BIOENERGETICS OF MEMBRANES

Developments in Bioenergetics and Biomembranes, Volume 1

BIOENERGETICS OF MEMBRANES

Proceedings of the International Symposium on Membrane Bioenergetics held on the Island of Spetsai, Greece, 10 - 15 July, 1977.

**Editors
Lester Packer
George C. Papageorgiou
Achim Trebst**

1977

ELSEVIER/NORTH-HOLLAND BIOMEDICAL PRESS

AMSTERDAM • OXFORD • NEW YORK

Published by:

Elsevier/North-Holland Biomedical Press
335 Jan van Galenstraat, P.O. Box 211,
Amsterdam, The Netherlands

Sole distributors for the USA and Canada:

Elsevier North-Holland Inc.
52 Vanderbilt Avenue,
New York, N.Y. 10017

ISBN Elsevier/North-Holland for this volume: 0-444-80016-6
ISBN Elsevier/North-Holland for the series: 0-444-80015-8

Library of Congress Cataloging in Publication Data

International Symposium on Membrane Bioenergetics,
    Spetsai, 1977.
    Bioenergetics of membranes.

    (Developments in bioenergetics and biomembranes ;
v. 1)
    Includes index.
    1. Membranes (Biology)—Congresses. 2. Bio-
energetics—Congresses. I. Packer, Lester.
II. Papageorgiou, George C. III. Trebst, Achim.
IV. Title. V. Series.
QH601.I537 1977   574.8'75   77-13215
ISBN 0-444-80016-6 (Vol. 1)
ISBN 0-444-80015-8 (Series)

Printed in The Netherlands

PREFACE

This volume contains contributions presented in the International Symposium on Membrane Bioenergetics, which was held on the Island of Spetsai, Greece, July 10-15, 1977.

This symposium was sponsored and supported by the Greek Ministry of Culture and Sciences, International Union of Biochemistry, International Union of Pure and Applied Biophysics, Federation of European Biochemical Societies, Nuclear Research Center at Demokritos and the National Hellenic Research Foundation. On this occasion it is worth adding a few words concerning the location of the meeting and the Spetsai tradition on international conferences. The Anargyrios and Korgialenios School of Spetsai (a boarding high school of the Island) where the symposium was convened, has established its international reputation in the field with a series of successful summer schools on molecular biology, molecular and cell biology and molecular and developmental biology which commenced in 1966. The symposium on membrane bioenergetics represent a departure from the Spetsai tradition and an enlargement of the scientific activities on the Island. Once again participants expressed their satisfaction for the arrangements of the scientific programme of the meeting. Senior scientists in the field have the opinion that they have participated in one of the most successful meetings in bioenergetics. Aside from the scientific merits of the place, Spetsai is the point where three Continents come close together (Africa, Asia and Europe) and indeed the conference was attended by persons from 16 countries. Since scientists and research workers are traditionally ranking among most suitable persons to act as spearheads for increasing understanding and international cooperation, it seems that in this global age of science, Spetsai scientific conferences may serve a dual purpose.

Published records of scientific symposia lose much of their value if publication is unduly delayed. In the present instance, we have chosen to proceed with publication at the expense of omitting the interesting discussion which occurred during the meeting and some of the articles which could not be ready by the time the volume went to press. This symposium volume should be of great value to scholars and educators in the fields of :

Structure and biogenesis of the three membrane systems: mitochondria, chloroplasts and sarcoplasmic reticulum.

Mechanisms of biological oxidation; iron sulphur proteins, interaction of iron sulphur proteins with quinones, photosynthetic electron transport.

Concepts of bioenergetics, ionophores, ionophoric proteins and transport mechanisms. Mechanism of action of bacteriorhodopsin (and rhodopsin) and ATP-synthetase-ATPase complex and their role in energy transduction or as ion pumps.

Some of the most recent advances in these fields are presented, with emphasis on membrane organization with respect to bioenergetic functions. These advances include the role of biological oxidations, electrical forces and ion gradients in mechanisms of energy coupling. In these areas there has been exceedingly rapid progress some of which is recorded here. The past few years almost as many redox components of the respiratory chain, as were known to exist before, have been identified by low-temperature EPR as iron sulphur proteins. These are now being characterized in terms of how they interact in flavoprotein dehydrogenases and with ubiquinone and the rest of the respiratory chain. Advanced technologies being used to exploit the understanding of membranes and bioenergetics as high resolution ^{31}P-NMR, ESR-spectroscopy, nano- and pico-second spectroscopy and the use of selective chemical probes for labelling, are well represented in this volume.

Most importantly, we thank each of the authors and other participants whose contributions made the Spetsai Symposium on Membrane Bioenergetics an intellectually stimulating and profitable occasion, finally, we thank one of the authors, Dr. J. Isaakidou for her invaluable endeavours for the success of the symposium.

A. E. Evangelopoulos

J. G. Georgatsos

L. Packer

G. C. Papageorgiou

A. Trebst

CONTENTS

MEMBRANE STRUCTURE AND BIOGENESIS

Bioenergetics of Membranes. L. Packer et al. ed.

THE MOLECULAR ORGANISATION OF CELL MEMBRANES

D.Chapman,
Chemistry Department,
Chelsea College,
University of London,
England.

INTRODUCTION

Considerable scientific work over the last ten years has increased our knowledge of the molecular organisation of cell membrane structure.   The original
idea suggested some 50 years ago that cell membranes are built upon a matrix of
a bilayer of lipid has been preserved.   This idea has now been strengthened
and enlarged to include knowledge of the mobility and asymmetry of the lipid
components.   Considerable information has also been obtained concerning the
protein arrangements associated with the lipid bilayer.   In some cases details
of the interior structure of the proteins themselves is now available.

It has been clear for some time that to understand the many functions associated with cell membranes that it would be necessary for us to understand their
detailed structure.   We now seem to be close to this important goal thereby
linking membrane structure with function.   Furthermore these studies are also
revealing the possibilities and the potential for simulating these functions
e.g. the simulation of energy transducer processes for technological
exploitation.

In this paper we attempt to briefly summarise the situation regarding membrane
structure so as to form an introduction and basis for the subsequent discussions
on Membrane Bioenergetics at this Conference.

CELL MEMBRANES AND FUNCTION

Biochemical and electron microscope studies by many scientists have revealed
the existence of many cell membranes including not only the outer plasma cell
membrane but also membranes associated with the various organelles within the
cell.   Associated with the membranes are various functions including the
control of permeability to organic molecules, water and ions.   Certain
membranes are also associated with the organisation of enzymes, visual pigments,
chlorophyll and other energy transducing molecules.

THE LIPID BILAYER MATRIX

We know that cell membranes consist of various lipid and protein components of varying ratios dependent upon the particular membrane chosen.   Other molecules such as cholesterol or chlorophyll occur in some membranes e.g. the former in appreciable quantities in the plasma membranes and chlorophyll in chloroplast membranes.   Carbohydrate groups also are found associated either with the lipid or protein structures.

The lipid component often consists of a variety of lipid classes e.g. lecithins, phosphatidyl-ethanolamines and associated with each of these lipid classes is a range of fatty acids of varying chain length and unsaturation[1]. Some lipids such as the lecithins are known to form spontaneously in water (above the lipid transition temperature) many bilayers each layer separated by water[2].   Various experiments e.g. spectroscopic and calorimetric studies have shown that the lipid bilayer is an important structural feature of many cell membrane structures.

(a) Molecular mobility and Bilayer fluidity

A range of physical techniques, calorimetry x-ray techniques and spectroscopic methods have given valuable information about the molecular mobility of the lipids within model lipid bilayer structures.[3]   This information has been useful in providing insight into the molecular processes involved within the lipid bilayer matrix of natural cell membranes and to develop the concept of membrane fluidity[4].

Studies of these model systems have shown that the lipids exhibit a phase transition at a characteristic temperature[5].   Below this temperature the lipid chains are in an ordered or crystalline arrangement.   Above this temperature the chains are "melted" and in a fluid condition.   In this state rotational isomerism occurs about the C-C bonds of the chain.   Spectroscopic studies including nmr, esr, ir and Raman spectroscopy have given information about the order parameter along the length of the chain.   These studies are consistent in showing that greater disorder occurs at the methyl end of the chain[6].

As well as this type of mobility within the chain it has also been shown that the lipid molecules (above the transition temperature) can readily diffuse within the plane of the bilayer[7].   On the other hand the movement of lipid molecules from one side of the bilayer to the other (sometimes termed flip flop) is a slow process[8].

(b) Asymmetric arrangement

A variety of experiments including chemical and enzyme treatments have now confirmed a suggestion made by Bretscher[9] that in the erythrocyte membrane

that the lipid classes are asymmetrically arranged. The lecithin and sphingo-
myelins are arranged on the outside half of the membrane and the phosphatidyl
ethanolamine and phosphatidyl serine molecules are arranged on the inner half
of the membrane, Experiments with some other membranes show that lipid
asymmetry occurs in other membrane systems. The biological significance of
lipid asymmetry is at present still uncertain.

(c) Cholesterol organisation

Many experiments using a variety of techniques have shown that the
presence of cholesterol within the bilayer matrix is to modulate the lipid fluidity[10].
When the lipid is above its transition temperature in the fluid condition the
presence of the cholesterol is to inhibit the rotational isomerism of the
methylene groups in the chain, causing the chain to become more rigid. The
presence of cholesterol also prevents the lipid chains from crystallising. At
high concentrations the cholesterol removes the lipid phase transition[11].
Essentially the presence of cholesterol is to cause the lipid to adopt an
"intermediate fluid" structure[12].

The organisation of the cholesterol molecules within the plane of the
bilayer has been discussed. Various types of complex have been postulated
including 1:1 and 2:1 lipid to cholesterol complexes[13,14]. An alternative
arrangement of <u>random</u> structure has also recently been discussed[15].

PROTEIN ARRANGEMENTS

For some time it had been considered that the proteins of cell membranes were
arranged on the outside of the lipid bilayer - the Danielli-Davson model. In
recent years it was realised that intrinsic protein may also occur within the
lipid bilayer matrix[16]. A variety of arrangements of the proteins can occur
even with a single membrane structure. A simple generalisation concerning the
protein organisation is not possible.

Some of the proteins (intrinsic) have been shown to span the bilayer using
chemical and electron microscope studies[17]. The latter using the techniques
of freeze-fracture exhibit these spanning proteins in the form of particles
seen on the fracture faces.[18] In many membranes these particles appear to be
randomly distributed within the plane of the bilayer. A detailed structure of
a membrane protein which spans the lipid bilayer has been revealed by the
technique of electron diffraction with the proteins of the purple membrane of
Halobacterium halobium[19]. This shows that the protein consists of seven helical

6

polypeptide chains folded back and forth across the membrane.   These helical
segments run approximately paralled to the plane of the membrane.

  (a) Protein mobility

      Evidence that proteins can rotate about an axis perpendicular to the
plane of the membrane was first obtained from studies of the visual pigment,
rhodopsin.   The highly regular parallel alignment of the disc membranes in
retinal rod outer segments has enabled the performance of detailed studies of
the arrangement of the major protein in these membranes[20].   Spectroscopic
studies showed that the retinal rod is strongly dichroic in that light polarised
perpendicular to the long axis of the rod is strongly absorbed compared with
light polarised parallel to the rod.   This indicates that the chromophore of
rhodopsin, retinal, is preferentially oriented parallel to the plane of the
disc membrane.   Initial studies of partial bleaching with short flashes of
polarised light lacked sufficient time resolution to demonstrate dichroism in
the plane of the membrane[21].   The rotational relaxation time of rhodopsin was
measured by Cone[22] using a flash photolysis apparatus capable of resolving
events in the microsecond time range.   He used lumi-rhodopsin as a tracer to
measure the (polarised) flash induced linear dichroism.   With the light
polarised parallel and perpendicular in output to the laser pulse he determined
the dichroic ratio as a function of time.   The rotational relaxation time of
the rhodopsin molecule was calculated and was of the order of 20 μsecs at room
temperature.

      From this result and making some assumption about the size of the protein
he was able to calculate the viscosity of the surrounding lipid matrix (about
2 poise).   This value is in good agreement with the value determined from
translational diffusion measurements for the same system[23].

      Similar studies have been reported for bacteriorhodopsin located in the
purple membrane of <u>Halobacterium</u> <u>halobium</u> in which a transient spectroscopic
chromophore centred at 410 nm was used to investigate rotational diffusion[24].
The absorbane has a lifetime of about 10 ms and was found to be strongly
dichroic following flash illumination with plane polarised light but in contrast
to rhodopsin this did not decay rapidly.   The rotational relaxation time was
found to be at least 10^3 times slower than rhodopsin consistent with the
crystal-like hexagonal packing of bacteriorhodopsin in purple membranes inferred
from X-ray analysis and unlike the liquid-like array of rhodopsin in the
thylakoid membrane.   Examination of the photolytic dissociation of cytochrome
a_3-CO complex in the inner mitochondrial membrane[25] and photoinduced dichroism
of chlorophyll a_1 in chloroplast membranes[26] has also suggested that these
proteins do not rotate rapidly about an axis perpendicular to the membrane.

Fluorescence techniques have also been applied to study protein rotation in membranes. The fluorescence polarisation of analino naphthalene sulphonate and dansyl chloride bound to electroplax membrane fragments, for example, have been resolved into two components; a rapid but partial decay, believed to be rotation of the probe at its binding site and the residual decay due to rotation of the protein which set a lower limit for rotational relaxation time[27] of 0.7 us.   Tryptophan fluorescence has been explored as an intrinsic probe to measure rotational kinetics of other membrane proteins and external probes such as eosin have been used successfully in model systems and erythrocyte membranes to study the rotation of proteins which have no natural chromophores[24,28]. In general, these studies indicate that rotational relaxation times for different membrane proteins vary considerably and probably reflect differences in the molecular environment surrounding each particular protein.

(b) Protein aggregation

Changes in the distribution of protein in the plane of the membrane was first noted with freeze-cleaved erythrocyte membrane preparations that had been quenched from acidic media.   Membrane-associated particles are distributed randomly on complementary inner fracture of these membranes at pH 7.5 or pH 9.5 but they aggregate when the pH is reduced to 5.5 or less.   Adjusting the pH again to 7.4 restores the original particle distribution so that the process is freely reversible[29].

Freeze-fracture techniques have been used to examine membrane-associated particle distribution of chloroplast membranes[30].   Particle distribution in these membranes appeared to be random in dark adapted chloroplasts but they became aggregated following illumination.   In other experiments, temperature dependent membrane-associated particle aggregation has been observed with the inner mitochondrial membrane[31].   Protein induced aggregation has also been reported in mitochondrial membranes and lipid reconstituted Ca^{2+} -activated ATPase of sarcoplasmic reticulum.[33]   Proteins of some bacterial membranes also remain dispersed even when the membranes are cooled to below the lipid phase transition temperature[34].   Erythrocyte glycophorin also appears to be dispersed both above and below the phase transition temperature[35].   Some interesting studies of reconstituted membranes containing rhodopsin have shown that dark adapted rhodopsin is aggregated below the lipid transition temperature but is dispersed randomly at higher temperatures.   When the chromophore is bleached, however, the rhodopsin molecules remain dispersed at all temperatures[36].

The relationship between membrane-associated particle aggregation and lipid crystallisation has been demonstrated convincingly in the micro-organism[37] *Mycoplasma mycoides* var. *capri*. This shows that in freeze-fracture replicas of a strain of the organism devoid of cholesterol two regions can be observed, one with a high density of particles, the other smooth and containing no particles. Similar replicas of the parent strain which contains cholesterol show that the proteins remain dispersed more or less randomly throughout the plane of the membrane. A simulation of the protein aggregation processes has been discussed[38].

Protein aggregation within the fluid lipid bilayer triggered by mechanisms other than lipid crystallisation has been related to various cell processes including pinocytosis, the stage of the cell growth cycle and to membrane interactions and fusion.

PROTEIN - LIPID INTERACTIONS

The fact that some proteins are embedded within the lipid bilayer has led to discussions of the perturbation which this produces on the neighbouring lipid[39,40]. Various terms have been used to describe this e.g. boundary layer lipid, halo lipid and annulus lipid. This has sometimes been associated with "residual lipid" i.e. the lipid which still appears to be tightly bound to the protein after extensive solvent extraction or the amount of lipid which is required to restore enzyme activity. The present situation is confused although it seems clear that proteins can affect the lipid chain gauche isomers. More experiments are required to clarify whether proteins in general have a tightly bound lipid annulus which they carry around with them within the fluid bilayer structure.

CATALYTIC HYDROGENATION AND MEMBRANE FLUIDITY

A new technique that we have been developing in our laboratory[41] may lead to the selective control of membrane fluidity. This is the technique of catalytic hydrogenation. Our recent experiments have shown that certain homogenous catalysts can be incorporated into phospholipid membrane systems and that indeed catalytic hydrogenation is observed. Thus the effect of hydrogenation of biological membranes on particular membrane functions will enable detailed and specific correlations to be established with saturation of particular double bonds of membrane phospholipids.

Genetic, nutritional and other methods of manipulating the fluidity of biological membranes will continue to make an important contribution to our understanding of how these membranes perform their various functions. Catalytic hydrogenation may play an equally important role in many future studies. Various catalysts will be investigated leading to selectivity of hydrogenation and fine modulation of membrane fluidity. Studies of enzymic and transport processes, excitable membranes, retinal, chloroplast and mitochondrial function, may become possible as a function of this hydrogenation process.

ACKNOWLEDGEMENTS

I wish to acknowledge the support of the Wellcome Trust, the Medical Research Council and the Science Research Council.

REFERENCES

1.   van Deenen, L.L.M. (1965) in Progress in the Chemistry of Fats and other Lipids ed Holman, R.T. (Pergamon Press, Oxford) Vol. VIII Part 1 pp 1-127.

2.   Chapman, D., Williams, R.M. and Ladbrooke, B.D. (1967) Chem. Phys. Lipids 1, 445 - 475.

3.   Oldfield, E., and Chapman, D. (1972) FEBS Lett. 21, 303 - 306.

4.   Chapman, D. Byrne, P. and Shipley, G.G. (1966) Proc.Roy.Soc.London Ser.A. 290, 115 - 142.

5.   Chapman, D. (1975) Q.Rev.Biophys. 8, 185 - 235.

6.   Seelig, A. and Seelig, J. (1974) Biochemistry 14, 2283.

7.   Kornberg, R.D. and McConnell, H.M. (1971) Proc.Nat.Acad.Sci. USA, 68, 2564.

8.   Rothman, J.E. and Lenard J. (1977) Science 195, 743 - 753.

9.   Bretscher, M.S. (1972) Nature New Biol. 236, 11 - 12.

10. (a) Chapman, D. and Penkett, S.A. (1966) Nature Lond. 211, 1304 - 1305.
    (b) Gally, H.V. Seelig, A and Seelig, J. (1976) Hoppe-Seyler's Z.Physical Chem. 357 1447 - 1450.

11.   Ladbrooke, B.D. Williams, R.M. and Chapman, D. (1968) Biochim. Biophys. Acta 150, 333 - 340.

12.   Williams, R.M. and Chapman, D. (1970) in Progress in the Chemistry of Fats and other Lipids ed. Holman, R.T. (Pergamon Press Oxford) Vol 2, p 1 - 79.

13.   Phillips, M.C. and Finer, E.G. (1974) Biochim.Biophys.Acta 356, 199 - 208.

14.   Engleman, D.M. and Rothman, J.E. (1972) J.Biol Chem. 247, 3694 - 3697.

15.   Cornell, B.A. Chapman, D. and Peel, W.E. (1977) Biochim.Biophys.Acta in press.

16.   Singer, S.J. and Nicholson, G.L. (1972) Science $\underline{175}$ 720 - 731.

17.   Bretscher, M.S. J.Mol.Biol. $\underline{58}$ 775 - 781 (1971).

18.   Branton, D. (1966) Proc.Natl.Acad.Sci. USA, $\underline{55}$ 1048.

19.   Henderson,R. and Unwin, P.N.T. (1975) Nature $\underline{257}$, 28 - 32.

20.   Daemen, F.J.M. (1973). Biochim.Biophys.Acta $\underline{300}$, 255 - 288.

21.   Hagins, W.A. and Jennings, W.H. (1959) Disc.Faraday Soc. $\underline{27}$ 180.

22.   Cone, R.A. (1972) Nature New Biology $\underline{236}$, 39 - 43.

23. (a) Brown, P.K. (1972) Nature New Biology $\underline{236}$, 35 - 38.

    (b) Poo, M. and Cone, R.A. (1974) Nature $\underline{247}$, 438 - 441.

24.   Razi-Naqvi, K. Gonsalez-Rodriquez, J. Cherry, R.J. and Chapman, D.
        (1973) Nature New Biology $\underline{245}$, 249 - 251.

25.   Junge, W. and Devault, D. (1975) Lasers in Physical Chemistry and
        Biology Elsevier.

26.   Junge, W. and Eckhof, A. (1973) FEBS Letts. $\underline{36}$ 207

27.   Wahl,P. Kàsai, M. Changeux, J.P. and Auchet, J.C. (1971) Eur. J.
        Biochem. $\underline{18}$, 332

28.   Cherry, R.J. and Schneider, G. (1976) Biochemistry $\underline{15}$ 3657 - 3661.

29.   Pinto da Silva (1972) J.Cell Biol. 777 - 787.

30.   Torres-Pereira, J. Mehlhorn, R. Keith, A.D. and Packer, L. (1974)
        Archives Biochem.and Biophysics 160, 90 - 99.

31.   Hackenbrock, C.R. Structure of Biological Membranes Nobel Symposium (1976)
        34. Plenum Press. p 199 - 234.

32.   Thompson, G.A. and Nozawa, Y. (1977) Biochim.Biophys.Acta $\underline{472}$, 55 - 92.

33.   Kleemann, W. and McConnell, H.M. (1976) Biochim.Biophys.Acta $\underline{419}$
        206 - 222.

34.   Haest, C.W.M. Verkleij, A.J. de Gier, J. Sheek, R. Ververgaert, P.H.J.
        van Deenen, L.L.M. (1974) Biochim.Biophys.Acta $\underline{356}$, 17 - 26.

35.   Kleemann, W. Grant, C.W.M. and McConnell, H.M. (1974) J.Supramol.
        Struct. $\underline{2}$, 609 - 616.

36.   Hong, K. and Hubbell, W.L. (1973) Biochemistry, $\underline{12}$, 4517 - 4523.

37.   Rottem, S. Yashouv, J. Ne'eman, Z. and Razin, S. (1973) Biochim.
        Biophys.Acta, $\underline{323}$, 495 - 508.

38.   Chapman, D. Cornell, B.A. and Quinn, P.J. (1977) Biochemistry of Membrane
        Transport FEBS Symposium No. 42.

39.   Jost, P.C. Griffith, O.H. Capaldi, R.A. and Vanderkooi, G. (1973) Proc.
        Nat.Acad.Sci.USA $\underline{70}$, 480 - 484.

40.   Vanderkooi, G. and Bendler, J.T. (1976) Nobel Symposium on Biological
        Membranes 34 Plenum Press.   p. 551 - 570.

41.   Chapman, D. and Quinn, P.J. (1976) Proc.Nat.Acad.Sci. $\underline{73}$ 3971 - 3975

USE OF APOLAR LABELS TO IDENTIFY

INTRINSIC MEMBRANE PROTEINS AND IONOPHORIC CHANNELS

Carlos Gitler

Department of Membrane Research
The Weizmann Institute of Science
Rehovot, Israel

INTRODUCTION

Membrane proteins can be divided according to most recent models into those
which span the bilayer and are therefore in contact with both the outer and the
inner aqueous phases. Those which appear to be embedded into the lipid bilayer
but which are in contact with only one aqueous phase, and those which are not in
contact with the lipid bilayer but which are probably attached to proteins of
either of the two previous categories. It would seem that the so-called intrinsic
or integral proteins include the first two types described above, while the last
group would constitute the extrinsic or peripheral proteins. With few exceptions,
little evidence is available to definitely classify a given protein as to whether
it is or not in contact with the lipid bilayer. Furthermore, with the exception
of glycophorin, no evidence is available as to the polypeptide regions of a given
protein which are in contact with the lipid bilayer.

The order of magnitude of the problem can be posed by some simple calculations.
If the liquid hydrocarbon region of the bilayer is some 30 to 35 Å wide, a peptide
of some 20-25 amino acids in length would, if present as an α-helix, suffice to
span the apolar region of the bilayer. Such a peptide would have a molecular
weight of some 3000 daltons. If the protein spanning the bilayer has a molecular
weight of some 100,000, then the problem is one of identifying 3% of the total
polypeptide. Even if more than one segment of the protein spans the bilayer, the
problem is one of identifying those segments which were in contact with the
phospholipids. Criteria for such identification could be similar to those used
with glycophorin, namely the presence of 20-25 adjacent apolar amino acids. However, if membrane proteins span the bilayer by forming multichain coiled-coils,
the segments spanning the bilayer might be identified by 3,4,3,4,... sequences of
polar groups. It is likely that, such identification would be extremely
difficult. Studies such as those of Unwin and Henderson[1,2] using low dose high
resolution electron microscopy might, when the definition is enhanced, allow
definite sequence assignments to those areas in contact with the lipid. However,

12

two dimensional ordered arrays of membrane proteins are as yet not easily
obtained.

From the above brief discussion, it is apparent that it would be desirable to
have more direct methods to identify the regions of the intrinsic membrane
proteins which exist in contact with the lipid bilayer. One alternative, which
was applied to glycophorin, would be to label those portions of the protein in
contact with the aqueous layers. The polypeptides between such areas in a
protein spanning the bilayer must include the segments burried in the bilayer. A
second approach, which is the subject of this presentation, is a more direct
attempt to label those polypeptides which penetrate the lipid bilayer. It is
based on the fact that apolar molecules will readily partition and dissolve in
the liquid hydrocarbon region of the phospholipid bilayer. Once located therein,
they can react directly with those components present therein. The first part of
this chapter deals with the use of apolar azides while the second section will
discuss the use of apolar carbodiimides.

APOLAR AZIDES AS REAGENTS TO DETERMINE THE PENETRATION OF PROTEINS INTO THE LIPID BILAYER

Two different approaches have been taken in the design of reagents which will
partition into the lipid bilayer and once located therein may be activated to
covalently label those components present therein. Klip and Gitler[3,4,5] synthesized aromatic azides which rapidly dissolve into the lipid and upon exposure to
light insert covalently both into proteins and into lipids. Chakrabarti and
Khorana[6], and Greenberg et $al.$[7], synthesized fatty acids containing azides which
could be incorporated into phospholipids and on exposure to light formed nitrenes
which might attach covalently both to proteins and to phospholipids.

AROMATIC APOLAR AZIDES

The use of apolar aromatic azides may be exemplified by the recent studies
performed with 5-[^{125}I]-iodonaphthyl-1-azide (INA)[8-11]. This compound can be
made highly radioactive by reacting 5-azidonaphthalene-1-diazonium chloride with
^{125}I-NaI. The iodine is easily added in the last step of the synthesis, the
precursor diazonium salt being stable in the cold. This is important because the
azides are easily destroyed by exposure to radiation during storage. When INA in
ethanol is added in the dark to a membrane suspension (< 1% final ethanol, 10^{-6} M
INA) it partitions within a few seconds following addition, almost quantitatively
(> 98%), into the membrane lipids. Its high molar extinction (ε_{310} = 21,400 M^{-1}
cm^{-1}) allows its conversion into the 5-[^{125}I]-iodonaphthyl-1-nitrene by short

periods of exposure to light (wild microscope lamp with an Osram HB 200 W mercury lamp). The use of the 314 nm line of the mercury lamp, together with Corning filters 7-60 and 0-54, allows the generation of nitrenes without damaging the membrane proteins or lipids. With membranes, such as rabbit skeletal muscle sarcoplasmic reticulum (SR)[12], intestinal microvillus membranes[11], a purified lipid bound Na^+, K^+-ATPase[13], or with hemoglobin-free erythrocyte membranes (HFE)[14], two minutes of exposure to light results in the complete conversion of the INA into the nitrene, even in suspensions containing 4 mg of membrane protein/ml. With a 10% suspension of human erythrocytes in PBS, 30 minutes are required for the complete generation of the nitrene. No hemolysis occurs during this period.

Once the nitrene is generated within the bilayer, it inserts covalently into the proteins and phospholipids present therein. Removal of non-covalently attached irradiation products can be accomplished by washing the membranes with BSA[3], by precipitating with cold acetone-10% water, by dialyzing against a BSA-containing buffer solution or alternatively, by the procedures used to fix and stain SDS-polyacrylamide gels used for electrophoresis of the proteins (SDS-PAGE). Very high efficiencies of covalent insertion of ^{125}I-INA products have been obtained. These range from 18-26% in HFE and erythrocytes to as high as 55% in SR and intestinal microvillus membranes.

DISTRIBUTION OF THE INA INSERTION PRODUCTS - INTRINSIC AND EXTRINSIC PROTEINS

The patterns of distribution of covalent ^{125}I-INA insertion products in SR and in erythrocyte membranes are shown in Figs. 1 and 2, respectively. It can be observed that in SR, the main labelled protein band is the intrinsic 102 KD Ca^{++}-ATPase (Band a, Fig.1). Some minor insertion also occurs into the intrinsic acidic proteins (Bands d-g, Fig.1) and perhaps into the proteolipid (Band h) and the lipids (k region). However, no radioactivity is present in the extrinsic or peripheral proteins, namely the high affinity Ca^{++}-binding protein or the calcequestrin (Bands c and d, respectively). In the case of the erythrocyte, little radio-activity is present in the spectrin region. The majority appears in the areas of bands 3, PAS 1-3 and 7. In order to determine whether extrinsic proteins are labelled, the HFE membranes were treated with "protein perturbants" as described by Steck and Yu[15]. Treatment with 0.1 N NaOH, 5 mM p-chloromercurybenzoate and 1.0 mM EDTA released 59.9, 46.2 and 36.9% of the membrane protein but only 7.8, 4.2 and 3.1% of the radioactivity. SDS-PAGE of such supernatants showed minor labelling only in the spectrin bands 1, 2 and none in the other released proteins (Bands 4.1, 4.2, 5 and 6). On the other hand, 0.5% Triton X-100 in 56 mM sodium borate, pH 8, which preferentially solubilizes the integral proteins[16] (Bands 3, PAS 1-3 and 7) released some 83% of the radioactivity. Purification of glyco-

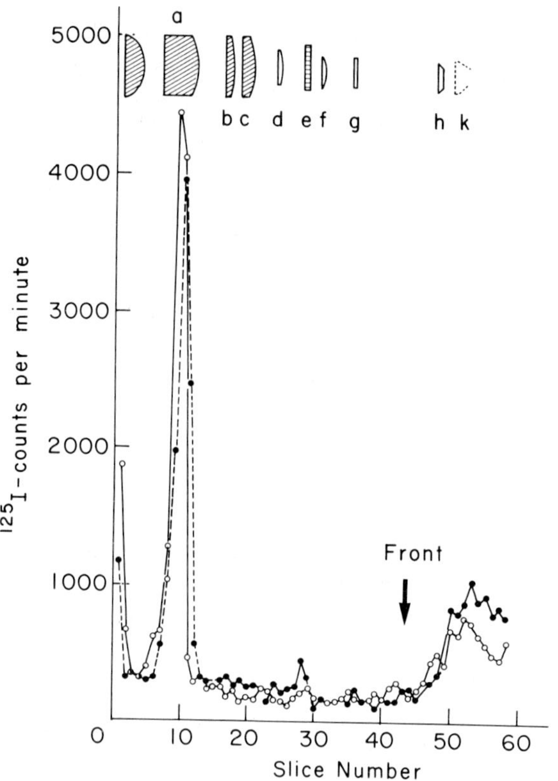

Fig.1 : Distribution of ^{125}I-INA covalent insertion products in the polypeptides of rabbit skeletal muscle sarcoplasmic reticulum separated by SDS-PAGE. Non-covalently attached INA irradiation products removed by BSA equilibration (——●——) and by acetone precipitation (——○——).

phorin[10,17] showed this protein to be significantly labelled. Trypsin digestion indicated that nearly all the radioactivity was associated with the trypsin insoluble peptide.

Studies with intestinal microvillus membranes labelled with INA showed that actin is not labelled even though preparations high in this protein were subjected to the INA labelling procedure.

All these studies clearly indicate that INA labels only integral or intrinsic proteins. Very little, if any, label inserts into those proteins which are operationally defined as extrinsic or peripheral proteins. It is likely that these latter proteins are not in contact with the lipid bilayer. Or if such contact exists, it prevents access to the INA-nitrene. Clearly these results support the contention that INA labels from within the lipid core.

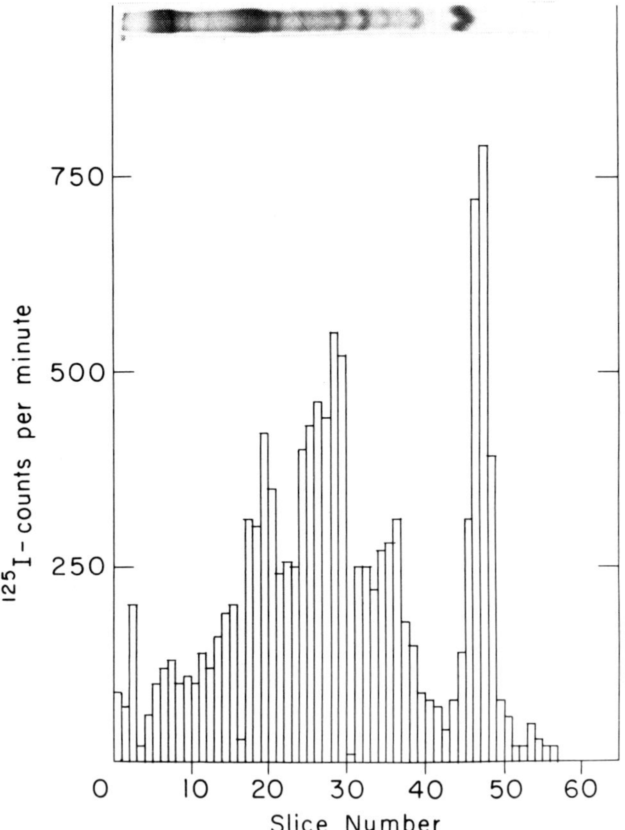

Fig.2 : Distribution of the ^{125}I-INA covalent insertion products in the polypeptides of human erythrocytes separated by SDS-PAGE.

DISTRIBUTION OF INA-INSERTION PRODUCTS - IDENTIFICATION OF LABELLED PEPTIDES

If the INA labels from within the bilayer, exhaustive proteolysis should
release those segments of the polypeptide chains exposed to the aqueous phases but
should not release those peptides immersed in the lipid bilayer. While not all
the peptides resistant to digestion necessarily reside within the bilayer, it is
likely that steric hindrance should protect those that are immersed.

A general finding with all the membranes studied is that digestion with trypsin,
acetyltrypsin, thermolysin or papain does not release more than a few percent of
the INA covalently inserted. As mentioned in the case of glycophorin, INA inserts
into the insoluble core remaining after trypsin digestion which includes the 23
apolar amino acid segment.

The Ca^{++}-ATPase is cleaved by acetyltrypsin to give two nearly equal fragments
of 52,000 and 46,000. Both fragments are nearly equally labelled. The activity

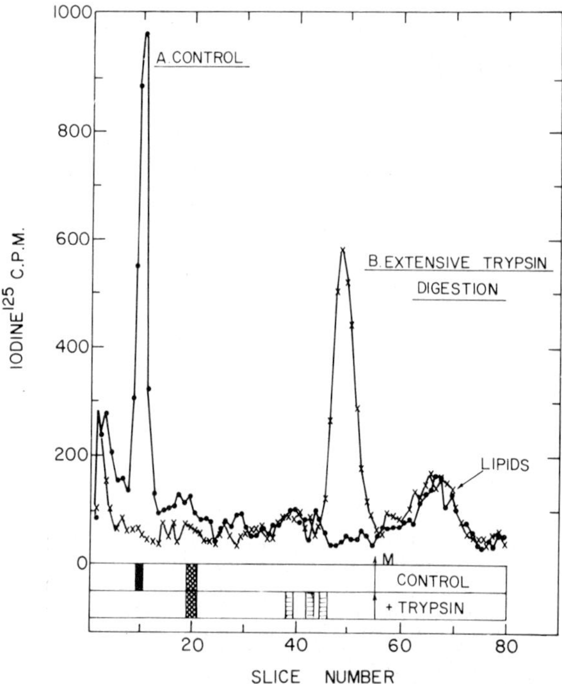

Fig.3 : Labelling of (Na^+, K^+) dependent ATPase with 5-[^{125}I]-iodonaphthyl-1-azide (A). Extensive trypsinolysis of labelled membranes (B).

of the Ca^{++}-ATPase is not affected by the incorporation of 0.5 moles of INA/mole of enzyme. Furthermore, the cleaved enzyme retains its activity. Digestion with higher amounts of acetyltrypsin results in the formation initially of radioactive fragments of mol. wt. 31K, 24K, 20K, 13K and ca. 6K. After prolonged incubation, only the 20K and 6K peptides remain. These contain the near total radioactivity present in the original 102K band.

A different behaviour emerges in the case of the purified, membrane-bound (Na^+, K^+) ATPase. Two polypeptide components remain in this purified preparation, a catalytic protein with mol. wt. 100 KD which constitutes 60-70% of the total protein and a smaller sialoglycoprotein of unknown function. On labelling with INA at roughly 0.5 moles INA/mole enzyme, the label was attached almost exclusively to the large chain (Fig.3A). Inhibition of the enzyme was less than 10%. The sialoglycoprotein was barely labelled and less than 20% of the radioactivity was associated with the lipid. Exhaustive digestion with trypsin or thermolysin resulted in nearly quantitative transfer of the radioactivity from the 100 KD band to a peptide(s) of 12 KD (Fig.3B).

After exposure of the enzyme at a much higher concentration of INA (roughly 25-50 moles INA per mole of enzyme) the lipids became labelled to a level 6-8 fold greater than the protein and incorporation was also seen in the glycoprotein. In these conditions the (Na$^+$, K$^+$) ATPase activity was almost completely inhibited and radioactivity equivalent to roughly one molecule of INA was incorporated into the large chain per mole of enzyme. Exhaustive trypsinization of these membranes resulted in essentially the same pattern of the label distribution as shown in Fig.3B.

The preferential insertion into the large chain at low concentrations of INA and inhibition of the (Na$^+$, K$^+$) ATPase could imply contrary to expectation, that the label binds specifically to catalytic areas of the protein. Such areas are often hydrophobic clefts in enzyme surfaces. This possibility is unlikely for the following reasons. All catalytic functions are destroyed by relatively limited proteolysis, but as seen above exhaustive proteolysis did not release the label. Furthermore, binding to the active site of a fluorescent ATP analogue, formycin triphosphate was unaffected by incorporation of sufficient iodonaphthyl groups to inhibit activity by 70%, and ATP did not alter incorporation of label into the protein when it was present during irradiation (not shown). Although the inhibition of the (Na$^+$K) ATPase activity by INA was not due to interference with ATP binding, other intermediate steps were not examined.

Labelling of the (Na$^+$, K$^+$) ATPase with INA in a K$^+$-containing medium[9] showed some 10-25% greater incorporation of radioactivity into the large chain when compared with labelling in the presence of Na$^+$-containing media. Furthermore, tryptic cleavage in the presence of KCl resulted in the formation of fragments of 58 KD and 46 KD containing 20% and 80% of the radioactivity, respectively. The 46K band was then split further to an unlabelled 36K segment and the 12K band shown in Fig.3B. In the presence of NaCl, tryptic cleavage gave an unlabelled 78K band and the terminal 12 KD peptide(s) from the onset of digestion. A minor fraction of the label (at best 2%) appeared in a 37 KD fragment. These results indicate that contrary to the Ca^{++}-ATPase, only one end of the (Na$^+$, K$^+$) ATPase is labelled[9].

In the case of the intestinal microvillus membrane, the INA-inserted was not removed by extensive papain digestion whereas chloramine-T iodinated membranes released appreciable radioactivity under the same conditions[11]. Neither glucose transport nor the activity of sucrase, alkaline phosphatase or leucine amino-peptidase, is affected by the labelling procedure.

The above results again support the contention that the INA labels proteins from within the bilayer. They suggest that an enzyme such as the Ca^{++}-ATPase is in contact with the bilayer in at least two segments of its polypeptide chain. These occur in each of the 2 halves of the protein formed during the initial

18

tryptic cleavage. The enzymatic activity is unaffected by the labelling. In the
case of the (Na^+, K^+) ATPase only one terminal segment of the protein is labelled.
Activity of the enzyme is lost during labelling but apparently not because the
label binds to the ATP binding site. It is possible that lipid-protein interaction might be affected by the labelling[18]. Furthermore, it is not clear whether
other polypeptides, not released by trypsin (Fig.3B) might not penetrate into the
bilayer. It is conceivable that all peptides inserting into the bilayer might
not be equally accessible to the INA. This might also explain the low incorporation of the INA into the 58K glycoprotein.

It is evident that much more work is required, however, those fragments
identified by their INA labelling may now be analyzed to determine their amino
acid content and sequence of amino acids. This might shed light in the manner
that peptides are inserted into the bilayer. Furthermore, such peptides are of
interest because in addition to a purely structural role, they might form parts
of the ion-conducting pathways.

APOLAR CARBODIIMIDES AND PROTEOLIPIDS

The second part of this chapter deals with the use of carbodiimides to identify
reactive carboxyl-groups which are present in apolar regions of membranes. It
also deals with the purification of proteolipids and mainly with work done in
collaboration with Hans and Kristine Sigrist and Natan Nelson in which a presump-
tive proton-translocating proteolipid from chloroplasts has been isolated essentially pure and reconstituted into liposomes.

Apolar Carbodiimides : These molecules are of interest because they react
with carboxyls (also with sulfhydryls and phenols). They have the general formula
R-N=C=N-R', where R and R' can either be an apolar group such as cyclohexyl in
dicyclohexylcarbodiimide (DCCD) or they can have one apolar and one polar group
such as 1-cyclohexyl-3-(2-morpholino ethyl)-carbodiimide. DCCD is very insoluble
in water and readily dissolves in solvents such as chloroform while those carbodiimides containing polar groups can dissolve readily in water and other polar
solvents. Thus, there are available reagents with the same reactive carbodiimide
but with very different solubilities and therefore they would be expected to
interact with different reactive groups present in membranes.

In various membrane systems, DCCD is capable of combining at low concentrations
(ca. 10^{-6}M) with discreet proteins to inhibit their function. The best studied
systems are those related to the energy transducing ATPase complexes of mitochondria chloroplasts and bacteria to be discussed below. However, it should be
pointed out that DCCD at low concentrations can also inhibit the activation by
isoproterenol of the adenyl cyclase of nucleated erythrocytes[19,20]. Preincubation
of turkey erythrocytes with 10^{-5}M DCCD for 5 minutes is sufficient to inhibit up

to 90% of the isopropylarterenol stimulation. Water-soluble carbodiimides show no
inhibitory effect even at 100 fold higher concentrations. A carboxyl group seems
to be involved because another carboxyl-activating series of compounds : N-ethoxy
carbonyl-2-ethoxy-1,2-dihydroquinoline (EEDQ) and the isopropyloxy-derivative
(IIDQ) are also inhibitory at low concentrations. Nothing is known as yet about
the mechanism of this inhibition although it appears to involve the adrenergic
receptor or its capacity to couple with the adenyl cyclase[21].

In various membrane systems, DCCD is reported to restore impaired respiratory
linked functions, presumably by decreasing proton permeability[22,23]. Using [14C]-
DCCD Beechey and coworkers[24] showed that at low concentrations, the [14C]-DCCD
binds covalently mainly to one protein band of low molecular weight. These
authors showed further that EEDQ is also inhibitory in this system. Furthermore,
water-soluble carbodiimides are poorly inhibitory in the intact membranes. Thus,
in these systems also a carboxyl(s) in an apolar environment seems to be involved.
Further evidence to this effect is derived from the fact that [14C]-DCCD covalently
labels the site to which it interacts. As shown in the following scheme, an
activated carboxyl intermediate (I) would not be stable in the presence of nucleophiles and of water. The radioactivity in these conditions would be liberated as
the dicyclohexylurea (II). Only in the absence of nucleophiles would it be
expected that the stable form III would be formed.

$$
\begin{array}{l}
\quad\qquad\qquad\qquad\qquad\qquad\qquad\qquad\ \ \overset{\textstyle H}{}\quad\overset{\textstyle H}{} \\
R_1\text{-N=C=N-}R_1 \ + \ Pr\ COOH \ \longrightarrow \ R_1\text{-N-C=}\overset{+}{N}\text{-}R_1 \\
\qquad\qquad\qquad\qquad\qquad\qquad\qquad\qquad\qquad\ |\ \\
\qquad\qquad\qquad\qquad\qquad\qquad\qquad\qquad\ O\text{-C-}Pr \\
\qquad\qquad\qquad\qquad\qquad\qquad\qquad\qquad\qquad\ \overset{\|}{O} \\
\qquad\qquad\qquad\qquad\qquad\qquad\qquad (I)
\end{array}
$$

$$
\begin{array}{l}
\qquad\qquad\qquad\qquad\qquad\qquad\qquad\ \ H\ \ O\ \ H \\
(I) \ + \ H_2O \ \longrightarrow \ Pr\ COOH \ + \ R_1\text{-N-}\overset{\|}{C}\text{-N-}R_1 \\
\qquad\qquad\qquad\qquad\qquad\qquad\ (II)
\end{array}
$$

$$
\begin{array}{l}
(I) \ + \ R\text{-NH}_2 \ \longrightarrow \ Pr\text{-CONH-}R \ + \ (II) \\[4pt]
\qquad\qquad\qquad\qquad\qquad\qquad\qquad\quad O \\
(I) \ \xrightarrow{\ (O\text{-}N)\ \text{acyl shift}\ } \ R_1\text{-N-}\overset{\|}{C}\text{-NH-}R_1 \\
\qquad\qquad\qquad\qquad\qquad\qquad\quad\ | \\
\qquad\qquad\qquad\qquad\qquad\qquad\ C\text{-}Pr \\
\qquad\qquad\qquad\qquad\qquad\qquad\quad\overset{\|}{O} \\
\qquad\qquad\qquad\qquad\qquad\qquad (III)
\end{array}
$$

where R_1 is the cyclohexyl residue and Pr COOH is a protein carboxyl group.

As mentioned, it was observed by Beechey and coworkers[24] that the main fraction
of the covalent bound [14C]-DCCD was associated with a small molecular weight
protein. These authors further made the important observation that this small
protein could be extracted as a proteolipid into methanol-chloroform. Other
similar reports have appeared[25-27].

Proteolipids : It has been known for some time that many membranes contain
proteins that may be extracted into organic solvents. In model systems it has
been possible to show that acidic phospholipids can form complexes with the amino

groups of proteins and that upon neutralization of the carboxyl groups by protons or Ca^{++} ions, complexes are formed which are sufficiently desolvated to be able to dissolve in isooctane or in decane[28,29].

The DCCD-binding proteolipid has been extensively purified by use of the methanol-chloroform procedure. However, this method has not apparently yielded a proteolipid which in the absence of DCCD could be reconstituted into liposomes and shown to retain some of its native properties. While searching for methods to purify membrane proteolipids in collaboration with H. Sigrist, a simple technique was devised based on the use of 1-butanol which allows the rapid and neat purification of several proteolipids[27]. The procedure differs from that of Maddy[30] in that water is kept at very low concentrations. Thus, addition of a membrane suspension in water or in water plus calcium or in buffer, to butanol such that the final water content is 2% results in the solubilization of the membrane lipids and proteolipids. All other proteins are rapidly and quantitatively precipitated. Evaporation of the butanol and addition of buffer and sonication allows the formation of liposomes containing the proteolipid and the original membrane lipids. Alternatively, ether can be added to the butanol extract to precipitate the proteolipid. Using this procedure, the proteolipids of SR, mitochondria and chloroplasts have been isolated and obtained as single bands. Fig.4 shows the SDS-PAGE patterns obtained during the purification of the DCCD-binding proteolipid of mitochondria[27].

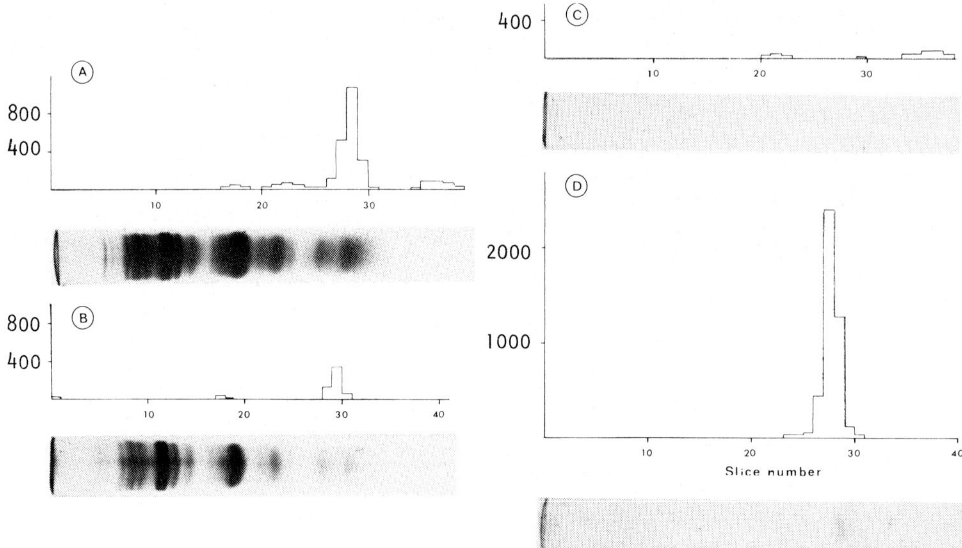

Fig.4 : Isolation of the ^{14}C-DCCD-binding protein from submitochondrial particles : distribution of protein and radioactivity in polyacrylamide gels.
A. ^{14}C-DCCD-treated submitochondrial particles (40 μg); B. Butanol precipitate (32 μg); C. Ether supernatant (equivalent to 4 μg); D. Ether precipitate (proteolipid) (5 μg).

RECONSTITUTION OF THE CHLOROPLAST DCCD-BINDING PROTEOLIPID, ACTIVE IN PROTON TRANSLOCATION

Application of the above 1-butanol procedure to lettuce chloroplasts, in collaboration with N. Nelson and colleagues, resulted in the purification of a single polypeptide of mol. wt. 8300 as the DCCD-binding proteolipid.

Liposomes were formed by evaporation of butanol from the butanol supernatant and sonication, following addition of an aqueous solution. Further sonication of the liposomes in the presence of bacteriorhodopsin led to the formation of liposomes showing light-induced proton uptake (Fig.5).

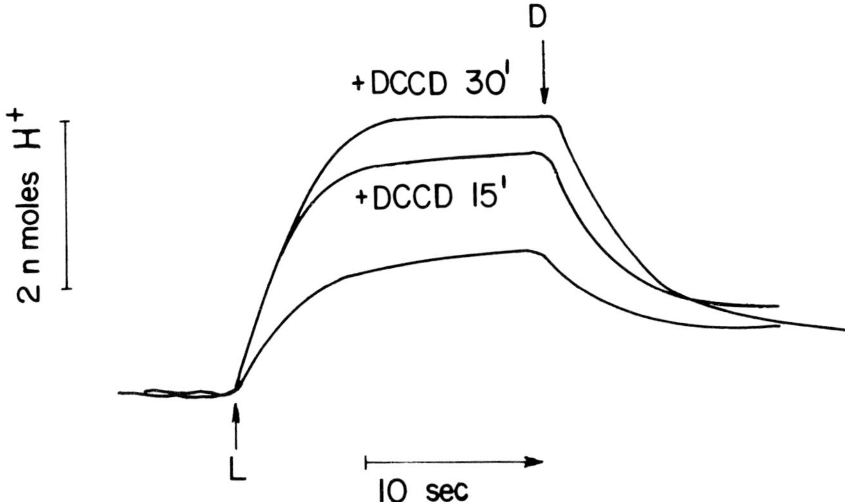

Fig.5 : Enhancement of light-induced proton uptake by DCCD in proteolipid-bacteriorhodopsin vesicles. After butanol extraction of 2 ml of chloroplast membranes (equivalent to 7.0 mg of chlorophyll) by the procedure described in results (omitting the ether precipitation step), 100 ml of the butanol supernatant was dried at low temperature and redissolved in 2 ml of 1-butanol. Aliquots (0.5 ml) were dried under nitrogen. Procedures for liposome formation and proton uptake measurements are described in Materials and Methods. DCCD, when present, was 20 μM. pH measurements were performed at 5 min intervals. L and D represent light on and off, respectively.

DCCD addition (20 μM) to these liposomes resulted in a progressive increase in the light-induced proton uptake. Fifteen minutes after addition of DCCD the proton uptake was found to be twice that measured in the absence of added DCCD. Comparison of the rates of proton uptake and release show that DCCD altered only the total magnitude of the uptake and not the observed rates. Addition of DCCD to liposomes identical to those discussed in Fig.5, but not containing proteolipid, effected no change in light-induced proton uptake or release in the dark even after prolonged incubation (data not shown).

22

Addition of [^{14}C]-DCCD to liposomes containing the proteolipid, chloroplast lipids, and bacteriorhodopsin resulted in the majority of the radioactivity being associated with the proteolipid band (Fig.6). In contrast, the bacteriorhodopsin protein band failed to show any enhancement in the binding of DCCD over background level.

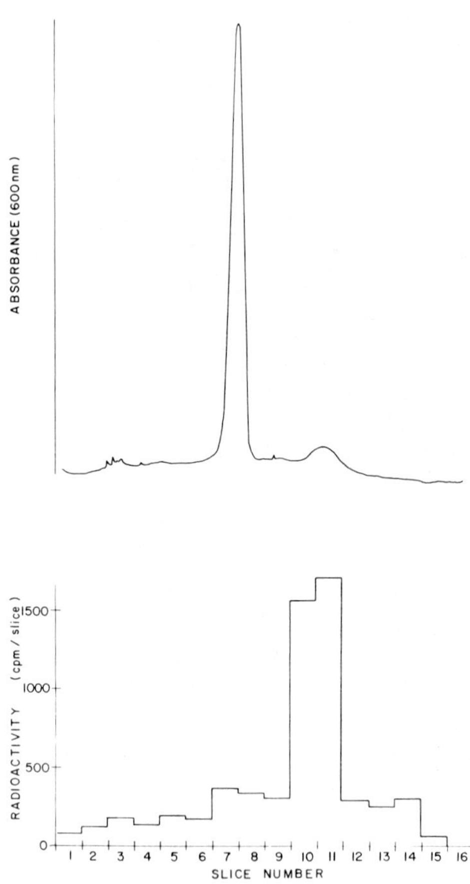

Fig.6 : [^{14}C]-DCCD labelling of the proteolipid-containing liposomes. Liposome formation and bacteriorhodopsin incorporation were performed as described in Materials and Methods, with the butanol supernatant containing approximately 50 μg of proteolipid protein. Upon incubation (room temperature, 1 hr) with [^{14}C]-DCCD (40 μM), the liposome suspension was solubilized (2% sodium dodecyl sulfate, 2% mercaptoethanol) and electrophoresed as described in Fig.1.

The isolated proteolipid fraction, when reincorporated into a liposomal system, appears to retain some of the functions ascribed to it *in situ*. Namely, it mediates proton translocation in the reconstituted systems, the H^+ leak being sensitive to DCCD. It should be emphasized that the system under study does not allow the direct measurement of the proton-translocating capacity of the proteolipid. Its implication in such a role is based on the increased accumulation of protons observed in the presence of DCCD and on the fact that this DCCD effect is due to its binding preferentially to the proteolipid. It is conceivable that the mere addition of protein to a liposome enhances its permeability to protons. Thus, addition of bacteriorhodopsin to liposomes leads, as expected, to accumulation of protons within the liposomes in the light, but a rapid leak ensues in the dark. However, such a leak is completely insensitive to added DCCD. Furthermore, as seen in Fig.6, there is little, if any, binding of DCCD to the bacteriorhodopsin moiety.

Even though, as noted, DCCD increased significantly the proton accumulation in the proteolipid-containing liposomes, it did not appear to modify the rates of translocation either in the light or in the dark (Fig.5). The most plausible explanation is that a mixture of liposomes is present. Some, without appreciable functional proteolipid, lead to the uptake observed in the absence of DCCD. The rest, containing the proteolipid, show no net uptake in the absence of DCCD, due to the proteolipid-induced leak. Upon addition of DCCD, the proteolipid would cease to function as a proton translocator and under these conditions all liposomes would behave as if no proteolipid were present. Preliminary experiments (N. Nelson, unpublished results) indicate that the two sets of vesicles indeed exist and can be separated by density centrifugation.

Results not presented indicate that liposomes containing only bacteriorhodopsin show a proton accumulation and release similar to that depicted in Fig.5 for liposomes in the presence of the DCCD-inhibited proteolipid.

The reported findings suggest that the native state of the proteolipid is preserved during the isolation and reconstitution experiments. The available evidence indicates that, in the chloroplast, the proteolipid resides within the membrane bilayer. Therefore, its native conformation might depend on its being surrounded by an apolar environment. Thus, during isolation, exposure to polar solvents should be minimized. Butanol is a weakly protic solvent that might accomplish this requirement. This is in contrast to the usual water/methanol/chloroform previously utilized for proteolipid isolation. Furthermore, the procedure utilized is rapid and therefore minimizes the time of exposure to the solvents.

In these results one surprising finding is the fact that DCCD does not react with all the carboxyls of the many proteins present in the membranes. It is likely that the DCCD partitions into the lipid phase and that it might react there with those carboxyls present therein[18]. A proteolipid might have such burried carboxyls. Alternatively, the DCCD might bind to a hydrophobic pocket of a membrane protein.

It is interesting to note that the cationic dye ethidium bromide binds to mitochondrial membranes with enhanced fluorescence and that it indicates few binding sites of high affinity[32]. Ethidium shows enhanced fluorescence only on binding to apolar-negatively charged surfaces. The binding of ethidium appears to be to the energy transducing ATPase based on its fluorescence behaviour in the presence of inhibitors[32]. Furthermore, ethidium diazide has been shown to covalently bind to the DCCD-sensitive proteolipid[33]. These findings indicate that the DCCD-binding proteolipid must be characterized by the presence of a carboxyl(s) in the vicinity of a hydrophobic region. This would be in accord with its function to interact with protons.

ACKNOWLEDGEMENTS

It is impossible in such a short article to give proper credit to the many observations on which the subject matter is based. However, these may be found in the references of those articles, mainly from our own work, which were cited.

REFERENCES

1. Unwin, P.N.T., and Henderson, R. (1975) J. Mol. Biol. 94, 425.

2. Henderson, R., and Unwin, P.N.T. (1975) Nature (London) 257, 28.

3. Klip, A., and Gitler, C. (1974) Biochem. Biophys. Res. Commun. 60, 1155.

4. Gitler, C., and Klip, A. (1974) in : Perspectives in Membrane Biology (Estrada, O.S., and Gitler, C., eds.), New York, Academic Press, p.149.

5. Klip, A., and Gitler, C. (1976) in : Mitochondria : Biogenesis, Structure and Function (Packer, L., and Gomez-Puyou, A., eds.), New York, Academic Press, p.315.

6. Chakrabarti, P., and Khorana, H.G. (1975) Biochemistry 14, 5021.

7. Greenberg, G.R., Chakrabarti, P., and Khorana, H.G. (1976) Proc. Nat. Acad. Sci. U.S.A. 73, 86.

8. Bercovici, T., and Gitler, C. (1977) Biochemistry, submitted for publication.

PROB
BY FLUORESCEIN DE

N.G.Oikonomako
The National H
48,

SUMMARY

The absorption
are altered during
proteins. These in
sites (eosin) or co
(FlNCS, FDMA, mercu
derivative-protein
ical solvent polari
glycogen phosphoryl
as optical probes v
the reaction rate o
glucose-1-P, and gl
enzyme and that P_i
of phosphorylase b
and that d) SDS den
cules of SDS were b
without affecting c

INTRODUCTION

An increasing re
cific sites on memb
environment of a pa
the recent advances
cular level[1]. Spect
ly bound or adsorbe
the chemical and ph
known that strong b
place predominantly

(*) Abbreviations u
phosphate α-glucosy
isothiocyanate, iso
mercurochrome, dibr
tetrabromofluoresce
mylated protein; SD

9. Karlish, S., Jorgensen, P.L., and Gitler, C. (1977) Nature (London), submitted for publication.

10. Kahane, I., and Gitler, C. (1977) in preparation.

11. Sigrist-Nelson, K., Sigrist, H., Bercovici, T., and Gitler, C. (1977) Biochim. Biophys. Acta, in press.

12. MacLennan, D.H. (1970) J. Biol. Chem. 245, 4508.

13. Jorgensen, P.L. (1974) Biochim. Biophys. Acta 356, 53.

14. Dodge, J.T., Mitchell, C., and Hanahan, D.J. (1963) Arch. Biochem. Biophys. 100, 119.

15. Steck, T.L., and Yu, J. (1973) J. Supramol. Struct. 1, 220.

16. Yu, J., Fishman, D.A., and Steck, T.L. (1973) J. Supramol. Struct. 1, 233.

17. Kahane, I., Furthmayr, H., and Marchesi, V.T. (1976) Biochim. Biophys. Acta 426, 464.

18. Jensen, J., and Ottolenghi, P. (1976) Biochem. J. 159, 815.

19. Gitler, C., Interiano de Martinez, I.A., Viso, F., Gasca, J.M., and Rudy, B. (1973) in : Membrane Mediated Information (Kent, P.W., ed.), Medical and Technical Publishing Company, Lancaster, England, Vol.2, p.129.

20. Delgadillo Gutierrez, H.J. (1977) Doctorate Thesis, Universidad Nacional Autonoma de Mexico, Mexico D.F.

21. Schram, M. (1976) J. Cyclic Nucleotide Res. 2, 347.

22. Racker, E., and Horstman, L.L. (1967) J. Biol. Chem. 242, 2547.

23. Patel, L., Schuldiner, S., and Kaback, H.R. (1975) Proc. Nat. Acad. Sci. U.S.A. 72, 3387.

24. Cattell, K.J., Lindop, C.R., Knight, I.G., and Beechey, R.B. (1971) Biochem. J. 125, 169.

25. Steckhoven, F.S., Waitkus, R.F., and van Moerkerk, H. (1972) Biochemistry 11, 1144.

26. Fillingame, R.H. (1976) J. Biol. Chem. 251, 6630.

27. Sigrist, H., Sigrist-Nelson, K., and Gitler, C. (1977) Biochem. Biophys. Res. Commun. 74, 178.

28. Gitler, C

29. Montal, M.

30. Maddy, A.H

31. Nelson, N.
    Gitler, C.

32. Gitler, C.
    193, 479.

33. deNobrega

strates, coenzymes and prosthetic groups in preference to other regions of the protein surface[5].

A number of fluorescein dyes have recently found useful in studying structure-function relationship of proteins; these studies include adsorption of the reagents on specific binding sites[6-8] or covalent modification of particular reactive groups[9-14].

We have attempted to study the spectroscopic changes of some fluorescein derivatives in a variety of solvents in order to be able to estimate the polarity of the protein environment around the fluorescein dye. Glycogen phosphorylase b was selected for our studies because it represents a favorable model enzyme for the identification of conformational changes resulting from the binding of substrates, effectors and denaturing agents[15].

MATERIALS AND METHODS

Phosphorylase b (three times recrystallized) was isolated from rabbit skeletal muscle according to the method of Fischer and Krebs[16]. Before use the enzyme solution was passed at 23° into a Sephadex G-25 column equilibrated with the appropriate buffer. Modified FlNHCS-phosphorylase b was prepared as described by Oikonomakos[17]. Glutamic-aspartic transaminase from pig hearts was isolated by the method of Jenkins et al.[18] as modified by Dimitropoulos et al.[9]

FlNCS, isomer 1, was a product of Sigma; FDMA, mercurochrome, and eosin were purchased from BDH. All organic solvents used were of the analytical reagent grade. Other materials of the highest available purity were products of Sigma (glucose-1-P, AMP, SDS, glycylglycine, bovine pancreas a-chymotrypsin, potato apyrase, jack beans urease, pancreas trypsin, bovine serum albumin), BDH (glucose-6-P), Merck (sodium b-glycerophosphate, 2-mercaptoethanol), Calbiochem. (african papaya papain) and Pharmacia (coarse grade Sephadex G-25).

Activity assays of phosphorylase b were carried out according to Illingworth and Cori[19] in the absence of 2-mercaptoethanol. Ultracentrifugation studies were performed in a MSE Centriscan 75 ultracentrifuge as described previously[20]. Absorption and difference spectra were recorded in 1 cm silica-semimicro cells (1.5 ml) by means of a Perkin-Elmer 356 Double Beam UV-visible recording spectrophotometer. Spectroscopic kinetics measurements were performed in an automatic Unicam SP 500 spectrophotometer.

RESULTS AND DISCUSSION

I. Estimation of Protein Binding Site Polarity

a). Solvent effects on the absorption spectrum of fluorescein derivatives.

As it has been shown the polarity of a solvent can be expressed
by the Z values, an empirical solvent polarity scale suggested by
Kosower[21]. Our studies have indicated that a decrease in solvent polarity caused a red shift in the absorption spectrum maximum (λ_{max})
of fluorescein dye. In addition we have demonstrated that a linear
relationship exists between the λ_{max} of the fluorescein derivative
in an organic solvent and the Z value (fig. 1). The observed correlation of λ_{max} with Z value indicates that the absorption process in
fluorescein dyes, like the one in 1-alkylpyridinium iodides, approximates a process in which a very polar state is changed to a neutral
one[4].

b). Fluorescein derivatives as probes of the microenvironment of
protein binding sites.

The λ_{max} of the four fluorescein derivatives bound to a number of
proteins, obtained from our studies as well as from a few other reports on the subject, are given in Table 1. The data of fig. 1 and

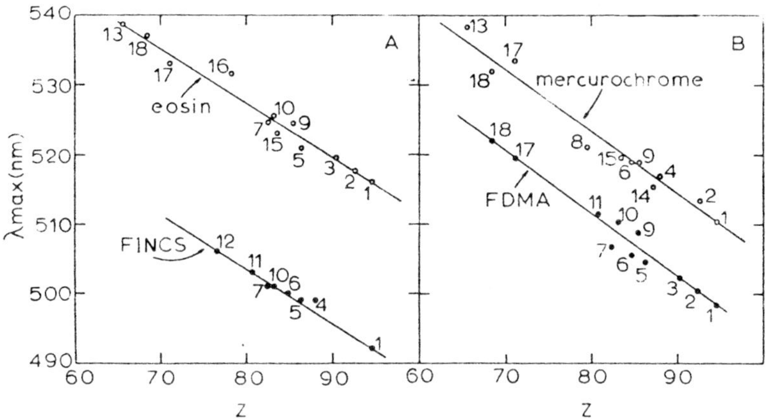

Fig. 1. Relationship between the wavelength of the absorption
maxima of fluorescein derivatives and solvent polarity expressed by the Z value. A concentrated solution of fluorescein
derivative in a 50 mM $NaHCO_3$-Na_2CO_3 buffer, pH 8.8 was diluted to a final concentration of 10 µM in several solvents
before recording the absorption spectrum. (A,o) is eosin,
(A,●) FlNCS, (B,o) mercurochrome and (B,●) FDMA. The numbers
on the figure indicate the solvents used: 1 is 50 mM $NaHCO_3$-
Na_2CO_3 buffer, pH 8.8, 2, 3, 4, 5, 6, 7, and 8 are 20, 40,
60, 70, 80, 90, and 100% of ethanol, 9, 10, 11, 12, 13 contained 60, 70, 80, 90, and 100% of acetone, 14 and 15 are 80
and 100% of methanol, respectively; 16 is n-propanol, 17 is
dimethylsulfoxide and 18 is dimethylformamide.

TABLE 1

ESTIMATION OF PROTEIN POLARITY SITES FROM THE λ_{max} OF BOUND FLUORESCEIN DERIVATIVES

The fluorescein derivative was titrated with increasing concentrations of protein until no further change in λ_{max} occurred. Titrations were carried out in a series of solutions containing 10 µM of dye in a 50 mM $NaHCO_3$-Na_2CO_3 buffer, pH 8.8 and increasing amounts of proteins. Their absorption spectra were recorded 2 hours later at 23°C.

Protein	Eosin		FlNCS		Mercurochrome		FDMA	
	λ_{max} (nm)	Z_{est}	λ_{max} (nm)	Z_{est}	λ_{max} (nm)	Z_{est}	λ_{max} (nm)	Z_{est}
Phosphorylase b	519	91	498	87	522	81.5	511	80.5
Glutamic-aspartic transaminase	520	88	497	88	520[11]	83.5	512[9]	79.5
a-Chymotrypsin	516	94.5	492	94.5	513	92	498	94.5
Apyrase	516	94.5	494	92	517	87.5	502	90.5
Urease	516	94.5	492	94.5	514	90.5	498	94.5
Trypsin	516	94.5	492	94.5	512	93	498	94.5
Papain	518	92.5	492	94.5	513	92	498	94.5
Albumin	529	76.5	501[10]	83	522	81.5	508	84
Yeast alcohol dehydrogenase	-	-	-	-	-	-	508[12]	84
D-Glyceraldheyde-3-P dehydrogenase	-	-	-	-	-	-	506[14]	86.5
Lactate dehydrogenase	-	-	-	-	-	-	512[13]	79.5

Table 1 can be used to estimate the Z value of the environment around the cojugated dye. These values (Z_{est}) are also included in Table 1.

The above Z_{est} values suggest: (1) that the microenvironment of the -SH groups of phosphorylase b, aspartate transaminase and the dehydrogenases which react with FDMA and mercurochrome is basically hydrophobic with a Z_{est} ranged between 79.5 and 86.5, (2) that the amino groups of phosphorylase b and aspartate transaminase which react with FlNCS hint a microenvironment with low polarity and comparable Z_{est} values, (3) that from the enzymes tested only phosphorylase b and aspartate transaminase (B_6 enzymes) showed hydrophobic adsorption sites for eosin and (4) that in addition to the above enzymes, albumin was also found to conjugate with the four fluorescein dyes in specific sites of very low polarity.

II. <u>Identification of Conformational Changes of Glycogen Phosphorylase b Induced by Ligands</u>

a). Reaction of phosphorylase b and mercurochrome.

It has been demonstrated that mercurochrome reacts rapidly with the -SH groups of aspartate transaminase[11] and phosphorylase b[22]. The effect of the substrate, glucose-1-P, and the activator, AMP, on the rate of the reaction between the enzyme and mercurochrome was investigated by following the enzyme activity and the change in the difference absorption maximum of mercurochrome at 533 nm[22] (fig. 2).

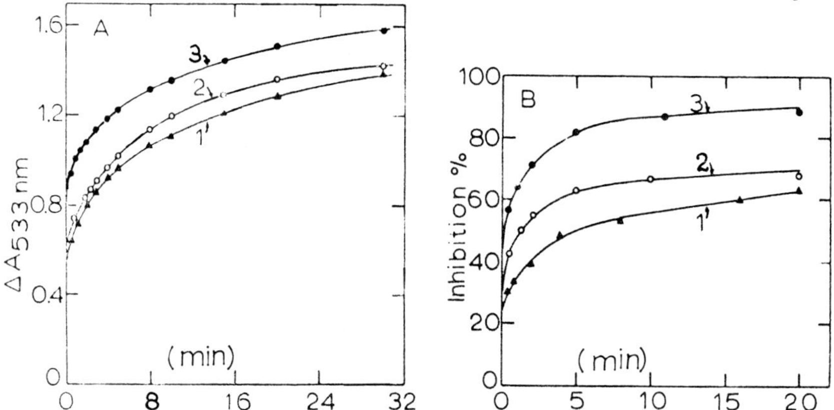

Fig. 2. Interaction of phosphorylase b with mercurochrome in the absence or presence of AMP or glucose-1-P. A). Effect of AMP and glucose-1-P on the reaction rate of phosphorylase b (4 µM) with mercurochrome (80 µM), as measured spectroscopically in a 50 mM Tris-HCl buffer, pH 8.8 at 30°C. (1) No addition, (2) with 5 mM AMP, (3) with 16 mM glucose-1-P. B). Effect of AMP and glucose-1-P on the rate of inactivation of phosphorylase b (4 µM) with mercurochrome (40 µM) in a 50 mM Tris-HCl buffer, pH 8.8 at 30°C. (1) No addition, (2) with 5 mM AMP, (3) with 16 mM glucose-1-P. Aliquots were removed at various time intervals and tested for activity.

32

The data of the fig. 2 demonstrate that the presence of glucose-1-P
and AMP increase the overall rate of the reaction. It is possible
that the enhanced reactivity arises from some conformational changes
around the -SH groups of the enzyme[23].

b). Effect of ligands on FlNHCS-phosphorylase b complex.

FlNCS, is one of the most efficient reagents for chemical modification of the $-NH_2$ groups of enzymes[17]. This is due to the fact that
the modifier exhibits a characteristic absorption spectrum with a
sharp λ_{max} at 492 nm, an area of the spectrum where neither enzymes,
substrates, cofactors or effectors absorb. So, whatever changes in
the absorption spectrum of FlNHCS-enzyme complexes in the presence
of ligands will reflect conformational changes of the examined protein molecule.

In the fig. 3 we present the effect of ligands on the spectroscop-

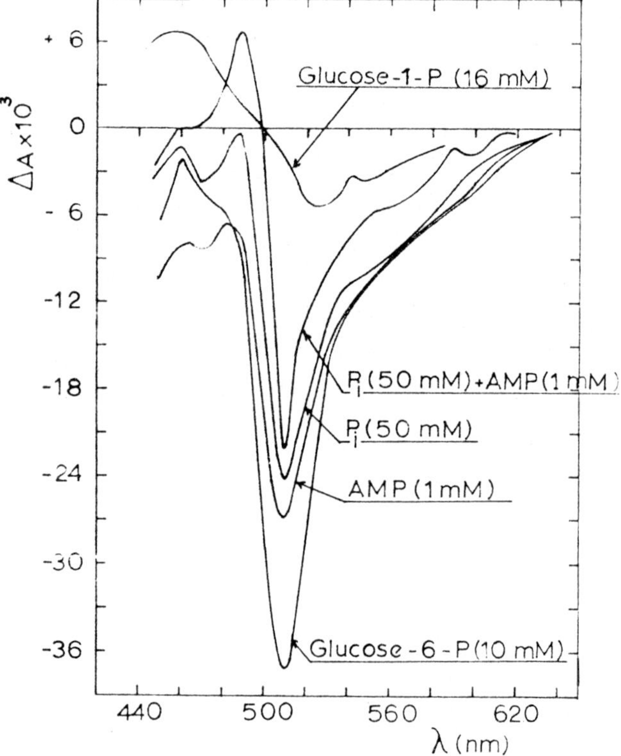

Fig. 3. Effect of various ligands on the difference spectrum of
FlNHCS-phosphorylase b complex. The difference spectra were taken
with 20 μM FlNHCS-enzyme (1.0 ε-NH_2 groups modified per dimer) in
the presence of the various ligands, against 20 μM FlNHCS-enzyme
complex in a 20 mM sodium b-glycerophosphate, 1.5 mM EDTA buffer,
pH 6.8 (23°C).

ical properties of FlNHCS-phosphorylase b complex. It can be seen that glucose-1-P, P_i, AMP or glucose-6-P cause a decrease in the difference absorption spectrum of FlNHCS-phosphorylase b conjugate. Glucose-1-P has a small effect, while P_i antagonizes the effect of AMP. The latter result is in agreement with the observations of Griffiths et al.[24]

III. <u>Identification of Conformational Changes of Glycogen Phosphorylase b Induced by Denaturing Agents</u>

a). Alkaline denaturation of phosphorylase b.

In order to study the denaturation process of phosphorylase b at high pHs we used a mercurochrome-phosphorylase b complex of a molar ratio of 2:1. The complex was prepared as described by Sotiroudis[22] and possessed 84% residual activity.

It has been reported that in contrast to acid denaturation, alkali induced denaturation of enzymes is time-dependent[25,26]. Alkaline denaturation kinetics shown in fig. 4, revealed two basic steps: the first step was very fast; an increase in the difference absorption spectrum at 522 nm was appeared instantaneously after the addition of NaOH, without any shift in the λ_{max} of the complex (522 nm). The second step exhibited a slow denaturation process as indicated by the decrease in the absorbance at 522 nm. A concomitant shift in the λ_{max} of the complex at 511 nm was recorded at the equilibrium.

Pseudo-first order rate plot of the decrease in the difference absorption spectrum at 530 nm (fig. 4) gave two linear regions with pseudo-first order rate constants of 7.5×10^{-3} and 4.9×10^{-3} min^{-1}, respectively (fig. 4, insert). The last linear region of the plot presumably represents the kinetics of alkaline denaturation of the more stable protein molecular species.

Livanova et al. have reported that the denaturation of phosphorylase b at high pHs is irreversible procedure; attempts to produce reassociation and reactivation by neutralization, resulted to protein precipitation[27]. In contrast to the previous observations, we found that when the alkali-denatured mercurochrome-phosphorylase b complex (λ_{max}=511 nm) was dialysed for 48 hours (at 4°C) against a 50 mM $NaHCO_3$-Na_2CO_3 buffer, pH 8.8, the λ_{max} of the complex returned to 522 nm. From the observed red shift during dialysis it is concluded that some refolding of the protein occurs, which restores the hydrophobic microenvironment of mercurochrome without precipitation.

b). SDS denaturation of phosphorylase b.

SDS have been used successfully for the estimation of the total number of the -SH groups in phosphorylase b or a[28]. The reagent

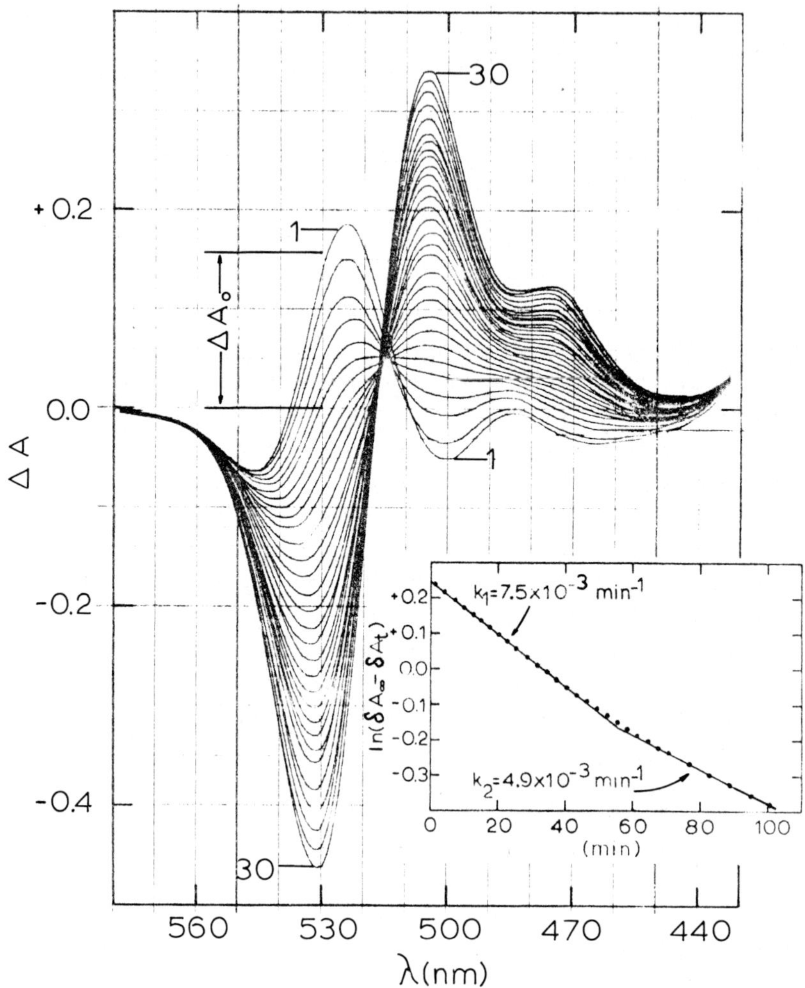

Fig. 4. Kinetics of alkali denaturation of phosphorylase b.
The absorption spectrum of mercurochrome-phosphorylase b complex
(37.5 μM mercurochrome + 18.7 μM enzyme) was recorded at various
time intervals after the addition of 0.5 N NaOH against the conjugate in a 50 mM $NaHCO_3$-Na_2CO_3 buffer, pH 8.8. Curve 1 is the difference spectrum 1.5 min and curve 30, 100.5 min after the addition of
NaOH. Insert: Estimation of the pseudo-first order rate constants
from the decrease in the difference spectrum at 530 nm.
$\delta A_t = \Delta A_o - \Delta A_t$, $\delta A_\infty = \Delta A_o - \Delta A_\infty$, where ΔA_t is the ΔA value at different times and ΔA_∞ is the ΔA value (at 530 nm) obtained at equi-
librium (24 hours).

causes dramatic changes on the tertiary and quaternary structure of
the enzyme molecule. These changes include dissociation to subunits,
unfolding of the polypeptides chains, and breakdown of the Schiff
base which unites pyridoxal-P with the apoenzyme[29,30]. It is apparent
that all these changes result in the complete inactivation of the
enzyme.

As shown in fig. 5, SDS causes characteristic changes in the absorption spectrum of FlNHCS-phosphorylase b complex, which include
a decrease in the λ_{max} of the complex (498 nm) and the appearance of
a new λ_{max} at 460 nm. It appears therefore that a FlNHCS-group in the
enzyme molecule permitts the detection of conformational changes of
the enzyme which take place during SDS denaturation.

Additional studies on the process of phosphorylase b denaturation

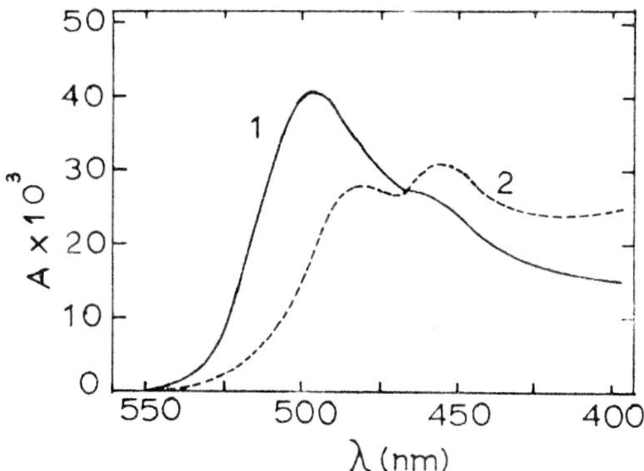

Fig. 5. Effect of SDS (2 mM) -curve 2- on the absorption spectrum
of FlNHCS-phosphorylase b complex -curve 1-. The concentration of
the modified enzyme (1.0 FlNHCS-groups bound per enzyme molecule)
was 0.84 µM in a 20 mM sodium b-glycerophosphate, 1.5 mM EDTA
buffer solution, pH 6.8 at 23°C.

by SDS were performed using eosin, a fluorescein derivative which
conjugates reversibly with phosphorylase b. The association of eosin
with the enzyme caused a red shift in the dye's spectrum (Table 1).
This perturbation on the absorption spectrum of eosin can be ascribed to the special properties of the enzyme binding sites. Difference spectra of the eosin-phosphorylase b complexes showed maxima
and minima at 536 nm and 516 nm, respectively, which can be used to
follow the titration of the enzyme with SDS.

As shown in fig. 6 the spectroscopic titration of eosin-phosphorylase b system with SDS revealed three basic steps in the denatur-

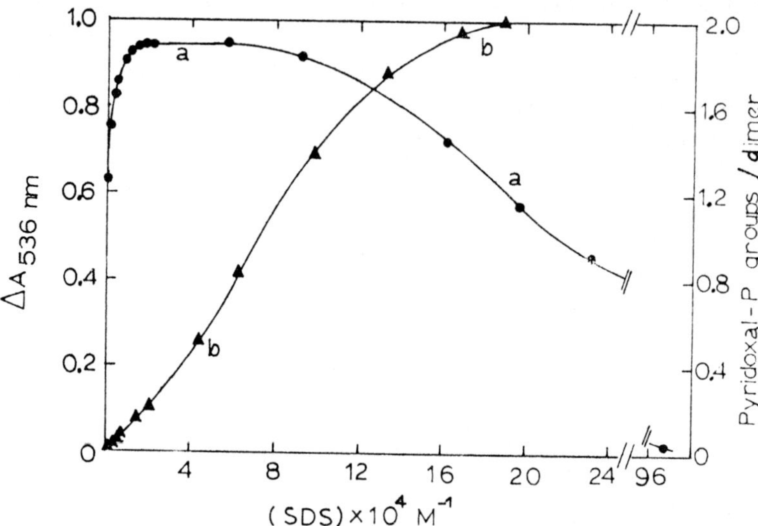

Fig. 6. Spectroscopic titrations of phosphorylase b (20 µM) with SDS in the presence (a) or absence (b) of eosin (15 µM) in a 50 mM glycyl-glycine, 5 mM 2-mercaptoethanol, 1 mM EDTA buffer, pH 6.8. Curve a was plotted from difference spectra measurements at 536 nm during SDS addition (enzyme+eosin+SDS against eosin+SDS). Curve b was plotted from absorbance measurements at 415 nm; 2.0 pyridoxal-P groups per enzyme dimer were released from the protein in equilibrium (A_{max} at 415 nm). SDS (1-20 µl) was added to the sample and the reference cell to compensate the effect of dilution. SDS has no effect on the absorption spectrum of eosin alone.

ation process. The first step (at 0-0.2 mM) is probably due to a conformational change leading to the enhancement of eosin binding or to a hydrophobic interaction between SDS and eosin at the binding sites of the dye. The second step (0.2-0.6 mM) represents a transition stage of the denaturation process, while the third step (0.6-100 mM) results in a gradual displacement of eosin from the enzyme. It has been reported that at this stage (equilibrium) one enzyme molecule binds about 700 molecules of SDS[31]. In the first titration step no significant changes in pyridoxal-P content of the enzyme were occurred (fig. 6).

As shown in fig. 7 the first molecules of SDS bind to the enzyme in a cooperative manner. A Hill coefficient (n_H) of 1.4 was esti-mated using a modified Hill equation[32],

$$\log \delta A/(\delta A_{max}-\delta A) = -\log K + n_H \log (SDS),$$

where K is the dissociation constant for SDS-enzyme complex and n_H is a measure of the strength of the interactions between SDS binding sites. Two molecules of SDS per enzyme molecule cause 70% of

the change at 536 nm and this means that   about two SDS molecules
(with a Km=19.5 µM) are bound per enzyme dimer at this stage. In parallel experiments under the same conditions it was found that the

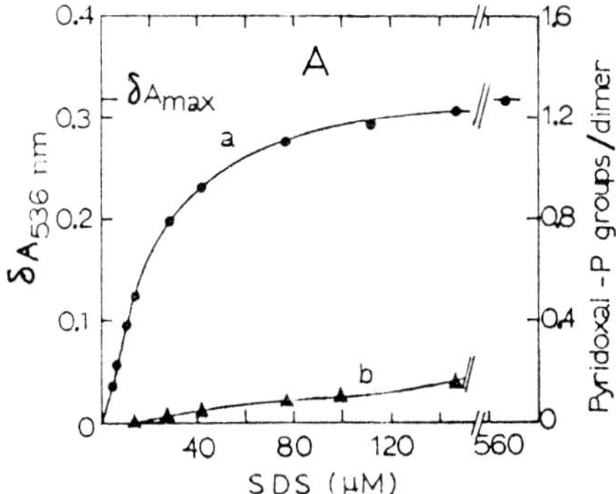
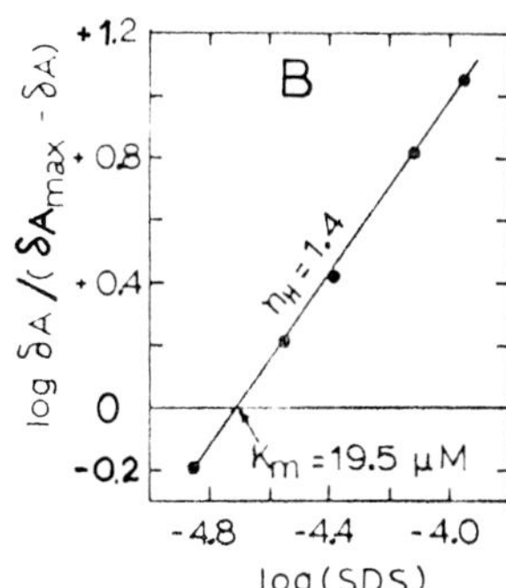

Fig. 7.   Binding of SDS to phosphorylase b. A). Replot of the first
step of titration curve shown in fig. 6. $\delta A_{536\ nm}$ represents
$\Delta(\Delta A_{536\ nm})$ -curve a-. Curve b is referred to pyridoxal-P
release (replotting again from fig. 6). B). Hill plot for phosphorylase b in respect to SDS binding; the data were taken from A.

sedimentation coefficient or the catalytic activity of phosphorylase
b does not change during the first step of SDS titration.

Supported by the National Hellenic Research Foundation.

REFERENCES

1. Carraway,K.L.(1975) Biochim.Biophys.Acta 415,379-410.

2. Horton,H.R. and Koshland,D.E.,Jr.(1967) Methods Enzymol.11,
   856-870.

3. Edelman,G.M. and McClure,W.O.(1968) Acct.Chem.Res. 1,65-70.

4. Turner,D.C. and Brand,L.(1968) Biochemistry 7,3381-3390.

5. Glazer,A.N.(1970) Proc.Natl.Acad.Sci.USA 65,1057-1063.

6. Wassarman,P.M. and Lentz,P.J.(1971) J.Mol.Biol. 60,509-522.

7. Jacobsberg,L.B.,Kantrowitz,E.R. and Lipscomb,W.N.(1975) J.Biol.
   Chem.250,9238-9249.

8. Somerville,L.L. and Quiocho,F.A.(1977) Biochim.Biophys.Acta
   481,493-499.

9. Dimitropoulos,C.G.,Oikonomakos,N.G.,Karni-Katsadima,I.A.,
   Kalogerakos,T.G. and Evangelopoulos,A.E.(1973) Eur.J.Biochem.
   38,537-552.

10. Andersson,L.-O.,Rehnström,A. and Eaker,D.L.(1971) Eur.J.Biochem.
    20,371-380.

11. Kalogerakos,T.G.,Oikonomakos,N.G.,Dimitropoulos,C.G.,Karni-Katsa-
    dima,I.A.and Evangelopoulos,A.E. Biochem.J (in press).

38

12. Heitz,J.R. and Anderson,B.M.(1968) Arch.Biochem.Biophys. 127, 637-644.

13. Markovich,D.S.,Umrikhina,N.V. and Vol'kenshtein,M.V.(1971) Molekul.Biol. 5,51-58.

14. Markovich,D.S. and Krapivinskii,G.B.(1973)Biokhimiya 38,106-112.

15. Fischer,E.H.,Pocker,A. and Saari,J.C.(1970) Essays Biochem. 6, 23-68.

16. Fischer,E.H. and Krebs,E.G.(1962) Methods Enzymol.5,369-373.

17. Oikonomakos,N.G.(1977) Ph.D. Thesis, Faculty of Physics and Mathematics, University of Athens, Athens, Greece.

18. Jenkins,W.T.,Yphantis,D.A.and Sizer,I.W.(1959)J.Biol.Chem. 234, 51-57.

19. Illingworth,B. and Cori,G.T.(1953) Biochem.Prep. 3,1-9.

20. Tzartos,S.J. and Evangelopoulos,A.E.(1977) Eur.J.Biochem. 74, 233-241.

21. Kosower,E.M.(1958) J.Am.Chem.Soc. 80, 3253-3260.

22. Sotiroudis,T.G.(1977) Ph.D. Thesis, Faculty of Physics and Mathematics, University of Athens, Athens, Greece.

23. Birkett,D.J.,Freedman,R.B.,Price,N.C. and Radda,G.K.(1970) in Chemical Reactivity and Biological Role of Functional groups in Enzymes (Smellie,R.M.S.,ed.) pp.147-155,Academic Press,London.

24. Griffiths,J.R.,Dwek,R.A.and Radda,G.K.(1976) Eur.J.Biochem. 61, 243-251.

25. Tachibana,A. and Murachi,T.(1966) Biochemistry 5,2756-2763.

26. Dyson,J.E.D. and Noltmann,E.A.(1969)Biochemistry 8,3533-3543.

27. Livanova,N.B.,Morozkin,A.D.,Pikhelgas,V.Ya. and Shpikiter,V.O. (1966) Biokhimiya 31,194-200.

28. Battell,M.L.,Smillie,L.B. and Madsen,N.B.(1968) Can.J.Biochem. 46,609-615.

29. Hedrick,J.L.,Smith,A.J. and Bruening,G.E.(1969) Biochemistry 8, 4012-4019.

30. Kleppe,K. and Damjanovich,S.(1969) Biochim.Biophys.Acta 185, 88-102.

31. Seery,V.L.,Fischer,E.H. and Teller,D.C.(1970) Biochemistry 9, 3591-3598.

32. Changeux,J.-P.(1963) Cold Spring Harbor Symp.Quant.Biol. 28, 497-504.

PHOTOAFFINITY BINDING OF CYTOCHROME C TO CYTOCHROME C OXIDASE

R. BISSON, Heidi GUTWENIGER, C. MONTECUCCO, R. COLONNA, A. ZANOTTI
and

A. AZZI

C.N.R. Unit for the Study of Physiology of Mitochondria and Institute
of General Pathology, University of Padova, Italy.

The chemical events that lead to the reduction of dioxygen to
water in the mitochondrial membrane are realized at the level of
cytochrome c oxidase. The electrons used in this process are provided by reduced cytochrome c, and energy conservation occurs simultaneously with the oxidation of the reduced oxidase. Besides
such important physiological events, cytochrome c oxidase and
cytochrome c represent also an interesting model system for studying the interaction of a multi-subunit intrinsic membrane protein
with a peripheral protein.

Cytochrome c oxidase from beef heart mitochondria has six
subunits which have been rather precisely located on the mitochondrial membrane: subunit II (mol. wt. 23,700), V (mol. wt. 11,100)
and VI (mol. wt. 8,800) on the external surface; subunits IV (mol.
wt. 14,200) and I (mol. wt. 34,000) within the membrane; and subunit III (mol. wt. 19,400) on the internal surface of the mitochondrial membrane[1,2]. Subunit II has been suggested to be the sum
of two different subunits with mol. wt. 25,000 and 21,000 respectively[2]. No indications are present however in the literature as
to the subunit of the enzyme involved in cytochrome c binding.
Birchmeier et al.[3] have proposed that subunit III of yeast cytochrome c oxidase may be involved in cytochrome c binding based
on affinity labeling studies.

The rational of the present study was to synthesize an arylazido derivative of cytochrome c with the active group located
in a position most probably involved in the binding of cytochrome
c with the oxidase, i.e. lysine 13. Such a derivative was expected to behave similarly to native cytochrome c[4] and therefore
to bind to the oxidase at the site of binding of native cytochrome
c. Light induced formation of the arylnitrene from the parent
azido derivative and insertion of the nitrene[5] at any CH group
in the surroundings should realize a covalent complex between
cytochrome c and the subunit(s) of oxidase involved in its binding.
Such a reaction scheme is shown in Fig. 1.

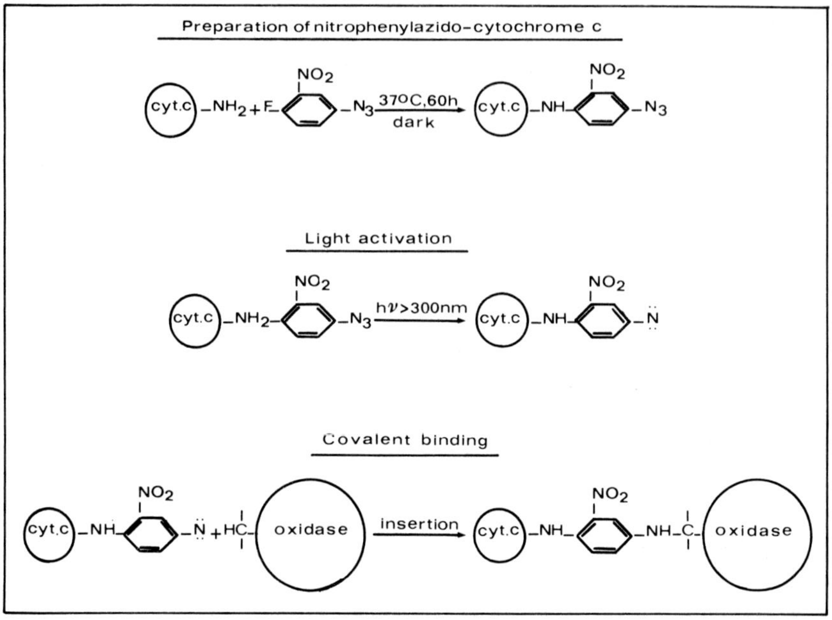

Preparation of nitrophenylazido–cytochrome c

$(cyt.c)-NH_2 + F-\langle\rangle-N_3 \xrightarrow[\text{dark}]{37^{\circ}C, 60h} (cyt.c)-NH-\langle\rangle-N_3$ (with NO$_2$ substituents)

Light activation

$(cyt.c)-NH_2-\langle\rangle-N_3 \xrightarrow{h\nu > 300nm} (cyt.c)-NH-\langle\rangle-\ddot{N}$ (with NO$_2$ substituents)

Covalent binding

$(cyt.c)-NH-\langle\rangle-\ddot{N} + HC-(oxidase) \xrightarrow{\text{insertion}} (cyt.c)-NH-\langle\rangle-NH-C-(oxidase)$ (with NO$_2$ substituents)

The derivatives of cytochrome c obtained after incubation with
fluoro nitrophenylazide were separated by Amberlite CG-50 column
chromatography into five fractions, the first eluted being named
P_4, followed by P_3, P_2, P_1 and P_0. The electrophoretic analysis of
the derivatives, together with their absorbance characteristics,
are reported in Table 1.

TABLE 1

CHARACTERIZATION OF ARYLAZIDO-CYTOCHROME C DERIVATIVES

Fraction	Electrophoretic migration %	Absorbance 480/408 nm	Trypsin sensitivity[..]	Labeled residues
native	100	0.07	yes	none
P_0	100	0.07	yes	none
P_1	88	0.105	yes	lys 22
P_2	86	0.105	no	lys 13
P_3	75	0.138	no	lys 13+22
P_4	–	0.28	no	lys 13+22+4a

[..] Trypsin digestion of 11-21 peptide.

The data indicate that the fraction named P_0 was identical to native cytochrome c, according to the above parameters. P_1 and P_2 appeared to be equally labelled (one residue per molecule) and P_3 had instead two residues per molecule.

It is well established that the first two lysine residues labeled by a number of nucleophilic reagents are lysine 13 and 22 (cf. Table 2). It is thus conceivable that the two fractions P_1 and P_2 eluted differently from the Amberlite column but with identical spectral and electrophoretic properties were positional isomers of cytochrome c with either lys 22 or lys 13 modified. To test this hypothesis, the hemeundecapeptide fragments (11-21) of the arylazido cytochrome c derivatives were prepared according to Harbury and Loach[6] by extensive pepsin digestion. The sequence of the hemeundecapeptide (11-21) contains only one lysyl residue, namely lys 13. It is expected that trypsin digestion of these peptides will depend upon the availability of the free ε-amino group of lys 13. Fractions P_0, P_1 as well as native cytochrome c peptides were fully digested while those from fractions P_2 and P_3 were not. Such an experiment indicates that fraction P_2 and P_3 had the ε-amino group of lysine 13 blocked. Therefore most probably fraction P_1 was labeled at the other mostly reactive lysine, namely lys 22, which was also most probably labeled in fraction P_3, besides lys 13. This conclusion was also supported by the activity data of cytochrome c derivatives (Table 2) consistent with the

TABLE 2

THE LABELING OF CYTOCHROME c LYSINE WITH CHEMICAL REAGENTS

Reaction	No. of residues modified	Residue No.	Increase in K_m %	Ref.
Acetylation	1	22	none	(7)
	2	22+13	+++	
Trinitrophenylation	1	13	300	(8)
	2	13+22	+++	
NBD-Cl	1	13	550	(9)
Trifluoroacetylation	1	22	10	(10)
	1	13	490	
Fluoro-nitrophenylazide	1	22	none	
	1	13	300	
	2	13+22	400	

+++ stands for an increase in K_m, not given in quantitative terms.

The data of fluoro-nitrophenylazide cytochrome c were determined in this study.

concept that fraction P_1 had the same K_m of native cytochrome c and fractions P_2 and P_3 had a 3-4 fold higher K_m, when measured with cytochrome c oxidase.

Illumination of cytochrome c oxidase in the presence of cytochrome c derivatives at lysine 22, 13 or both, resulted in the formation of covalent complexes between the two proteins, as established after their purification in Amberlite CG-50 chromatography or ammonium sulfate precipitation. The repetion of the illumination procedure in the presence of freshly added cytochrome c derivatives resulted in an increase of complex formation. It should be mentioned that, in the absence of added cytochrome c, the complexes were inactive, as measured from the oxygen consumption in the presence of ascorbate plus TMPD. The binding data can be compared instead with the activity of the complexes in transferring electrons from reduced cytochrome c, added in excess, to oxygen.

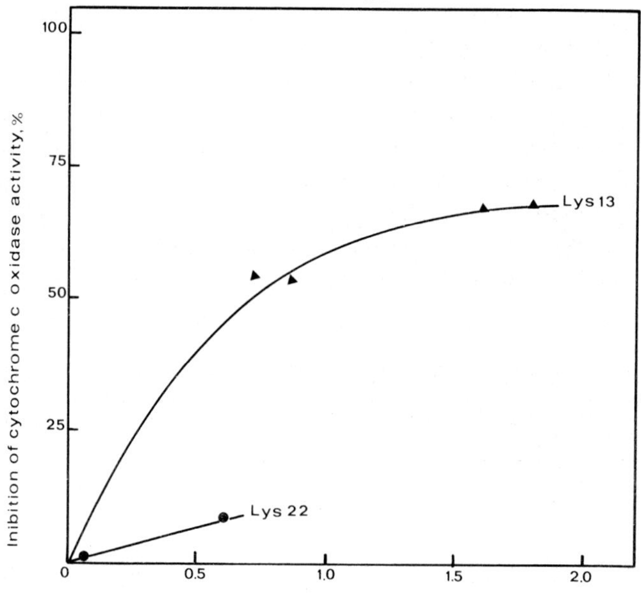

Fig. 2 – Catalytic activity of cytochrome c–cytochrome oxidase complexes. The catalytic activity of the complexes between cytochrome oxidase and cytochrome c derivatives (or native cytochrome c) was measured in a polarographic apparatus using a Clark type oxygen electrode, in the following assay system: 50 mM K-phosphate, pH 6.0, 1 mM EDTA, 20 mM Na-Ascorbate, 0.7 mM tetramethyl-p-phenylene-diamine, 0.1 μM cytochrome oxidase (or its complexes), 0.6 mg asolectin in a total volume of 3.5 ml. Bound cytochrome c was measured from the absorbance at 550-540 nm.

From the data of Fig. 2 it appears that when the covalent complex was established through lys 22 little if any activity was lost.
However, when the complex was established through lysine 13 a
decrease in activity could be observed, which became higher at
higher amounts of cytochrome c bound. The lack of a direct propor-
tionality could be attributed to some non specific binding of cytochrome c to phospholipids which was also evident from the polyacrylamide gel electrophoresis in the presence of Na-dodecylsulfate.

The conclusion which can be drown from the experiments described
above is consistent with a covalent binding of arylazido-13-cyto-
chrome c to, or close to, the catalytic site of the enzyme, but in
a somewhat distorted situation, preventing electron transfer from
bound ferro-cytochrome c to oxygen and inhibiting at the same time
binding of catalytically active, native cytochrome c. This situation was specific, since arylazido-22-cytochrome c was far less
efficient in inhibiting electron transfer. A polypeptide analysis
of the complexes formed between oxidase and arylazido cytochrome c
derivative was carried out by polyacrylamide gel electrophoresis in
the presence of Na-dodecylsulfate.

In Fig. 3 the polypeptide profile obtained under standard conditions after staining with amido black and scanning the colored
gel at 610 nm is shown. The illuminated oxidase in the presence of
native cytochrome c or in the presence of arylazido-22-cytochrome
c were not different neither in terms of the number of polypeptides nor in their apparent relative proportions or molecular weights.
Thus the small binding observed to occur between this derivative
and the oxidase may have occurred to sites other than protein, and
therefore most probably to phospholipid. The polypeptide profile
of the oxidase illuminated in the presence of arylazido-13-cyto-
chrome c derivative or (not shown), of the arylazido-13,22-deriva-
tive resulted in clear alterations of the polypeptide pattern. The
most evident change appeared to be a large decrease of the polypeptide of mol. wt.23,700 (band II) which, it should be recalled,
is one of the three polypeptides located on the outer surface of
the inner mitochondrial membrane. The second, less evident, but
always present, was a substantial increase in the area occupied by
the polypeptide of mol. wt. 34,000 (band I). The two phenomena may
be interpreted in terms of a summation polypeptide formed between
cytochrome c and subunit II, whose molecular weights added together
are approximatly 36,000. Such a polypeptide would migrate, in the

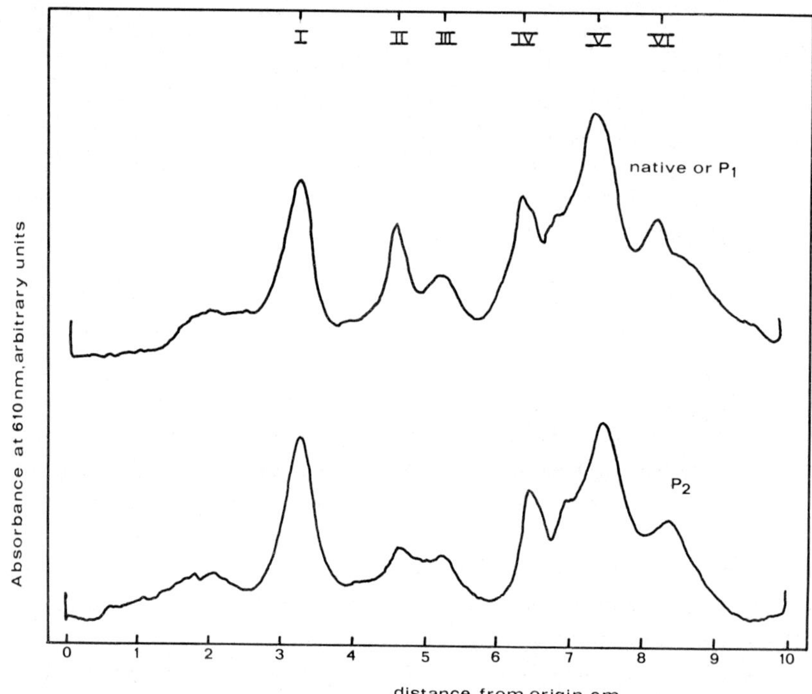

Fig. 3 - Polypeptide analysis of cytochrome oxidase-cytochrome c
covalent complexes. The interaction of cytochrome c derivatives
with cytochrome c oxidase was studied under the following conditions: 2 µM cytochrome c oxidase was dissolved in 0.5 mM EDTA-tris
pH 6.5 containing 0.1 % tween 80 in the presence of 12 µM cytochrome c or its derivatives. The samples in a 4 mm thick layer were
illuminated by a 100 W ultraviolet lamp shielded with a glass filter
for 45 min at 0 C under stirring at a distance of 10 mm. After illumination, the samples (2 ml) were rapidly passed through a Pasteur
pipette (1 x 80 mm) filled with Amberlite CG-50 equilibrated with
50 mM ammonium acetate pH 7.2. The eluate was lyophilized, dissolved in a small volume and used for spectral analysis and gel electrophoresis. The protein content of the samples was 7 mg/ml and
fraction of 20 µl were applyed to the gel. Dissociation of the pro
tein complex was obtained by treatment with 25 mM tris-acetate pH
8.2 containing 3% Na-dodecylsulfate, 1.5 % β-mercaptoethanol and 2
mM ethylene-diamino tetracetate. Electrophoresis was carried out in
a 1 mm thick gel of 100 x 140 mm dimension of 12% acrylamide.

gel technique employed,at the same approximate rate of band I.

Scanning of the unstained gel with 410 nm light, revealed (uniquely

for the arylazido-13/and arylazido-13,22-cytochrome c derivatives),

an absorption at mol. wt. 35,000, in correspondence of band I. The

three data, when considered together, strongly support the idea

that, when arylazido derivatives of cytochrome c at position 13 are

reacted with the oxidase, a covalent complex is established, which

involves subunit II.

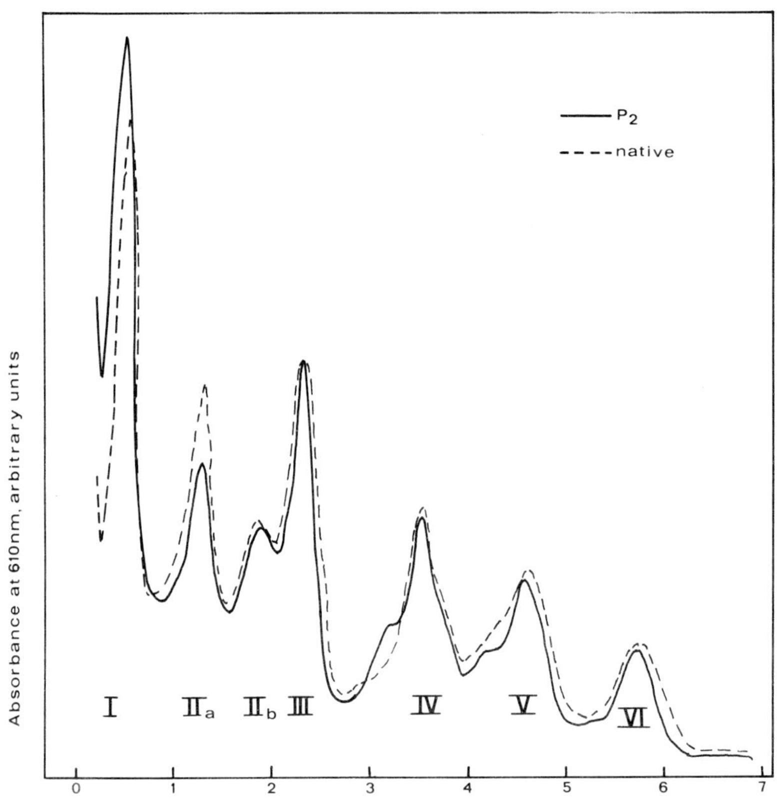

Fig. 4 - Polypeptide analysis of cytochrome oxidase cytochrome c covalent complexes in the presence of urea. The experimental conditions were as in Fig. 3. The analysis was carried out according to ref. 2.

Under the conditions described by Downer et al.[2] resolution of cytochrome oxidase in seven polypeptides was obtained either with the native enzyme or that illuminated in the presence of arylazido cytochrome c derivatives. In Fig. 4 an experiment is shown in which the oxidase was illuminated in the presence of cytochrome c or of its arylazido-13-derivative. From the two traces it is possible to conclude that the illumination in the presence of arylazido-13-cytochrome c resulted in a diminution of band IIa (the one of apparent higher mol. wt.) without affecting band IIb (the one of apparent lower mol. wt.). Thus polypeptide IIa was the one more directly involved in cytochrome c binding, since no substantial changes of the other bands could be seen. Arylazido-22-cytochrome c derivative was not effective in producing any polypeptide change under the above conditions.

The data reported in this study may be summerized in a model shown in Fig. 5.

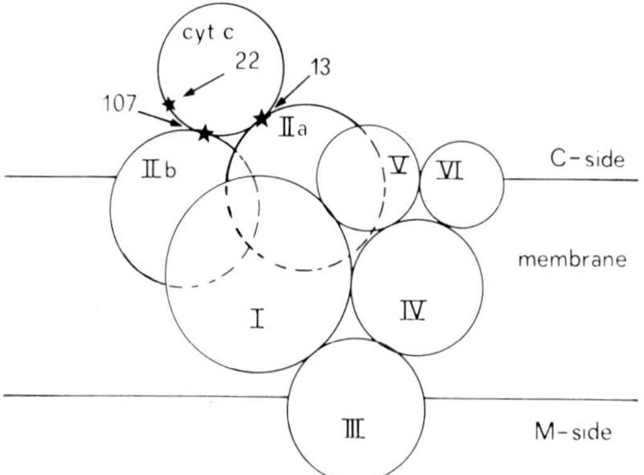

Fig. 5 – Model of the binding of cytochrome c to the oxidase.

The interaction of cytochrome c with subunit IIa through lysine 13 appears in the model, as inferred from the present study. Another adjacent subunit (IIb, for beef-III for yeast) may be however responsible for forming part of cytochrome c binding site as concluded from the study of Birchmeier et al.[3] Lys 22 should not interact with the protein, but rather with the lipid. Such a picture is apparently consistent with the intramolecular distances and location of the different residues of cytochrome c as obtained from X-ray diffraction analysis of the crystalline protein.

REFERENCES

1. Eytan, G.D., Carrol, R.C., Schatz, G. and Racker, E. (1975) J. Biol. Chem. 250, 8598-8603

2. Downer, N.W., Robinson, N.C. and Capaldi, R.A. (1976) Biochemistry 15, 2930-2936

3. Birchmeier, W., Kohler, C.E. and Schatz, G. (1976) Proc. Natl. Acad. Sci. U.S.A. 73, 4334-4338

4. Dickerson, R.E. and Timkovich, R. (1975) in The Enzymes (Boyer, P.D., ed.) vol. 2, pp. 397-547, Academic Press, New York

5. Knowles, J.R. (1972) Acc. Chem. Res. 5, 155-160

6. Harbury, H.A. and Loach, P.A. (1960) J. Biol. Chem. 235, 3640-3645

7. Wada, K. and Okunuki, K. (1969) J. Biochem. (Tokyo) 66, 263-271

8. Wada, K. and Okunuki, K. (1969) J. Biochem. (Tokyo) 66, 249-262

9. Margoliash, E., Ferguson-Miller, S., Tulloss, J., Kang, C.H., Feinberg, B.A., Brautigan, D.L. and Morrison, M. (1973) Proc. Natl. Acad. Sci. U.S.A. 70, 3245-3249

10. Staudenmayer, N., Smith, M.B., Smith, H.T., Spies, F.K. and Millet, F. (1976) Biochemistry 15, 3198-3205

ROTATIONAL DIFFUSION OF VARIOUS MEMBRANE BOUND PROTEINS AS DETERMINED BY SATURATION TRANSFER EPR SPECTROSCOPY .

Philippe F. Devaux, Anne Baroin, Alain Bienvenue
Edith Favre, Annie Rousselet and David D. Thomas
Institut de Biologie Physico-Chimique
(ERA-CNRS 690 : Sondes Moléculaires dans les Biomembranes)
13, rue P. et M. Curie - 75005 Paris

SUMMARY

Saturation transfer spectroscopy permits the determination of the rotational correlation time τ_2 of spin labeled proteins for $10^{-6} < \tau_2 < 10^{-3}$ sec. By application of this technique we have shown that, while rhodopsin in the disc membranes rotates with a correlation time τ_2 of $5 \ 10^{-5}$ sec. at 4°C, the cholinergic receptor protein in Torpedo marmorata membrane fragments is completely immobilized at the same temperature ($\tau_2 > 10^{-3}$ sec). Interpretation of the spectra obtained with spin labeled acyl-atractylosides in rat heart mitochondria suggests that the atractyloside binding protein could interact with some other protein unit in the mitochondrial membrane.

INTRODUCTION

Intrinsic membrane bound proteins are embedded in a lipid matrix. Experimental evidence suggests that, in most membranes, the lipid matrix forms a fluid bilayer where most, if not all the phospholipids diffuse laterally. If these proteins are attached to the membrane by simple hydrophobic interactions with the lipids, they should also diffuse laterally and be randomly distributed in the plane of the membrane. There is however experimental evidence suggesting protein organization in many membranes. Also on cell surfaces, under certain conditions, it has been shown that some proteins cannot diffuse (1). These restrictions to the motion of membrane bound proteins certainly have biological significance and are important to be detected. To study them it is necessary to be able to measure the parameters associated with protein motion. The rotational correlation time τ_2 is the relevant parameter that characterises the rotational diffusion.

The use of fluorescence labels as well as spin labels has been moderately successful in the past for the determination of the rotational correlation times of membrane bound proteins, simply because their motion is slow. In 1972 Cone (2) using transient photodichroism was able to measure the rotational correlation time of membrane-bound rhodopsin. More recently Cherry (3) showed that the bacterio-rhodopsin has a very low rotational diffusion constant. In the present paper we describe the application of a new EPR technique referred to as saturation transfer spectroscopy. It will be shown that it is possible to measure by this technique rotational correlation times as slow as $\tau_2 \simeq 10^{-3}$ sec and therefore to

48

discriminate between different types of membrane bound proteins whose correlation times for rotation are in the range of 10^{-5} to 10^{-3} sec. For example, it is possible to determine if a protein is freely "floating" in a lipid bilayer or not. Considering that the lateral mobility is clearly associated with rotational mobility, the determinations of τ_2 give also clues on the lateral mobility of the proteins considered.

<u>Saturation transfer spectroscopy</u>

Classical EPR spectroscopy of nitroxide radicals has been widely used over the past decade to study the motion of lipids in biological membranes (see Spin labeling, L. Berliner ed. Acad. Press 1976). The sensitivity of the technique covers the range of correlation times : 10^{-10} to 10^{-7} sec which is very appropriate for the study of chain motion. However the motion of membrane bound proteins such as rhodopsin which is known to diffuse freely in the lipids, is already too slow to be measured with the classical EPR technique.

Recently a new technique (saturation transfer spectroscopy) was developped by J. Hyde, L. Dalton and D. Thomas (4, 5). Technically the saturation transfer spectra are obtainable with a classical EPR spectrometer provided with the second harmonic display and a precise tuning of the phase on the phase detector. Indeed saturation transfer spectra are recorded with the second harmonic at 90° out-of-phase, with a high power (H_1 = 0.25 G). Figure 1 shows some reference spectra with the corresponding correlation times (5).

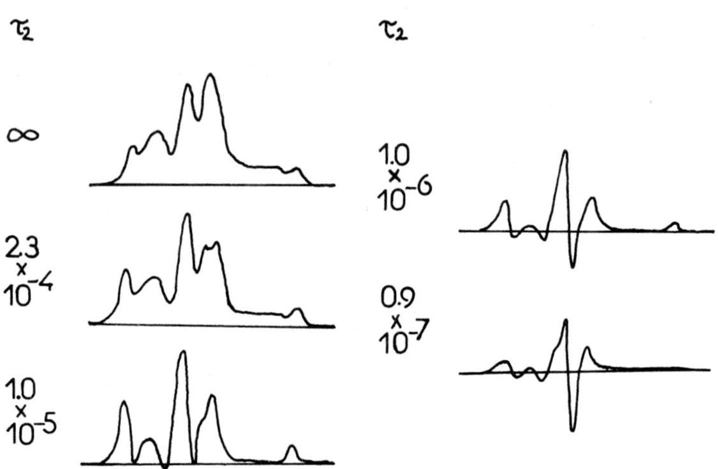

Fig. 1 Maleimide spin labeled hemoglobin : reference spectra (reproduced from D.D. Thomas et al. (4)).

D.T. Thomas et al (6) have applied successively this technique to study a biological system, namely subfragment 1 in Myosin. We have extended this technique to study slow anisotropic motions experienced by membrane bound proteins such as rhodopsin, the cholinergic receptor of Torpedo marmorata and finally the ADP carrier in rat heart mitochondria.

It will be shown that even if lipids have an apparent high degree of fluidity, the motion of the proteins in two different membranes can be very different. Lipids therefore do not entirely determine the mobility of the proteins.

<u>Rotational diffusion of rhodopsin and fluidity of the lipid annulus around rhodopsin.</u>

We have spin labeled membrane bound rhodopsin with different maleimide derivatives. A short chain derivative (Syva, Palo-Alto) :

SPIN LABEL I 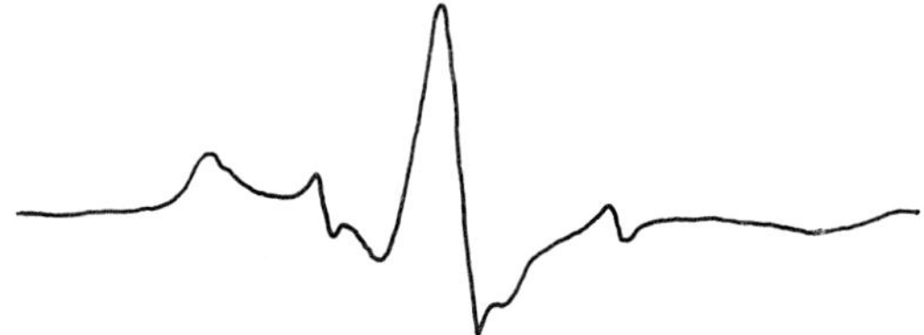

shows by classical EPR evidence of a strong immobilization (apart from 10% of the signal due to a weakly immobilized component (Fig. 2).

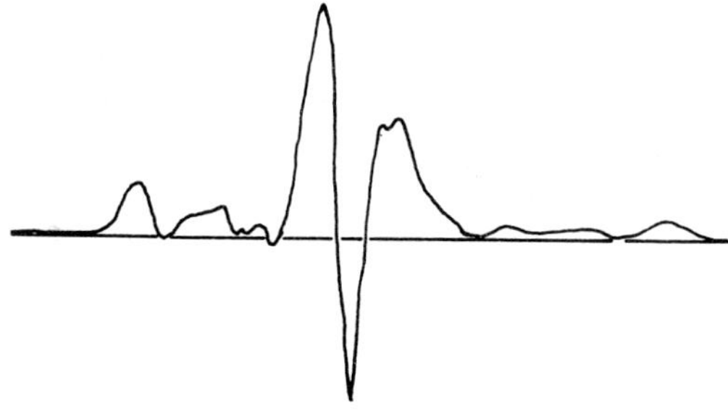

Fig. 2 : First harmonic EPR spectrum of membrane bound rhodopsin labeled with spin label I (4°C).

The second harmonic out-of-phase recording gives the spectrum of Fig. 3

Fig. 3 : Second harmonic out-of-phase spectrum of membrane bound rhodopsin labeled with spin label I.

The analysis of this spectrum according to the criteria of Thomas et al. (5), enables us to estimate the correlation time of the protein to be 50 μ sec at 4°C, which is very close to Cone's results (20 μ sec at 20°C) (2). The addition of glutaraldehyde (5%) results in a complete immobilization since the saturation transfer spectrum corresponds now to a correlation time $\tau_2 > 10^{-3}$ sec. In a similar way, delipidation completely stops the motion of the probe (50% decrease of the phospholipid to protein ratio by treatment with phospholipase A_2). Rhodopsin was also labeled with the following components :

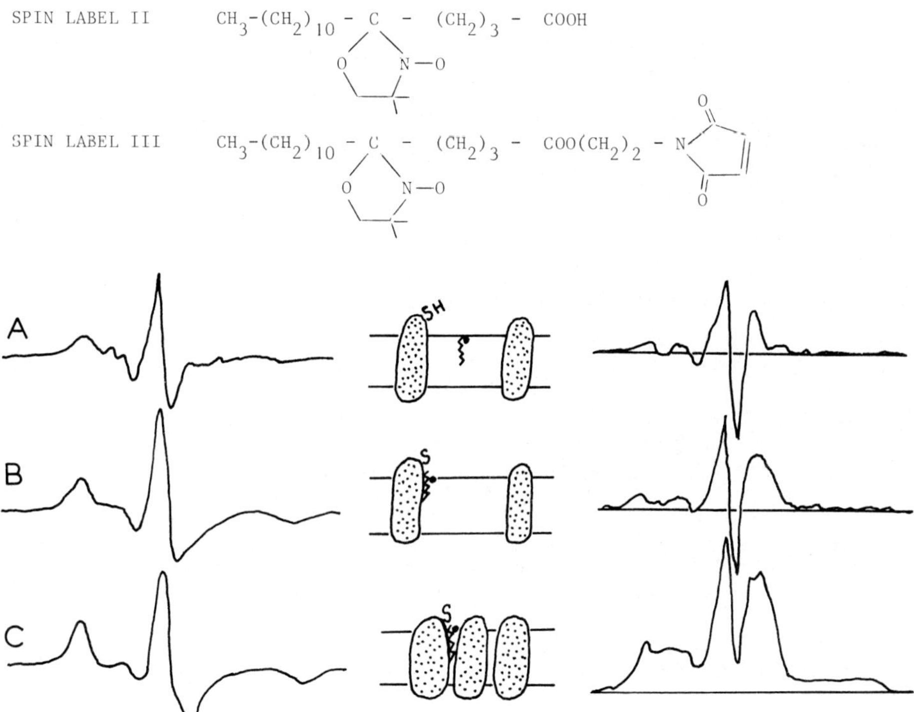

Fig. 4 : Left 1st harmonic spectra. Right 2nd harmonic out-of-phase spectra.
(A) Rod outer segments membrane fragments labeled with spin label II.
(B) Membrane bound rhodopsin labeled with spin label III. (C) same as (B) but in delipidated membrane fragments.

Spin label II is simply added to the membrane preparation and is dissolved in the lipids. When labeling the membranes with spin label III, the unreacted maleimide derivatives were removed with B.S.A.. Only a fraction of these spin labels binds to available SH groups. It can be shown that the binding sites

where the maleimide reacts are directly exposed toward the aqueous phase. This is
demonstrated by systematic studies with ascorbate, a classical reducing agent of
nitroxides.

Figure 4 shows on the left the usual EPR spectra that can be obtained with
spin labels II and III in normal and delipidated membranes. The right side of
Figure 4 shows the same set of spin labels recorded with the second harmonic
out-of-phase display. One striking feature of these saturation transfer spectra
is the difference obtained with the fatty acid chain covalently attached to the
protein between normal membranes and delipidated membranes. This difference
demonstrates that the lipid annulus has a viscosity depending on the presence
of the bulk lipids and therefore suggests a continuity between the lipid annulus
and the bilayer.

Absence of motion of the cholinergic receptor in Torpedo Marmorata membrane
fragments.

The cholinergic receptor protein can be studied in its membranous state with
Torpedo membrane fragments (7). We used two different spin labels. The maleimide
derivative, called spin label I in the previous paragraph, and the following
specific spin labeled ligand (8) :

SPIN LABEL IV $\qquad CH_3-(CH_2)_7 - \underset{\underset{|}{\overset{/\overset{O}{\diagdown}\ \overset{N-O}{\diagup}}{\diagdown\diagup}}}{C} - (CH_2)_6 - COO-(CH_2)_2 - N^+(CH_3)_3$, $\qquad I^-$

Both spin labels show by saturation transfer that the cholinergic receptor
proteins are completely immobilized ($\tau_2 > 10^{-3}$ sec), see figure 5.

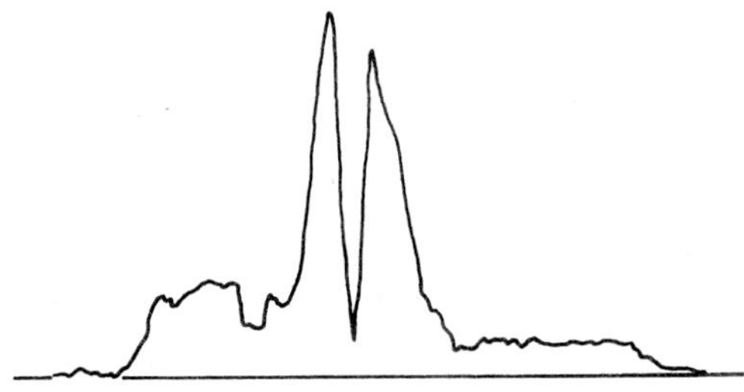

Fig. 5 : Second harmonic out-of-phase spectrum of cholinergic receptor rich
membrane fragments of Torpedo marmorata labeled with spin label I (4°C).

Therefore it seems that strong protein-protein interactions exist between adjacent proteins of the post synaptic membrane.

The ADP carrier in mitochondria

It has been demonstrated that,next to the ADP carrier in mitochondria, fluid lipids exist (9). However there are some indications that the ADP carrier is associated, at least during phosphorylation, with the ATPase (10). Quantitative knowledge of the rotational freedom of the atractyloside binding protein, assumed to be the ADP carrier, would help to decide about this possible protein-protein association. We have studied rat heart mitochondria with spin labeled long chain acyl-atractyloside, a very specific probe of the ADP carrier (9) :

SPIN LABEL V $\quad$ $CH_3-(CH_2)_{10} - C - (CH_2)_3 - COO - Atractyloside$

This probe gives rise by classical EPR to a completely immobilized signal. Figure 6 shows the spectrum one can obtain by saturation transfer.

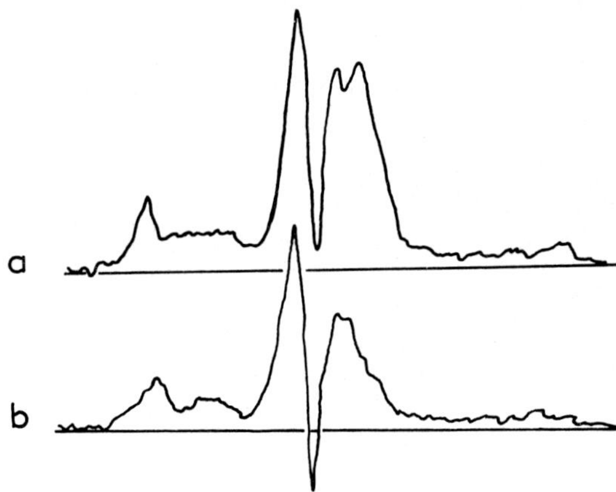

Fig. 6 : Second harmonic spectra of rat heart mitochondria labeled with spin label V. (a) spin label V attached to the ADP carrier. (b) spin label V displaced from the ADP carrier by atractyloside 10^{-3} M.

Top spectrum corresponds to the protein bound spin label (concentration of spin labels about half of that of the sites) and lower spectrum corresponds to the displaced spin label (atractyloside is added and the concentration of free spin labels increased). The spectrum of the displaced probe reflects its motion when in the lipid phase.

The protein bound spectrum of figure 6 corresponds according to the criteria put forward by Thomas et al. (5) to a correlation time of about 10^{-4} sec. However the probe is rather far from the binding site and may reflect more motion than what is actually experienced by the protein ($\tau \simeq 10^{-3}$ sec ?). With a molecular weight of about 50 000 (10) for the atractyloside binding protein one would anticipate a correlation time very close to the value obtained for the rhodopsin (m.w. 40 000). If the value of 10^{-3} sec is to be taken seriously, then the atractyloside binding protein should be associated with some other component that slows down its motion.

CONCLUSIONS

Saturation transfer spectroscopy seems to be a promising technique to study the rotational motions of molecules in the correlation times range of $5 \times 10^{-7} <$ $\tau_2 < 10^{-3}$ sec. This corresponds to the appropriate range for the rotation of membrane bound proteins. More theoretical work has to be done to simulate the spectra corresponding to anisotropic motions. Simulations of saturation transfer spectra corresponding to cylindrical motions are in progress in our laboratory. We are convinced that the orientation of the membranes would make the inter-pretations simpler and therefore is an important goal to achieve.

We have already shown by this technique that fundamental differences between membrane bound proteins can be detected. Namely we can differentiate between proteins "floating" in a fluid lipid bilayer and proteins tightly anchored or so tightly packed that no motion occurs.

ACKNOWLEDGEMENTS :

This work was supported by grants from the "Centre National de la Recherche Scientifique" (ERA n° 690; "Sondes moléculaires dans les biomembranes") and the "Délégation Générale à la Recherche Scientifique et Technique, commission : Membranes Biologiques".

REFERENCES

1. Nicholson, G.L. (1976) Biochim. Biophys. Acta, 457, 57-108.

2. Cone, R.A. (1972) Nature N.B., 236, 39-43.

3. Cherry, R.J. (1975) Febs Lett. 55 ,1-7.

4. Hyde, J.S. and Dalton, L.R. (1972) Chem. Phys. Lett. 16, 568-572.

5. Thomas, D.D., Dalton, L.R. and Hyde, J.S. (1976) J. of Chem. Phys. 65, 3006-3024.

6. Thomas, D.D., Seidel, J.C., Hyde, J.S. and Gergely, J. (1975) Proc. Nat. Acad. Sci. USA 72, 1729-1733.

7. Changeux, J.P., Benedetti, L., Bourgeois, J.P.; Brisson, A., Cartaud, J., Devaux, P.F., Grünhagen, H., Moreau, M., Popot, J.L., Sobel, A. and Weber, M. (1976) Cold Spring Harbour Symposia on Quantitative Biology, Vol XL, 211-230.

8. Brisson, A.D., Scandella, C.J., Bienvenüe, A., Devaux, P.F., Cohen, J. and
    Changeux, J.P. (1975) Proc. Nat. Acad. Sci. USA $\underline{72}$, 1087-1091.

9. Lauquin, G.L., Devaux, P.F., Bienvenüe, A., Villiers, C. and Vignais, P.V.
    (1977) Biochemistry, $\underline{16}$, 1202-1208.

10. Vignais, P.V. (1976) Biochim. Biophys. Acta, $\underline{456}$, 1-38.

SURFACE POTENTIAL CHANGES ON ENERGIZATION OF MITOPLASTS AND CHLOROPLASTS

A.T. Quintanilha and L. Packer

Membrane Bioenergetics Group, Lawrence Berkeley
Laboratory, and the Department of Physiology-
Anatomy, University of California, Berkeley,
California 94720 USA

INTRODUCTION

The inner mitochondrial and chloroplast membranes generate transmembrane electrochemical proton gradients. This gradient is divided into an electrical, a measure
of the potential difference, and a concentration, related to the pH difference,
term (cf. [1]). The contribution of these terms for different membranes has recently
been reviewed[1]: respiring or ATP energized mitochondria pump out protons and
establish small pH differences estimated to be $\leq$ 1 pH unit, but large transmembrane
potentials as high as 180 mV[2-4], whereas illuminated chloroplast membranes pump in
protons and establish pH differences as large as 3 pH units, and small transmembrane
potentials estimated to be of the order of 10 mV[5]. However, a more meaningful description of the role of electrical forces at each membrane interface in the control
of energization would be derived from knowledge of surface potential changes at
each membrane interface. Since pH and ionic strength influence the surface potential of charged phospholipid membranes[6], changes in the transmembrane potential
could modify surface potentials of the inner mitochondrial and chloroplast membranes. More positive surface potentials inhibit both the H^+ pump[7] and respiration[8]
in mitochondria.

ESR studies of the partition of impermeable charged paramagnetic amphiphiles have
shown that the surface potential of charged phospholipid vesicles depends on the
ionic aqueous concentration in a way similar to that predicted by the Gouy equation[9]. This technique could, in principle, be used to measure the surface potential of biological membranes[10] and we have recently applied it to show that[11] the
inner membrane of mitochondria becomes more negatively charged upon energization
with ATP. Here we report on the use of impermeable charged paramagnetic amphiphiles
to show that energization of mitoplasts (inner membranes + matrix) with ATP and
chloroplasts by light changes the surface potential of the outer half of these
membranes.

MATERIALS AND METHODS

<u>Preparations and assays</u>: Rat liver mitochondria were prepared in a 0.25 M
sucrose, 1 mM Tris base and 1 mM EDTA medium at pH 7.6[12] and mitoplasts prepared
as described[13], and then suspended in a 0.25 M sucrose and 1 mM Tris base (pH 7.6)
medium at 20 mg protein/ml. Two incubation media were used: a) 0.24 M sucrose

56

and 10 mM Tris base (pH 7.6) and b) 0.14 M KCl and 10 mM Tris base (pH 7.6). Mitoplasts were energized by ATP and de-energized by oligomycin. Oxidizable substrates
were not used to energize the mitoplasts because electron transport reduced the
spin label[14].

Chloroplasts were prepared in a 0.4 M sucrose, 10 mM NaCl and 50 mM Tris base
medium at pH 8 as described[15] and suspended in the same medium at chlorophyll concentrations of 6 mg/ml. The incubation medium used was: c) 30 mM Tricine (pH 8),
40 mM NaCl, 10 mM $MgCl_2$ and 1 mM Na phosphate buffer (pH 8).

Chloroplasts were energized by heat filtered saturating light intensities from
a 40 W Tiyoda lamp and de-energized by NH_4Cl or carbonyl cyanide p-trifluoromethoxy-
phenylhydrazone (FCCP).

Surface potentials: Changes in the partition between the membrane and the aqueous environment of the positively-charged spin-labeled detergent 4-(dodecyl
dimethyl ammonium)-1-oxyl-2,2,6,6-tetramethyl piperidine bromide, CAT_{12} (Fig. 1A)
and of the uncharged spin labels 2,2-dimethyl-5,5-methylnonyl-N-oxazolidinyloxyl,
$2N11$[11] and 2,2,-dimethyl-5,5-methylheptyl-N-oxazolidinyloxyl, 2N9 (Fig. 1B) synthesized by Dr. R.J. Mehlhorn of our laboratory, were recorded in a Varian E109 Spectrometer at modulation amplitude of 1G, time constant of 0.128 s and microwave
power of 10 mW at room temperature.

Mitoplast assays were carried out in media a) and b) at protein concentrations
of 3 mg/ml and spin label concentrations $\leq$ 15 nmoles/mg protein. Chloroplast assays
were carried out in medium c) at chlorophyll concentrations of 0.6 mg/ml and spin
label concentrations $\leq$ 300 nmoles/mg chlorophyll and illuminated in a quartz EPR
tube of 1 mm internal diameter. The concentrations of spin label used were such
that they did not affect the rate of respiration in mitoplasts or the Hill reaction
by chloroplasts.

The partition P is defined as the ratio of the spin label concentration in the
aqueous medium to the concentration on the membrane. To determine the changes in
surface potential $\Delta\psi_s$, in mV, as a function of changes in partition P, we follow
the procedure described[9] viz.:

$$\Delta\psi_s = \frac{RT}{ZF} \ln \frac{P_1}{P_2} \tag{I}$$

where P_1 is the partition in state 1, P_2 the partition in state 2, Z is the charge
on the spin label, F the Faraday constant, R the universal gas constant and T the
absolute temperature. At 25°C and for a positive charge of unity, $RT/ZF \sim 26$ mV.

The determination of P is done by estimating the decrease in the spin label
aqueous EPR signal upon addition of the mitoplasts or chloroplasts to the respective
media. The value of P in the de-energized state was determined in the presence of
very small amounts of $K_3Fe(CN)_6$ to ensure that all the spins were oxidized but no
paramagnetic quenching was occurring. Thereafter changes in P and not its absolute
value were only of interest.

RESULTS

Mitoplasts: The effect of ATP and oligomycin on the partition P of CAT_{12} (14 nmoles/mg protein) in the sucrose or KCl medium is shown in Table I. Identical results are obtained in non-ionic or ionic buffered media a) and b). Control studies indicate that: 1) the changes in partitioning upon addition of GTP, ITP or cyclic-AMP (not shown) are oligomycin insensitive and that therefore part of the partition change on addition of ATP is probably due to binding of CAT_{12} to ATP; this is also observed in the absence of mitoplasts; and 2) the partition of the uncharged spin label 2N11 is not altered by ATP or oligomycin. When the values of the partition coefficient (Table I) are used in equation I, the potential of the outer surface of the mitoplasts is found to be $\sim$ 20 mV more negative when energized in sucrose whereas in KCl medium the corresponding value is $\sim$ 17 mV[11].

TABLE I

DEPENDENCE OF CAT_{12} SPIN LABEL PARTITIONING ON ADENINE NUCLEOTIDES

Additions	Partition coefficient (P)	
	Sucrose medium	KCl medium
None	0.11(1)	0.31(8)
ATP	0.03(9)	0.14(8)
ATP + oligomycin	0.08(3)	0.24(4)
GTP or ITP	0.08(3)	0.23(9)

Adenine nucleotides (1.33 mM) and oligomycin (3 μg/mg mitoplast protein) were tested in 0.25 M sucrose and 0.15 M KCl media.

Chloroplasts: The effect of energization of chloroplasts (in the light) and de-energization (light + 30 mM NH_4Cl) on the partition P of CAT_{12} is shown in Table II. Figure 1 shows that both spin labels used, CAT_{12} and 2N9 are reduced in the light. However, in the case of CAT_{12} in the absence of uncouplers, the aqueous signal increases when the light is turned off (Fig. 1A).

TABLE II

CAT_{12} SPIN LABEL PARTITIONING IN CHLOROPLASTS

	Partition coefficient (P)
Dark	0.25(2)
Light	0.19(0)

Medium: 30 mM Tricine, 40 mM NaCl, 10 mM $MgCl_2$, 1 mM Na phosphate (pH 8).

Control studies indicate that: 1) the uncharged spin label 2N9 which, like CAT_{12}, partitions mainly in the membrane shows no recovery of the aqueous signal in the coupled state once the light is turned off (Fig. 1B), and 2) in the presence of 30 mM NH_4Cl or 1 mM FCCP (not shown) there is no recovery of the aqueous signal of CAT_{12} when the light is turned off. This indicates that only CAT_{12} responds to changes in the energization state of chloroplasts.

When the values of the partition coefficient of CAT_{12} after 10 s of illumination and after 60 s in the dark are used in equation I, the surface potential of the outer surface of the chloroplasts is found to be $\sim$ 7 mV more negative when illuminated.

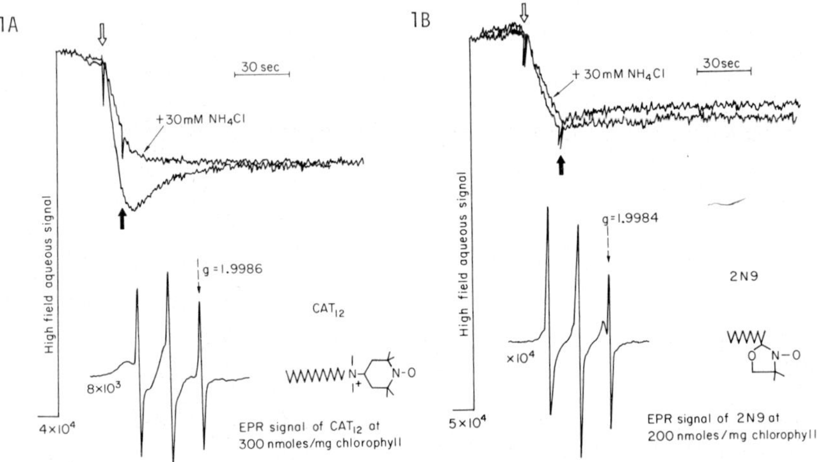

Fig. 1. Structure and EPR signal of CAT_{12} and 2N9 in chloroplasts. A) CAT_{12}-medium: 30 mM Tricine, 40 mM NaCl, 10 mM $MgCl_2$ and 1 mM Na phosphate (pH 8), 300 nmoles/mg chlorophyll at 0.6 mg chlorophyll/ml. Changes in the aqueous signal at g = 1.9986 in the light and dark are indicated by the arrows. B) 2N9-medium: as above with 200 nmoles/mg chlorophyll at 0.6 mg chlorophyll/ml. Changes in the aqueous signal at g = 1.9984 in the light and dark (arrows).

DISCUSSION

In general the direct determination of changes in surface charge at membrane interfaces is difficult to measure. Using the positively-charged amphipathic spin-label probe CAT_{12}, which partitions between membrane and aqueous domains as an indicator of changes in surface charge[10,11] we found in the present study that the surface potential of the outer surface of mitoplasts becomes about 17-20 mV more negative when energized with ATP and that the outer surface of chloroplasts becomes $\sim$ 7 mV more negative when energized by light. In other studies, electrophoretic methods have shown that whole mitochondria and chloroplasts have a negative ζ

potential and therefore carry a net negative charge within their surface of shear[16-22]; there have also been reports that energization of mitochondria with ATP[21,22] or chloroplasts by light[20] decreases the ζ potential. The ζ potential measures the potential at a certain distance from the surface of the membrane and indicates how strongly ions in the aqueous medium, screen the surface charge. Thus, ζ potentials and surface potentials are different and, indeed, it is possible for ζ potentials to even have a different sign from the surface potential.

Experiments are being undertaken in our laboratory to extend the use of spin probes to try to clarify several interesting questions raised by the studies reported here, viz. whether conformational changes are involved in surface charge changes; why it is that in spite of the fact that chloroplasts and mitoplasts pump protons in opposite directions when energized, in both cases their outer surface becomes more negative; and how surface charge may regulate energy transductions in these membranes.

ACKNOWLEDGMENTS

The spin labels were a gift of Dr. R.J. Mehlhorn of this laboratory. Research supported by ERDA.

REFERENCES

1. H. Rottenberg, J. Bioenergetics, 7 (1975) 61.

2. P. Mitchell and J. Moyle, Eur. J. Biochem., 7 (1969) 47.

3. E. Padan and H. Rottenberg, Eur. J. Biochem., 40 (1973) 431.

4. D.G. Nicholls, Eur. J. Biochem., 50 (1974) 305.

5. H. Rottenberg, T. Grunwald and M. Avron, Eur. J. Biochem., 25 (1972) 54.

6. S.G.A. McLaughlin, G. Szabo and G. Eisenman, J. Gen. Physiol., 58 (1971) 667.

7. G. Schäfer and G. Rowohl-Quisthoudt, J. Bioenergetics, 8 (1976) 73.

8. R.J. Mehlhorn and L. Packer, Biochim. Biophys. Acta, 423 (1976) 382.

9. J.D. Castle and W.L. Hubbell, Biochemistry, 15 (1976) 4818.

10. R.J. Mehlhorn and L. Packer , Methods in Enzymology, Academic Press, New York, (1977) in press.

11. A.T. Quintanilha and L. Packer, FEBS Letters, 78 (1977) 161.

12. R.C. Stancliff, M.A. Williams, K. Utsumi and L. Packer, Arch. Biochem. Biophys., 131 (1969) 629.

13. C. Schnaitman and J.W. Greenawalt, J. Cell Biol., 38 (1968) 158.

14. A.T. Quintanilha and L. Packer, Proc. Nat. Acad. Sci. USA, 74 (1977) 570.

15. M. Avron, Anal. Biochem., 2 (1961) 535.

16. H.W. Douglas, M.V. Laycock and D. Boulter, J. Exp. Bot., 14 (1963) 198.

17. T.E. Thompson and B.D. McLees, Biochim. Biophys. Acta, 50 (1961) 213.

18 F.V. Mercer, A.J. Hodge, A.B. Hope and J.D. McLean, Aust J. Biol. Sci., 8 (1955) 1.

19. J.A. Gross, M.J. Becker and A.M. Shefner, _Experientia_, 20 (1964) 261.
20. P.S. Nobel and H.C. Mel, _Arch. Biochem. Biophys._, 113 (1966) 695.
21. N. Kano, M. Muratsugu, K. Kurihara and Y. Kobatake, _FEBS Letters_, 72 (1976) 241.
22. T. Aiuchi, N. Kano, K. Kurihara and Y. Kobatake, _Biochemisty_, 16 (1977) 1626.

BIOGENESIS OF CHLOROPLAST MEMBRANES IN ALGAE

Itzhak Ohad
Department of Biological Chemistry
The Hebrew University of Jerusalem
20 Mamilla Road, Jerusalem, Israel

INTRODUCTION

Formation of photosynthetic membranes is considered to occur by integration of polypeptides, chlorophyll and lipids into preexisting membranes which grow within the plastid to a predetermined amount and degree of organization. The process of membrane growth, as it appears from the study of several algal systems, presents two main features: it consists of a stepwise or simultaneous addition of newly synthesized components, and it is the result of a close cooperation between the chloroplast and nuclear genetic potential as well as associated protein synthesizing machinery (1). As such, photosynthetic membrane development is of interest to the investigator of bioenergetics and to the cell biologist. To the first, it offers a variety of experimental systems in which one can study the dynamics of composition-structure-function relationships; to the latter, the opportunity to study the intricate process of eucaryotic gene expression, its regulation and evolution of subcellular DNA containing organelles. In the time and space allowed for this presentation I will try to focus the attention on those properties of the chloroplast developmental systems in algae which seem to be of a general character.

Intracellular origin and possible role of chloroplast membrane polypeptides: The powerful and yet relatively simple technique of SDS-polyacrylamide gel electrophoresis was used in order to analyze the polypeptide composition of chloroplast membranes in a variety of algae. Combined with radioactive labeling, use of protein and RNA synthesis inhibitors and selection of mutants, this technique has permitted the identification of several polypeptide bands in *Chlamydomonas'* chloroplast membranes which are correlated with the ability to perform the following functions: H_2O splitting activity of PSII; fluorescence induction and diphenylcarbazide (DPC) photooxydation, which are a measure for the presence of PSII reaction center; presence of P700 complex and methylviologen mediated oxygen uptake, which are considered to represent PSI activity; and formation of the chlorophyll-protein complexes CPI and CPII (2-5).

The polypeptides associated with the binding of chlorophyll a and b and formation of the CPII complex which is considered to represent the light harvesting chlorophyll, have been repeatedly shown to be synthesized by the 80S type cytoplasmic ribosomes, while all the other above mentioned polypeptides seem to be

Table 1: STEPWISE SYNTHESIS AND INTEGRATION OF CHLOROPHYLL AND POLYPEPTIDES OF CYTOPLASMIC AND CHLOROPLASTIC ORIGIN INTO DEVELOPING THYLAKOIDS

Order of events		Final state of organization				Experimental systems	References
First step	Second step	CPII	CPI	PSII	PSI		
1. Cytoplasmic polypeptides and chlorophyll	(a) Chlorop. polypeptides	+	–	+	+	*Clamydomonas* y-1 greening in presence of CAP and restoration of activity	1, 15, 19
	(b) Chlorop. polypeptides and chlorophyll	+	+	+	+		
2. Cytoplasmic polypeptides	(a) Chlorophyll	+	–	–	–	*Chlamydomonas* y-1 grown in the dark in presence of CAP, greened in presence of CAP or cycloheximide;	12
	(b) Chlorophyll and chlorop. polypeptides	+	+	+	+	*Chlorella* ts mutant	11
3. Cytoplasmic and some of the chloroplastic polypeptides, chlorophyll	(a) Chlorop. polypeptides	+	–	+	+	*Chlamydomonas* ts mutant T_4 grown at 37°C and then incubated or grown at 25°C	2
	(b) As in (a) + chlorophyll	+	+	+	+		
4. Cytoplasmic and chloroplastic polypeptides, chlorophyll	(a) Chlorop. polypeptides	–	–	+	+	*Euglena* dark grown and then greened in alternate light-dark-light with addition of CAP or cycloheximide	6, 20
	(b) Some of chloroplastic polypeptides	–	–	–	+		
	(c) As in (b) + missing polypeptides and chlorophyll	±	+	+	+		

CPII, CPI, chlorophyll-protein complexes, as detected by SDS-acrylamide gel electrophoresis.
PSII, PSI, activity measured as oxygen evolution, DCIP, photoreduction and fluorescence induction for photosystem II and oxygen uptake in presence of DCMU with Ascorbate-DCIP as electron donors and MV as electron acceptor for photosystem I.

synthesized by the 70S type chloroplast ribosomes. The possibility that one of
the polypeptides required for the formation of the reaction center of PSII and one
associated with the function of PSI, are coded by nuclear DNA, has been considered
(2, 5). Chloroplastic origin of polypeptides required for PSII activity has been
demonstrated also in *Euglena* (6). The origin of the chlorophyll-binding polypeptides forming a CPII-like complex in *Euglena* appears also to be cytoplasmic (6,7).

Stepwise addition of components to developing membranes: Chloroplast membranes can be formed in non-dividing or dividing cells without concomitant synthesis of protein by the chloroplast ribosomes. Moreover, in non-dividing cells,
inhibition of chloroplast RNA synthesis by rifampicin does not prevent the process
of chlorophyll synthesis and membrane formation (8). However, membranes formed
without the participation of chloroplast synthesized proteins are impaired or completely devoid of photosynthetic activity (1). Restoration of photosynthetic activity can be obtained if chloroplast protein synthesis is reestablished. The
synthesis of the missing polypeptides does not require concomitant synthesis of
chlorophyll. Nevertheless, their complete integration into functional units
might be enhanced by photosynthetic electron transfer activity (2, 5).

Chloroplast membranes are not formed without synthesis of the polypeptides of
cytoplasmic origin which can be prevented by specific protein synthesis inhibitors
or altered by specific mutations. Thus, the only possible sequence of polypeptide
addition to the growing membranes seems to be: (a) polypeptides of cytoplasmic
origin; (b) polypeptides of chloroplastic origin. Since synthesis of soluble proteins in the chloroplast can occur in absence of concomitant protein synthesis in
the cytoplasm (9), the regulatory role of the cytoplasmic polypeptides on chloroplast membrane formation seems to be specific for the membrane and acts at the
post-transcriptional level. This could be explained if one assumes that the insertion of the chloroplast made polypeptides into the membrane requires binding of
polyribosomes (10), which might depend on the presence of the polypeptides of cytoplasmic origin in the membrane.

Formation of chlorophyll-protein complexes: As mentioned above, chlorophyll
synthesis can be dissociated from that of the membrane polypeptides of chloroplastic origin, but apparently not from that of the polypeptides of cytoplasmic
origin. This concept was based so far on the fact that cells which are unable
to synthesize chlorophyll in the dark or mutants which have lost this ability
following single gene mutation, are not able to accumulate the above mentioned
polypeptides in the dark. Support for this view was also found in the fact that
inhibition of protein synthesis by the 80S cytoplasmic ribosomes prevented formation of both chlorophyll and membrane polypeptides (1). Recently, a temperature
sensitive *Chlorella* mutant was described (11) in which membranes containing polypeptides, tentatively identified as of cytoplasmic origin, are formed at 30°C in

64

absence of chlorophyll synthesis. The synthesis of small amounts of these polypeptides could also be demonstrated in non-dividing *Chlamydomonas* cells pretreated in the dark with chloramphenicol (CAP) (1). Moreover, dissociation between
the synthesis of the membrane polypeptides of cytoplasmic origin and chlorophyll
can be achieved in *Chlamydomonas* y-1 cells when grown in the dark in the presence
of CAP (12). In such cells the above polypeptides are synthesized in absence of
chlorophyll synthesis in substantial amounts and can be subsequently used for the
formation of active, normally organized membranes. This will occur if the cells
freed of CAP are further incubated in the light in presence of cycloheximide (CHI)
which in this case does not prevent chlorophyll synthesis but does prevent additional synthesis of cytoplasmic proteins. The chlorophyll synthesized under such
conditions is properly integrated into the membranes and forms a CPII complex.
Thus, polypeptides and chlorophyll synthesized independently can be integrated
into a CPII complex, as detected by the SDS-acrylamide gel electrophoresis (12).
However, the formation of CPI complex appears to require simultaneous synthesis
and integration of chlorophyll and the related polypeptides. This conclusion is
based on the following observations: Membranes present in the y-1 mutant of
Chlamydomonas grown in the dark for 3-4 generations contain the polypeptides related to the CPI complex but not the complex itself. The complex is not formed
when chlorophyll is synthesized by such cells in absence of *de novo* synthesis of
the CPI polypeptides (incubation in the light in presence of CAP). The complex
is still not formed if cells are now allowed to synthesize additional amounts of
the CPI polypeptides in absence of concomitant chlorophyll synthesis (incubation
in fresh medium in the dark) (4). Simultaneous synthesis of the polypeptides and
chlorophyll are not necessarily sufficient for the formation of the CPI complex.
Thus, in the temperature sensitive mutant T_4 grown at 37°C and lacking polypeptides required for PSII activity, the CPI major polypeptides are present in normal amounts, as well as chlorophyll and PSI activity, yet the CPI complex is not
detectable by the regular SDS-acrylamide gel electrophoresis technique (2). The
possible order of addition of membrane polypeptides and chlorophyll to the growing membrane and ensuing formation of chlorophyll-protein complexes, as well as
PSI and PSII activities are summarized in Table 1.

Correlation between presence of cytoplasmic polypeptides, CPII complex and
stacking: The organization of the major proteins or chlorophyll-protein complexes in the membrane plane has been the subject of extensive investigation by
means of chemical modification, use of covalently bound membrane impermeable reagents, controlled proteolysis and quantitative electron microscopical analysis
of freeze fractured membranes. From such studies a model of membrane organization has emerged which, in its main features, accounts for the presence of the
intramembrane particles, their distribution, size and possible correlation with

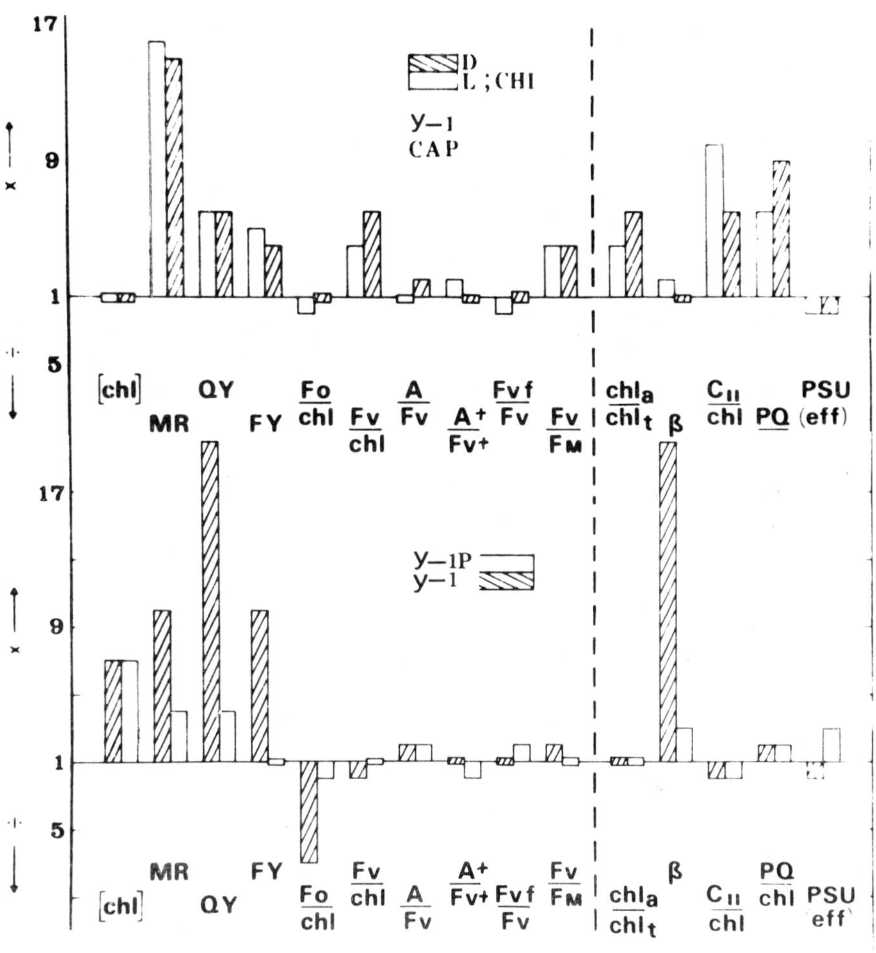

Fig. 1: Histogram showing changes (increase or decrease in various measurable (left hand of the vertical broken line) and derived (right hand of the vertical broken line) parameters. Upper figure: restoration of photosynthetic activity in the y-1 mutant greened in presence of CAP and further incubated in the dark or light with addition of CHI. Lower figure: greening of the *Chlamydomonas* y-1 and y-1p mutants. [Chl], chlorophyll/cell; MR, maximal rate of DCIP photoreduction with H_2O as electron donor; QY, FY, quantum yield and flash yield for the same reaction; Fo, Fv, Fv^+, intrinsic, variable fluorescence and variable fluorescence in presence of DCMU respectively; A, A^+, area above the fluorescence curve without or with addition of DCMU; Fvf, FM, the fast phase and maximal extent of variable fluorescence; Chl_a/Chl_t, the fraction of chlorophyll active in DCIP reduction; β, the degree of connectivity between the PSII reaction center and the rest of the electron transfer chain; CII/Chl, the fraction of chlorophyll organized into PSII reaction center; PQ/Chl, the fraction of active plastoquinone; $PSU_{(eff)}$, the effective size of PSII units. The data are taken from refs. 18, 19. A detailed description of the model on the basis of which the derived parameters were calculated, is given in ref. 18.

polypeptide composition of the membranes, as well as distribution of various enzymatic activities along the surface of the stacked and unstacked thylakoids (13). According to this model, the stacking of thylakoids and formation of grana is related to the organization of the chlorophyll-protein complex CPII. Thylakoid stacking is a reversible process regulated by ionic interaction. The ability of the CPII complex to move in the membrane plane (lateral diffusion) and reversibly associate with PSII complexes, has been considered as one of the major causes for the observed Mg^{++}-dependent rise in induced fluorescence (14). However, several examples, in which stacking does not appear to be simply correlated to the presence of a CPII chlorophyll-protein complex, have been described. In *Chlamydomonas* y-1 mutants thylakoids formed in presence of CAP and thus lacking polypeptides of chloroplastic origin and corresponding activities, the CPII complex is present (4) but the membranes are unstacked (15). On the other hand, the thylakoids are extensively stacked in the T_4 mutant grown at non-permissive temperature (37°C) and lacking only the PSII activity, but exhibiting a normal content of the CPII complex (2). Stacking is observed also in residual thylakoids formed in *Chlamydomonas* y-1 cells in the dark in presence of CAP and lacking both chloroplast synthesized polypeptides and the CPII complex, but containing polypeptides of cytoplasmic origin (12). A similar situation has been reported also for the *Chlorella* mutant described by Galling (10) which lacks chlorophyll when grown at 30°C but exhibits stacked thylakoids. Apparently, when stacking occurs, the polypeptides of cytoplasmic origin are present either organized in a detectable CPII complex or not. However, the presence of these polypeptides, although required, might not be sufficient for the formation of stacked thylakoids. Possibly, changes in local relative concentration of various ions, as a result of electron transfer activity or changes in the redox state of electron carriers, might affect the stacking process. Such changes are expected to be influenced by the state of organization of the membrane which, in its turn, is an expression of its polypeptide, lipid and pigment composition.

Development of photosynthetic activity: The various sequences of addition of polypeptides and chlorophyll to the growing membrane found in different systems implies also that development of photosynthetic activity might exhibit different patterns. Sequential addition and/or interconnection of various segments of the electron transfer chain can be detected by measurements of overlapping partial reactions using natural or artificial electron donor and acceptors. As an example, the measurement of maximal rate of electron transfer combined with that of intrinsic and variable fluorescence, quantum yield, flash yield and effective photosynthetic unit size, provide information on the presence of PSII reaction centers, their connection to the electron donor and acceptor sites and connection with the light harvesting chlorophyll, presence of unorganized chlorophyll and interconnection between the two photosystems. This approach has been used for the study of development of photosynthetic activity in *Scenedesmus* (16, 17), *Chlamydomonas* (1, 18, 19) and *Euglena* (5, 6, 20).

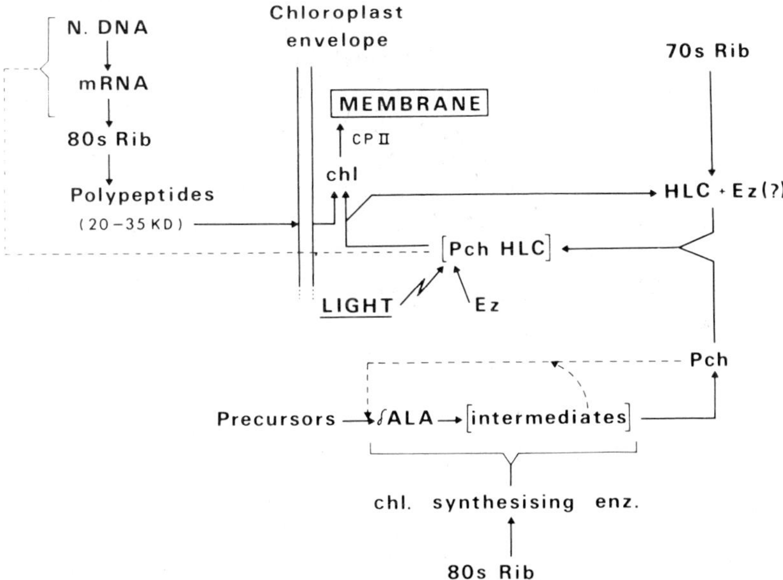

Fig. 2: Schematic representation of the interaction between chloroplast and cytoplast, and the effect of light on the process of chloroplast membrane formation; solid lines, synthesis, transport or assembly; broken lines, control exerted by various intermediates in chlorophyll biosynthesis on the synthesis of δALA and the membrane polypeptides of cytoplasmic origin. The synthesis and integration of polypeptides of chloroplastic origin are not shown in this scheme, simplified from ref. 12. For additional information, see text. Pch, protochlorophyll(ide); HLC, holochrome.

Some of the possible patterns for the development of PSII activity in *Chlamydomonas* mutants are shown in Fig. 1. Thus, in the pale green mutant y-1P, derived from the y-1 mutant, photosynthetic membranes are not formed during growth of the cells in the dark (19). The resultant yellow cells green in the light very slowly doubling the amount of membranes at each division. On a chlorophyll basis the developing photosynthetic activity is practically constant, as shown by measurements of the maximal rate of electron transfer, quantum yield, flash yield, intrinsic fluorescence and variable fluorescence. Only a small rise is observed in the effective size of the photosynthetic unit (19). On the other hand, in the parent cell y-1, which greens very fast and in which preexisting PSII components become integrated into the electron transfer chain before

68

chlorophyll and other components are synthesized, the maximal rate of electron
transfer, quantum yield and flash yield increase considerably, while the intrinsic fluorescence decreases as chlorophyll becomes organized. The relative amount
of PSII reaction centers on a chlorophyll basis remain practically constant, but
the degree of interconnection between PSII units and the electron transfer chain
increases considerably (18). A third example can be found in the process of
restoration of photosynthetic activity in the y-1 cells greened initially in
presence of CAP and thus devoid of polypeptides of chloroplastic origin. The
chloroplast membranes thus formed contain mostly polypeptides of cytoplasmic
origin and chlorophyll organized into a CPII complex, but not the reaction centers of PSII (4). When such cells are incubated in fresh medium in the dark or
light with addition of CHI, only the missing chloroplast polypeptides are synthesized and integrated into the membrane, and an active PSII complex is formed (4,
15). In this case there is no change in the chlorophyll content, but the maximal
rate of electron transfer increases considerably as well as the quantum yield,
flash yield and variable fluorescence per chlorophyll. The fraction of chlorophyll already present and organized in PSII units increases, as well as the active pool of plastoquinone, while the effective size of the PSII units decrease
slightly (19).

Using similar techniques, it was possible to show that in *Euglena* cells greened in alternate light-dark-light cycles, inhibition of protein synthesis in the
chloroplast causes a reduction in the ability to use H_2O as an electron donor for
PSII (6) and disconnection between the PSII reaction center and PQ pool (20). Re-
sumption of chloroplast protein synthesis brings about restoration of oxygen evolution and increase in the degree of connectivity between PSII and PQ pool (20).
These phenomena are associated with *de novo* synthesis and integration of chloroplast synthesized polypeptides into the thylakoids, as detected by electrophoretic
analysis and radioactive labeling of polypeptides (6). The major conclusion of
this discussion is that one should not consider a certain pattern of development
of photosynthetic activity, as exhibited by a certain system, as representative.
There are no rules except those imposed by the sequence of addition of polypeptides and chlorophyll to the developing membranes. Thus, increase or decrease in
the variable fluorescence, quantum yield, maximal rate, effective or apparent photosynthetic unit size, as observed in so many different systems (16-20), should
not be considered as contradictory, but rather the expression of the degree of
freedom in the pathways of membrane formation.

<u>Chloroplast-cytoplast interactions and control of membrane formation</u>: The
role of light and the control of cytoplasm on the synthesis of chloroplast membranes has been the subject of extensive investigation in algae (1, 16, 17). Action spectra for the differentiation of the proplastid or development of the

photosynthetic capacity of plastids have been obtained in *Scenedesmus* (16, 17), *Euglena* (21) and *Chlamydomonas* (1). Two major effects have been established: one, dependent on blue light, is connected with the process of proplastid differentiation proper and triggers synthesis of RNA, proteins, as well as mobilization of substrates for energy production due to degradation of carbolydate reserve; the second light effect, exerted also by red light, operates at the level of photoconversion of protochlorophyll(ide) into chlorophyll(ide) and is mandatory in cells in which chlorophyll synthesis does not occur in the dark (1). Once the differentiation process has been initiated, chloroplast membranes' synthesis proceeds at a constant rate and is subject to control by light mostly at the photoconversion level. Based on recent findings that in *Chlamydomonas* chlorophyll synthesis requires participation of enzymes of both cytoplasmic and chloroplastic origin (12) and on the fact that the synthesis of δALA and most of the subsequent intermediates in the chlorophyll biosynthesis pathway appears to be related to activity of enzymes of cytoplasmic origin (1, 22), one could consider as a working hypothesis the possibility that the level of the protochlorophyll-holochrome complex controls the synthesis of the membrane polypeptides of cytoplasmic origin required for the formation of the membranes. In addition, the formation of this complex is supposed to depend on the presence of polypeptides of chloroplastic origin (12).

The scheme shown in Fig. 2 presents the main features of such a control system which might account for most of the available experimental results, as follows: Polypeptides of cytoplasmic origin could be synthesized and integrated into the membranes in cells grown in the dark in presence of CAP or other conditions preventing activity of 70S ribosomes; δALA will not accumulate in such cells nor membrane polypeptides of chloroplastic origin; inhibition of δALA synthesis will have a similar effect to that of inhibition of protein synthesis in the chloroplast which -- if continued for long periods of time during cell division -- will result in the loss of the ability to synthesize chlorophyll, unless 70S translation in the chloroplast is resumed; concomitant synthesis of chlorophyll and cytoplasmic polypeptides might facilitate their final integration and organization into the developing membranes. This model (12) serves as a basis for further experimental work in our laboratory intended to clarify the mechanism of transport of the cytoplasmic polypeptides across the outer and inner chloroplast envelope, as well as the mechanism of negative control on the synthesis of these polypeptides by the chloroplast.

ACKNOWLEDGEMENTS

This lecture is based largely on results obtained through close cooperation with Dr. S. Bar-Nun, Dr. D. Cahen, Mr. J.M. Gershoni, Mr. M. Gurevitz, Dr. N. Lavintman, Dr. S. Malkin and Ms. S. Shochat. I wish to thank them all for their assistance, criticism and ingenuity in devising experiments, and suggesting the interpretations as brought here.

The work unpublished yet (refs. 12, 20) has been supported by grants from the Israeli Academy of Science and Stiftung Volkswagenwerke No. AZ 11.2882.

REFERENCES

1. Ohad, I. (1975) in "Membrane Biogenesis, Mitochondria, Chloroplasts and Bacteria" (A. Tzagaloff, ed.), Plenum Press, N. Y., p. 279.

2. Kretzer, F., Ohad, I. and Bennoun, P. (1976) in "Genetics and Biogenesis of Chloroplasts and Mitochondria" (T. Bücher, W. Neupert, W. Sebald and S. Werner, eds.), North Holland Publishing Co., p. 25.

3. Bar-Nun, S. and Ohad, I. (1977) Plant Physiol. $\underline{59}$, 161.

4. Bar-Nun, S., Schantz, R. and Ohad, I. (1977) Biochim. Biophys. Acta $\underline{459}$, 451.

5. Ohad, I. (1977) J. Cell Biol. (in press).

6. Gurevitz, M., Kratz, H. and Ohad, I. (1977) Biochim.Biophys.Acta (in press).

7. Bingham, S. and Schiff, J.A. (1976) *ibid* ref. 2, p. 79.

8. Jennings, R.C. and Ohad, I. (1972) Arch. Biochem. Biophys. $\underline{153}$, 79.

9. Gershoni, J.M. and Ohad, I. (1976) Colloque Intern. C.N.R.S., "Acides Nucleiques et Synthese des Proteins chez les Vegetaux", $\underline{261}$, 447.

10. Chua, N.H., Blobel, G., Siekevitz, P. and Palade, G.E. (1973) Proc. Nat. Acad. Sci. USA $\underline{70}$, 1554.

11. Galling, G. (1976) *ibid* ref. 2, p. 53.

12. Gershoni, J.M. and Ohad, I. (1977) (in preparation).

13. Staehelin, A., Armand, P.A. and Miller, K.R. (1976) in "Chlorophyll-Proteins, Reaction Centers and Photosynthetic Membranes" (J.M. Olson and G. Hind, eds.) p. 278.

14. Arntzen, J.C., Briantais, M.J., Burke, J.J. and Novitzky, P.W. (1976) *ibid* ref. 13, p. 278.

15. Hoober, J.K., Siekevitz, P. and Palade, G.E. (1969) J. Biol. Chem. $\underline{244}$, 2621.

16. Bishop, N.I. and Senger, H. (1972) Plant & Cell Physiol. $\underline{13}$, 937.

17. Oh-Hama, T. and Senger, H. (1975) Plant & Cell Physiol. $\underline{16}$, 395.

18. Cahen, D., Shochat, S., Malkin, S. and Ohad, I. (1976) Plant Physiol. $\underline{58}$, 257.

19. Cahen, D., Malkin, S. and Ohad, I. (1977) Plant Physiol. (in press).

20. Cahen, D., Gurevitz, M. and Ohad, I. (1977) (in preparation).

21. Egan, J.M., Jr., Dorsky, D. and Schiff, J.A. (1975) Plant Physiol. $\underline{56}$, 318.

22. Feierabend, J. (1977) Planta (in press).

Bioenergetics of Membranes. L. Packer et al. ed.

BIOGENESIS OF THE PHOTOSYNTHETIC MEMBRANES
IN HIGHER PLANTS

George Akoyunoglou
Biology Department, Nuclear Research Center
"Demokritos", Athens, Greece

INTRODUCTION

Plants grown in complete darkness have no chloroplasts nor Chl[+], but etioplasts
and PChl. The etioplasts contain a highly ordered structure, known as prolamellar
body. The etioplast to chloroplast differentiation in angiosperms depends on
light. The differentiation consists essentially in the transformation of the prolamellar body to primary thylakoids, and their subsequent growth and development
into grana and stroma thylakoids. The growth and development of the chloroplast
membrane is a multi-step process. It involves the biosynthesis and incorporation
of both structural and functional components (pigments, lipids and proteins), as
well as the assembly and integration of these components into functional units.
Chl $\underline{a}$ is the essential pigment for photochemical activity, and it is formed by the
photoreduction of the bound PChl. The protein moiety of the PChl-Holochrome acts
as an enzyme, PChl being the substrate[1,2]. The peptides of the membrane are of
nuclear transcription-cytoplasmic or chloroplast translation, and chloroplast tran-
scription-chloroplast translation. There are no specific growing sites in the membrane, but the newly synthesized Chl $\underline{a}$ is inserted throughout the expanding thylakoid[3].

Under continuous illumination the differentiation of etioplast to chloroplast
proceeds normaly, i.e., after a lag-phase, which depends on the age of the etiolated tissue[4-6], the different components are synthesized in a parallel way and integrated into the developing membrane, and in a short time the characteristic features of the mature chloroplasts are observed[6,7]. It has been found that it is
possible to distinguish the different steps in thylakoid development, by changing
the mode of illumination[8,9]. For example, in periodic light (2 min light-98 min
dark) the etioplasts are transformed to protochloroplasts. The protochloroplasts
can be transformed to chloroplasts only after transfer of the plants to continuous
light[9]. In spite of the incomplete development of the protochloroplast thylakoid
they are photosynthetically very active[6,10].

Making use of this system some aspects of chloroplast membrane biogenesis and
development have been studied. In this paper emphasis will be put on the develop-

---

+Abbreviations: Chl, chlorophyll; PChl, Protochlorophyll(ide); DPC, 1,5-diphe-
nylcarbazide; DCIP, dichlorophenol indophenol; DCMU, 3-(3,4-dichlorophenyl)-1,1-
dimethylurea; PS, photosystem; CP, Chl-protein Complex; LDC, light dark cycles.

ment of the photosynthetic units and the establishment of the photosynthetic electron flow.

MATERIALS AND METHODS

Etiolated bean leaves _Phaseolus vulgaris_ (red kidney var.) were used for the study. The growth and handling of the bean seedlings were done as previously described[9]. Active plastids were isolated in dim green light from the developing bean leaves as previously described[11]. PS II activity, fluorescence measurements, _in vivo_ $^{14}CO_2$-fixation experiments, and the 518 nm absorbance change measurements were done as previously described[10-12].

RESULTS AND DISCUSSION

Composition and structure of the developing protochloroplasts. Under periodic light the main Chl molecules synthesized are those of Chl _a_. There is no lag-phase in its formation, and the final concentration reached is almost 1/6th of that of mature green leaves[9]. Chl _b_, which is formed from Chl _a_ on the developing membrane[3,13], accumulates only after transfer of the periodic-light leaves to continuous light[9]. The rate of carotenoid synthesis increases slightly under periodic light, and at the same time a change in their composition takes place, so that the ratio xanthophyll/β-carotene decreases as the time of intermittent illumination increases, and finally reaches the value of mature chloroplasts[14,15]. The carotenoids formed are inserted into the developing membrane in an heterogeneous way. The heterogeneous distribution of carotenoids is controlled by the development of PS I and PS II, and not by the structural differentiation of the membrane into stroma and grana thylakoids[14,15]. The thylakoids formed under intermittent light are unstacked[9], they do not contain the Chl _b_-rich CP II[16], and the polypeptides of the 25-30 Kdaltons range of lipid-free thylakoids, probably derived from CP II[18,19]. Moreover, they do not have the ability for cation-induced stacking and cation-induced Chl _a_ fluorescence rise[20].

Transfer of the periodic-light leaves to continuous light induces the gradual synthesis of the 25-30 Kdaltons polypeptides[13], Chl _b_[9], and CP II[16,17], and the appearance of grana stacks, which are formed by the stacking of preexisting primary thylakoids[3]. At the same time the ability for cation-induced stacking and cation-induced Chl _a_ fluorescence rise appears[18,21]. It is obvious that the synthesis of all or some of the above components are controlled by continuous illumination.

Development of the photosynthetic activity: Development of PS II. It is known, that the PS II activity ($H_2O \longrightarrow$ DCIP or DPC $\longrightarrow$ DCIP) appears very early during greening when 5-day old etiolated bean leaves are exposed to continuous light. The activity on a Chl basis increases with the time of illumination and after 4 h a plateau is reached. The value at the plateau is almost equal to the value of mature chloroplasts[6]. This shows that in continuous illumination the complete PS

TABLE I

ONSET OF PHOTOSYSTEM II ACTIVITY IN DEVELOPING PHOTOCHLOROPLASTS FROM 5-DAY OLD ETIOLATED BEAN LEAVES EXPOSED TO LIGHT-DARK CYCLES (LDC) AND THEN TRANSFERRED TO CONTINUOUS LIGHT (CL)

Sample	Chl $\underline{a}$	Chl $\underline{b}$	μmoles DCIP/mg Chl/h		μmoles DCIP/g fresh wt/h	
	μg/g fresh wt		$H_2O \longrightarrow$ DCIP	DPC $\longrightarrow$ DCIP	$H_2O \longrightarrow$ DCIP	DPC $\longrightarrow$ DCIP
6 LDC	45	0	150	350	7	16
14 LDC	110	1	500	750	55	83
42 LDC	230	7	750	1150	173	272
28 LDC	313	12	950	1050	309	341
28 LDC+2h CL	452	78	480	540	254	236
28 LDC+4h CL	638	156	320	350	254	278
Chloroplasts	1980	640	175	190	67	73

II unit is formed from the beginning of greening, and that prolonged illumination increases only the number of the units.

The development of the PS II activity under periodic light is shown in Table I. The rate on a Chl basis increases with time and after 2-3 days it reaches a plateau. The rate at the plateau is almost 5-times higher than that of mature chloroplasts. The $H_2O \longrightarrow$ DCIP reduction is lower than that of DPC $\longrightarrow$ DCIP but the difference decreases as the time of greening increases. The increase in the rate on a Chl basis observed in the beginning of greening indicates the reduction of the apparent size of the units with time, and it is due to the presence of unorganized Chl in a high proportion, which decreases as the time of illumination increases. During the remainder of the greening process under periodic light the activity per gram fresh weight increases, but on a Chl basis remains constant, indicating that primary thylakoids of the same composition and organization are formed. After transfer of the plants to continuous light the rate of Chl $\underline{a}$ and Chl $\underline{b}$ increases, and the PS II activity on a Chl basis decreases; after some time the value of mature chloroplasts is reached. These results, together with the observation that high light intensities are needed for maximum photosynthetic activity[6], suggest that (a) the PS II unit is formed in a step-wise manner; (b) the Chl $\underline{a}$ formed under periodic light is used for the formation of small PS II units; (c) the Chl formed after transfer of the plants to continuous light are used as antenna and light-harvesting pigments of the growing units; (d) the lower rate observed for $H_2O \longrightarrow$ DCIP reduction reflects the step-wise formation of the units, i.e., the reaction center is assembled first and then the H_2O-splitting enzymes are added. This is supported also from the results of Sironval and coworkers[22], who found no oxygen evolution in leaf segments greened under msec flashes.

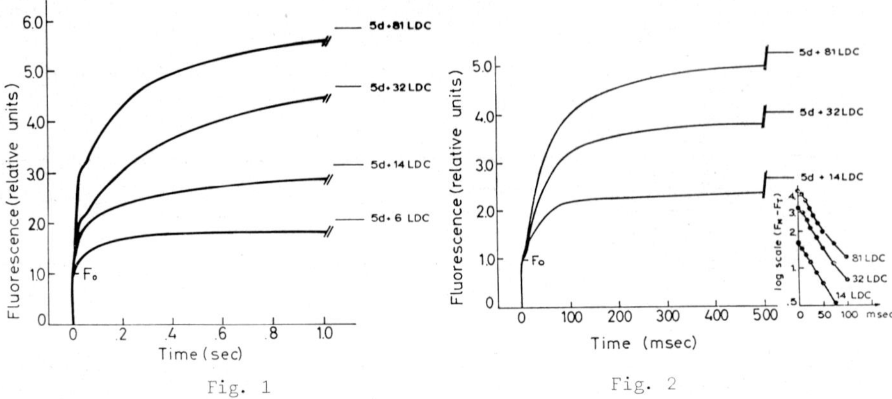

Fig. 1   Fig. 2

Fig. 1. Fluorescence induction of developing protochloroplasts from etiolated bean leaves exposed to LDC. Room temperature; 5 µg Chl/ml; exciting light intensity I_o = 2610 µW/cm^2. The curves were taken from actual oscilloscope traces and normalized to the same F_o.

Fig. 2. Fluorescence induction in the presence of 12 µM DCMU of developing protochloroplasts. Exciting light intensity I_o = 575 µW/cm^2. Insert: log plot of F_M-F_T for the same curves. Other experimental conditions as in Fig. 1.

Development of the PS II unit: The development and interaction of the PS II units has been also studied by fluorescence emission spectroscopy[11]. Figures 1 and 2 show the normalized at F_o fluorescence rise kinetics curves of developing protochloroplasts in the presence or absence of DCMU, and Table II some of the fluorescence kinetics values. The F_{max} level increases as the number of light--dark cycles increases and values twice as high as those of mature chloroplasts are reached. Addition of NH$_2$OH increases the F_{max} even higher. The half-rise time in the presence or absence of DCMU is larger (at least 7-fold) than the corresponding of mature chloroplasts. In the presence of DCMU the slow rise kinetics is exponential in the beginning of greening, but later becomes sigmoidal. Transfer of the periodic-light leaves to continuous light results in a dramatic decrease of the half-rise time and of the F_{max}/F_o ratio (see Table II). According to the Duysens and Sweers hypothesis[23], the fluorescence rise reflects the photoreduction of Q, a fluorescence quencher, to the non-quenching form Q$^-$. If the electron flow from Q$^-$ is not inhibited, then Q$^-$ can reduce the secondary electron carriers and finally the reaction center of PS I[24]. In the presence of an inhibitor (e.g DCMU), the rate of the fluorescence rise represents the rate by which Q is reduced. The rate of Q reduction depends on the rate by which the absorbed photons reach the reaction center. Thus, it depends on the actinic light, the number of antenna Chl per PS II unit, and the efficiency of the units to utilize the absorbed photons[25] With the same actinic light, and assuming that the efficiency remains constant, the difference in half-rise time in different samples reflects the difference in the

TABLE 2

FLUORESCENCE INDUCTION OF PROTOCHLOROPLASTS AND CHLOROPLASTS FROM 5-DAY OLD ETIO-
LATED BEAN LEAVES EXPOSED TO LIGHT-DARK CYCLES (LDC) AND THEN TRANSFERRED TO
CONTINUOUS LIGHT (CL)

Sample	Chl $\underline{a}$	Chl $\underline{b}$	F_{max}	F_{NH_2OH}	Half-rise time of fluorescence	
	$\mu g/g$ fresh wt		F_O	F_O	$I=296\ \mu W/cm^2$ +DCMU	$I=2610\ \mu W/cm^2$ -DCMU
					(ms)	
14 LDC	105	0	3.15	3.80	110	120
41 LDC	315	12	5.30	6.20	95	105
81 LDC	495	42	5.90	6.45	70	80
28 LDC	230	7	4.30	4.95	98	110
28 LDC+3.5h CL	594	124	3.30	3.30	20	21
23 LDC+14h CL	1210	378	3.30	3.30	13	14
Chloroplasts	1930	640	3.30	3.30	13	14

size of the PS II unit. In protochloroplasts the half-rise time of the fluorescence rise is 7-times larger than that of mature chloroplasts, suggesting, therefore, that the size of their PS II unit is several times smaller than that of the
mature chloroplast.

The 5-fold increase in the rate of the fluorescence rise in only 3.5 h of continuous illumination after the 28 LDC (see Table II) indicates a fast increase in
the size of the PS II unit. This suggests that the Chl formed during the first
hours of continuous illumination is inserted into preexisting units, increasing
thus their size.

The sigmoidal shape of the fluorescence induction has been attributed to the
intersystem II unit-unit interaction[26]. The change, therefore, from exponential
to the sigmoidal shape can be considered as an indication of the development of
energy transfer and connection between the PS II units. It seems, therefore, that
at the beginning of greening under periodic light the system II units are structurally isolated but later they come into contact. This unit interaction may result either from unit aggregation or from insertion of newly synthesized membrane
components in between the units rather, than from the increase in unit size. Since
the aggregation takes place even when the PS II units are small in size, it seems
that the reaction centers of the connected units are close to each other, so that
a small number of Chl, already packed around each unit, is adequate for the onset
of the excitation energy transfer.

It is accepted that at room temperature only Chl $\underline{a}$ of PS II fluoresces. According to Kitajima and Butler[27], in isolated chloroplasts the non-variable (F_o) and

the variable ($F_{max}-F_o=F_v$) fluorescence emission are both emitted from the bulk Chl $\underline{a}$ of PS II. Structural changes of the thylakoids induce changes of the F_v level. For example, suspension of granal chloroplasts in low salt medium dissociates the grana, and decreases the F_v level. Addition of salts reverses the process. Protochloroplasts have single thylakoids, and the presence or absence of salts has no effect on the F_{max} level in the presence of DCMU[20,21,28]. They have, however, very high F_v, and the F_{max}/F_o is almost twice as high as that of mature chloroplasts. This implies that either in mature chloroplasts all F_o is not originate from the bulk Chl of PS II[27], or that the intrinsic F_o is lower in protochloroplasts. In the latter case it indicates that in protochloroplasts the bulk Chl is organized in a more efficient way. The change observed in the ratio F_{max}/F_o with time in periodic light reflects the decrease in the relative concentration of the unorganized Chl with time, which contributes only to the F_o level.

The decrease in the F_{max}/F_o observed after transfer of the periodic light leaves to continuous light reflects the different changes that take place in continuous illumination; i.e., formation of new Chl which increases the relative concentration of unorganized Chl (effect on F_o); increase in size of the PS II units, which has an effect on both the F_o and the F_{max} level; and finally, formation of grana with the concomitant appearance of the Mg^{++} effect on the spillover phenomena, which affect the F_v level.

Development of the heterogeneous PS II units: It has been shown[29], that the area over the fluorescence induction curve of isolated chloroplasts is a measure of the total number of charge separations that occur at PS II during the fluorescence induction period. A first order kinetic analysis of the area growth in the presence of DCMU revealed that it occured in two distinct linear phases, a fast and a slow phase. This suggested a functional differentiation in PS II, and the existence of two types of PS II centers, the efficient α-centers and the relatively less efficient β-centers[30]. The rate constants K'_α and K_β of the two types of centers can be calculated from the slope of the lines corresponding to the two phases of the logarithmic plot of the area growth. Similarly, the relative concentration of the α and β-centers can be calculated from the intercept of the line of the slow phase with the logarithmic ordinate at zero time[30]. The differential effect of Mg^{++} ions on the function of the two types of PS II centers led to the hypothesis that the chloroplast grana are the loci of the efficient α-centers while the stroma thylakoids are the loci of the β-centers[30].

The development of the two types of units (α and β) has been studied[31] in an effort to understand how they develop, what's their interelationship, and, moreover, if the heterogeneity of the photochemical centers of the PS II units is an endogeneous property of the chloroplast lamellae or a consequence of its differentiation into grana and stroma thylakoids.

It was found that in continuous illumination the first photosynthetic membranes formed contain both the α and β-centers. During the first hours of continuous light the relative concentration of β-centers is high, but gradually decreases, and in 6-3 h it reaches the steady-state value of mature chloroplasts (about 65%). The magnitute of the photochemical rate constants, K'_α and K_β increases with time in continuous light and it reaches the steady-state value of the mature chloroplast. Most of the increase is accomplished within 4 h of continuous light. The K'_α and K_β depends among other things and on the size of the PS II unit. Under our experimental conditions the change of the value of K'_α and K_β during greening gives an idea of the change in the size of the PS II unit. Obviously, under continuous light there is a fast growth of the units around the α-and β-centers.

Figures 3 and 4 show the change in the relative concentration of the α- and β-centers, and their respective K'_α and K_β during greening under periodic light. It

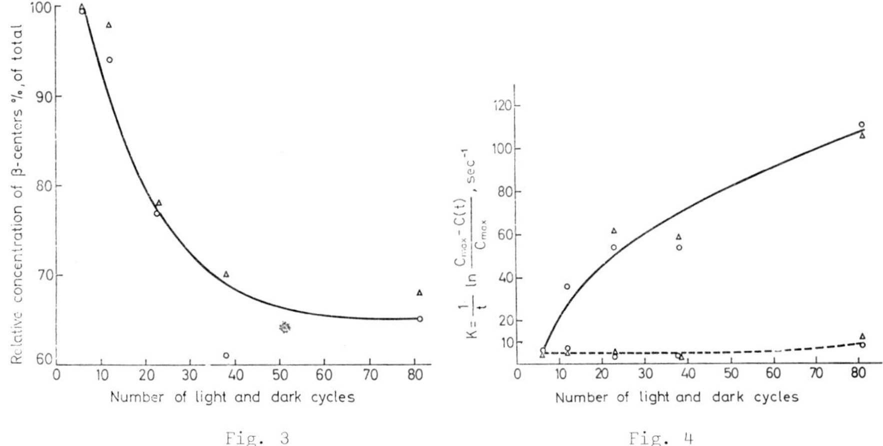

Fig. 3

Fig. 4

Fig. 3. Intermittent illumination-induced change of the relative concentration of β-centers in developing protochloroplasts; 12 µM DCMU(o), 12 µM DCMU and 1 mM $NH_2OH(\Delta)$, 4 µg Chl/ml, exciting light intensity I_o = 2610 µW/cm^2.
Fig. 4. Intermittent illumination-induced change of the photochemical rate constants K'_α (—) and K_β(---) in developing protochloroplasts. Other conditions as in Fig. 3. For mature chloroplasts, K'_α = 480 sec^{-1}, K_β = 44 sec^{-1}.

is seen that the originally high relative concentration of the β-centers decreases, and finally it reaches the steady-state value of the green control (65%). In the beginning the values of both K'_α and K_β are almost equal. Subsequently, however, there is a dissimilar dependence of K'_α and K_β on the number of light-dark cycles. The value of K_β remains constant, while the value of K'_α increases steadily with time in periodic light, but none ever reaches the value of the green control. Transfer of the periodic-light leaves to continuous light induces a fast and parallel increase of K'_α and K_β up to the value of the green control. The results of a representative experiment, where etiolated leaves were exposed first to 14 LDC and

78

then to continuous light, showed that in 1 h the K'_α and K_β increase almost 5-ti-
mes, and in 3 h they both reach the value of mature chloroplasts.

These results indicate that, (a) the α- and β-centers are both present in the
photosynthetic membranes from the beginning of the light-induced greening; (b)
the expression of the α-center does exist in protochloroplasts, and the differentiation between the photochemical centers is already in an advanced state; (c) the
α-centers and the photosynthetic units associated with them develop independently
of the β-centers; (d) the α-centers do not share the same pigment bed with the   .
β-centers, otherwise, there should have been a parallel change in K'_α and K_β under
periodic light; (e) the presence of Chl b, CP II and grana is not a prerequisite
for the formation of the α-centers; and (f) the heterogeneity of the photochemical
centers in PS II is an endogenous property of the chloroplasts lamellae.

The 518 nm change: Photosynthetic organisms show a light-induced absorbance
change near 518 nm, which has a spectrum with a maximum at 518 nm and a minimum
at 475 nm[32]. In spite of the wide variety of explanations, the origin of the
518 nm absorbance change is generally accepted to be closely related to the photosynthetic activity and the energized state of the photosynthetic apparatus. The
appearance and development of the 518 nm absorbance change during the etioplast
to protochloroplast and protochloroplast to chloroplast differentiation has been
studied in an effort to correlate changes in composition, structure and function
of the developing membrane with changes in absorbance at 513 nm, and, moreover,
as an indication of the appearance and development of the photosynthetic activity,
and the energized state of the developing plastid[12].

Figure 5 shows the kinetics of the 518 nm and 480 nm absorbance change of a
green leaf (1A), and an etiolated leaf which was exposed to 25 LDC (1B). As it is
evident, periodic-light leaves do show the 518 nm change, but contrary to the
green leaves, there is a decrease in absorbance at 513 nm. The change at 480 nm
is similar to the green leaves. The signals are biphasic in both kinds of leaves, but the half-time of the slow phase is larger in the periodic-light leaves
than in the green ones. Moreover, the dark decay is monophasic in the green leaves, but biphasic in the periodic-light leaves. A second and subsequent illuminations after the dark decay induces the same biphasic signal in the periodic-light
leaves, but monophasic in the green leaves. The inversion in the sign of the
change at 518 nm is observed at all wavelengths in the region 510-540 nm; in the
region 470-510 nm the change is similar to that of green leaves (Compare Fig. 6,A
and 6,I).

Since the periodic-light leaves show a light-induced decrease in absorbance at
518 nm, while the green leaves a light-induced increase, an inversion in sign is
expected to take place after transfer of the periodic-light leaves to continuous
light. Figure 6 shows the spectra of the fast component of the change at 518 nm
of etiolated leaves exposed first to 25 LDC (6,A) at then transferred to conti-

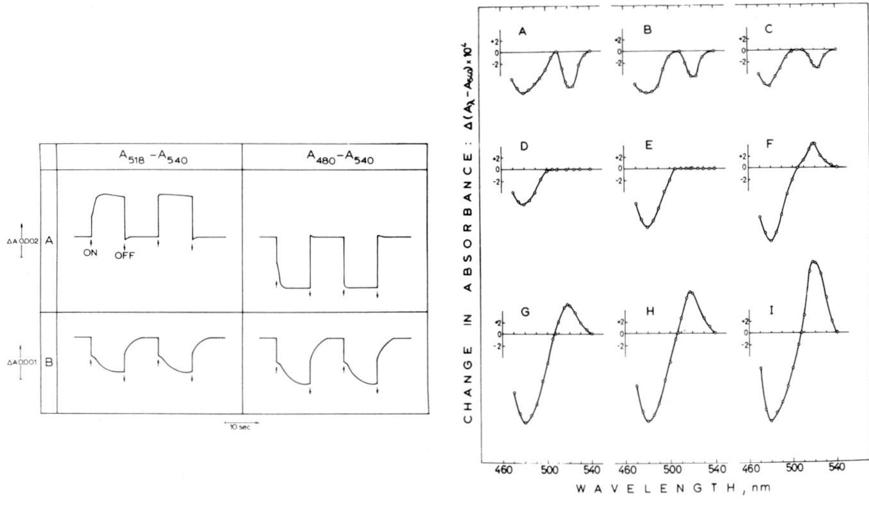

Fig. 5   Fig. 6

Fig. 5. Time course of the light-induced absorbance change at 518 nm and 480 nm in green leaves (A) and in 5-day old etiolated bean leaves exposed to 25 LDC (B). Fig. 6. Spectra of the fast component of the light-induced absorbance change in 5-day old etiolated bean leaves exposed first to 25 LDC (A), and then transferred to continuous light for 0.5, 1, 2, 3.5, 5, 9, 12 and 33 h (B-I).

nuous light for different intervals of time. As it is evident, the 510-540 nm region of the spectrum changes from negative to zero, and then to positive during the continuous illumination, while there is no change in the 470-510 nm region. As long as the change at 518 nm is negative, the kinetics at 518 nm and 480 nm is similar to periodic-light leaves, but when the change becomes positive then the kinetics is similar to green leaves. The kinetics of the samples in the intermediate stage (Fig. 6, D-F), where the change in absorbance from negative to positive takes place, is really striking. In the 5 h (Fig. 6,F) sample, for example, the kinetics in the 510-540 nm region is similar to normal green leaves, but in the 470-510 nm region is similar to periodic-light leaves[12].

Periodic-light leaves are the only plant material, so far, which show a light-induced decrease in absorbance both in the 510-540 nm and 470-510 nm region. All photosynthetic processes which have been shown to be related with the 518 nm change take place in these leaves. The only difference resides in their structure and composition, i.e., the periodic-light leaves lack Chl b, CP II and grana, and they have small PS units. The absence of grana can not be responsible for this abnormal behaviour, since both isolated grana and stroma lamellae show normal 518 nm change[33]. The difference in pigment composition can not explain this behaviour, since the electrochromic spectra of Chl _b_ show a large decrease in absorbance at 470-500 nm, and a small increase at 500-540 nm[34] the electrochromic spectra of carotenoids, and to a smaller degree of Chl _a_, show an increase in the whole 480-540 nm region[38]. So, in the absence of Chl _b_ the net result should be expected to be an increase in absorbance throughout the 470-540 nm region.

Recent studies[34,35] have shown that the 518 nm absorbance change depends on the degree of orientation of the chloroplast pigments within the membrane. If this is correct, then it is possible that the orientation of the pigments which are close to the reaction center is different from the orientation of the rest antenna pigments. However, the question remains of why the difference in orientation would affect only the 510-540 nm region.

The observation that the kinetic behaviour of the 518 nm absorbance change depends on the developmental stage of the chloroplast, suggests that it has to be regarded as a characteristic property of the "518 nm effect". Our results, therefore, suggest that the "518 nm effect" is probably the net result of more than one phenomena that take place in the chloroplast.

Establishment of the photosynthetic electron flow between PS II, PS I.

The establishment of the photosynthetic electron flow between PS II and PS I, and, moreover, the appearance and development of the overall photosynthetic activity has been studied by measuring the rate of $^{14}CO_2$-fixation[10]. The results showed that soon after exposure to LDC the leaves are capable of photosynthetically fixing $^{14}CO_2$. The rate on a Chl basis increases with time and after 35-45 LDC it reaches a plateau. The activity per gram fresh weight, however, continuous to increase. The rate on a Chl basis at the plateau is 5-6 times higher than that of green control. The results indicate that the electron flow from PS II to PS I is established early during greening in LDC, and, moreover, that during the growth of the membrane units of the same size and organization are formed, which are several times smaller than that of green leaves.

Effect of salts on primary thylakoids: A variety of phenomena are affected by cations in isolated chloroplasts, such as light scattering, light-dependent shrinkage, grana stacking and Chl _a_ fluorescence rise. In most of these changes divalent cations are more effective than monovalent[36]. Protochloroplasts have single thylakoids, and they do not have the ability for cation-induced stacking and cation-induced Chl _a_ fluorescence rise. Moreover, by suspending them in low-salt Tricine buffer they lose easily their ionic content[28]. Using, therefore, protochloroplasts, the effect of salts on the structure and function of the single thylakoid can be studied. We especially studied the effect of salts on PS II activity and on the fluorescence induction.

Suspension of protochloroplasts in low-salt Tricine buffer inactivates the H_2O-splitting enzyme, and a decrease in the rate of the $H_2O \rightarrow DCIP$ reduction by more than 60% is observed. Addition of NaCl (10 mM) increases the rate to almost its initial value. This effect is similar to that observed when Cl^- ions are added to Cl^--depleted chloroplasts[37]. This is really a Cl^- effect, since addition of $NaNO_3$ or $Mg(NO_3)_2$ does not increase the rate. Moreover, the rate of DPC $\longrightarrow$ DCIP reduction is not affected by the presence or absence of NaCl.

The fluorescence rise kinetics of protochloroplasts suspended in low-salt Tricine buffer are shown in Fig. 7. In the absence of salts (curve A) the fluorescence yield is low and the half-rise time of the fluorescence kinetics high. Addition of NaCl (curve C) increases the fluorescence yield, which reaches the F_{max} level, and decreases the half-rise time of the fluorescence kinetics. The

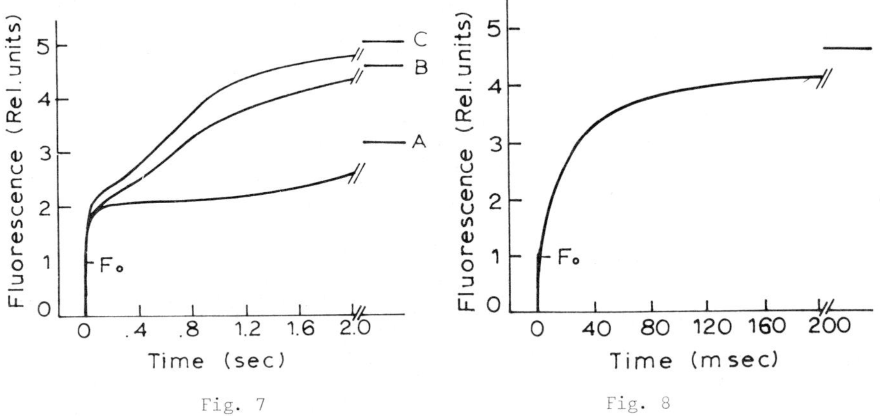

Fig. 7

Fig. 8

Fig. 7. Effect of salts on the fluorescence induction of protochloroplasts, from etiolated leaves exposed to 42 LDC, suspended in Tricine, 20 mM-Sucrose, 0.3 M, buffer, pH 8.0. A,-salts; B, + 50 mM $NaNO_3$; C, + 50 mM NaCl. Other conditions as in Fig. 1.

Fig. 8. Fluorescence induction of protochloroplasts (42 LDC) in the presence of 12 µM DCMU, suspended in Tricine, 20 mM-Sucrose, 0.3M, buffer, pH 8.0, ± 50 mM NaCl. Other conditions as in Fig. 1.

same gross effect is also found with $NaNO_3$ (curve B). The fluorescence yield, according to Duysens and Sweers hypothesis[23], determines the reduction/oxidation level of the electron acceptor Q. At the fluorescence steady state the rates of reduction and oxidation of Q are equal. Therefore, the effect of salts on the fluorescence yield indicates their effect on the redox state of Q. The NaCl effect on the fluorescence induction correlates well with the Cl^- effect on the PS II activity. However, the $NaNO_3$ effect on the fluorescence induction indicates that in addition to the Cl^- effect on the oxidizing site of PS II, there is also a Na^+ effect. Since $NaNO_3$ does not increase the rate of the $H_2O \longrightarrow$ DCIP reduction, but it increases the rate of the fluorescence rise kinetics and the fluorescence yield, it indicates that Na^+ must affect some structural features either of the PS II unit, or of the secondary electron carrier components. In order to check this hypothesis, we measured the effect of salts on the fluorescence kinetics of

DCMU-poisoned protochloroplasts suspended in low-salt Tricine buffer. The results are shown in Fig. 8. As it is evident, there is no Na^+ effect on the fluorescence rise kinetics, and the fluorescence yield. These results indicate that the Na^+ effect must be on the reducing site of PS II and it seems to monitor the intra-membrane structure of the thylakoid, and especially the structure of the secondary electron carrier components. An other conclusion that can be drawn is that the intersystem II unit-unit interaction, and the relative concentration of α- and β-units are not affected by the presence or absence of salts in protochloroplasts.

This cation effect is pH dependent, and monovalent cations are much more effective than divalent (Mg^{++}). The concentration of Na^+ necessary for maximum reconstitution depends on pH,, and it increases as the pH increases from 6.0 to 8.0, and, moreover, it depends on the time the protochloroplasts were left in the low-salt medium, i.e., on the degree of disorganization.

Suspension of protochloroplasts in low-salt Tricine buffer affects also the light-induced 518 nm absorbance change, i.e., it induces a large decrease in the magnitude of the slow phase of the decay, and a small decrease in the slow phase of the signal. Addition of salts eliminates this effect.

CONCLUSIONS AND SUMMARY

The growth of the chloroplast membrane is a multi-step process, and some of the steps are controlled by light. With different modes of illumination (i.e., periodic light) the various steps in the development of the membrane can be clearly distinguished. The main Chl molecules synthesized under periodic light are those of Chl a, and they are used for the formation of the reaction centers, the core of the PS units, and CP I. Chl b, which is formed from Chl a on the developing membrane, is synthesized in continuous light together with the 25-30 Kdaltons polypeptides. CP I and and CP II develop independently of each other. Carotenoids are formed and inserted into the developing membrane in an heterogeneous way, which is controlled by the development of PS I and PS II. There are no specific growing sites in the membrane, but the newly synthesized components are inserted throughout the expanding thylakoid.

The development of the PS II unit is also a step-wise process. It does not require the formation of CP II. During the biogenesis and growth of the PS II unit in periodic light the reaction center is formed first with a few antenna Chl closely packed around. The unit is thus very efficient for photochemistry. This structure represents the core of the PS II unit. At the same time the H_2O-splitting enzymes are formed, which are light induced. As the number of small units increases they form clusters which have the reaction centers very close to each other. The aggregation of the units is controlled by the structural development and organization of the membrane and not by the concentration and type of Chl. The heteroneity of the PS II units is an intrinsic property of the membrane.

The α-units and β-units develop independently of each other from the beginning of the light-induced greening. They do not share the same pigment bed. The presence of grana, Chl b and CP II is not a prerequisite for the formation or development of the α-centers.

Further greening in continuous light results in the formation of Chl b and the 25-30 Kdaltons polypeptides, and more Chl a and carotenoids. The Chl, which is formed with a high rate, is inserted into preexisting units increasing their size, and making them more efficient for absorbing the incident light. At the same time the stacking component CP II is formed, which has as a consequence the formation of grana, and the development of the ability for cation-induced stacking and cation--induced Chl a fluorescence rise. Grana are formed by stacking or folding of preexisting and growing primary thylakoids. CP II is involved in both the regulation of the excitation energy distribution between the two photosystems and the stacking process. The orientation of the pigments which are close to the reaction center seems to be different from that of the rest of the pigments. Cations regulate the intramembrane structure of the thylakoid and the stacking of grana, and control the rate of electron flow from H_2O to NADP and the distribution of the excitation energy between the two photosystems.

Acknowledgements: Parts of this work have been conducted in collaboration with Dr. A. Melis (Heterogeneous PS II units), and Mrs. Tsimilli-Michael (518 nm change). I wish to thank Mr. S. Daoussis for his excellent technical assistance.

REFERENCES

1. Griffiths, W.J. (1974) FEBS Letters 49, 196.

2. Manetas, Y. and Akoyunoglou, G. (1976) Plant Physiol. 58, 43.

3. Argyroudi-Akoyunoglou, J.H., Kondylaki, S. and Akoyunoglou, G. (1976) Plant Cell Physiol. 17, 939.

4. Akoyunoglou, G. and Siegelman, H.W. (1968) Plant Physiol. 43, 66.

5. Akoyunoglou, G. and Argyroudi-Akoyunoglou, J.H. (1969) Physiol. Plant. 22, 288.

6. Akoyunoglou, G. and Michelinaki-Maneta, M. (1975) in: Proc. 3rd Internatl. Congress on Photosynthesis (Avron, M. Ed.). North-Holland, Amsterdam, p. 1885.

7. Boardman, N.K., Anderson, J.M., Hiller, R.G., Kahn, A., Roughan, R.G., Treffy, T.E. and Thorne, S.W. (1972) in: Proc. 2nd Internatl. Congress on Photosynthesis (Forti, G., et al., Eds). Dr. W. Junk N.V., The Hague, p. 2265.

8. Akoyunoglou, G., Argyroudi-Akoyunoglou, J.H., Michel-Wolwertz, M.R. and Sironval C. (1966) Physiol. Plant. 19, 1101.

9. Argyroudi-Akoyunoglou, J.H. and Akoyunoglou, G. (1970) Plant Physiol. 46, 247.

10. Akoyunoglou, G. and Argyroudi-Akoyunoglou, J.H. (1972) in: Proc. 2nd Internatl. Congress on Photosynthesis (Forti, G. et al., Eds). Dr. W. Junk N.Y., The Hague, p. 2427.

11. Akoyunoglou, G. (1977) Arch. Biochem. Biophys. (In press).

12. Akoyunoglou, G. and Tsimilli-Michael, M. (1977) Plant Sci Letters (In press).

13. Akoyunoglou, G., Argyroudi-Akoyunoglou, J.H., Michel-Wolwertz, M.R. and Sironval, C. (1967) Chimika Chronika 32A, 5.

14. Tsimilli-Michael, M. and Akoyunoglou, G. (1976) Biochem. Biophys. Newsl. 8, 20.

15. Tsimilli-Michael, M. and Akoyunoglou, G. (In preparation).

16. Argyroudi-Akoyunoglou, J.H., Feleki, Z. and Akoyunoglou, G. (1971) Biochem. Biophys. Res. Comm. 45, 606.

17. Argyroudi-Akoyunoglou, J.H. and Akoyunoglou, G.(1973) Photochem. Photobiol. 18, 219.

18. Argyroudi-Akoyunoglou, J.H. (1977) This volume, p.

19. Anderson, J.M. and Levine, R.P. (1974) Biochim. Biophys. Acta 357, 118.

20. Argyroudi-Akoyunoglou, J.H. and Akoyunoglou, G. (1977) Arch. Biochem. Biophys. 179, 370.

21. Davis, D.J., Armond, P.A., Gross, E.L. and Arntzen, C.J. (1976) Arch. Biochem. Biophys. 175, 64.

22. Dujardin, E., DeKouchkovsky, Y. and Sironval, C. (1970) Photosynth. 4, 223.

23. Duysens, L.N.M. and Sweers, H.E. (1963) in: Studies on Microalgae and Photo-synthetic Bacteria (Jap. Soc. Plant Physiol., Ed.). p. 353.

24. Malkin, S. and Kok, B. (1966) Biochim. Biophys. Acta 126, 413.

25. Dubertret, G. and Joliot, P. (1974) Biochim. Biophys. Acta 357, 399.

26. Joliot, A. and Joliot, P. (1964) C.R. Acad. Sci. (Paris) 258, 4622.

27. Kitajima, M. and Butler, W.L. (1975) Biochim. Biophys. Acta 376, 105.

28. Akoyunoglou, G. (1976) 10th Internatl. Biochem. Congress. Abstr. Book, p. 330.

29. Murata, N., Nishimura, M. and Takamiya, A. (1966) Biochim. Biophys. Acta 120, 22.

30. Melis, A. and Homann, P.H. (1976) Photochem. Photobiol. 23, 343.

31. Melis, A. and Akoyunoglou, G. (1977) Plant Physiol. (In press).

32. Duysens, L.N.M. (1954) Science 120, 353.

33. Murata, N. and Fork, D.C. (1971) Biochim. Biophys. Acta 245, 356.

34. Reich, R., Scheerer, R., Sewe, K.-U. and Witt, H.T. (1976) Biochim. Biophys. Acta 449, 235.

35. Breton, J. and Mathis, P. (1974) Biochim. Biophys. Res. Comm. 58, 1071.

36. Murakami, S., Torres-Pereira, J. and Packer, L. (1975) in: Bioenergetics of Photosynthesis (Govindjee, Ed.), p. 555.

37. Hind, G., Nakatami, H.Y. and Izawa, S. (1969) Biochim. Biophys. Acta 172, 277.

Bioenergetics of Membranes. L. Packer et al. ed.

DEVELOPMENT OF THE CATION-INDUCED STACKING CAPACITY

IN HIGHER PLANT THYLAKOIDS DURING THEIR BIOGENESIS

J.H. Argyroudi-Akoyunoglou
Department of Biology
Nuclear Research Center "Demokritos"
Greek Atomic Energy Commission, Athens, GREECE

We studied the capacity of pea thylakoids to form grana stacks in the presence
of added monovalent or divalent cations to Tricine "low-salt" disorganized plastids during their greening. Grana stacking was monitored by the yield of heavy
subchloroplast fractions separated by differential centrifugation of digitonin disrupted plastids(Argyroudi-Akoyunoglou, Arch. Biochem. Biophys. 176, 267, 1976).
The primary thylakoids of the agranal protochloroplasts formed in periodic light
do not show the cation-induced stacking capacity of mature green chloroplast thylakoids. Similarly the effect saturates at lower cation concentration in mature
chloroplasts than in plastids of the early stages of greening. The capacity for
cation-induced stacking and for saturation at low cation concentrations appear
gradually after exposure of the plants to continuous light. This capacity develops in parallel with the formation of Chl $\underline{b}$, the polypeptides of the 25-30
Kdalton range of lipid-free thylakoids, and polypeptides of low isoelectric point.
The thylakoid peripheral stroma proteins RuDP carboxylase and the coupling factor
are not involved in the cation induced stacking since their removal (Strotmann et
al, Biochim. Biophys. Acta, 314, 202, 1973) does not affect the thylakoid aggregation. The good correlation found between the extent of grana reconstitution
and the increase in the Chl $\underline{a}$ variable fluorescence yield (F_{DCMU}) caused by cation addition to low-salt chloroplasts (Argyroudi-Akoyunoglou and Akoyunoglou,
Arch. Biochem. Biophys. 179, 370, 1977) point out to a parallel development of
the cation control on the spillover of the excitation energy between the two
photosystems.

INTRODUCTION

The most characteristic feature of a green algal or higher plant chloroplast
is the occurence of grana, densely packed thylakoid regions, joined by unstacked
stroma lamellae. Even though the mechanism by which thylakoid stacking occurs is
under intense investigation, we still do not know when, why and how it occurs.

Abbreviations: Chl, Chlorophyll; CP, Chlorophyll-protein Complex; LDC, Light-Dark
cycles; CL, Continuous light; MV, methyl viologen; DPC, 1,5-diphenyl carbazide;
DCIP, dichloroindophenol; DCMU, 3-(3',4'-dichlorophenyl)-1,-1-dimethyl urea; PS,
photosystem.

Isolated stroma and grana lamellae differ in composition and activities; the stroma lamellae fractions are enriched in Chl $\underline{a}$, CP I, group I polypeptides and have PS I activity; the grana fractions contain both Chl $\underline{a}$ and Chl $\underline{b}$, are enriched in CP II, group II polypeptides and have both PS I and PS II activities[1-5]. The question was therefore raised[6] whether stacking of thylakoids occurs at specific binding sites, i.e. whether some granal components are needed for stacking. To answer this question in the past we made use of the Izawa and Good finding that granal chloroplasts lose their grana by suspension in low-salt media, while cation addition restores grana like structures[7]. Making use of digitonin disruption to separate the grana from the stroma lamellae we studied the low-salt dissociation cation-reconstitution process and found that the reconstituted grana recover the composition and activities of grana obtained from control chloroplasts[6-9]. This showed that the cation-induced stacking of thylakoids is a specific binding leading to grana reconstitution. In addition, the good correlation found between the cation-induced reconstitution of grana and the cation-induced increase in the Chl $\underline{a}$ fluorescence yield[9] suggested that the cation induced control on the excitation energy spillover between the two photosystems[10] is controlled by the structural changes within the thylakoid leading to grana stacking[9].

In protochloroplasts, agranal plastids obtained from etiolated leaves exposed to periodic light, the presence of cations did not affect any of these parameters[6-9]. This suggested that components necessary for stacking are missing from the primary thylakoids and that the stacking capacity develops gradually during thylakoid maturation. Since the primary thylakoids lack Chl $\underline{b}$[11] and are deficient in CP II[5,12], while both components are synthesized after transfer of the plants to continuous light[5,11,12], we proposed that they may be involved in grana formation as prerequisites or consequence of stacking[6,11]. Indeed it was shown recently that the onset of the Mg^{2+} control on spillover occurs parallel to the appearance of CP II in plants exposed to periodic light-dark cycles and then transferred to continuous light[13].

Adopting the low-salt dissociation-cation reconstitution approach we studied how the cation-induced stacking capacity of thylakoids develops during their biogenesis and in what way it is correlated with the biosynthesis of the growing thylakoid components.

MATERIALS AND METHODS

Chloroplasts were isolated from spinach (<u>Spinacea oleracea</u>), bean (<u>Phaseolus vulgaris</u>) or pea (<u>Pissum sativum</u>). Beans and peas were grown in the laboratory[11]; etiolated plants were exposed to periodic light-dark cycles (2 min light-98 min dark) or to continuous light[11]. Plastids were isolated[1,6,8] as pellets fcr 10 min at 6,000xg (periodic light plastids), 3,000xg (continuous light plastids) or 1,000xg (mature chloroplasts). They were washed once with the homogenization buffer (0.3 M sucrose-0.01 M KCl-0.05 M phosphate buffer, pH 7.2).

Disorganization of the grana was done by treatment with Tricine-NaOH, 0.05 M, pH 7.2[8] (40 ml/pellet of 5 g leaves). The plastids were repelleted and resuspended in Tricine to have 100 or 400 μg Chl/ml, and were kept for 30 min at 0°C. $MgCl_2$ or NaCl were added and reconstitution proceeded for 30 min at 0°C. Digitonin disruption followed as described before[1,8]. Isolation of subchloroplast fractions was done by differential centrifugation first for 10 min at the centrifugal force used for plastid isolation (1K, 3K, or 6K) and then for 30 min at 10,000xg (10K). Chl was determined[14] in pellets and supernatant, and based on recovered Chl which was usually 90%, we determined the distribution of the Chl in the fractions. Spinach subchloroplast fractions were isolated as previously described[1,8,9].

The RuDP carboxylase and the coupling factor protein were removed from the thylakoids according to Strotmann[15]; the thylakoids were then suspended in Tricine and reconstitution and digitonin disruption followed as above.

For electrophoresis and isoelectric focusing of the thylakoids, the plastids were washed with Tricine as above, were delipilized, and either dissolved in SDS and electrophoresed[16] or in 2% Triton X-100 and electrofocused. The gel composition for isoelectric focusing was: 5% acrylamide, 0.2% bisacrylamide, 4% glycerol, 0.16% TEMED, 1% Servalyte (pH 2-11), 2% Triton X-100, and 0.06% ammonium persulphate. The catolyte was 0.01% H_3PO_4, the anolyte 3% ethylene diamine. Staining was done with 5% sulphosalicylic-5% TCA: methanol: 1% Coomassie in methanol, 100:20:3; destaining was done with 10% acetic acid in 25% methanol. Densitometer tracings were obtained in a Joyce Loeble Chromoscan with a 620 nm filter. CP I and CP II were separated by gel electrophoresis of lipid rich lamellae[8]; the gels were scanned without staining; the CP II to CP I ratio was determined by integration of the area under the peaks.

The fluorescence and photochemical activity measurements were done as previously described[6,8,9].

RESULTS

<u>Effect of cations on stacking and on the Chl <u>a</u> fluorescence yield</u>: The dissociation of grana in low-salt media and their reconstitution by added cations is correlated with the formation in high yield of light or heavy subchloroplast fractions, respectively, produced after digitonin disruption of chloroplasts[6,8]. Figure 1A shows the effect of cation addition to low-salt agranal chloroplasts and protochloroplasts on thylakoid aggregation. Figure 1B shows the effect of cations on the Chl <u>a</u> fluorescence yield. Both parameters increase drastically after cation addition to the Tricine disorganized spinach chloroplasts. Divalent cations are more effective than monovalent. None of these parameters is affected in protochloroplasts after cation addition.

<u>Specificity in the cation-induced stacking</u>. Table I shows results of experi-

88

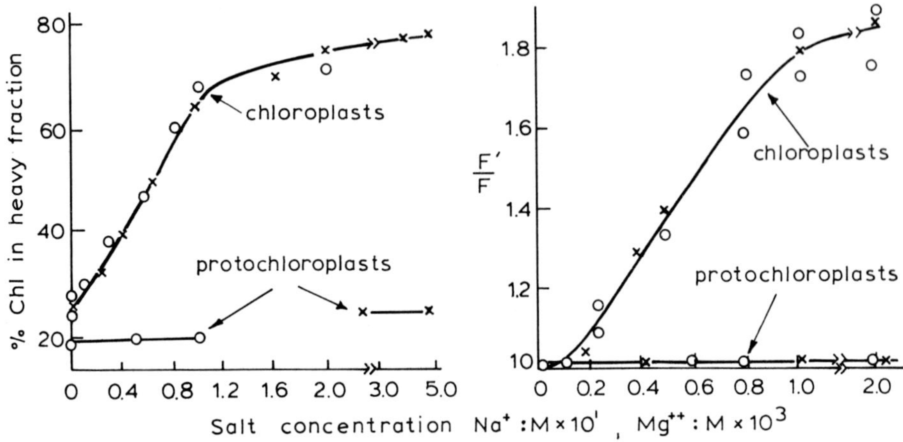

Fig. 1. The effect of Na$^+$(o) and Mg^{++}(x) on A. thylakoid aggregation and B. the
Chl a fluorescence yield, in Tricine disorganized spinach chloroplasts and bean
protochloroplasts A. Heavy subchloroplast fractions obtained by differential centrifugation of digitonin disrupted plastids represent 1,000g+10,000g pellets
(spinach) or 5,000g+10,000g pellets (beans).
B. F', the fluorescence yield in the presence of cations; F, in the absence of
cations. Intensity of excitation light: 574 µW/cm^2 (chloroplasts), 2,610 µW/cm^2
(protochloroplasts). Chl, 10 µg; DCMU, 10 mM. Chloroplast incubation with the
cation for 5 min prior to measurement.

ments which suggest that the cation-induced stacking of thylakoids occurs at specific binding sites rather than between random parts of the membranes. As shown by
the activities and composition of the subchloroplast fractions produced after digitonin disruption of the low-salt chloroplasts, all fractions are mixtures of grana and stroma thylakoids. If an unspecific binding occured, after cation-addition,
one would expect all fractions produced to have activities and composition similar
to those found in the low-salt sample. On the contrary, after cation reconstitution, the heavy and light fractions produced recover to a high extent the activities and composition of the respective fractions derived from stroma and grana
lamellae, obtained from control non disorganized chloroplasts.

Development of the cation-induced stacking capacity: Figure 2 shows the rate
of Chl biosynthesis in etiolated pea leaves exposed directly to continuous light
or to periodic light and then transferred to continuous light. Similar to the results with beans[11] in periodic light only Chl a is formed; Chl b is synthesized
after transfer of the plants to continuous light, and a Chl a to Chl b ratio of
3.0 is found at the plateau. A very small amount of Chl b can be also detected
after 60 light-dark cycles. Assuming that a high Chl a to Chl b ratio is indicative of single lamellae, these results suggest that single lamellae predominate in
the protochloroplasts and that grana start to be formed after transfer to continuous light, in aggreement with the results with bean leaves[17].

TABLE I

ACTIVITIES AND COMPOSITION OF SUBCHLOROPLAST FRACTIONS OBTAINED AFTER DIGITONIN DISRUPTION OF CONTROL GRANAL, LOW-SALT AGRANAL AND CATION-RECONSTITUTED SPINACH CHLOROPLASTS

Subchloroplast fraction	Chl a / Chl b	CP II / CP I	PS II[a] +DPC	PS II[a] -DPC	PS I[b]	PS II + PS I[b]
Control granal chloroplasts						
10 K	2.5	4.3	80	62	75	70
50 K	5.6		12	0	250	0
144 K	5.8	0.7	0	0	300	0
Low-salt agranal chloroplasts						
10 K	3.0	3.0	89	80	95	57
50 K	3.3		86	80	130	35
144 K	3.0	2.6	37	11	190	15
Cation-reconstituted chloroplasts						
10 K	2.5	3.9	60	60	110	80
50 K	4.7		17	10	380	0
144 K	5.2	0.8	4	0	380	0

[a]Rate of DCIP reduction, in the presence or absence of DPC[6]; the values show percent of chloroplast rates.

[b]Rate of methyl viologen mediated oxygen uptake[8,9]. Values show percent of chloroplast rates. PS II: $DCIPH_2$-MV reaction, PS II+PS I: H_2O-MV reaction.
Control medium: 0.05 M phosphate-0.01 M KCl, pH 7.2. Low-salt medium: 0.05 M Tricine-NaOH, pH 7.2. Reconstitution medium: 0.05 M Tricine-NaOH-0.1 M NaCl, pH 7.2. Digitonin disruption as described[8].

Figure 3 shows the development of the capacity of thylakoids to stack after cation addition to Tricine suspended plastids in leaves that received 42 light-dark cycles and were then transferred to continuous light. Only 10 to 30% of the Chl is pelletable as a heavy fraction in the Tricine samples, and this depends on the time of exposure. As exposure to light is prolonged the effectiveness of Tricine washing to disorganize the grana formed is diminished. The cation-induced stacking capacity increases with time in continuous light and appears in parallel with the synthesis of Chl _b_.

In Figure 4 comparison is made of the cation effect development in plastids of leaves transferred to continuous light with that of plastids remaining in periodic light. A small cation effect is observed after 130 light-dark cycles. In this figure the net value of the cation effect is shown, i.e., after subtracting the respective values found in the Tricine samples. The biphasic type of curve shown in Figure 3 is no longer noticed.

90

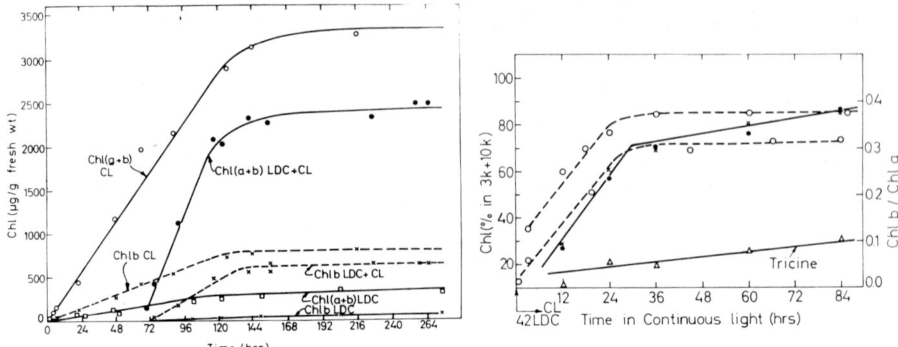

Fig. 2. Rate of Chl formation in 8-
day old etiolated pea leaves exposed
to continuous light (CL) periodic light
(LDC) or first to 42 LDC and then transferred to continuous light.

Fig. 3. Development of the cation induced stacking in continuous light
after 42 light-dark cycles. Prior to
cation addition: $\triangle$--$\triangle$ After 100 mM Na$^+$.
$\bullet$--$\bullet$, or 5 mM Mg^{2+}: x--x. Chl b/Chl a:
o--o, upper curve in chloroplasts, lower
curve in leaves.

Table II shows the appearance and development of the cation stacking effect in
plastids greening directly in continuous light. The effect develops gradually as
in the case of the leaves transferred to continuous light after periodic light
exposure.

The gradual appearance of the effect, brought up the question whether these results reflect a different saturation pattern between the immature and the mature
thylakoid. Indeed, as shown in fig. 5, the effect saturates at lower cation con-

TABLE II

DEVELOPMENT OF THE CATION-INDUCED THYLAKOID STACKING IN ETIOLATED PEA LEAVES EXPOS-
ED TO CONTINUOUS LIGHT

Exposure time (h)	(% Chl in 3K+10K)$_{+cation}$ - (% Chl in 3K+10K)$_{-cation}$		
	5 mM Mg^{2+}	100 mM Mg^{2+}	100 mM Na$^+$
15	19	40	26
16	22	–	–
25	23	52	39
48	33	54	33
66	52	73	53
89	65	73	65
Lab. green	63	67	61

The Chl concentration before digitonin disruption was in the order shown:
91, 175, 218, 250, 380, 407 and 410 µg/ml.

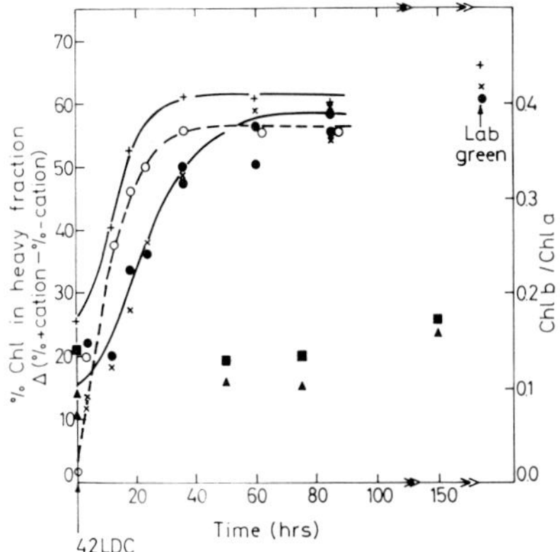

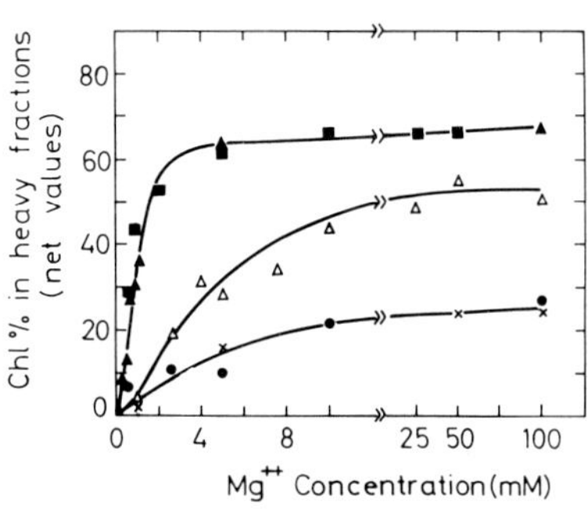

Fig. 4. Development of the cation-induced stacking in continuous or periodic light. Plastids of leaves remaining in LDC: 5 mM Mg^{2+} (▲), 100 mM Na^+ (■). Plastids of leaves transferred to CL: 5 mM Mg^{2+} (×), 100 mM Na^+ (●), or 100 mM Mg^{2+} (+). Chl $\underline{b}$/Chl $\underline{a}$: o--o. Lab. green: 19-day plants, day-night conditions.

Fig. 5. Effect of the thylakoid developmental stage on the Mg^{2+} concentration dependence. Plastids of leaves exposed to 40 (●) or 72 (×) LDC transferred to CL: for 18 hours after 43 LDC (Δ), or 144 hours after 43 LDC (■), grown in laboratory day-night conditions for 19 d (▲).

centrations in mature chloroplasts (1-2 mM Mg^{2+}, as in spinach) than in plastids of the early stages of greening. This may reflect the ease with which the immature thylakoids lose their ionic content. Another possibility is that grana

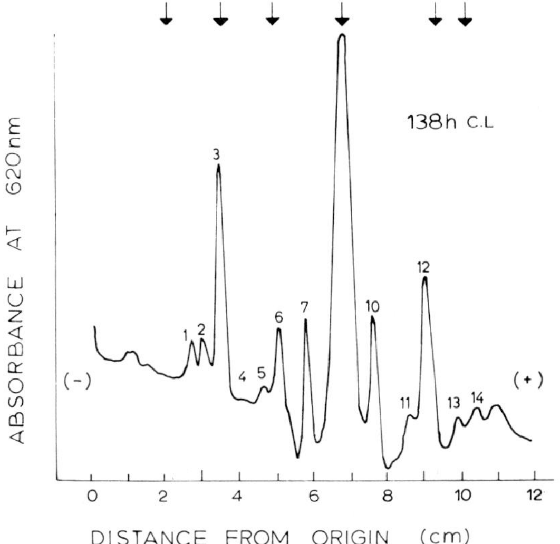

Fig. 6A. SDS-gel electrophoresis run of thylakoids of leaves exposed to CL for 138 h. Marker proteins from left to right: γ-globulins, 160 Kd, Bovine albumin 67 Kd ovalbumin 45 Kd, chymotrypsinogen 25 Kd, myoglobin 17.8 Kd, cytochrome c 12.4 Kd.

92

formation occurs gradually in plants transferred to continuous light after periodic light exposure, so that grana at different developmental stages are present,
i.e. from incompletely developed to fully developed ones. The fully developed
grana would be pelletable at low cation concentrations, the less developed at
higher. If this is so, the various polypeptides involved in stacking seem to have
different cation binding constants and that their appearance is reflected by the
difference in the saturation pattern.

 Appearance of polypeptides. Figs. 6 and 7 show the electrophoretic and isoelectric focusing analyses of thylakoids at various stages of development. The
SDS gel electrophoresis pattern of the mature pea thylakoids (Fig. 6A) resembles

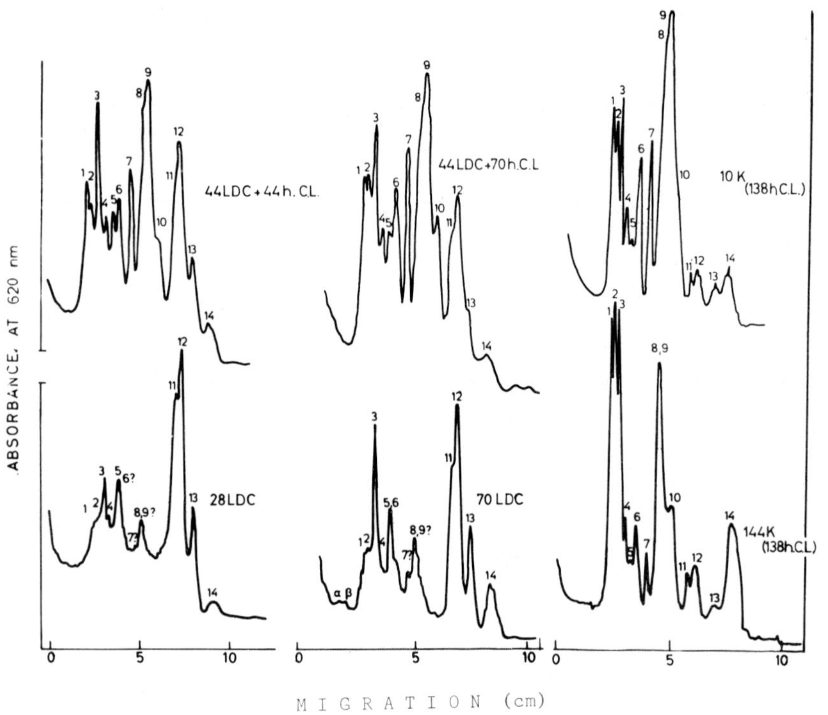

Fig. 6B. Densitometer tracings from SDS-gel electrophoresis runs of thylakoids and
subchloroplast fractions obtained from leaves at various stages of development.

that obtained from lipid-free lamellae of a variety of sources[16,18-21]. A marked
difference in the polypeptide region marked 8-10 and corresponding to 25-30 Kd
molecular weight is evident between the primary thylakoids and the mature green
ones. These peptides appear gradually with greening and finally they predominate
(Fig. 6B). This group of polypeptides predominates also in the 10K grana fraction
obtained from mature chloroplasts, while it is present in lower amount in the 144K

stroma lamellae fraction, in accordance with earlier results[21]. In the later fraction the group 1-3 polypeptides predominates. The thylakoid peripheral proteins,
RuDP carboxylase and the CF_1 protein seem to be responsible at least partly for
the polypeptides appearing at 52 and 14 Kd (large and small subunit of RuDP carboxylase) and 56 and 53 Kd (the a and b subunits of the coupling factor protein)
polypeptides[22]. They may therefore add to the group 1-3 polypeptides and much
more so in the stroma lamellae, which have a larger surface exposed to the stroma,
than the grana.

Tricine disorganization of the thylakoids, as performed in our experiments, was
expected to remove the RuDP-carboxylase since it is released at low osmolarity. A
residual amount of this protein, however, could still be present. On the other

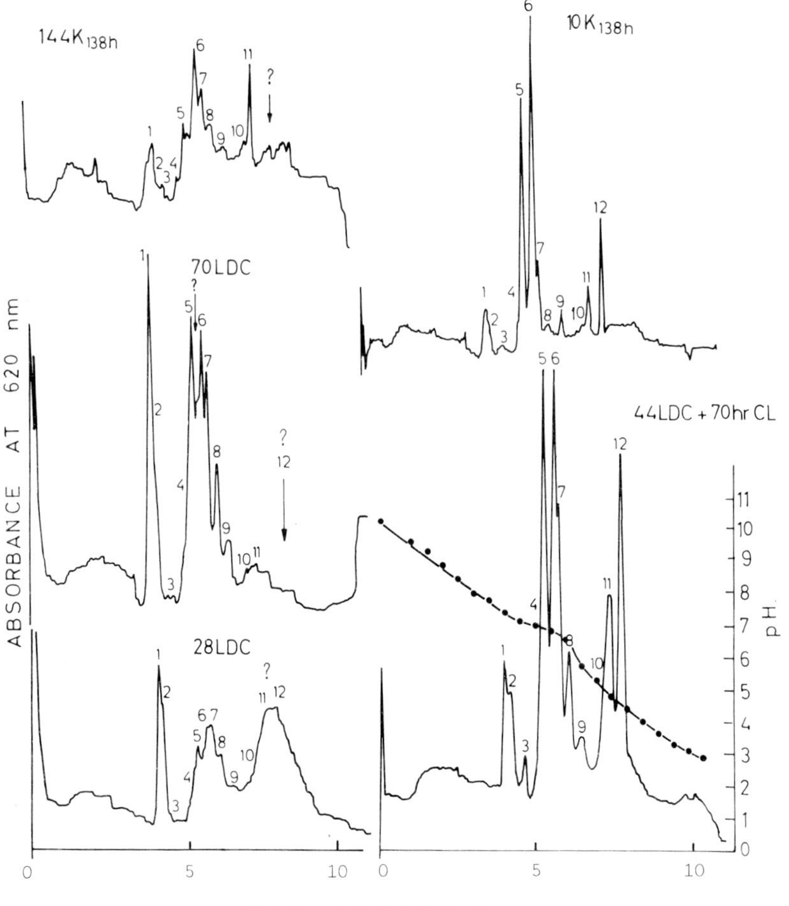

Fig. 7. Densitometer tracings and pH-gradient of thylakoids and subchloroplast
fractions at various developmental stages after polyacrylamide gel isoelectric
focusing.

TABLE III

CATION-INDUCED STACKING CAPACITY OF THYLAKOIDS FROM WHICH THE RuDP CARBOXYLASE
AND THE COUPLING FACTOR PROTEIN WERE REMOVED.

	% Chl in heavy subchloroplast fraction (1K + 10K)		
	Tricine washed	Pyrophosphate washed	Pyrophosphate-sucrose washed
Tricine	30	53	52
+ 25 mM Mg^{2+}	95	100	100

The RuDP carboxylase and the coupling factor protein were removed by pyrophosphate
washing or pyrophosphate followed by sucrose washing, respectively, according to
Strotmann et al[15]. Plastids isolated from leaves exposed to 144h CL after 43 LDC.

hand the coupling factor protein, which is released only at high osmolarity and in
the absence of salts, was expected to be present.

To see whether these peripheral proteins are involved in the cation-induced
stacking we removed them from mature thylakoids[15] and determined the cation-induced stacking capacity of the remaining thylakoids. As shown in Table III, their
removal does not affect the ability of thylakoids to stack. The stacking capacity,
therefore, resides within the thylakoid. The parallel appearance of the cation effect and the 25 Kd polypeptide group, on one hand, and the fact that these polypeptides are integral parts of the thylakoid (structural protein) suggest strongly
that they may be involved in stacking. The isoelectric focusing experiments, although still preliminary, have shown the gradual appearance of low isoelectric
point polypeptide, which are missing from the primary thylakoids, during greening
in continuous light. These polypeptides (pI of about 4.5) no 11 and 12 in fig. 7,
are also present in high amount in the 10K grana fraction. Their correspondence to
the SDS gel electrophoresis polypeptides is still not known.

DISCUSSION

Cations are generally believed to bind on the fixed negative charges of the thylakoids; the isoelectric point of chloroplasts is 4.7[23]. Stacking of adjacent membranes is thought to be facilitated by divalent cation salt bridge formation, or
by monovalent cation suppression of the electrostatic repulsion, through neutralization of the charges. Electrolytes are also expected to act by lowering the water
solubility of the membrane enhancing the hydrophobic forces between adjacent granal
membranes[24]. Although both lipids and proteins may contribute to these charges,
proteins are more strongly implicated in the cation binding. Berg et al[25], for
example, have shown that modification of the free carboxyl groups eliminates all
the thylakoid surface charges and induces multiple membrane associations in low-
salt chloroplasts. This resembles the effect of salts on low-salt mature chloroplasts.

The parallel appearance of the 25-30 Kd polypeptides with the development of the cation-induced stacking capacity in thylakoids during greening in continuous light, suggests that these components may be required for stacking. A variety of other finding has implicated this group of polypeptides in the stacking process. These polypeptides, also called by others group II[4,20,26], are believed to be derived from CP II after delipidation[20]. CP II is missing, or is present in minor amounts, in agranal blue green algae or agranal plastids of higher plants[5,12,27,28], but it is a major component of thylakoids of granal plants and algae[5,12,27]. Its gradual appearance in developing chloroplasts during the transformation of proto-chloroplasts to mature chloroplasts, coincides with grana formation[12,17] and the onset of the cation control on the spillover of the excitation energy[13]. Subchloroplast fractions derived from grana and stroma lamellae are enriched in CP II and CP I or in group II and group I polypeptides, respectively[1-5,21]. The unstacked membranes of the mixotrophically grown ac-5 mutant of C reinhardi lack the group II polypeptides in contrast to the stacked membranes of the phototrophically grown cells[29]. Our data also suggest that the 25-30 Kd polypeptides are involved in the cation-induced stacking; however, we can not say whether the parallel appearance of thylakoid components and of the stacking property shows a direct relationship, nor which one of these polypeptides is more specifically involved. Similarly, the parallel appearance of Chl b may imply that its formation is a consequence of the formation of some of these polypeptides. The appearance of Chl b, in a way, monitors the formation of CP II, since it is needed for stabilization of the pigment binding on the protein of CP II[5,30]. In thylakoids which, for some reason, lack Chl b, but not the 25 Kd peptides, it might be expected that cation induced stacking may still occur. The Chl b-less barley mutant, for example, shows low-extent of cation-induced changes in the Chl a fluorescence yield and a low extent of lamellar appressions[31]. These thylakoids have reduced amount of some of the group II polypeptides[26]. The parallel appearance of the stacking capacity of the thylakoids and the synthesis of thylakoid components may, therefore, reflect the development of a certain organization within the thylakoid, which offers the necessary charge distribution and hydrophobic environment needed for stacking. The results of the electrofocusing experiments indeed suggest that continuous light exposure induces the formation of polypeptides of low isoelectric point. We do not know yet whether these polypeptides correspond to the 25-30 Kd ones. It seems certain, however, that the stacking property resides within the thylakoid and is not affected by the removal of the stroma peripheral proteins.

The main question is therefore still open; what kind of forces capable to induce grana stacking are induced by continuous light, and why these forces are not formed in periodic light.

ACKNOWLEDGEMENTS. Part of this work has been conducted in collaboration with Mr. S. Tsakiris. I wish to thank also Mr. S. Koussoulakos and Mr. J. Lambris for their assistance in the electrofocusing experiments.

REFERENCES

1. Anderson, J.M. and Boardman, N.K. (1966) Biochim. Biophys. Acta, 112, 403.

2. Sane, P.V., Goodchild, D.J. and Park, R.B. (1970) Biochim. Biophys. Acta 216,162

3. Goodchild, D.J. and Park, R.B. (1971) Biochim. Biophys. Acta, 226, 393.

4. Levine, R.P., Burton, W.G. and Duram, H.A. (1972) Nature New Biol., 237, 176

5. Argyroudi-Akoyunoglou, J.H. and Akoyunoglou, G. (1973) Photochem. Photobiol., 18, 219.

6. Akoyunoglou, G. and Argyroudi-Akoyunoglou, J.H., (1974) FEBS Lett., 42, 140.

7. Izawa, S. and Good, N.E. (1966) Plant Physiol., 41, 544.

8. Argyroudi-Akoyunoglou, J.H. (1976) Arch. Biochem. Biophys., 176, 267.

9. Argyroudi-Akoyunoglou, J.H. and Akoyunoglou, G. (1977) Arch. Biochem. Biophys., 179, 370.

10. Murata, N. (1969) Biochim. Biophys. Acta, 189, 171.

11. Argyroudi-Akoyunoglou, J.H. and Akoyunoglou, G.(1970) Plant Physiol., 46, 247.

12. Argyroudi-Akoyunoglou, J.H., Feleki Z. and Akoyunoglou G. (1971) Biochem. Biophys. Res. Commun. 45, 606.

13. Davis, D.J. et al. (1976) Arch. Biochem. Biophys., 175, 64.

14. Mackinney, G. (1941) J. Biol. Chem., 140, 315.

15. Strotmann, H., Hesse, H. and Edelmann, K. (1973) Biochim. Biophys. Acta, 314,202

16. Hoober, K. (1970) J. Biol. Chem., 245, 4327.

17. Argyroudi-Akoyunoglou, J.H. et al. (1976) Plant Cell Physiol., 17, 939.

18. Remy, R. (1971) FEBS Lett., 13, 313.

19. Klein, S.M. and Vernon, L.P. (1974) Photochem. Photobiol., 19, 43.

20. Anderson, J.M. and Levine, R.P. (1974) Biochim. Biophys. Acta, 357, 118.

21. Henriques, F. and Park, R.B. (1974) Plant Physiol., 54, 386.

22. Henriques, F. and Park, R.B. (1976) Arch. Biochem. Biophys., 176, 472.

23. Dilley, R.A. and Rothstein, A. (1967) Biochim. Biophys. Acta, 135, 427.

24. Murakami, S. and Packer, L. (1971) Arch. Biochem. Biophys., 146 337.

25. Berg, S. et al. (1974) Plant Physiol., 53, 619.

26. Anderson, J.M. and Levine, R.P. (1974) Biochim. Biophys. Acta, 333, 378.

27. Argyroudi-Akoyunoglou, J.H. (1976) 7th International Photobiology Congress, Rome, Italy, Abstracts, p. 87.

28. Brown, J.S. et al. (1974) In Proceedings 3rd Intern. Congress on Photosynthesis, Rehovot, Israel, (Avron, M. ed) p. 1951, Elsevier Amsterdam, Holland

29. Levine, R.P. and Duram, H.A. (1973) Biochim. Biophys. Acta, 325, 565.

30. Genge, S., Pilger, D. and Hiller, R.G. Biochim. Biophys. Acta, 347, 22.

31. Boardman, N.K. and Thorne, S.W. (1976) Plant Sci. Lett., 7, 219.

DISCUSSION

D. Chapman
Chemistry Department
Chelsea College
University of London, England

A discussion took place concerning protein-lipid interactions within cell membranes and particularly on the nature of the boundary layer. D. CHAPMAN pointed out that it seems common for some scientists to draw a tightly packed immobilised layer of lipid molecules around the protein e.g. an "annulus" of lipid and to envisage that this lipid is carried around by the protein within the plane of the membrane. He raised the question as to whether this picture allowed for any direct protein-protein contacts or whether this was not allowed on this model. He also asked whether this model was considered to be a general one for proteins within cell membranes and the evidence for this.

J. METCALFE said that with the sarcoplasmic reticulum ATPase that he had evidence from saturation transfer ESR experiments that the rate of exchange of the lipids from the annulus was about two orders of magnitude lower than the rate of exchange between phospholipids in a lipid bilayer. In his view the lipid annulus corresponds to the minimum amount of lipids required to retain biological activity.

P. DEVAUX said that in his experiments that when lipids are progressively removed from a membrane system that (a) the rotational correlation time of the protein decreases regularly and (b) that the state of the lipids including the lipids specifically attached to the protein (in his case spin-labelled fatty acids bound covalently to r proteins) are progressively modified i.e. a decrease of mobility occurs. This he had shown using saturation transfer experiments. These experiments demonstrated the influence of the presence of the bulk lipids on the protein mobility. He also pointed out that some recent experiments by M. BROWN *et al.* indicated from proton NMR studies with retinal rods that rapid exchange of the phospholipid occurs with the bulk and boundary lipid. DEVAUX said that more experiments were required than were now published to support the concept of an immobilised lipid annulus which only slowly exchanged with the bulk lipid.

D. CHAPMAN mentioned some of his recent experiments with Gramicidin A in liposome systems. This showed using Raman spectroscopy that this hydrophobic polypeptide increases the trans content of the lipid chains by its presence somewhat similar in effect to the action of cholesterol. At very high content of the polypeptide to lipid using ESR spin labels an "immobilised" component was observed.

This is similar to the spectral feature attributed to the boundary layer or annulus layer. In the case of the polypeptide however the presence of this component was associated with random arrangements of the lipid and polypeptide where polypetide-polypeptide contacts could occur. This allowed lipid to contact not only one but two and three polypeptides. The immobilised ESR component was considered then to arise from the spin label being contacted or "trapped" by multiple polypeptides rather than with only the single contact. He suggested therefore that perhaps the presence of this ESR immobilised component might not be a good measure of the annulus lipid rigidity. He further pointed out that in the model outlined by METCALFE for the sarcoplasmic reticulum protein of a cylinder surrounded by thirty tightly packed lipids that this led to great packing problems and to holes appearing in the resultant structure. He had shown this in a slide during his formal lecture presentation. He also indicated that he and his co-workers had shown that the co-ordination number in a random arrangement of large cylinders and small cylinders is always much less than the maximum number of small cylinders which can be arranged around the large cylinder.

In general it was agreed that the presence of a protein inside the lipid bilayer will perturb the lipid chain rotational isomerism in the boundary layer. Furthermore the existence of a lipid annulus exchanging slowly or not at all with the bulk lipid may also occur in some specific cases (e.g. with specifically attached covalent or electrostatic linkages) but the evidence for this requires careful examination and more experiments are required to strengthen the case for and against the existence of this highly immobilised lipid layer certainly as a general phenomenom.

BACTERIORHODOPSIN

LIGHT-DRIVEN PROTON AND SODIUM ION TRANSPORT IN BACTERIORHODOPSIN-CONTAINING PARTICLES

S. Roy Caplan, Michael Eisenbach, Shulamit Cooper, Haim Garty, Gil Klemperer and Evert P. Bakker

Department of Membrane Research, The Weizmann Institute of Science
Rehovot, Israel

SUMMARY

The light-induced pH changes in a suspension of bacteriorhodopsin-containing proteoliposomes were followed by measuring the fluorescence changes of the impermeant indicator FITC-Dextran. The kinetics of the alkalization by the proteoliposomes as well as the kinetics of the light-driven acidification by KCl-loaded sub-bacterial particles of *Halobacterium halobium* fitted a sum of two exponentials, corresponding to two simultaneous processes, one rapid and one slow.

The observation that external effects such as pH and temperature mainly influenced the extent of the slow process led to the development of a model. According to this model the slow process represents net proton transport across the membrane while the rapid process represents the dissociation (in sub-bacterial particles) or association (in proteoliposomes) of protons from or to the membrane as a result of a light-induced conformational change. Further evidence is given to support this model.

A similar phenomenon was observed with NaCl-loaded sub-bacterial particles, provided that the light intensity was below 45 W/m^2. Above this light intensity an additional phase, this time of alkalization, followed the acidification. This alkalization is the result of a H^+/Na^+ antiport, *i.e.* an influx of protons accompanied by an efflux of Na^+ ions. Illuminating the particles for four hours in order to deplete them of Na^+ caused the disappearance of the alkalization phase during the subsequent re-illumination. By following the Na^+ flux driven by a light-induced membrane potential difference, it is shown that the H^+/Na^+ antiport is electrogenic.

INTRODUCTION

Illumination of a suspension of *Halobacterium halobium* under anaerobic conditions results in complex pH changes in the suspending medium, *i.e.* a combination of acidification and alkalization processes, the net result of which is acidification[1-4]. This is associated with the formation of a proton electrochemical potential difference across the membrane ($\Delta\tilde{\mu}_H{+}$), which is composed of a ΔpH component ($pH_{in} > pH_{out}$) and a $\Delta\psi$ or membrane potential component (inside negative)[5,6]. In simpler preparations derived from *H. halobium* where processes affecting the pH cannot occur, less complicated pH changes are observed. Sub-bacterial particles, *i.e.*, cell envelopes prepared from intact bacteria by sonication and oriented inside-in, acidify the suspending medium upon illumination[7-9]. On the other hand, reconstituted bacteriorhodopsin-containing proteoliposomes are oriented inside-out and alkalize the suspending medium upon illumination[9-11]. Most of the experiments which are described below were carried out with these two types of particles, where their internal and external contents could be pre-determined and changed at will. In these experiments the internal and external solutions contained NaCl or KCl and buffer only.

It is shown in this communication that both the alkalization by proteoliposomes and the acidification by sub-bacterial particles are composed of two first order (or pseudo first order) processes occurring simultaneously, and a model is developed to explain this behaviour. Differences in behaviour are observed also between NaCl-loaded and KCl-loaded sub-bacterial particles and explained on the basis of a H^+/Na^+ antiport model[12,13].

MATERIALS AND METHODS

The preparation of *H. halobium* (M-1 strain), sub-bacterial particles, purple-membrane fragments, and bacteriorhodopsin-containing proteoliposomes was carried out as described before[9,13]. The measurements of external pH were performed either by a combined pH electrode (as described elsewhere[9,13]) or fluorometrically using FITC-Dextran* from Pharmacia Fine Chemicals (MW = 20,000). In the latter case the light-induced pH changes were followed by a Perkin-Elmer MPF 44 A model spectrofluorometer (the photomultiplier was protected by narrow band interference filter Baird Atomic B10). The indicator in its basic form emits light at 520 nm after being excited at 480 nm. As a consequence of its high molecular weight the indicator is impermeant through the membrane and therefore measures the external pH only. In the experiments with the fluorometer, the sample was illuminated

---

*Abbreviations : FCCP - carbonyl cyanide *p*-trifluoromethoxyphenylhydrazone; FITC - fluorescein isothiocyanate; MES - 2-(N-morpholino)ethanesulfonic acid; MOPS - morpholinopropane sulfonic acid; PIPES - piperazine-N,N'-bis(2-ethane-sulfonic acid); TAPS - tris(hydroxymethyl)methylaminopropane sulfonic acid; TPB^- - tetraphenyl boron; $TPMP^+Br^-$ - triphenylmethylphosphonium bromide.

through a Corning 3-68 "cut-on" filter (only part of the cuvette was exposed to the illuminating beam). The measurements of ΔpH and $\Delta\psi$ in sub-bacterial particles were performed by following the accumulation of [14C] propionic acid (50 µM) and [3H] TPMP$^+$ (10 µM), respectively. For both measurements filtration was used to separate particles from medium. This was carried out as described previously[13], with the exception that the filter was washed 2-3 times only within a time period of 15 s. An attempt to correct for the absorption of TPMP$^+$ by the membrane[5,6] was made by subtracting the accumulated TPMP$^+$ in lysed particles (obtained by 200-fold dilution in water) from the accumulation measured in intact particles. This correction yielded $\Delta\psi$ = 0 in the dark, but since the TPMP$^+$ accumulation might not correspond to the *exact* value of $\Delta\psi$, we expressed the results as $(RT/F)\ln([TPMP^+]_{in}/[TPMP^+]_{out})$ rather than $\Delta\psi$. Na$^+$ efflux was measured using ^{22}Na$^+$ as described previously[13].

RESULTS AND DISCUSSION

KCl-loaded particles and proteoliposomes

We have shown in previous communications that the kinetics of the light-induced pH-changes in sub-bacterial particles of *H. halobium* and in bacteriorhodopsin-containing proteoliposomes (acidification in the former and alkalization in the latter) could be described as a sum of two exponentials[9,14,15], *i.e.* as two first-order (or pseudo first-order) processes occurring simultaneously. The pH in these experiments was measured by a pH-electrode, and the possibility of a perturbation of the observed rate constants by the pH-electrode seemed unlikely but could not be excluded[14]. To bypass this difficulty we used FITC-Dextran in the following experiment as an impermeant fluorescent pH indicator for the external pH.

Fig.1 shows the pH-changes measured fluorometrically in bacteriorhodopsin-containing proteoliposomes (A) and their kinetic analysis (B) after switching the light on and off (following 5-15 min of dark period preceded by pre-illumination, so that most of the bacteriorhodopsin was in its light-adapted form, *cf*. Ref.16). The kinetic analysis was performed by plotting the logarithm of the difference $(H^+_\infty - H^+_t)$, *i.e.* the total extent of proton uptake minus the quantity taken up at time t, *vs*. time. It can be seen that even when a pH electrode is not used the kinetics of both the "on" and "off" reactions may be described as a sum of two exponentials. The slow process may be observed alone at the end of the reaction as a straight line on the semi-logarithmic plot, while the fast process gives rise to the straight line obtained by subtracting the extrapolated line of the slow process from the experimental points (Fig.1B). The rate constants of the fast and slow processes (k_1 and k_2, respectively) are calculated from the slopes of these

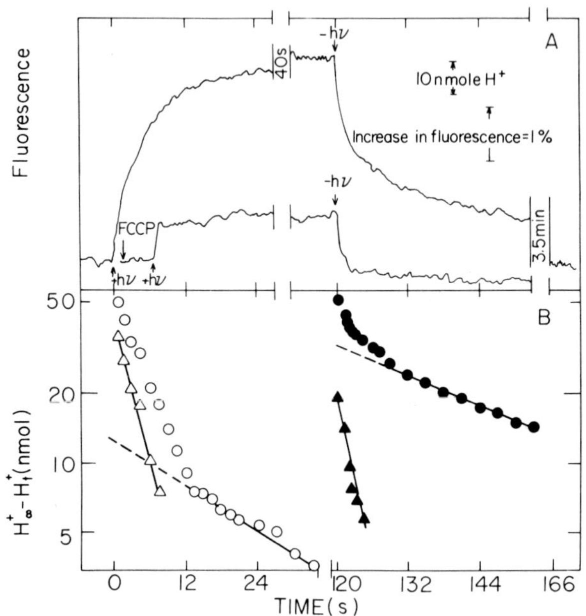

Fig.1 : Light-induced pH changes and their kinetic analysis in bacteriorhodopsin-containing proteoliposomes. Proteoliposomes, prepared at 17 nmol/ml in 1 M KCl by sonication for 75 min (final pH = 6.3), were diluted to 15 nmol bR/ml at 25°C with a solution containing 1 mg FITC-Dextran (3 ml final volume). A, the light-induced pH changes (I = 75 W/m^2) as followed by recording the change of the indicator fluorescence (pK $\simeq$ 6.5) at the wavelength pair 480/520 nm. The lower trace was recorded in the presence of 20 μM FCCP. B, the semi-logarithmic plot of the upper trace in part A. o,• - experimental points taken from the upper trace; Δ,▲ - differences between the experimental points and the extrapolated line. The open and closed symbols are for the "on" and "off" reactions, respectively; k_1(on) = 0.26 s^{-1}; k_2(on) = 0.046 s^{-1}; k_1(off) = 0.29 s^{-1}; k_2(off) = 0.025 s^{-1}.

straight lines, and they are comparable to those obtained with a pH-electrode under the same light intensity, pH, and temperature[15]. The extents are calculated from the semilogarithmic plot by extrapolation of the straight lines to t = o. The uncoupler FCCP abolished the slow phase and decreased the extent of the rapid phase as in our earlier experiments using a pH-electrode. These results justify our use of the pH electrode in routine kinetic studies of this system.

The kinetics were examined at different wavelengths, and also at various light intensities, pH's, and temperatures and in the presence of various ions and iono-

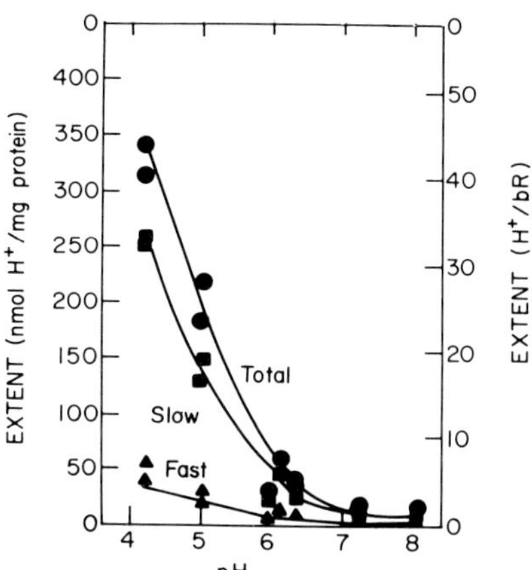

Fig.2 : Effect of initial pH on the extent of the pH changes. Portions of KCl-loaded sub-bacterial particles (4 mg protein) were suspended in 2.5 ml of 4 M KCl, 0.2 mM TAPS, lmM MES, and 2.5 mM citrate. (These buffers were required to keep the external pH as close as possible to the original value during the illumination. This was especially important at pH's below 6, where otherwise the external pH could drop by 0.5 pH units.) The pH was adjusted by addition of small amounts of NaOH or HCl and the suspensions were incubated overnight at 4°C. After incubation and pre-illumination the light-induced pH changes were followed by a pH-meter. ● - total extent; ■ - slow process; ▲ - fast process. The dependence of the extent on the pH was similar in the "on" and "off" reactions, and since the phases could be separated more easily in the "off" reaction, the points are taken from this reaction (From Ref.14).

phores[14,15,17]. The extents of both phases were maximal when the illuminating wavelength was in the vicinity of 570 nm - the absorption peak of bacterio-rhodopsin. The remaining parameters showed a significant effect : for the most part they influenced the extent of the slow process, with little effect on that of the rapid one.

As shown in Fig.2, decreasing the pH of a suspension of KCl-loaded sub-bacterial particles caused a drastic increase in the extent of the slow process, while that of the rapid process increased only slightly.

106

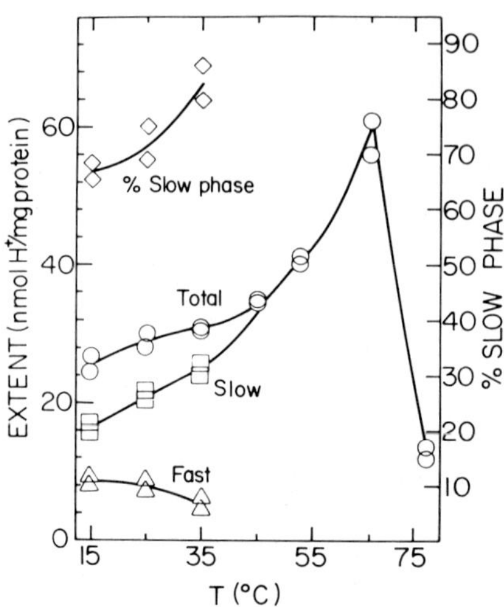

Fig.3 : Effect of temperature on the extents of light-induced pH changes. KCl-
loaded sub-bacterial particles (3.9 mg protein) were suspended in 2.5 ml 4 M KCl
(pH 6.3), pre-illuminated and followed by a pH-meter. The experiment was
initiated after 30 min incubation at each temperature. o - total extent; Δ -
extent of the fast process; ☐ - extent of the slow process; ◊ - extent of the
slow process as a fraction of the total extent (From Ref.14).

Similarly, Fig.3 shows that increasing temperature led to an increase in the
extent of the slow process while the extent of the rapid process decreased. In a
suspension of proteoliposomes the extent of the slow process similarly increased
with increasing temperature or decreasing pH, while that of the rapid process
decreased (not shown)[*]. Furthermore, in the presence of the permeant cation $TPMP^+$
which decreases $\Delta\psi$[19], only the extent of the slow phase increased. A similar
increase in the extent of alkalization was observed with proteoliposomes in the
presence of $TPMP^+$ or the permeant anion NO_3^-. Under these conditions we would
expect the predominant effect to be an increase in H^+ transport[14], and may there-

---

[*]The extents obtained with proteoliposomes were usually smaller than those
obtained with KCl-loaded sub-bacterial particles, probably because their volumes
are several orders of magnitude smaller than those of the latter (compare Ref.18
with Ref.13), and because of their lower internal buffer capacity (*cf*. Ref.15).

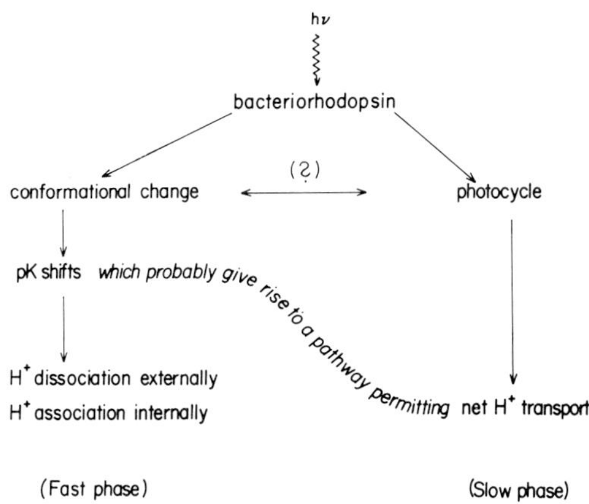

Fig.4 : A schematic representation of a model explaining the biphasic acidification in a suspension of sub-bacterial particles.

fore conclude that the rapid phase does not represent net H^+ transport. This conclusion is supported by the low extents of the rapid phase (generally 1 and at most 3 protons per bacteriorhodopsin molecule at very low pH) as compared to the higher extents of the slow process in sub-bacterial particles (above 60 protons per bacteriorhodopsin molecule below pH 5). Furthermore, the rapid process operates for a short while only (less than 15 seconds), while a transport process would be expected to operate for longer periods. Our suggested model is shown in Fig.4, where the arrows signify the sequence of operations. Excitation of bacteriorhodopsin by light leads to a photocycle[20-24] during which protons are taken up and released by the membrane as shown by flash photometry, and to conformational changes of the protein[25-28]. These processes may be dependent on each other. According to the model, the experimental expression of the pumping by the photocycle is the slow phase of the acidification, leading to net H^+ transport. The conformational changes bring about changes in the number of exposed groups on both sides of the membrane, the result of which in sub-bacterial particles is H^+-dissociation externally and H^+-association internally. This is seen as the rapid phase in the light-induced pH change. In proteoliposomes, because of their inside-out orientation, association occurs externally and dissociation internally. The groups involved might well serve as proton carriers between the photocycle and the external solutions. The main supporting evidence for this view can be summarized as follows : (1) Under conditions where transport alone can occur, *e.g.* when $TPMP^+$ is added during the steady state of proton transport in the light, only the slow process is observed[14]. (2) Under conditions where transport cannot occur, *e.g.* in a suspension of purple-membrane fragments where both faces of the membrane are exposed to the same medium, only the fast process is observed[14]. (3) With these fragments one expects to observe almost simultaneous proton release (by the exterior) and binding (by the interior), as was demonstrated by following the flash-induced pH change of the medium on a millisecond scale in suspensions of purple-membrane fragments, proteoliposomes and sub-bacterial particles[29]. (4) Because of differences in the groups exposed on either side of the membrane, the extent of acidification by one side would be expected, according to our model, to be different from the extent of alkalization by the other, and hence the *net* acidification *or* alkalization observed on a macro-scopic time scale should be dependent on the pH. We have shown recently[30] that either acidification or alkalization is observed with purple-membrane fragments, depending on the pH. (5) In sub-bacterial particles the K^+ ionophore monactin had a similar effect to $TPMP^+$ in the presence of KCl, *i.e.* it increased the extent of the slow process without affecting the rapid process[17]. Similarly, in proteo-liposomes valinomycin, which is another K^+-ionophore, increased the rate constant of the slow process by an order of magnitude while that of the rapid process increased only slightly[17].

It should be mentioned that with intact bacteria, under conditions such that the alkalization is not observed (*e.g.* at low pH), similar biphasic kinetics characterize the acidification[4].

These results are important for our understanding of the proton-transfer mechanism in illuminated suspensions of different preparations of *H. halobium*, and especially for re-evaluating old results describing light-induced pH changes with purple-membrane fragments. According to the present model, conclusions regarding proton transport drawn either from the initial rate of the light-induced pH changes with any preparation, or from experiments with purple-membrane fragments (where no transport can be observed), might not be valid. From such experiments the only legitimate conclusions that can be drawn, if our model is correct, refer to proton dissociation-association since the proton pump can be evaluated from the slow process only.

NaCl-loaded particles

The kinetics of the light-induced pH changes in NaCl-loaded proteoliposomes were identical to those observed with KCl-loaded proteoliposomes (not shown). However, with NaCl-loaded sub-bacterial particles different kinetics were observed, unless low light intensities ($I \leq 45$ W/m^2) were used. Fig. 5 compares the light-induced pH changes in KCl-loaded sub-bacterial particles with those in NaCl-loaded particles at $I = 400$ W/m^2. It is apparent that an additional phase, this time of

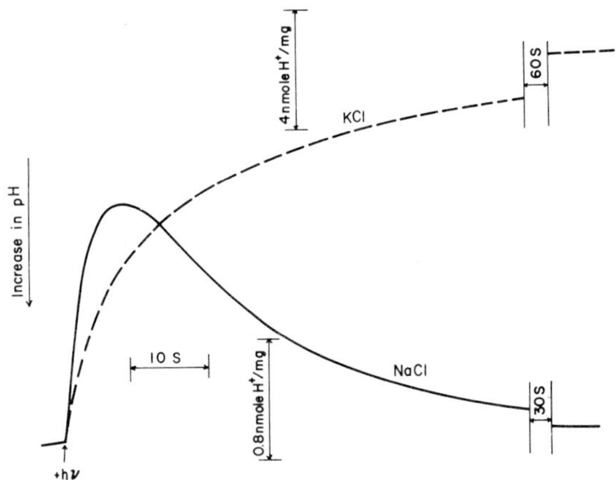

Fig.5 : A comparison between the light-induced pH changes in KCl-loaded and in
NaCl-loaded sub-bacterial particles. The KCl- and NaCl-loaded vesicles were
prepared simultaneously, and suspended respectively in either 4 M KCl or 4 M NaCl
plus 1 mM MOPS (pH 7.2) to a concentration of 2.4 or 4.2 mg/ml, respectively.
I = 400 W/m^2. Temperature, 25°C. (From Ref.13.)

alkalization, is observed with NaCl-containing particles. Since the only
difference between the types of particles examined is their Na$^+$ content, and since
the additional phase was not observed with NaCl-loaded proteoliposomes, we conclude that in the cell envelope there might exist a pathway for the Na$^+$-dependent
uptake of protons. Studies of Na$^+$ transport, using ^{22}Na$^+$ as a marker, showed a
light-induced Na$^+$ efflux (even against its own electrochemical potential gradient)
accompanied by a H$^+$ influx. Our conclusion is further supported by the experiment
described in Fig.6. NaCl-loaded sub-bacterial particles were illuminated and the
trace of the pH-changes was recorded (A). Another sample of the same batch was
illuminated for 4 hours (a period longer than the time needed for the system to
reach "static head", *i.e.* a point where no *net* Na$^+$ transport is observed),
followed by 5 min in the dark and then re-illuminated (B). With sample B no
alkalization was observed, showing that efflux of Na$^+$ is obligatory for influx of
H$^+$. As was suggested previously[12,13], a H$^+$/Na$^+$ antiport could account for the
observed results. These additional results can be summarized as follows :
(1) The rate of Na$^+$ loss, like the rate of proton uptake, is a function of light
intensity. (2) ATP does not appear to be involved in Na$^+$ efflux, since inhibitors
of ATP synthesis do not abolish Na$^+$ efflux, and loading the vesicles with ADP plus
P$_i$ does not stimulate Na$^+$ efflux. (3) Uncouplers which destroy or decrease

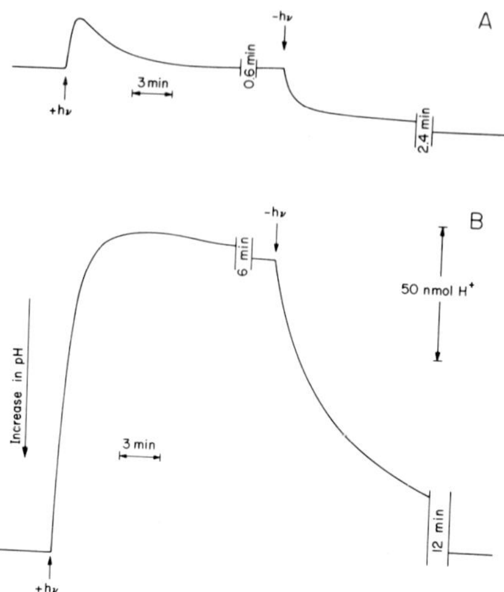

Fig.6 : Light-induced pH changes in NaCl-loaded and NaCl-depleted sub-bacterial particles after long illumination. NaCl(3M)-loaded sub-bacterial particles were suspended in 3M NaCl to a concentration of 2.5 mg protein/ml. The suspension was pre-illuminated (not shown) and then re-illuminated through an N-500 "cut-on" filter (Baird-Atomic) at light intensity of 380 W/m2 (trace A). Another sample of the same batch was illuminated for 4 hours, followed by 5 min in the dark and then re-illuminated (trace B). The initial pH in both samples was 6.7. Temperature, 25°C.

$\Delta\tilde{\mu}_{H^+}$ also reduce Na^+ efflux. (4) Imposition of a Na^+ gradient causes a movement of protons in a direction opposite to the direction of the Na^+ gradient. (5) Na^+ efflux in absence of illumination (and consequently membrane potential difference) can be induced by acidification of the medium, whereas Na^+ uptake occurs upon its alkalization. Thus the direction of Na^+ transport is dependent on the direction of the pH gradient in the absence of illumination.

The following mode of operation is suggested :

(1) When the light is turned on, the proton pump extrudes H^+, and a $\Delta\tilde{\mu}_{H^+}$ is formed (the medium is acidified). At a certain stage protons re-enter the vesicle in exchange for vesicular Na^+, resulting in an alkalization of the medium and an extrusion of Na^+.

(2) When the light is turned off, the proton pump ceases to operate, and protons enter the vesicle (because of back diffusion down their electrochemical potential gradient) with a resultant alkalization of the medium. With the entry of H^+, $\Delta\tilde{\mu}_{H^+}$ decreases. Since the internal Na^+ concentration is less than the external, there is now a tendency for medium Na^+ to exchange with vesicular H^+ leading to reacidification of the medium.

Details of the stoichiometry of the H^+/Na^+ exchange have not yet been established, although previous results of Lanyi and MacDonald[12] were in favour of an electrogenic H^+/Na^+ antiport. Such an electrogenic model predicts that the value of $\Delta\psi$ will affect the rate of Na^+ extrusion. A neutral H^+/Na^+ antiport model, on the other hand, excludes any effect of $\Delta\psi$ on the rate of Na^+ efflux. Measuring Na^+ extrusion in the presence of $TPMP^+$ or TPB^- (not shown) showed an inhibition of Na^+ efflux (*cf*. Ref.31), which is in favour of an electrogenic anti-port model ($H^+/Na^+ > 1$). However, the possibility of direct interaction between $TPMP^+$ and the antiport could not be excluded[13]. We bypassed this difficulty by preparing a system where a pre-determined ΔpH opposed the direction of operation of the H^+/Na^+ antiport; with this ΔpH the operation of the anitport could be driven by $\Delta\psi$ only. Fig.7 shows the results of such an experiment. NaCl-loaded sub-bacterial particles were prepared with a ΔpH so oriented as to oppose Na^+ extrusion ($pH_{in} < pH_{out}$). Because of the high buffer capacity on both sides of the membrane, illumination decreased the ΔpH only slightly, but stimulated a rapid increase in $\Delta\psi$ (externally positive). As shown in the figure the extrusion of Na^+ followed the development of $\Delta\psi$, indicating that $\Delta\psi$ drives the H^+/Na^+ antiport against the ΔpH gradient. This supports the suggestion of Lanyi and MacDonald[12] that the stoichiometry between the inward flow of H^+ and the outward flow of Na^+ is greater than 1. The above experiment enabled us to reach this conclusion without using high concentrations of $TPMP^+$ to swamp $\Delta\psi$, which could have interacted with the antiport system directly.

This mechanism could account for the relatively low internal concentration of Na^+ inside the intact bacteria (0.5-2M, depending on the growth phase[32]), while the external Na^+ concentration is maintained at 4.3M.

ACKNOWLEDGEMENT

This study was supported by grants from the U.S.-Israel Binational Science Foundation (BSF), Jerusalem, Israel, and from the National Council for Research and Development, Israel, and the KFA Jülich, Germany. One of us (M.E.) is grateful to "The B. De Rothschild Foundation for the Advancement of Science in Israel" for a research grant. The authors are deeply indebted to Prof. R.M. Johnstone

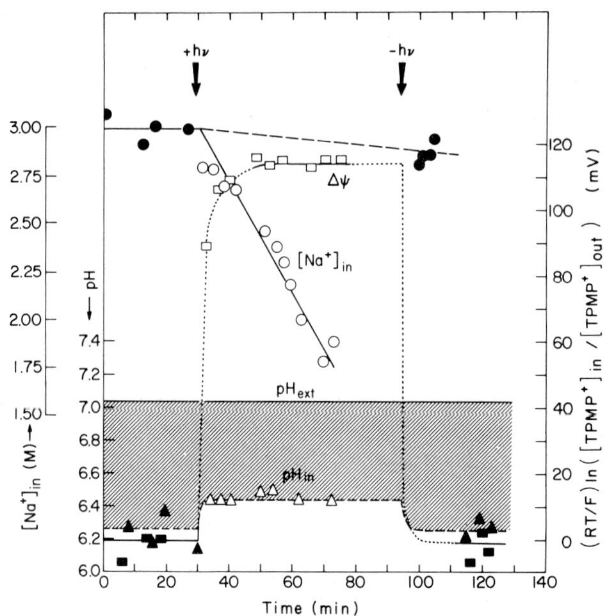

Fig.7 : Na^+ flux driven by light-induced membrane potential difference. NaCl(3M)-loaded sub-bacterial particles were suspended in a solution of 3M NaCl plus PIPES and MES (35 mM of each; pH 6.5) to a concentration of 7.1 mg protein/ml. The suspension was divided into 3 portions : one was equilibrated for 4 days with ^{22}NaCl, and to the other portion 57 µM (final concentration) [^{14}C]propionic acid (for internal pH measurement) or 10 µM [^{3}H]TPMP$^+$ (for Δψ measurement) were added, respectively. The external pH was simultaneously followed by a pH-meter. In all the samples the pH was adjusted to 7.0 before the initiation of the experiment, thus giving rise to an unfavourable ΔpH (the shaded region). All the samples were simultaneously measured. I = 480 W/m^2. Temperature, 25°C. o, ●, internal concentration of Na$^+$; □, ■, TPMP$^+$ accumulation as a Δψ indicator; △, ▲, internal pH. The open and closed symbols are for the light and dark samples, respectively. The upper dashed line represents the behaviour of a part of the ^{22}NaCl portion which was not exposed to light.

and Prof. H. Rottenberg for their help and discussions, to Prof. S. Malkin for suggesting and guiding us in using his spectrofluorometer, and to Mrs. Charlotte Weissmann for technical assistance.

REFERENCES

1. Oesterhelt, D., and Stoeckenius, W. (1973) Proc. Nat. Acad. Sci. USA 70, 2853-2857.

2. Danon, A., and Caplan, S.R. (1974) in : Proceeding of the Third International Congress on Photosynthesis (M. Avron, ed.) pp.2163-2170, Elsevier, Amsterdam.

3. Bogomolni, R.A., Baker, R.A., Lozier, R.H., and Stoeckenius, W. (1976) Biochim. Biophys. Acta 440, 68-88.

4. Wagner, G., and Hope, A.B. (1976) Aust. J. Plant Physiol. 3, 665-676.

5. Bakker, E.P., Rottenberg, H., and Caplan, S.R. (1976) Biochim. Biophys. Acta 440, 557-572.

6. Michel, H., and Oesterhelt, D. (1976) FEBS Lett. 65, 175-178.

7. MacDonald, R.E., and Lanyi, J.E. (1975) Biochemistry 14, 2882-2889.

8. Kanner, B.I., and Racker, E. (1975) Biochem. Biophys. Res. Commun. 64, 1054-1061.

9. Eisenbach, M., Bakker, E.P., Korenstein, R., and Caplan, S.R. (1976) FEBS Lett. 71, 228-232.

10. Racker, E., and Stoeckenius, W. (1974) J. Biol. Chem. 249, 662-663.

11. Kayushin, L.P., and Skulachev, V.P. (1974) FEBS Lett. 39, 39-42.

12. Lanyi, J.K., and MacDonald, R.E. (1976) Biochemistry 15, 4608-4614.

13. Eisenbach, M., Cooper, S., Garty, H., Johnstone, R.M., Rottenberg, H., and Caplan, S.R. (1977) Biochim. Biophys. Acta 465, 599-613.

14. Eisenbach, M., Garty, H., Rottenberg, H., and Caplan, S.R. (1977) submitted.

15. Bakker, E.P., and Caplan, S.R. (1977a) submitted.

16. Oesterhelt, D., Meentzen, M., and Schuhmann, L. (1973) Eur. J. Biochem. 40, 453-463.

17. Garty, H., Eisenbach, M., Shuldman, R., and Caplan, S.R. (1977) submitted.

18. Bakker, E.P., and Caplan, S.R. (1977b) submitted.

19. Renthal, R., and Lanyi, J.K. (1976) Biochemistry 15, 2136-2143.

20. Stoeckenius, W., and Lozier, R.H. (1974) J. Supramol. Struct. 2, 769-774.

21. Slifkin, M.A., and Caplan, S.R. (1975) Nature 253, 56-58.

22. Lozier, R.H., Bogomolni, R.A., and Stoeckenius, W. (1975) Biophys. J. 15, 955-962.

23. Dencher, N., and Wilms, M. (1975) Biophys. Struct. Mechanism 1, 259-271.

24. Kung, M.C., DeVault, D., Hess, B., and Oesterhelt, D. (1975) Biophys. J. 15, 907-911.

25. Oesterhelt, D., and Hess, B. (1973) Eur. J. Biochem. 37, 316-326.

26. Stoeckenius, W., Bogomolni, R.A., and Lozier, R.H. (1975) in : Molecular Aspects of Membrane Phenomena (Kaback, H.R., Neurath, H., Radda, G.K., Schwyzer, R., and Wiley, W.R., eds.) pp.306-315, Springer-Verlag, Berlin, Heidelberg, N.Y.

27. Konishi, T., and Packer, L. (1976) Biochem. Biophys. Res. Commun. 72, 1437-1442.

28. Trissl, H.-W., and Montal, M. (1977) Nature 266, 655-657.

29. Lozier, R.H., Niederberger, W., Bogomolni, R.A., Hwang, S.-B., and Stoeckenius, W. (1976) Biochim. Biophys. Acta 440, 545-556.

30. Eisenbach, M., Garty, H. Klemperer, G., Weissmann, C., Tanny, G., and Caplan, S.R. (1977) this book.

31. Lanyi, J.K., Renthal, R., and MacDonald, R.E. (1976) Biochemistry 15, 1603-1610.

32. Ginzburg, M., Sachs, L., and Ginzburg, B.Z. (1970) J. Gen. Physiol. 55, 187-207.

LIGHT-DEPENDENT CO_2 FIXATION IN ANAEROBIC *HALOBACTERIUM HALOBIUM*

Arlette Danon and S. Roy Caplan
Department of Membrane Research
The Weizmann Institute of Science
Rehovot, Israel

ABSTRACT

Anaerobic cells of *Halobacterium halobium* containing bacteriorhodopsin assimilate CO_2 by means of a light-dependent mechanism; the major product was found to be succinate.

INTRODUCTION

Proton gradients serve as intermediate energy pools between electron transport and ATP synthesis[1,2]. Proton gradient dependent reverse electron flow has been demonstrated in chloroplasts[3,4]. Indirectly related to the proton gradient is the CO_2 fixation cycle which utilizes both ATP and reducing power (Ferredoxin$_{red}$ and NADPH)[3,4] generated by the backward flow of electrons.

In *Halobacterium halobium* photoactivation of the bacteriorhodopsin results in the transfer of protons from the interior to the exterior of the cells. A proton gradient is thus established across the cell membrane which drives ATP synthesis and is not dependent on the electron transport chain[5]. This proton gradient could in principle also promote a light-dependent reverse electron flow. An indirect way of testing this possibility is to search for light-dependent CO_2 fixation in cells kept under strictly anaerobic conditions.

MATERIALS AND METHODS

Halobacterium halobium R_1 was used. The growth medium and conditions were as described previously (Danon and Stoeckenius, 1974). A suspension in basal salt was kept in the dark for one hour before starting an experiment by addition of $[^{14}C]NaHCO_3$. The assay for CO_2 fixation has been reported before[6]. Identification of the products of the CO_2 fixation was attempted by autoradiography of the ^{14}C-labelled compounds subjected to co-electrophoresis and chromatography with authentic markers.

Abbreviations : DCCD N,N'-dicyclohexyl carbodiimide, NQNO 2-n-nonyl-4-hydroxy-quinoline-N-oxide, CCCP carbonyl cyanide m-chlorophenyl hydrazone, DSPD disalicylidene propanediamine, PIPES piperazine-N-N'-bis[2-ethanesulfonic acid], MES 2-[N-morpholino]-ethanesulfonic acid.

RESULTS AND DISCUSSION

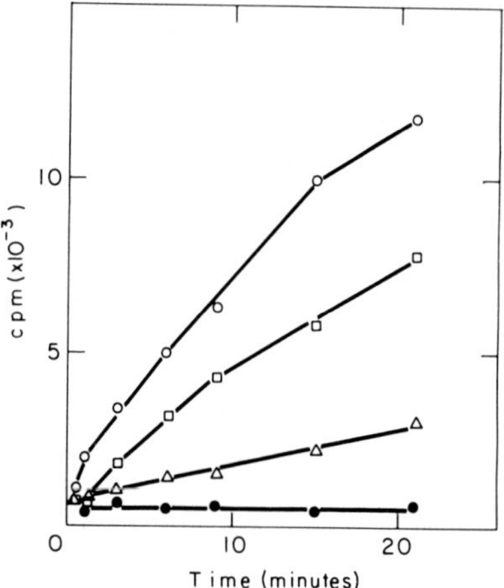

Fig.1 : Fixation of CO_2 into acid stable products. ($\bullet$——$\bullet$) $NaH^{14}CO_3$ added at time 0 in the dark, ($\triangle$——$\triangle$) $NaH^{14}CO_3$ added in the dark after 3 min pre-illumination, ($\square$——$\square$) $NaH^{14}CO_3$ added in the light without pre-illumination, ($\circ$——$\circ$) $NaH^{14}CO_3$ added in the light after 3 min pre-illumination. (From Ref.6)

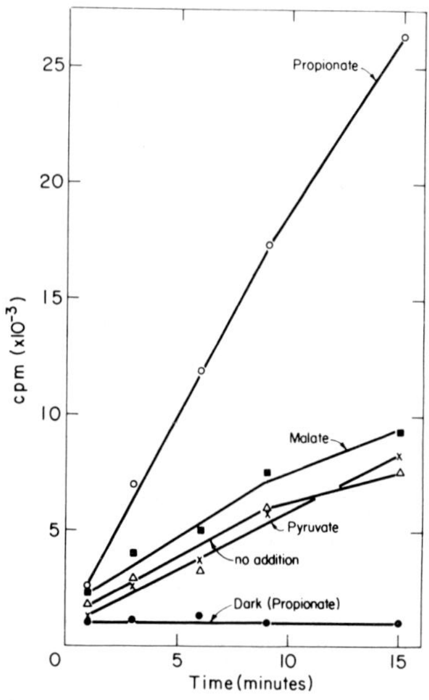

Fig.2 : Effect of different substrates on CO_2 fixation. All the substrates were added at 10 mM concentration. The cells were pre-incubated in the dark before adding the labelled bicarbonate. (From Ref.6)

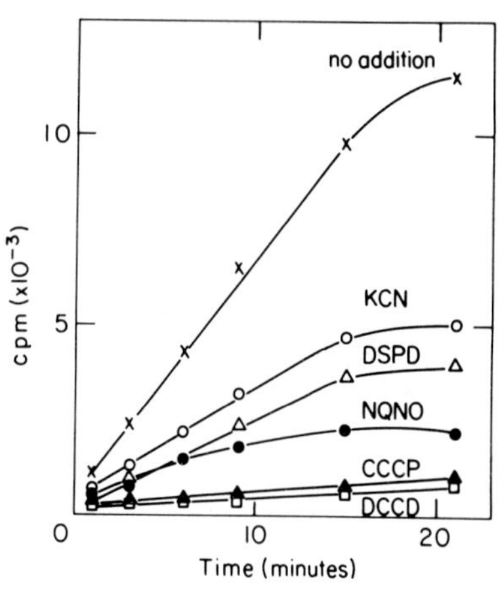

Fig.3 : Effect of different inhibitors on CO_2 fixation. KCN, 1 mM; DCCD, 10^{-4}M; DSPD, 1 mM; CCCP, 5×10^{-5}M; NQNO, 10^{-5}M. (From Ref.6)

From the results shown in Figs. 1, 2, and 3 the following points can be deduced. Anaerobic cells in the dark do not fix CO_2. When the light is switched on, fixation starts with a lag of about one minute or more. Pre-illumination suppresses the lag, but is not sufficient to allow CO_2 fixation if the light is off at the time of the [14C] bicarbonate addition. Propionate has a pronounced stimulating effect. Besides the substrates shown in Fig.2, succinate, citrate, and fumarate were tested. None had a substantial stimulatory effect over and above the level attained with basal salt alone.

The energy transfer inhibitor, DCCD, and the proton-transporting uncoupler, CCCP, which were both shown to strongly inhibit photophosphorylation (Danon and Stoeckenius, 1974) also inhibit CO_2 fixation. NQNO, KCN, and sodium azide, which have no effect on the photophosphorylation, decreased the assimilation of CO_2 by at least 50%. DSPD is known to prevent the photoreduction of NADP by inhibition of the catalytic function of ferredoxin[7,8]. In our system, DSPD did not prevent photophosphorylation but did impair CO_2 fixation.

Analysis of the products indicates that the earliest and major compound formed is succinate, whether [14C] bicarbonate or [14C] propionate is used as labelling substrate. Other compounds found were malate, citrate, fumarate, glutamate, and aspartate. Traces of keto-acids, probably keto-glutarate and keto-butyrate were also found by their reaction with 2,4-dinitro-phenylhydrazine.

CONCLUSIONS

Anaerobic *Halobacterium halobium* assimilates CO_2 in the light. The reaction requires ATP, a continuous flow of protons, and the integrity of the electron transport chain. The following model is suggested[9,10] :

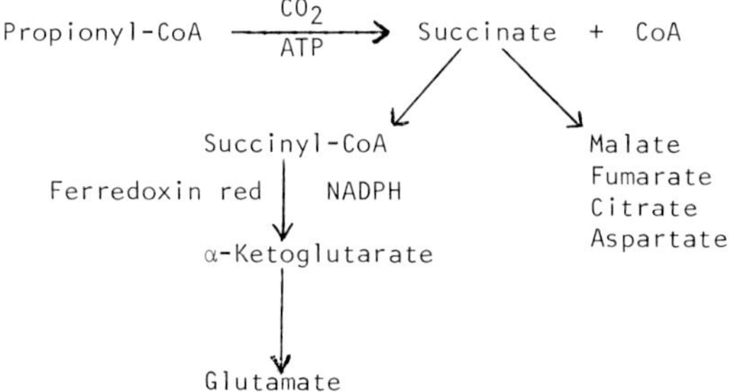

ACKNOWLEDGEMENT

This study was supported by grants from the U.S.-Israel Binational Science Foundation (BSF), Jerusalem, Israel, and from the National Council for Research and Development, Israel, and the KFA Jülich, Germany.

REFERENCES

1. Avron, M., Pick, U., Shahak, Y., and Siderer, Y. (1976) Structure of Biological Membrane, pp.25-391.

2. Mitchell, P. (1968) Glynn Research Ltd., Bodmin, U.K.

3. Shahak, Y., Hardt, H., and Avron, M. (1975) FEBS Lett. 54, 151-154.

4. Shahak, Y., Pick, U., and Avron, M. (1976) Proc. 10th FEBS Meeting, Vol.40, pp.305-314.

5. Danon, A., and Stoeckenius, W. (1974) Proc. Nat. Acad. Sci. USA 71, 1234-1238.

6. Danon, A., and Caplan, S.R. (1977) FEBS Lett. 74, 255-258.

7. Ben-Amotz, A., and Avron, M. (1972) Plant Physiol. 49, 244-248.

8. Avron, M., and Gibbs, M. (1974) Plant Physiol. 53, 140-143.

9. Knight, M. (1962) Biochem. J. 84, 170.

10. Buchanan, B.B., and Evans, M.C.W. (1965) Proc. Nat. Acad. Sci. USA 54, 1218-1221.

Bioenergetics of Membranes. L. Packer et al. ed.

LIGHT-INDUCED pH CHANGES IN PURPLE-MEMBRANE
FRAGMENTS OF *HALOBACTERIUM HALOBIUM*

Michael Eisenbach, Haim Garty, Gil Klemperer, Charlotte Weissmann, Gerald Tanny
and S. Roy Caplan

Department of Membrane Research, The Weizmann Institute of Science,

Rehovot, Israel

SUMMARY

With the goal of examining the validity of the hypothesis which suggests that
illuminated vesicles from *Halobacterium halobium* dissociate or associate protons
in addition to net transport (*cf.* Caplan *et al.*, this book), purple-membrane
fragments (where no net transport can be observed) were subjected to external
effects - pH, temperature, and electrical field. Increasing the pH or decreasing the temperature favoured the proton dissociation process, leading to net
acidification. However, decreasing the pH or increasing the temperature not only
decreased the acidification but was also capable of bringing about net alkalization.

The transport capability of purple-membrane fragments was examined by inserting them into a copolymer of acrylic acid:acrylamide (2:3) and attempting to
orient them electrically during the polymerization. The success in their orientation was shown by the ability of the resulting gel to pump protons preferentially
from one side to the other (in contrast to an "unoriented" gel), proving that
each face of the purple-membrane fragment carries a different fixed charge.

INTRODUCTION

Illumination of sub-bacterial particles of *Halobacterium halobium* or bacterio-
rhodopsin-proteoliposomes acidifies or alkalizes, respectively, the suspending
medium. The kinetics of these processes in both preparations appear to be biphasic, *i.e.*, two first-order (or pseudo first-order) processes occur simultaneously[1-3]. The rapid process can be observed for 10-15 s following illumination
and the slower process is observable for several minutes. A recently developed
model[1-3] suggested that the slow phase represents net proton transport while the
rapid phase is the result of dissociation from or association to the external
surfaces of particles or liposomes, respectively. This association-dissociation
is regarded as a consequence of light-induced conformational changes[4-6] of the

protein, leading either to pK-shifts of exposed groups or to a change in their number.

Support for this model was obtained from experiments with purple-membrane fragments[2], where both faces are exposed to the same medium. With such fragments no net transport can be observed, and thus any light-induced pH changes must be the result of dissociation or association (corresponding to the rapid phase of the light-induced pH change in sub-bacterial particles or proteoliposomes). According to the suggested model, the light-driven conformational change is expressed in these fragments by release of protons on one side of the membrane and proton binding on the other side[2]. Supporting evidence of an indirect nature in favour of this concept was observed by Lozier *et al.*[7], who measured milli-second scale flash-induced pH changes in a purple-membrane suspension. The flash was followed by a transient acidification, indicative of proton dissociation from the fragments with subsequent reassociation. This observation, together with the observed flash-induced acidification by sub-bacterial particles (which are oriented "inside-in") and alkalization by proteoliposomes (which are oriented "inside-out") led Lozier *et al.* to suggest that the release of protons is from one side of the purple-membrane (the external side of the bacterium) and the association is to the other. On a time scale of seconds, the individual process cannot be observed and only the difference between acidification and alkalization is expected to be seen.

In light of the suggested hypothesis, the purpose of this communication is to effect a closer examination of these macroscopic light-induced pH changes in purple-membrane fragments.

MATERIALS AND METHODS

Purple-membrane fragments were prepared from *H.halobium* M-1 strain according to Oesterhelt and Stoeckenius[8]. The illumination and the light-induced pH changes (measured by a combined pH electrode) were carried out as described else-where[9,10]. The bacteriorhodopsin-loaded gel was formed according to the procedure reported previously[11].

RESULTS AND DISCUSSION

(a) <u>Association-Dissociation</u>

If the hypothesis described in the Introduction is correct, the *net* light-induced pH change should be highly sensitive to the pH of the suspension.

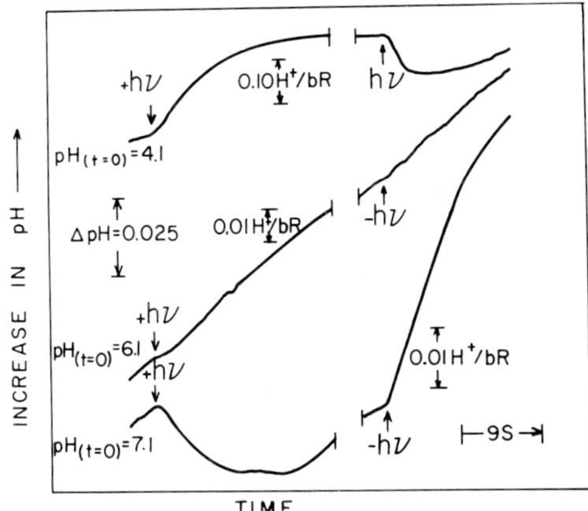

Fig.1 : Light-induced pH changes of purple-membrane fragments at different pH's. Purple-membrane fragments, containing 30 nmol bacteriorhodopsin (bR), were suspended in 2 ml of 1 M KCl solution at the indicated pH's. The suspensions were preilluminated (not shown) and then reilluminated through an N-500 "cut on" filter (Baird Atomic) at I = 140 W/m^2 and followed using a pH-meter. The pH scale for all the traces is identical, but because of the change in the buffer capacity with the pH the calibration for protons is written separately for each pH. This calibration was performed by adding small quantities of standardised HCl solution and measuring the pH change. The traces in the figure are all of the same sample of purple-membrane fragments. Temperature, 14°C.

Fig.1 shows the effect of the pH of the purple-membrane suspension on the light-induced pH changes. At pH 7.1 (lower trace) illumination caused acidification, which was reversible in the absence of light. At pH 4.1, on the other hand, illumination brought about a reversible alkalization. In between these pH's, at pH 6.1, neither net release of protons nor binding of protons could be observed. Evidently at this pH the number of protons released equals the number of protons associated. This dependence of the magnitude, initial rate, and direction of the light-driven pH changes on the pH of the suspension was examined over a wide range of pH and with several batches of purple-membrane fragments. The same phenomenon was observed for all these batches, with the exception that the pH of transition from acidification to alkalization varied from batch to batch, but was usually in the region of pH 5-7. Above this pH only acidification was observed, and its extent and initial rate further increased with increasing pH; below this pH only alkalization was observed which increased with decreasing pH[12]. The transition from light-induced acidification at high pH to alkalization at low pH (and vice versa) was fully reversible upon returning to the original pH.

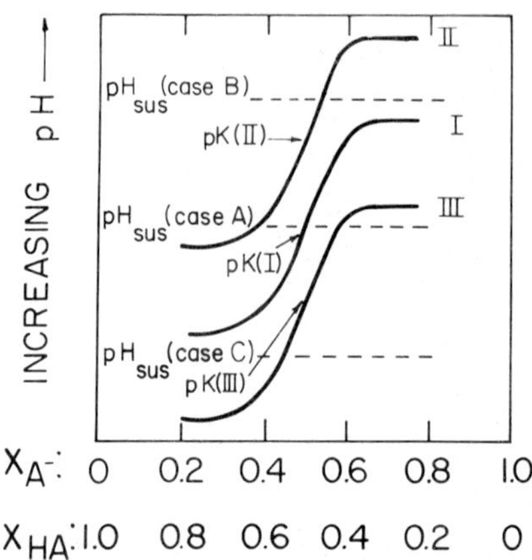

Fig.2 : Hypthetical titration curves of a light-sensitive species. X_{A^-} and X_{HA} are the mole fractions of the dissociated and undissociated forms of the acid HA, respectively. Curve I with its associated pK represents the behaviour in the dark. According to our model for purple membrane, illumination causes an upward pK shift [pk(I) $\xrightarrow{h\nu}$ pK(II)] on the side which was internal in the bacterium (leading to association of protons), and to a downward pK shift [pK(I) $\xrightarrow{h\nu}$ pK (III)] on the other side of the membrane (leading to dissociation of protons). Three principal cases may be distinguished according to the pH of the suspension (pH_{sus}) : (A) $pH_{sus} \simeq pK(I)$. At this pH both dissociation and association of protons will take place, and a small acidification or alkalization (or no net change in the pH) is expected. (B) $pH_{sus} \gg pK(I)$. The light-induced pK shift pK(I) $\xrightarrow{h\nu}$ pK(II) will lead to association of protons on one side of the membrane and to no effect on the other side where pK(I) $\xrightarrow{h\nu}$ pK(III). (C) $pH_{sus} \ll pK(I)$. Under these conditions pK(I) $\xrightarrow{h\nu}$ pK(II) will not lead to association because all the groups are already associated, but pK(I) $\xrightarrow{h\nu}$ pK(III) will cause dissociation of protons from the membrane.

The present results do not exclude the possibility of different conformations of bacteriorhodopsin at various pH's which give rise to different light-induced pH changes. However, it is not necessary to propose such a mechanism, because the model provides a simpler explanation. In principle, the light-induced conformational change may lead *either* to exposure (or masking) of new groups, *or* to changes in the pK values of the groups already exposed to the medium. The latter explanation may be excluded because it predicts more acidification at low pH and more alkalization at high pH (see Fig.2), which contradicts the data shown in Fig.1. The first possibility, on the other hand, is in accord with Fig.1. Thus, if the pH of the suspension is below the pK of the groups which change their location, most of these groups are protonated. Incorporation of protonated groups into the interior of the membrane or exposure of protonated groups from the interior of the membrane does not change the concentration of free hydrogen ions in the suspension. However, exposure of unprotonated groups may lead to their protonation, with a resultant decrease in the free proton concentration in the suspension. If the pH is above the pK, most of the groups are unprotonated. Hence, while incorporation or exposure of these groups does not alter the external pH, exposure of protonated groups to the high pH of the medium leads to deprotonation with a resultant increase in the proton concentration in the suspension. Experiments designed to establish the pK values of the groups on the membrane are presently being carried out in our laboratory.

It has been shown in our previous communications[1-3] that an increase in temperature led to a decrease in the extent of the rapid process, *i.e.* the dissociation of protons from the membrane of sub-bacterial particles or the association of protons to the membrane of proteoliposomes. Since the effect of temperature on the acidification by sub-bacterial particles[2] was more drastic than on the alkalization by proteoliposomes[3], one may expect to observe a temperature effect on the net pH changes with purple-membrane fragments.

Fig.3 shows the effect of temperature on both the extents and initial rates of the light-induced pH changes in a suspension of purple-membrane fragments at pH 7.6. An increase in the temperature led to a decrease both in extent and initial rate to a value of zero at about 30°C (typical for this batch at this pH). A further increase in the temperature caused limited alkalization with an increase in initial rate of alkalization. Generally, an increase in the temperature is expected to increase the initial rate of a single process. The observation in Fig.3, that the initial rate decreased with increasing temperature to a value of zero and then, with a further increase in the temperature, began to increase, indicates that (a) the apparent pH-changes are indeed a difference between single processes of acidification and alkalization, and (b) the activation energy of the alkalization is higher than that of the acidification.

The combined effects of pH and temperature on the light-induced pH changes are presented in Fig.4, which shows the extent of the light-induced pH changes (of a different batch of purple-membrane fragments) at various pH's and at 3 representative temperatures. It is seen that at low temperatures (*e.g.* at 5°C) the extent of net acidification was high, and the pH range where net acidification was observed was wide (*i.e.* the transition from net acidification to net alkalization occurred at a relatively low pH). Though an increase in the temperature decreased the extent of net acidification, the transition-pH from acidification to alkalization as well as the extent of alkalization at lower pH's were unaffected by the temperature, provided that the temperature did not exceed 30°C. Above 30°C both the pH range and the extent of alkalization increased. These data are indicative of two different effects of the temperature, *i.e.* one below 30°C and the second

124

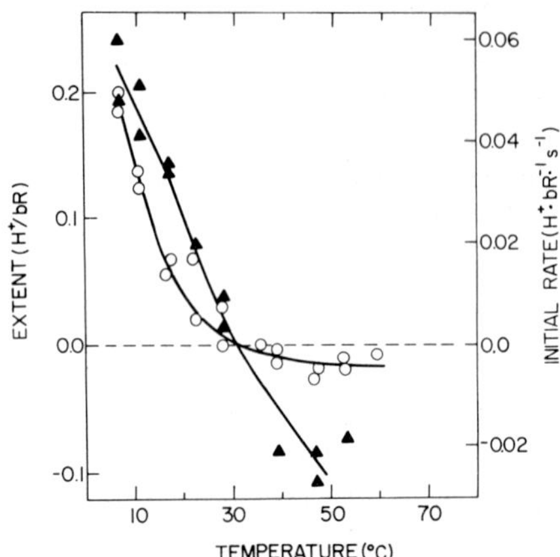

Fig.3 : Effect of temperature on the light-induced pH changes of purple-membrane
fragments. Purple-membrane fragments (from a different batch than in Fig.1),
containing 25 nmol bacteriorhodopsin, were incubated for 30 min at each temperature in 2 ml of 1 M KCl solution (pH 7.6), and treated as in the legend to Fig.1.
o, extent; ▲, initial rate. The figure is taken from Ref.12.

above this temperature. An explanation for the above effects may be proposed by
assuming that the light-induced pK-shifts which lead to acidification are tempera-
ture-sensitive, while those which lead to alkalization are relatively temperature-
insensitive. In this way one could explain the above phenomena at temperatures
below 30°C. At 30°C a phase-transition of the purple-membrane is known to occur[13],
thus above this temperature bacteriorhodopsin may be in a different conformation
where more dissociable groups are buried inside the membrane and thereby rendered
unavailable for dissociation.

This explanation for the observed phenomenon above 30°C is supported by
temperature-reversibility experiments. Increasing the temperature to a higher
value (but below 30°) and decreasing it again to the original value showed full
reversibility, *i.e.* the extent of light-induced pH change after the heating-cool-
ing cycle was the same as the original at the same temperature. However, if the
temperature were raised above 30°C and then reversed to the original value, the
new extent of the light-induced pH change was always smaller or even negative.
Apparently, conformational changes which below the phase transition are limited
by the highly viscous environment within the membrane[14,15], are permitted to occur
above 30°C, leading to an irreversible alteration in the extent of the light-
induced pH change.

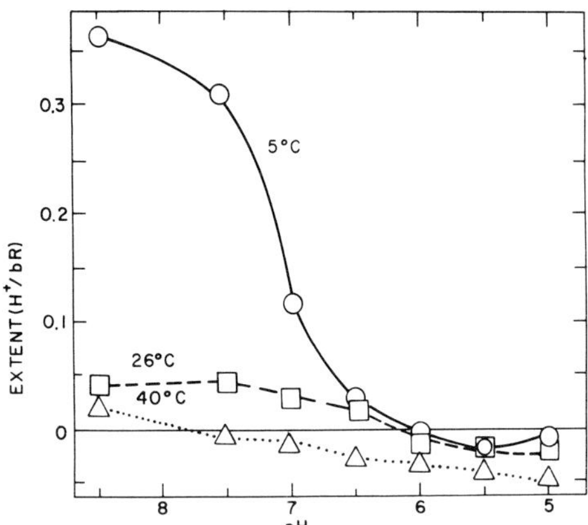

Fig.4 : Combined effects of temperature and pH on the extent of light-induced pH changes of purple-membrane fragments. Three samples of the same batch of purple-membrane fragments as in Fig.1 were treated as described in the legend to Fig.1. Each sample was incubated for 30 min at the indicated temperatures before the measurement. o, 5°C; □, 26°C; △, 40°C.

(b) Transport

The ability of the purple membrane to function as a proton pump must imply an inherent asymmetry. Consequently it would be expected that at a given pH the electric charges on each surface of the membrane are different, even in the dark. The fact that the two surfaces of the purple membrane are structurally different from one another has already been shown by electron microscopy[16]. However, no information relating to the electric charge of the membrane surface has been obtained, except for the observation (Weissmann and Sherman, unpublished results)   that purple-membrane fragments are negatively charged at neutral pH since they migrate towards the positive electrode in an electric field. Evidence for the existence of different electric charges on each surface of the purple membrane was obtained from studies in our laboratory directed toward the development of bacteriorhodopsin-loaded charged synthetic membranes which utilize light energy to generate electric current[11]. In these studies electrically oriented purple-membrane fragments were incorporated into a gel as described below.

A gel constructed of 6% (w/w) acrylic acid and 9% (w/w) acrylamide was found to meet the following requirements : (a) cation selectivity, (b) transparency, and (c) compatibility with the proton pump activity of bacteriorhodopsin. The purple-membrane fragments were added to the monomer solution at pH 4 (final concentration 15 µM bacteriorhodopsin) and the polymerization was then initiated by addition of ammonium persulfate (final concentration 4%). The progress of the exothermic polymerization was measured by a thermocouple which recorded the temperature of the suspension. When the suspension appeared viscous enough to prevent the migration of purple-membrane fragments, but fluid enough to permit their rotation, an electrical field of 100 $V \cdot cm^{-1}$ was applied for 1-2 min in order to orient the fragments. The resultant bacteriorhodopsin-loaded copolymer disc (6.5 mm in width and 3 cm in diameter) was introduced between the two half cells of an Ussing-type lucite chamber. This was equipped with salt bridges for current-carrying and potential-sensing electrodes and connected to a voltage clamp. Illumination of the gel at pH 7 produced a current (measured under short-circuit conditions) of 150 µA (equivalent to 21 $\mu A \cdot cm^{-2}$). A summary of the above steps is shown schematically in Fig.5.

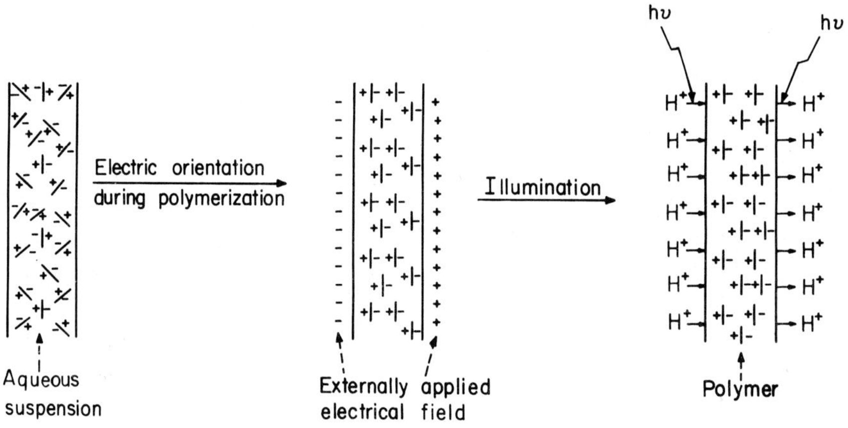

Fig.5 : A schematic representation of the main steps in polymerizing the gel and orienting the purple-membrane fragments in it. The short lines represent purple-membrane fragments. The + signs do not necessarily represent positive electrical charges, but may also stand for charges less negative than those of the sides marked by a minus symbol. The scheme is a simplification of the real situation : it might be the case that only a small percentage of the purple-membrane fragments are oriented, and that their orientation is not exactly perpendicular to the electrical field as shown. For further details see text.

No current could be measured with an identical gel which was not previously exposed to an electrical field. The fact that an electrical field can orient the fragments asymmetrically, allowing them to pump protons cooperatively, indicates that in the fragments there is a difference in charge between both faces (Fig.5).

The direction of the current generated was always the same (independent of the direction of illumination) : the solution on that side of the gel connected to the positive electrode during the electrical orientation step became positive upon illumination (Fig.5). Since the more negative face of the purple-membrane fragments is oriented toward this direction, and since this side is the one which extrudes protons, and since protons are extruded from the external side of the bacteria[7], we may conclude that the external side of the bacterial membrane is the more negative side.

In conclusion, our results illustrate the two processes associated with the functioning of the purple membrane. First, there is a light-induced proton association-dissociation phenomenon, the magnitude and direction of which can be altered by changes in pH and temperature. Second, there is a light-driven proton transport which can be observed after electrical orientation of purple membrane fragments incorporated in an appropriate planar matrix. These results support the model previously proposed by our group[1,2].

ACKNOWLEDGEMENT

This study was supported by grants from the U.S.-Israel Binational Science Foundation (BSF), Jerusalem, Israel, and from the National Council for Research and Development, Israel, and the KFA Jülich, Germany. One of us (M.E.) is grateful to "The B. De Rothschild Foundation for the Advancement of Science in Israel" for a research grant.

REFERENCES

1.   Caplan, S.R., Eisenbach, M., Cooper, S., Garty, H., Klemperer, G., and Bakker, E.P. (1977) this book.

2.   Eisenbach, M., Garty, H., Rottenberg, H., and Caplan, S.R. (1977) submitted.

3.   Bakker, E.P., and Caplan, S.R. (1977) submitted.

4.   Oesterhelt, D., and Hess, B. (1973) Eur. J. Biochem. 37, 316-326.

5.   Konishi, T., and Packer, L. (1976) Biochem. Biophys. Res. Commun. 72, 1437-1442.

6.   Trissl, H.-W., and Montal, M. (1977) Nature 266, 655-657.

7.   Lozier, R.H., Niederberger, W., Bogomolni, R.A., Hwang, S.B., and
     Stoeckenius, W. (1976) Biochim. Biophys. Acta 440, 545-556.

8.   Oesterhelt, D., and Stoeckenius, W. (1974) Methods in Enzymology 31,
     667-678.

9.   Eisenbach, M., Bakker, E.P., Korenstein, R., and Caplan, S.R. (1976) FEBS
     Lett. 71, 228-232.

10. Eisenbach, M., Cooper, S., Garty, H., Johnstone, R.M., Rottenberg, H., and
     Caplan, S.R. (1977) Biochim. Biophys. Acta 465, 599-613.

11. Eisenbach, M., Weissmann, C., Tanny, G., and Caplan, S.R. (1977) FEBS Lett.,
     in press.

12. Garty, H., Klemperer, G., Eisenbach, M., and Caplan, S.R. (1977) submitted.

13. Korenstein, R., Sherman, W.V., and Caplan, S.R. (1976) Biophys. Struct.
     Mechanism 2, 267-276.

14. Esser, A.F., and Lanyi, J.K. (1973) Biochemistry 12, 1933-1939.

15. Bakker, E.P., Eisenbach, M., Garty, H., Pasternak, C., and Caplan, S.R.
     (1977) J. Supramol. Struct., in press.

16. Blaurock, A.E., and Stoeckenius, W. (1971) Nature New Biol. 233, 152-155.

Bioenergetics of Membranes. L. Packer et al. ed.

LIGHT-INDUCED TRANSPORT IN HALOBACTERIUM HALOBIUM

CELL ENVELOPE VESICLES

Janos K. Lanyi

Ames Research Center, National Aeronautics and
Space Administration, Moffett Field, California 94035

Active transport in cells may be defined as a process by which solutes are
accumulated or extruded against their electrochemical gradients. It is clear that
for active transport there is a requirement for energy. In many systems the
energy is supplied either by terminal oxidation, for example of NADH or succinate,
or by the hydrolysis of ATP. For some bacterial membranes the means of coupling
transport to these energy-yielding reactions has been shown to be the transmembrane gradient of H^+ generated[1,2]. Thus, according to the chemiosmotic hypothesis[3,4] this kind of energy-transduction requires an enclosed compartment, and
the circulation of protons between the internal and external space through the
respiratory chain and ATPase, and through the solute transporters. The former act
generally to extrude H^+, causing the cell interior to become alkaline and electrically negative, while the latter allow the inflow of H^+, coupled to the inflow
of substrates.

The discovery of purple membrane in Halobacterium halobium[5-7] provided a
dramatic and unambiguous demonstration of the principle of chemiosmotic energy-
coupling. These membranes, which are differentiated regions of the cytoplasmic
membrane, contain a crystalline array of a retinal-protein complex, termed
bacteriorhodopsin. There is abundant evidence to show that the function of this
membrane is to translocate protons during illumination (for a review see ref. 8),
and that the resulting protonmotive force can be utilized for the synthesis of
ATP[9], for the transport of Na^+ and K^+ [10-15] and for the transport of amino
acids[11, 16-20]. Some of these processes have been studied to best advantage in
cell envelope vesicles rather than in intact cells. The vesicles are prepared
from cells by mechanical breakage[16,21], and are topologically analogous to cells.
They can be loaded with appropriately chosen NaCl-KCl solutions and appear inert
for transport unless gradients are created by the loading process or by illumination. Energization can also be accomplished by adding oxidizable substrates,
such as dimethyl phenylenediamine[22].

Illumination of the vesicles in 3 M NaCl or in 3 M KCl causes the extrusion of
protons and the rapid ($<$ 1 min) development of a steady-state, where the interior
becomes more alkaline and electrically negative. The protons in the external
medium can be measured with a fast glass electrode[23], as since their number
exceeds the number of bacteriorhodopsin molecules by a factor of up to 50, it

was assumed[12] that corresponding pH changes of the opposite sign take place in the vesicle interiors. More recently this point was confirmed (Helgerson and Lanyi, manuscript in preparation) by following ^{14}C-acetate uptake with the flow dialysis method[24]. Similarly, the electrical potential was measured previously by a method using a fluorescent cyanine dye[23], but now has been determined by following ^{3}H-TPMP[+] (triphenylmethylphosphonium ion) uptake. The results show that the pH gradients during illumination are much smaller in NaCl than in KCl. On the other hand, significant membrane potential is obtained only in NaCl. Gramicidin, which allows rapid Na[+] and K[+] movement across the membranes, abolishes the light-induced membrane potential and increases Δ pH in NaCl. The pH difference is unaffected by this antibiotic in KCl solution, presumably because K[+] permeability is already considerable in these membranes[15,25]. Interestingly, the value of ΔpH achieved in the presence of gramicidin does not reach the ΔpH found in KCl, as would be expected if Na[+] and K[+] influx regulated H[+] extrusion in a simple manner. A typical flow dialysis experiment is illustrated in Figure 1.

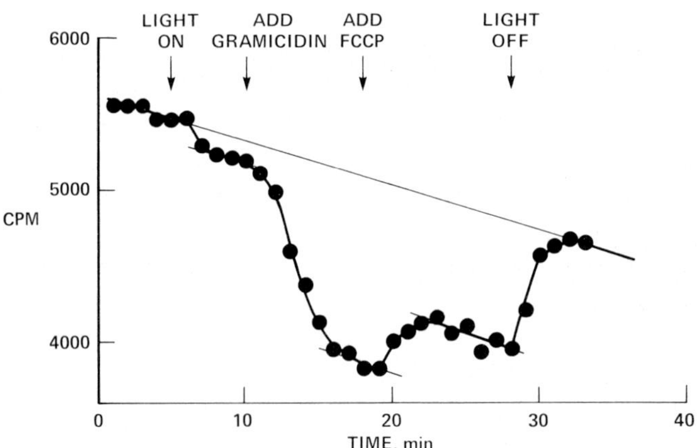

Figure 1. Flow dialysis experiment demonstrating the light-induced formation of transmembrane pH difference in H. halobium vesicles. The apparatus used and the principle of the method are explained in ref. 24. The buffer and the vesicles contained 3 M NaCl. Light initiates the alkalinization of the vesicle interior, resulting in uptake of radioactive acetate (and its disappearance from the external medium). Addition of gramicidin causes a further increase in acetate uptake, while the uncoupler FCCP causes its decrease. The values of ΔpH at each stage can be calculated from the external pH and the amount of acetate taken up. (Helgerson and Lanyi, unpublished work).

The size of the light-induced membrane potential (approx. 80-90 mV) is relatively unaffected by external pH. In vesicles, as in intact cells, the pH difference is greatly dependent on external pH, however, decreasing in magnitude with increasing pH. No detectable ΔpH is observed at and above pH 7.5 in either NaCl or KCl. This pH dependence is not caused by decreased Na^+ or K^+ permeability at higher pH, because the pH dependence pattern is not altered in the presence of gramicidin. Furthermore, direct measurements of passive K^+-Na^+ exchange in these vesicles show that the permeabilities increase, rather than decrease with pH[26]. The pH dependence must thus reflect requirements for the primary process, the translocation of protons. Very similar dependence on pH has been described for Escherichia coli vesicles[24], where ΔpH arises from lactate oxidation.

When both NaCl and KCl are present inside the vesicles the pH changes and membrane potential show complex time-course[12]. With 0.1 M NaCl (plus 2.9 M KCl) inside there is an initial, smaller pH change at the beginning of the illumination, of approx. 30 sec duration, followed by the establishment of a ΔpH value normally observed in KCl-filled vesicles. With 1 M NaCl inside (plus 2 M KCl) the initial smaller pH change exhibits a much longer duration, approx. 5 mins. The length of these Na^+-dependent pH anomalies is determined also by light-intensity, and they become progressively lengthened at decreased intensities. These results could be explained[12] if a) the initial decreased ΔpH is caused by influx of H^+, which is dependent on Na^+ inside the vesicles, and b) the internal Na^+ is depleted during illumination. Indeed, light-dependent efflux of Na^+ in H. halobium vesicles has been demonstrated in two laboratories[12,14]. Since during illumination Na^+ moves _against_ its electrochemical gradient, the sodium flux must be an actively driven process. If it is driven by the circulation of protons, through sodium-proton exchange (or _antiport_), the electrical potential should increase during Na^+ exit, since the net effect is the removal of positive charges from the vesicles. If Na^+ efflux is linked to the movement of any other ion, the potential should be smaller initially, during Na^+ exit, since then it would be a dissipative process. In the latter case the membrane potential should increase only once the Na^+ is depleted from the vesicles.

The measurement of changes in membrane potential, with the time-resolution required, is not straight-forward. Flow dialysis, as presently used, provides information only about the steady-state distribution of radioactive substances across the vesicle membranes[24]. For kinetic measurements we used the fluorescent dye, 3,3'-dipentyloxadicarbocyanine (di-O-C_5), described by Sims et al[27]. The molecule is somewhat hydrophobic and contains a permanent positive charge. It associates with membranes, and when the vesicle interior becomes negative, it accumulates there and aggregates, resulting in the quenching of the fluorescence. Illumination of H. halobium vesicles causes such reversible quenching of the dye fluorescence[23] and K^+-diffusion potentials of known magnitude were used to calibrate the effect. Fluorescence decrease is linear with membrane potential up to - 110 mV.

The inclusion of NaCl in the vesicles causes a complex pattern of changes for light-induced membrane potential, reciprocal to the pH changes observed: the potential is initially large, approaching values observed in 3 M NaCl, in the absence of KCl, and decreases with time in a manner analogous to the increase in ΔpH[12]. The findings for membrane potential are thus consistent with the H^+/Na^+ antiport model for sodium efflux, and explain the reduced ΔpH in terms of Na^+-efflux dependent H^+-influx. For various reasons it was suggested[11,12] that the antiporter functions electrogenically and thus can be driven by electrical potential alone.

Net efflux of measurable quantities of Na^+ can only take place if another ion is also displaced. Since light-induced K^+ uptake by H. halobium vesicles and cells has been demonstrated[10,15], it seems likely that the counterion to Na^+ is K^+. The net result of illumination would be the exchange of intravesicle Na^+ for K^+, a process consistent with the physiology of this organism. The gradients produced by the redistribution of Na^+ and K^+ during illumination collapse only very slowly[26]. A model for these ion movements is given in Figure 2, which links proton circulation to Na^+ circulation and to the potential-driven K^+ influx.

A general description of the light-driven amino acid transport systems in H. halobium is available[19,20]. The results show that nineteen out of twenty commonly occurring amino acids are actively accumulated by cell envelope vesicles. The most revealing experiments were those with glutamate. Although this amino acid is not chemically altered during transport, it is taken up by the vesicles in a virtually irreversible fashion[18]. This irreversibility, while not completely understood, permitted the designing of experiments to reveal the kinetic properties of the driving force for the transport[11].

At low light-intensities the transport rates for glutamate are not greatly reduced, but instead a lag period appears. This lag becomes very pronounced when larger and larger amounts of NaCl are included in the vesicles. These observations suggested that the transport of glutamate may be energized not directly by proton-circulation but by Na^+-circulation. If so, glutamate transport would require Na^+ in the external medium. This was indeed found[18], for glutamate as well as for the other amino acids transported[19,20]. The requirement for Na^+ is absolute, and the kinetic constant, K_m, for transport is not altered by NaCl[18].

If a concentration difference for Na^+ across the membrane (out > in) is established by appropriately loading the vesicles with KCl, it should be possible to obtain glutamate transport in the dark. Under these conditions rapid glutamate uptake takes place[18], and the kinetic properties measured are similar to those

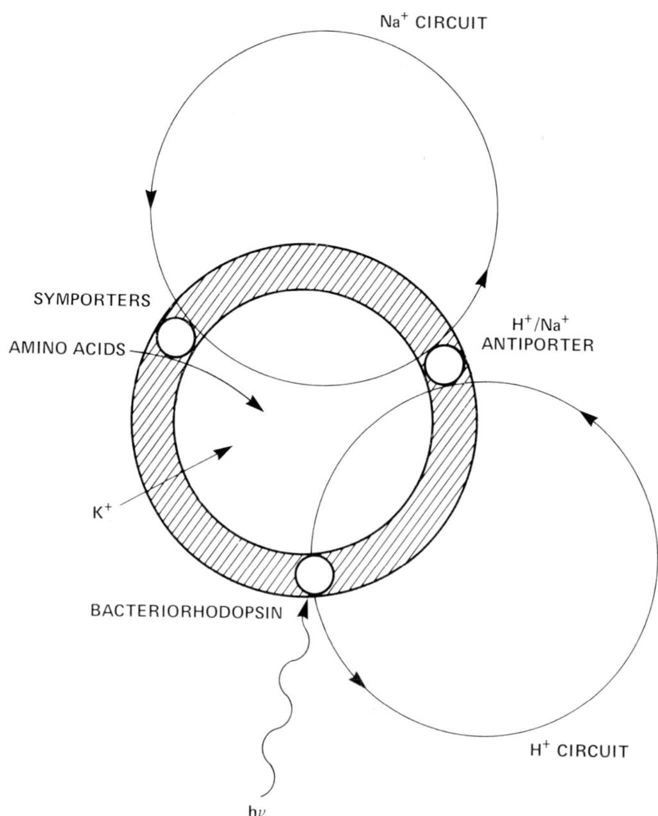

Figure 2. Scheme of energy transduction in H. halobium envelope vesicles. The absorption of photons initiates proton movement in bacteriorhodopsin, resulting in ΔpH (inside alkaline) and membrane potential (inside negative). The circulation of protons drives sodium ion efflux through proton-sodium antiport, and a gradient for sodium (out > in) is formed. The circulation of sodium ions drives active amino acid uptake by co-transport (symport).

obtained during illumination. When small amounts of NaCl are included in the vesicles, so that the size of the Na^+-gradient is lowered, glutamate transport is very much diminished, indicating that transport depends on the size of the gradient.

The kinetics of the generation and the decay of the Na^+ concentration-gradient during and after illumination are much slower than those for the H^+-gradient. When vesicles are first illuminated and then radioactive glutamate is added in

the dark at various times, the energized state for transport is found to persist
for 15-20 mins[11]. Similarly, it takes 5-10 mins of illumination to establish
maximal transport activity, detectable after the illumination, in the dark. The
kinetics of energizing the vesicles can thus be separated from the changes in
protonmotive force, which arises and decays in less than a minute at the
beginning and after the illumination. The observations indicate, when taken together, that the transport of glutamate can take place in the absence of a
gradient for H^+, but not in the absence of a gradient for Na^+. For such reasons
we suggested that glutamate and Na^+ are co-transported in H. halobium[11]. Because
glutamate transport does not respond to membrane potential, only to a sodium ion
concentration difference, this amino acid is probably transported as an anion, so
that the co-translocation across the membrane does not result in the net movement
of a charge. In this respect glutamate transport is different from the transport
of the other amino acids in H. halobium, since these respond to both membrane
potential and sodium ion concentration difference[19,20]. Nevertheless, the involvement of sodium ion circulation in all of the amino acid transport systems of this
organism is evident. The general scheme for energy coupling suggested, shown in
Figure 2, thus includes sodium-amino acid symport. Recent results indicate that
in H. halobium calcium transport is also linked to Na^+-circulation (Belliveau and
Lanyi, manuscript in preparation).

At least two questions arise concerning amino acid transport energetics in H.
halobium: a) are there _any_ H^+-circulation linked translocation systems, and b)
does the stoichiometry between Na^+ and amino acid exceed one for the neutral
amino acids? Evidence available at this time tends to suggest that the answer to
both question is in the negative. Since efflux as well as influx of methionine
requires Na^+ on the _cis_ side of the membrane to the translocation, transport in
either direction must involve a Na^+-dependent carrier (Helgerson and Lanyi, manuscript in preparation). Hence, at equilibrium the electrochemical gradient of the
amino acid should be equal and opposite to the electrochemical gradient of the
driving ion. In the presence of NaCl only (which eliminates the contribution of a
sodium ion concentration difference) the gradients (in mV) of methionine and
serine were found to be equal to the membrane potential, and a contribution from
ΔpH was missing (Helgerson and Lanyi, manuscript in preparation). These results
are consistent with Na^+ symport only, and with a stoichiometry of one.

Although previous reports of Na^+-linked transport in bacteria have been in-
frequent[28], it has become apparent very recently that such systems are, in fact,
present in many bacteria[29-32]. Amino acid transport in H. halobium thus presents
a useful general model for the energetics of Na^+ circulation.

REFERENCES

1. Simoni, R.D. and Postma, P.W. (1975) Annu.Rev.Biochem. 44, 523-554

2. Harold, F.M. (1977) in Current Topics in Bioenergetics (Sanadi, R.D., ed.) vol. 6, Academic Press, New York, pp. 83-149

3. Mitchell, P. (1969) in Theoretical and Experimental Biophysics, (Cole, A. ed.,) vol. 2, Marcel Dekker, New York, pp. 160-216

4. Mitchell, P. (1970) Symp.Soc.Gen.Microbiol. 20, 121-166

5. Oesterhelt, D. and Stoeckenius, W. (1971) Nature, N.B. 233, 149-152

6. Blaurock, A.E. and Stoeckenius, W. (1971) Nature, N.B. 233, 152-155

7. Oesterhelt, D. and Stoeckenius, W. (1973) Proc.Nat.Acad.Sci. U.S.A. 70, 2853-2857

8. Lanyi, J.K. in Membrane Proteins in Energy Transduction (Capaldi, R.A., ed.) Marcel Dekker, New York (in press)

9. Danon, A. and Stoeckenius, W. (1974) Proc.Nat.Acad.Sci. U.S.A. 71, 1234-1238

10. Kanner, B.I. and Racker, E. (1975) Biochem.Biophys.Res.Comm. 64, 1054-1061

11. Lanyi, J.K., Renthal, R. and MacDonald, R.E. (1976) Biochemistry 15, 1603-1610

12. Lanyi, J.K. and MacDonald, R.E. (1976) Biochemistry 15, 4608-4614

13. Lanyi, J.K. and MacDonald, R.E. (1977) Federation Proc. 36, 1824-1827

14. Eisenbach, M., Sprung, S., Garty, H., Johnstone, R., Rottenberg, H. and Caplan, S.R. (1977) Biochim.Biophys. Acta 465, 599-613

15. Garty, H. and Caplan, S.R. (1977) Biochim.Biophys. Acta 459, 532-545

16. MacDonald, R.E. and Lanyi, J.K. (1975) Biochemistry 14, 2882-2889

17. Hubbard, J.S., Rinehart, C.A. and Baker, R.A. (1976) J.Bact. 125, 181-190

18. Lanyi, J.K., Yearwood-Drayton, V. and MacDonald, R.E. (1976) Biochemistry 15, 1595-1603

19. MacDonald, R.E. and Lanyi, J.K. (1977) Federation Proc. 36, 1828-1832

20. MacDonald, R.E., Green, R.V. and Lanyi, J.K. (1977) Biochemistry (in press)

21. Lanyi, J.K. and MacDonald, R.E. in Methods in Enzymology (Fleischer, S. and Packer, L., eds.) Academic Press, New York (in press)

22. Belliveau, J.W. and Lanyi, J.K. (1977) Arch.Biochem.Biophys. 178, 308-314

23. Renthal, R. and Lanyi, J.K. (1976) Biochemistry 15, 2136-2143

24. Ramos, S., Schuldiner, S. and Kaback, H.R. (1976) Proc.Nat.Acad.Sci. U.S.A. 73, 1892-1896

25. Wagner, G. and Oesterhelt, D. (1976) Ber.Deutsch.Bot.Ges. 89, 289-292

26. Lanyi, J.K. and Hilliker, K. (1976) Biochim.Biophys. Acta 448, 181-184

27. Sims, P.J., Waggoner, A.S., Wang, C.-H. and Hoffman, J.F. (1974) Biochemistry 13, 3315-3330

28. Stock, J. and Roseman, S. (1971) Biochem.Biophys.Res.Comm. 44, 132-138

29. Tsuchiya, T., Raven, J. and Wilson, T.H. (1977) Biochim.Biophys.Res.Comm. 76, 26-31

30. Tokuda, H. and Kaback, H.R. (1977) Biochemistry 16, 2130-2136

31. Dunn, S.D. and Snell, E.E. (1977) J.Supramol.Struc. Suppl. 1, p. 136

32. MacDonald, R.E., Lanyi, J.K. and Green, R.V. (1977) Proc.Nat.Acad.Sci.
     U.S.A. (in press)

TRANSIENT PHOTOVOLTAGES GENERATED BY CHARGE DISPLACEMENTS
IN INTERMEDIATES OF THE BACTERIORHODOPSIN PHOTOREACTION CYCLE

San-Bao Hwang, Juan I. Korenbrot, and Walther Stoeckenius[*]
Departments of Physiology and Biochemistry and
[*]Cardiovascular Research Institute
University of California School of Medicine
San Francisco, California 94143

INTRODUCTION

Bacteriorhodopsin, the only protein component of the purple membrane of
Halobacterium halobium, contains a retinal chromophore covalently bound through
a Schiff base.[1,2] Protonation of the Schiff base[3] and non-covalent interactions
between protein and chromophore result in a characteristic visible absorption
spectrum for bacteriorhodopsin with a maximum absorbance near 570 nm, BR_{570}.
Upon absorption of light, bacteriorhodopsin molecules proceed through a cyclic
sequence of at least five conformational intermediates.[4,5,6] Each of these
intermediates is characterized by a distinct visible absorption spectrum and
defined by a unique wavelength of maximum absorbance. Thus, light absorbed
by bacteriorhodopsin molecules, BR_{570}, converts them into the intermediate K of
590 nm λ_{max}, (K_{590}). The K_{590} intermediate decays at room temperature in the
dark through intermediates L_{550}, M_{412}, N_{520} and O_{640} and finally returns to
BR_{570}.[4,5,6] The intermediates are in thermal equilibrium with each other, and
BR_{570} and K_{590} can be interconverted by light.[4] The rate at which any
intermediate state proceeds to the next one is a function of temperature,[4,7,8]
isotopic form of hydrogen present in solution,[7,8] pH[9] and relative humidity.[10]
At room temperature and physiological pH, the half-time of the photoreaction
cycle in an aqueous suspension of purple membrane is about 10 msec.[9]

In each photocycle, bacteriorhodopsin molecules act as a transmembrane proton
pump[11,12,13] by successively releasing protons on the extracellular surface of
the purple membrane and taking them up on the cytoplasmic surface.[9,14,15] The
proton release on the extracellular membrane surface occurs in the time scale
of conversion of the M_{412} to the N_{520} intermediate,[9,16] and formation of M_{412} is
associated with deprotonation of the Schiff base linkage between retinal and the
protein.[3] Kinetic isotope effects on the photocycle of bacteriorhodopsin
indicate that a proton transfer step occurs in the protein in the formation of
M_{412} and O_{640} intermediates.[7,8] Because the Schiff base linkage is not directly
accessible from the aqueous medium,[1,17] it is possible that the released proton
does not come directly from the Schiff base, but rather from a chain of proton
binding sites in the protein through which protons can cross the membrane. If
so, it may be expected that charge displacements occur in the bacteriorhodopsin
molecules upon illumination, reflecting the movement of protons along the chain of

binding sites. In addition, the light-induced protein conformational changes which underlie the spectral changes in the photoreaction cycle may also produce molecular-charge displacements as the result of changes in the distribution of net charges and/or dipole moment in the protein molecules. We report here the measurements of light-induced charge displacements in bacteriorhodopsin molecules specifically assocated with the photoexcitation of several of the intermediates in the photoreaction cycle.

METHODS OF MEASUREMENT

Light-induced charge displacements in bacteriorhodopsin were detected through the measurement of fast photovoltages generated by intense flash illumination of model cells consisting of multilayers of dry and oriented purple membrane fragments and lipids sandwiched between two metal electrodes. To construct these cells, a thin layer of Pd (30% to 60% transmittance) was deposited on an optically flat glass slide (2.5 cm diameter, 0.5 cm thickness) by vacuum deposition. Onto this metallic surface, air-water interface films consisting of oriented purple membrane fragments surrounded by a lipid monolayer were then transferred. The interface films were formed on a subphase of 4 times distilled water by spreading a suspension of sonicated purple membrane fragments in hexane, 0.5 mg/ml, prepared as previously described,[10] except that no excess soya phosphatidyl-choline was added. As reported in detail elsewhere,[10] in these interface films, approximately 85% of the purple membrane fragments are oriented with their intracellular surface towards the aqueous subphase. Under the experimental conditions reported here, some membrane overlap occurs on the water surface. In these interface films bacteriorhodopsin is spectroscopically[10] and functionally[15] intact. The bacteriorhodopsin interface films were transferred to the Pd covered glass slide by repeatedly inserting and withdrawing the slide across the interface film (2 cm/min), while maintaining constant surface pressure at the collapse pressure of the film (47 dynes/cm). Film transferred to the Pd-glass slide exclusively on withdrawal across the interface. This is most important since it resulted in the formation of multilayer stacks in which the purple membrane fragments remained highly oriented. The solid-support emerged wet and was allowed to thoroughly air-dry between transfers. We were able to deposit up to 30 successive layers on the Pd-glass support. After the last interface layer was transferred and allowed to dry, a second Pd metallic layer was vacuum deposited. A mask was used such that the two metal electrodes had a 0.7 ± 0.07 cm^2 area of overlap (Fig. 1). The overheating resulting from the deposition of metal most probably resulted in the irreversible damage of bacteriorhodopsin in the top 4 to 6 layers of the stack. Electrical contact with the Pd metal surfaces was made through thin phosphorbronze shims (0.3 mm thick).[18] The voltage difference between the metal surface was measured with a high input

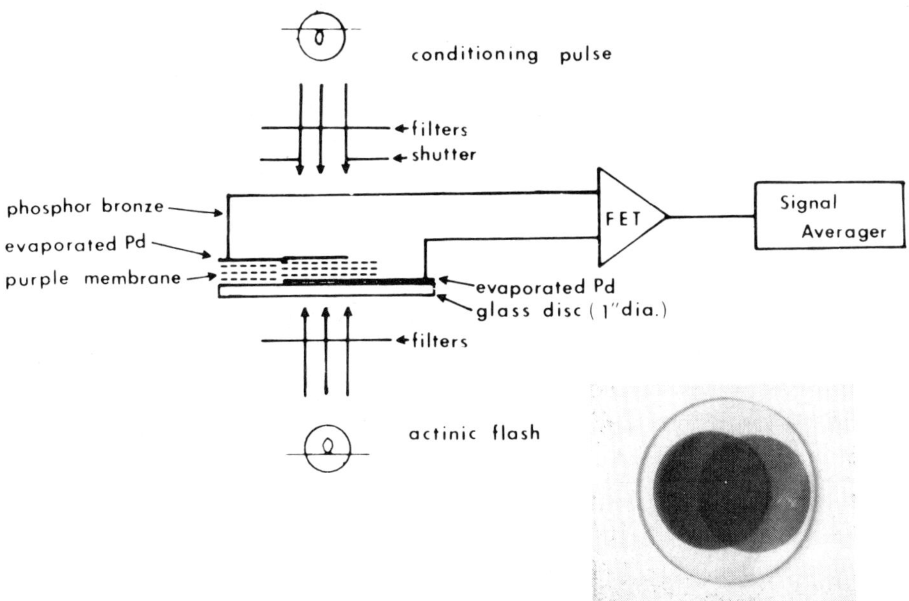

Fig. 1. Purple membrane-lipid multilayer sandwich cell. The glass slide was thoroughly cleaned in dichromate-sulfuric acid before metal deposition. Pd was deposited by evaporation from a tungsten basket at 10^{-6} torr. The potential difference between the metal electrodes was recorded through a unity gain, high input impedance differential amplifier and further amplified through a high gain linear amplifier.

impedance ($R_{in} > 10^{13}$ Ω) differential amplifier with a recording bandwidth of DC to 50 KHz (Winston Electronics, Model 1090). The photovoltage waveforms were averaged with a Nicolet 1074 signal averager, which was also used to reject light switching artifacts. The actinic light flash, produced by a Strobonar 880 unit (Honeywell, Inc.) was delivered through a 1.25 cm diameter and 50 cm long fiberglass light-pipe to the area of metal electrode overlap, avoiding any illumination of the Pd-phosphor bronze contact. The conditioning step of light, produced by a 250 W quartz-iodine source, was delivered through a system of lenses to the same area illuminated by the flash, but from the opposite direction (Fig. 1). One-quarter inch IR absorbing filters were used and calibrated neutral density (Balzers), colored glass (Corning), and interference filters (Baird-Atomic, 15 nm half bandwidth) controlled the wavelength and intensity of the light as required.

RESULTS

The BR_{570} Photovoltage

In the dark, there is no potential difference between the metal electrodes in the bacteriorhodopsin sandwich cell. The electrical resistance across the sandwich cell, measured by recording the current in response to applied voltages, varies between 1000 and 2000 Ω for sandwich assemblies of between 5 and 30 interface layers. The capacitance of the cell, measured at 1000 HZ with an AC impedance bridge, is less than 1 nF in all cases. These values for both resistance and capacitance are low when compared with the values reported for metal sandwich cells containing pure lipid monolayers[19, 20] or multilayers,[21] indicating that the purple membrane multilayers are not as well preserved structurally as pure lipid multilayers, and therefore allow low resistance paths between the electrodes. Furthermore, in the purple membrane sandwich cells the resistance is independent of the number of layers in the stack. In contrast, the resistance of lipid multilayers increases linearly with the number of layers.[21] In our case, however, the low values of the passive electrical parameters of the sandwich cell are advantageous since they result in a short time constant ($\tau \leq 10^{-6}$ sec) for the cell, which allows resolution of fast transients.

The photovoltage measured in response to flash illumination of the bacteriorhodopsin-metal sandwich cell in the dark at room temperature is shown in Fig. 2. The polarity of the photovoltage indicates that the extracellular surface of purple membrane becomes positive with respect to the intracellular surface. The time course and waveform of the photovoltage are identical to those of the stimulus flash, regardless of stimulus intensity. Since the first photoproduct of bacteriorhodopsin recognized at room temperature, K_{590}, forms on the order of picoseconds,[22] the rate of photoexcitation of bacteriorhodopsin molecules under our conditions is limited simply by the relatively slow rate of photon delivery by the actinic flash. Charge displacements concomitant with this photoexcitation should follow the kinetics of the actinic flash. That the photovoltage exactly follows the kinetics of the actinic flash, simply reflects the fact that the kinetics of the photoexcitation (msec time scale) is slow compared with the electrical bandpass characteristics of the sandwich cell ($\tau \leq 10^{-6}$ sec).

The maximum amplitude of the photovoltage measured is linearly proportional to the intensity of the actinic flash over the range tested here, as shown in Fig. 2. This linearity is observed regardless of the stimulus wavelength tested. The maximum amplitude of photovoltages generated by constant intensity flashes of different wavelengths plot an action spectrum for the photovoltages which matches the absorption spectrum of BR_{570} measured in multilayer stacks[10] (Fig. 2). These characteristics indicate that the recorded photovoltage results from a molecular

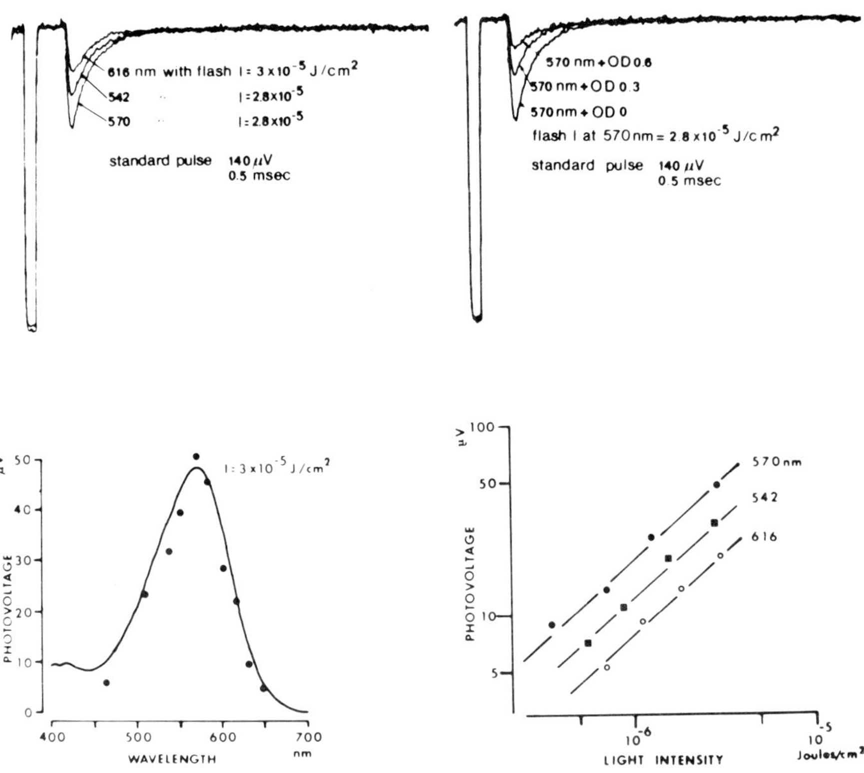

Fig. 2.   Transient voltages generated by photoexcitation of BR_{570}.   Each tracing is the average of 64 photoresponses to successive single flashes presented at 20 sec intervals.   Switching artifacts were rejected in the computer by subtracting from each averaged signal an average waveform obtained by flashing the sandwich cell an equal number of times under conditions where light was completely blocked.   A calibration standard pulse was also signal averaged.   In our convention downward deflections indicate that the extracellular surface of the purple membrane is positive with respect to the intracellular surface.   The waveform of the photovoltage is the same as that of the stimulus flash.   The integrated intensity of the stimulus flash at different wavelengths was measured with a calibrated photodiode (Optometer 40A, United Detector Technology).   The peak amplitude of the photovoltage is linear with intensity and its action spectrum matches the absorption spectrum of bacteriorhodopsin measured in the multilayers (continuous line).   The multilayers in this experiment consisted of 17 interface layers.

event closely associated with the photoexcitation of the BR_{570} state of the bacteriorhodopsin molecules.

The M_{412} and N_{520} Photovoltages

To identify whether charge displacements occur in conjunction with photoexcitation of other intermediates in the photoreaction cycle of bacteriorhodopsin,

142

we took advantage of the fact that the kinetics of the photocycle vary with relative humidity. The M_{412} intermediate in aqueous suspensions of purple membrane at room temperature exhibits a half-time decay of about 10 msec.[9] In contrast, the half-time of decay of M_{412} in the air-dried purple membrane multilayers described here is about 1000 msec.[10] Therefore, illumination of the sandwich cell of the dry purple membrane multilayers with 200 msec duration steps of light absorbed by BR_{570} should result in the formation of a photomixture containing BR_{570}, M_{412} and possibly other slow intermediates which follow M_{412} in the photoreaction cycle. Fig. 3 illustrates the action spectrum of the maximum

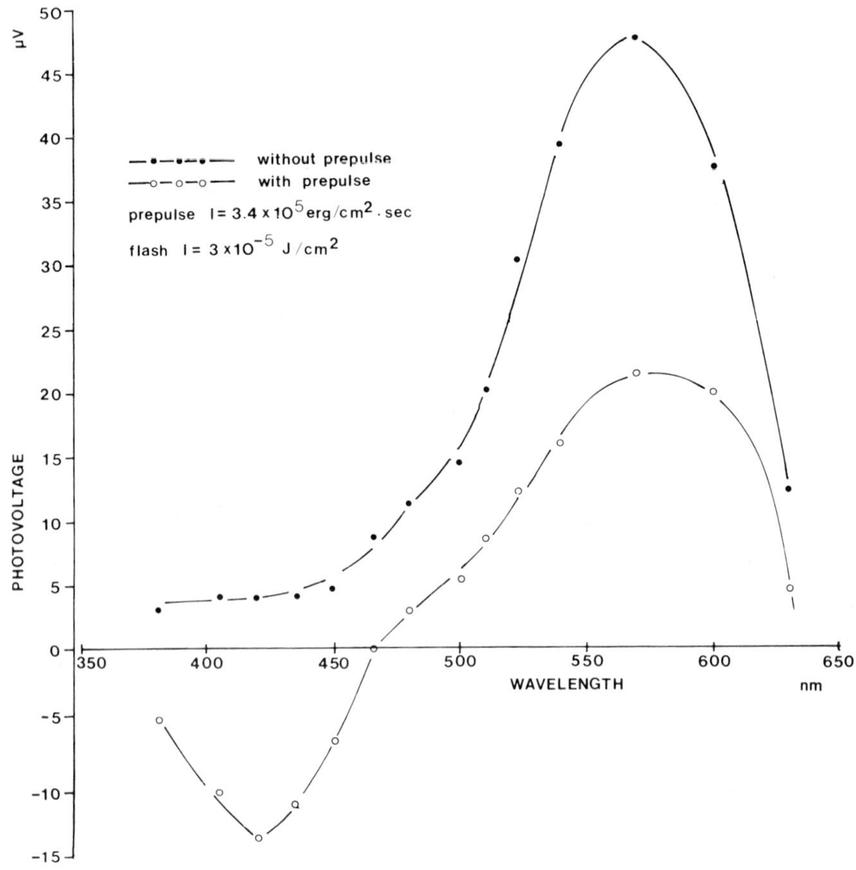

Fig. 3. Action spectrum of the peak amplitude of photovoltages generated by flash illumination of constant intensity. Closed circles are data points measured in the absence of a conditioning pulse. Open circles are data points generated by flashes presented 2 msec after the end of a 200 msec duration conditioning pulse. The test flashes are nearly monochromatic (15 nm half bandwidth interference filter). The conditioning pulse is filtered to pass light of wavelength longer than 540 nm (Corning OG-4). The multilayers in this experiment consist of 29 interface layers.

amplitude of photovoltages produced by constant intensity flashes presented either alone or 2 msec after a 200 msec conditioning pulse of light of wavelength longer than 540 nm. In the absence of a prepulse, positive potentials are recorded throughout the spectral range tested and the action spectrum matches the absorption spectrum of BR_{570}. In contrast, following the conditioning pulse smaller positive potentials are recorded in the red region of the spectrum and negative potentials in the blue region. Moreover, the action spectrum cannot be uniquely matched by the absorption spectrum of any one photointermediate. Clearly a mixture of several photointermediates has been produced, and each can generate independent photovoltages.

We first attempted to separate the contribution of BR_{570} to the photovoltages recorded following the conditioning pulse. Because the half-time of formation of M_{412} at room temperature in the dry multilayers is the same as in aqueous suspensions of purple membrane (about 30 μsec), and is fast compared with the duration of the conditioning pulse we assumed that in addition to BR_{570}, only the slow decay intermediates M_{412}, N_{520} and O_{640} could contribute to the photovoltages recorded after the conditioning pulse. Through independent experiments we have found that photoexcitation of O_{640} in dry multilayers produces a transient voltage detectable beginning about 200 msec after the conditioning pulse. Since according to Lozier et al[4] and Becher and Ebrey[23] neither M_{412} nor N_{520} absorb light at 600 nm, the photovoltage elicited by the 600 nm flash, 2 msecs after the conditioning pulse, arises nearly exclusively from photoexcited BR_{570}. The ratio of photovoltages produced by 600 nm light with and without prepulse is a measure of the fraction of BR_{570} molecules excited by the conditioning pulse and, therefore, a measure of the fraction of molecules which BR_{570} represents in the photomixture. The contribution of the BR_{570} photovoltage to the action spectrum following the prepulse in Fig. 3 can thus be calculated and subtracted. Fig. 4 illustrates the action spectrum obtained after subtraction of the BR_{570} contribution. The photovoltage polarity is such that the extracellular surface of purple membrane becomes transiently negative with respect to its intracellular surface. Furthermore, two distinct spectral peaks are measured, one maximum near 410 nm and the other near 520 nm. The data points recorded in the far blue reach a peak near 420 nm and match well the absorption spectrum of M_{412} as given by Becher and Ebrey.[23] A line drawn through the data points which reach a peak near 520 nm is similar to the calculated absorption spectrum of N_{520} as given by Lozier et al.[4] Fig. 4 also illustrates that the corrected maximum amplitude of the M_{412} photovoltage is linear with intensity of the actinic flash over the range tested and that the waveform of the M_{412} photovoltage is identical to that of the actinic flash. In conclusion, photoexcitation of the conformational intermediates M_{412} and N_{520} also produces photovoltages in the cell,

144

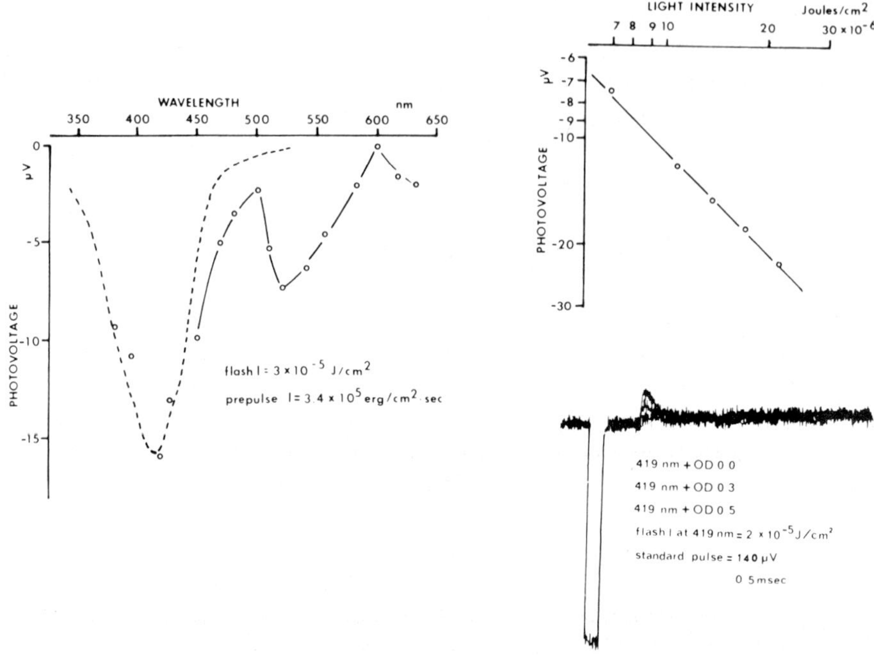

Fig. 4. M_{412} and N_{520} photovoltages. Data points were obtained from Fig. 3, by subtracting the contribution of the BR_{570} photovoltages (see text for method). Data points reach peak at two different wavelengths. The discontinuous line is the absorption spectrum of M_{412} as given by Becher and Ebrey[23] scaled to fit the points. The continuous line peaking near 520 is drawn to join the data points and closely matches the spectrum of N_{520} inferred by Lozier et al.[4] The M_{412} photovoltage is linear with light intensity.

but these are of an electrical polarity opposite to those produced by photoexcitation of BR_{570}.

We have argued above from direct spectrophotometric data that the conditioning 200 msec pulse creates a photomixture of BR_{570}, M_{412} and N_{520} in the multilayer sandwich cell. The ratio of photointermediates in the mixture should therefore depend on the light intensity of the conditioning pulse. Fig. 5 demonstrates that this is indeed the case. The amount of BR_{570} or M_{412} in the photomixture was determined by simply measuring the amplitude of their respective photovoltages produced by adequate stimulus flashes presented always 2 msecs after a 200 msec conditioning pulse of varying intensity. As the intensity of the conditioning pulse increases, the photomixture contains progressively less BR_{570} and more M_{412}. This observation is significant in that a purely spectrophotometric expectation has been confirmed directly through the use of photovoltages. That is, the photovoltaic responses of a given intermediate parallel its spectrophotometric responses.

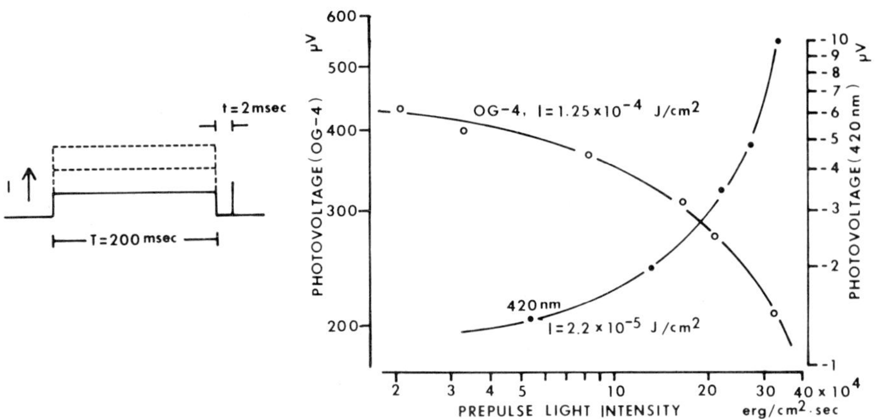

Fig. 5. The effect of varying conditioning pulse intensity on the BR_{570} and M_{412} photovoltages. Peak amplitude of the BR_{570} and M_{412} photovoltages generated by a constant intensity stimulus flash presented 2 msec after a conditioning pulse, were measured for various intensities of the conditioning pulse. The BR_{570} photovoltage was elicited by light of wavelength longer than 540 nm (Corning, OG-4). The M_{412} photovoltage was generated by nearly monochromatic light of 420 nm. The conditioning pulse was delivered through filters which passed light longer than 540 nm.

DISCUSSION

The fast photovoltages in bacteriorhodopsin we have described here share many characteristics with fast photoelectric effects described in other pigmented systems, most notably, the rhodopsin-containing membranes of vertebrate[24] and invertebrate photoreceptors.[25] In all cases, fast photovoltages have been found to be weak, linearly proportional to incident light intensity and to persist in the absence of ionic batteries or, in our case, in the absence of electrolytic solutions. Also Trissl and Montal[26] have recently reported the measurement of capacitative photovoltages associated with the photoexcitation of bacterio-rhodopsin. Fast photoelectric effects have generally been explained to arise from two possible physical mechanisms:[24,27] 1) a thermoelectric effect, where light is converted to heat and the resulting temperature gradient produces a flow of charge, and 2) molecular charge displacements. Molecular charge displacements may arise from various sources such as redistribution of net charges in the molecule, change in magnitude and/or orientation of permanent dipoles in the molecule, transfer of charged particles in or along the molecule. In the case of photoreceptor membranes, the existence of a small thermoelectric contribution to fast photovoltages has been demonstrated by Hagins and McGaughy.[25] In the purple membrane sandwich cells described here, potential thermoelectric contributions would be of a different physical origin than those in intact photoreceptor cells.

In any case, if present, they must be small since we were able to detect both small positive and negative photovoltages (~ 1 μV) and thermoelectric effects would not be expected to change polarity. Furthermore, control sandwich cells formed with lipid multilayers without purple membrane exhibited no photovoltages even in the absence of IR filters in the actinic beam. The observations that the photovoltages described here: 1) occur in the absence of electrolytic solutions; 2) follow the waveform of the actinic flash; 3) occur in parallel with spectrophotometric changes; and 4) are produced by at least three intermediates in the photoreaction cycle of bacteriorhodopsin independently of one another, all strongly suggest that they arise from molecular charge displacements.

The detailed molecular origin of the charge displacement which results from photoexcitation of the spectral intermediates of bacteriorhodopsin remains to be established. Moreover, this molecular origin may be different for different intermediates. For example, if the proton transfer steps suggested by Sherman, Korenstein and Caplan[7,8] both for M_{412} and O_{640} intermediates occurred along the protein molecule following photoexcitation, they could generate a photovoltage. Petersen and Cone[28] have observed light dependent changes in the permanent dipole moment of vertebrate rhodopsin and have suggested that this may be the source of the fast photovoltage associated with the decay of the Meta I intermediate. If similar changes occurred in photoexcited BR_{570} or any of the intermediates, a photovoltage would be generated.

In summary, photoexcitation of bacteriorhodopsin and of the M_{412}, N_{520} and O_{640} intermediates in the photoreaction cycle generates a transient voltage. This photovoltage arises almost certainly from molecular charge displacements in or near the protein molecule as it undergoes photochemical conversion in response to the absorption of light.

ACKNOWLEDGMENTS

SBH was supported by a Sloan Foundation Postdoctoral Fellowship. JIK is an Alfred P. Sloan Research Fellow. Research was supported by NIH grants EY 01586 and HL 06285 and NASA grant NSG-7151.

REFERENCES

1. Oesterhelt, D. and Stoeckenius, W. (1971) Nature New Biol. 233, 149.

2. Bridgen, J.A. and Walker, I.D. (1976) Biochemistry 15, 792.

3. Lewis, A., Spoonhower, J., Bogomolni, R.A., Lozier, R.H. and Stoeckenius, W. (1974) Proc. Natl. Acad. Sci. (USA) 71, 4462.

4. Lozier, R.H., Bogomolni, R.A. and Stoeckenius, W. (1975) Biophys. J. 15, 955.

5. Kung, M., DeVault, D., Hess, B. and Oesterhelt, D. (1975) Biophys. J. 15, 907.

6. Dencher, N. and Wilms, M. (1975) Biophys. Struct. Mech. 1, 259.

7. Sherman, W.V., Korenstein, R. and Kaplan, S.R. (1976) Biochim. Biophys. Acta 430, 454.

8. Korenstein, R., Sherman, W.V. and Kaplan, S.R. (1976) Biophys. Struct. Mech. $\underline{2}$, 267.

9. Lozier, R.H., Niederberger, W., Bogomolni, R.A., Hwang, S.B. and Stoeckenius, W. (1976) Biochim. Biophys. Acta $\underline{440}$, 545.

10. Hwang, S.B., Korenbrot, J.I. and Stoeckenius, W. (1977a) J. Membrane Biol. $\underline{36}$, in press.

11. Oesterhelt, D. and Stoeckenius, W. (1973) Proc. Natl. Acad. Sci. (USA) $\underline{70}$, 2853.

12. Racker, E. and Stoeckenius, W. (1974) J. Biol. Chem. $\underline{249}$, 662.

13. Drachev, L.A., Frolov, V.N., Kaulen, A.D., Liberman, E.A., Ostrumov, S.A., Plakunova, V.G., Semenov, A.Y. and Skulachev, V.P. (1976) J. Biol. Chem. $\underline{251}$, 7059.

14. Hwang, S.B. and Stoeckenius, W. (1977) J. Membrane Biol. $\underline{33}$, 325.

15. Hwang, S.B., Korenbrot, J.I. and Stoeckenius, W. (1977b) J. Membrane Biol. $\underline{36}$, in press.

16. Chance, B., Porte, M., Hess, B. and Oesterhelt, D. (1975) Biophys. J. $\underline{15}$, 913.

17. Peters, J., Peters, R. and Stoeckenius, W. (1976) FEBS Letters $\underline{61}$, 128.

18. Tang, C.W. and Albrecht, A.C. (1975) J. Chem. Phys. $\underline{62}$, 2139.

19. Mann, B. and Kuhn, H. (1971) J. Appl. Phys. $\underline{42}$, 4398.

20. Procarione, W.L. and Kauffman, J.W. (1974) Chem. Phys. Lip. $\underline{12}$, 251.

21. Buchwald, C.E., Gallagher, D.M., Haskins, C.P., Thatcher, E.M. and Zahl, P.A. (1938) Proc. Natl. Acad. Sci. (USA) $\underline{24}$, 204.

22. Kaufman, K.J., Rentzepis, P.M., Stoeckenius, W. and Lewis, A. (1976) Biochem. Biophys. Res. Commun. $\underline{68}$, 1109.

23. Becher, B. and Ebrey, T.G. (1977) Biophys. J. $\underline{17}$, 185.

24. Cone, R.A. and Pak, W.L. (1971) Handbook of Sensory Physiology. Vol. 1, pp. 345-382, W.R. Lowenstein, ed., Springer-Verlag, New York.

25. Hagins, W.A. and McGaughy, R.E. (1967) Science $\underline{157}$, 813.

26. Trissl, H.W. and Montal, M. (1977) Nature $\underline{266}$, 655.

27. Hagins, W.A. and Ruppel, H. (1971) Fed. Proc. $\underline{30}$, 64.

28. Petersen, D.C. and Cone, R.A. (1975) Biophys. J. $\underline{15}$, 1181.

Bioenergetics of Membranes. L. Packer et al. ed.

USE OF LIPID IMPREGNATED MILLIPORE FILTERS FOR THE DIRECT MEASUREMENT OF PHOTO-
POTENTIALS ACROSS ENVELOPE VESICLES OF HALOBACTERIUM HALOBIUM AND FOR ASSAY OF
IONOPHORE ACTIVITY OF THE OLIGOMYCIN BINDING SUBUNIT 9 OF THE YEAST MITOCHONDRIAL
ATPase

L. Packer, P.K. Shieh, J.K. Lanyi and R.S. Criddle
University of California, Berkeley, CA 94720; NASA-Ames
Research Center, Moffett Field, CA 94305 and the University
of California, Davis, CA 95616 USA

INTRODUCTION

The central role of ion gradients and membrane potentials in biological energy
transduction has gained wide acceptance. However, the direct measurement of membrane
potential in cell organelles, bacterial cells and vesicles, liposomes, containing
suspected ionophore proteins, etc., is usually difficult because these structures
are usually too small for the insertion of microelectrodes. Thus, estimation of
electrical potentials in such membrane systems has been mainly indirect as through
determinations of the transmembrane distribution of radioactive membrane-permeable
ions [1-3] or fluorescent dyes[4]. A promising new technique [5,6] is based on the
divalent cation induced association of lipid vesicles or proteoliposomes with
planar lipid membranes separating two aqueous compartments with the measurement of
electrical properties of this system. This method combines the ease of incorpora-
ting proteins into liposomes with the possibility of direct measurement of the
resulting electrical properties in the membranes formed. For example, illumination
generates electrical potentials of considerable size across the compartments when
liposomes containing bacteriorhodopsin are associated with a planar membrane [7-9].
Lipid-impregnated filters show great stability [10] but behave functionally like the
more fragile thin black lipid films[11].

This chapter will cover three topics:

I. Preparation of the lipid impregnated millipore filter technique itself,
    and the application of this technique to address two questions:

II. What is the magnitude of the photopotentials that can be developed across
    intact envelope vesicles of Halobacteria, and do such measurements agree
    with results obtained with indirect methods? The significance of these
    studies is that if the agreement is good, the planar membrane technique
    would offer a rapid and sensitive way to characterize ion antiport
    systems in intact cell membranes; and

III. How can the photopotentials generated by bacteriorhodopsin across the
     lipid impregnated filter membranes be used to study ionophoric activity
     of proteins such as the hydrophobic protein fraction subunit 9 of the

yeast mitochondrial ATPase? The significance of these measurements is that this proteolipid has been proposedto be important in linking ATP synthesis with the creation of a proton gradient by electron transport. The availability of a rapid and sensitive assay for ionophore activity would be of value in characterizing the protein responsible for such activity and in elucidating the action of oligomycin and dicyclohexyl-carboamide (DCCD) which bind to subunit 9.

I. PREPARATION OF LIPID IMPREGNATED MILLIPORE FILTERS

Millipore filters, pore size 0.3 μm (type PHWP 01300) were found to be the best. A template is used to cut a section of the filter into a circle 1.2 cm in diameter, which is glued to a 0.5 cm^2 area hole located on the wall of the teflon cup (Fig. 1A). The cup is then immersed into a solution of soybean phospholipid, 20 mg lipid per ml of decane, for 15 h at 22°C. The cup is then rinsed with distilled water several times, then acetone is used to carefully remove excess lipid from the entire area except on the filter itself. Lipid incorporated into the filter can be measured by weighing a dry filter before and after immersion in the lipid-decane solution. During incubation in the lipid-decane solution, lipid content increases progressively until the filter is saturated at approximately 500 μg lipid/cm^2 after 15 h of storage at 22°C. As shown in Fig. 1, two compartments are formed by placing the teflon cup with the lipid-impregnated filter covering the hole on its surface, inside of a 20 ml glass cup. Both compartments are then filled with bathing solution, and electrical properties are measured with calomel electrodes.

Envelope vesicles of H. halobium contain bacteriorhodopsin in its natural orientation and, upon illumination, pump protons into the surrounding medium. Such vesicles are prepared as described by MacDonald and Lanyi[12,13] by a procedure which basically involves mechanical breakage of the cell envelope and separation of the vesicles by differential centrifugation. High salt concentration is obligatory as the vesicles disintegrate in low salt. The two compartments are filled with a bathing solution containing 4 M NaCl, then 0.5 ml aliquot of envelope vesicles is added to the inner compartment, together with $CaCl_2$, to a final concentration of 100 - 500 mM.

Illumination of the membrane was with a tungsten lamp and a Hoya HA30 filter, usually at a light-intensity of 18 mW/cm^2 (measured at the surface of the teflon cup).

Fig. 1. Schematic of apparatus for photoelectric activity measurement. Diagram shows: A) configuration of the teflon cup (inner compartment), the filter membrane, and the glass container (outer compartment).

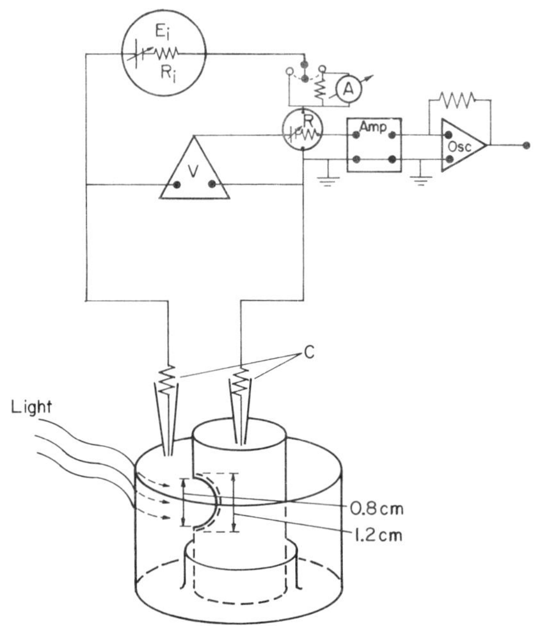

B) Schematic shows: V, Keithley electrometer 610; A, 441 Keithley micro-ammeter; R, Electronik 19 Honeywell; Amp, Electro Instruments Amplifier, Model A20B-3; Osc, Tektronix oscilloscope, Type 3A72; E_i, power supply (battery); R_i, external resistors; and C, calomel electrodes.

152

II. ASSAY OF PHOTOPOTENTIALS ACROSS ENVELOPE VESICLES OF <u>H</u>. <u>halobium</u>

Maximal photopotentials of 20-30 mV develop after 50 min of incubation of the assembled apparatus containing <u>H</u>. <u>halobium</u> vesicles and $CaCl_2$ (Fig. 2). The time dependence is analogous to previous findings with liposomes, where it has been found to be related to the time required for the vesicles to become associated with the planar membrane. The magnitude of the photopotential depends both on the amount of vesicles added and on the $CaCl_2$ concentration (increasing with the latter up to 500 mM). The kinetics of the rise of the photopotential and its decay in the dark (upper inset Fig. 2) are symmetrical and contain no transients, unlike those obtained with liposomes reconstituted with bacteriorhodopsin[6]. Its direction is opposite to the photopotentials recorded with purple membrane reconstituted liposomes: positive in the inner compartment, as expected from the opposite membrane orientation of the two kinds of vesicles[12,14,15]. Since the membrane resistance is lowered to about 10^7 ohm.cm^2 in 4 M NaCl solution, the photopotential obtained is 3-10 fold lower than the value determined by other means in the vesicles[16,17].

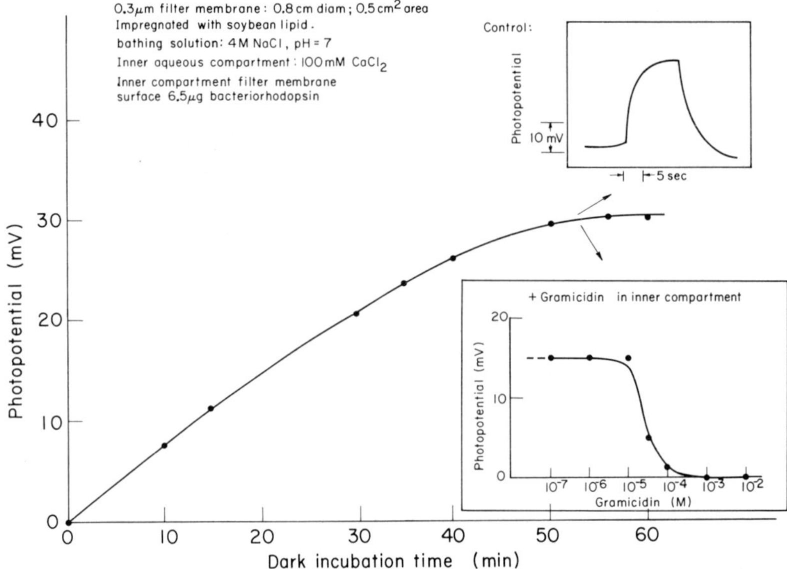

Fig. 2.   Photopotential development by <u>H</u>. <u>halobium</u> envelope vesicles with lipid-impregnated filter membranes. With increasing time of dark incubation, the magnitude of the photopotential becomes larger until a maximal value is reached in about 50 min.

<u>Effect of ionophores and triphenylmethylphosphonium ion on the photopotential</u>.
In much the same way as is found with liposomes containing bacteriorhodopsin,
addition of gramicidin collapses the photopotential. Half-maximal effect is obtained at 2×10^{-5} M gramicidin (lower inset, Fig. 2). Valinomycin similarly
causes the collapse of the photopotential, dependent on added KCl.

Triphenylmethylphosphonium ion ($TPMP^+$), a membrane-permeable cation which has
been shown to discharge membrane potential in <u>H</u>. <u>halobium</u> vesicle membranes[16],
decreases the electrical resistance of the filter membrane and causes the collapse
of the photopotential. The half-maximal effect is seen at $1-2 \times 10^{-4}$ M $TPMP^+$, in
remarkable agreement with results from measurements of the fluorescence of a cyanine
dye[16].

<u>Effect of Na^+-K^+ Transport on Photopotential</u>. Earlier it was shown that during
the illumination of <u>H</u>. <u>halobium</u> vesicles a proton-sodium antiport system is activated[17,18]. According to the model proposed, when Na^+ is present inside the
vesicles the proton-sodium antiporter will complete the pathway of proton circulation through bacteriorhodopsin, and the net result is the exit of Na^+. Owing to
the loss of Na^+ from the vesicles, large membrane potentials are achieved in the
absence of added K^+. When KCl is added, however, this cation is accumulated by the
vesicles, diminishing the photopotential, until virtually all of the internal Na^+
is replaced by K^+ [19]. At equilibrium the light-generated proton-motive force
maintains a fairly large Na^+-gradient (outside >> inside), and the value of the
membrane potential is kept low [17].

If the electrical properties of the filter membrane system depend on processes
taking place in the vesicle membranes, rather than in the planar membrane component,
the light-induced Na^+/K^+ -exchange due to H^+/Na^+ antiport will result in a gradual
K^+-dependent loss of the photopotential during illumination, corresponding to the
depletion of Na^+ from the vesicles. Figure 3 shows traces of electrical potential
measured in the presence and absence of KCl in the inner compartment. As trace <u>a</u>
indicates, in the absence of KCl the photopotential is stable and repeated illuminations cause only minor reductions in its magnitude. When 4 M NaCl in the inner
compartment is replaced by 3.8 M NaCl plus 0.2 M KCl, however, the photopotential
shows a time-dependent decrease, reaching values near zero (trace <u>b</u>) after a series
of prolonged illuminations (total time about 40 min). Trace <u>c</u> in Fig. 3 illustrates
the gradual diminution of the photopotential with repeated illuminations, and the
partial recovery of the system during the dark periods between successive illumination. The above described K^+ dependent effect on the photopotential should be
absent in liposomes reconstituted with bacteriorhodopsin since this system contains
no other proteins, which would catalyze the efflux of Na^+. The lower trace in Fig.
3 shows that in this case, repeated illumination has no cumulative effect on the
photopotential even when KCl is added.

154

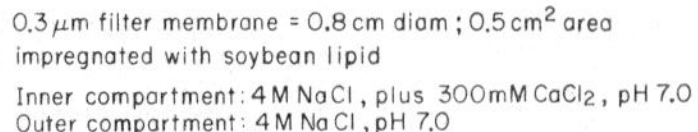

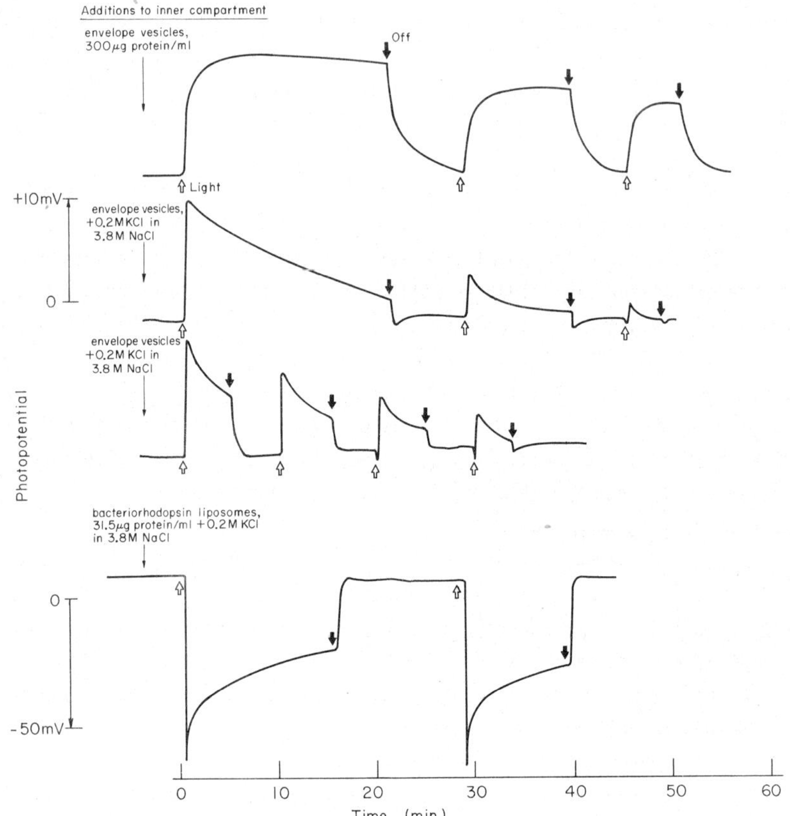

Fig. 3. Effect of K$^+$ on photopotential. The outside compartment contained 4 M
NaCl in all cases. Upper trace: inner compartment contained 4 M NaCl,
300 mM CaCl$_2$ and envelope vesicles (300 μg/ml protein). Second and third
trace: as above, but 3.8 M NaCl plus 0.2 M KCl in inner compartment.
Lower trace: as above but instead of envelope vesicles, liposomes reconstituted with bacteriorhodopsin (0.63 mg/ml lipid, 31.5 μg/ml protein)
were added.

The addition of KCl causes the gradual loss of the photopotential, presumably because $\underline{H}$. $\underline{halobium}$ membranes are permeable to K^+ [19-21]. Another experiment showed that membrane resistance was reduced to 2.5×10^7 ohm cm^2 by 0.5 M KCl addition compared with 2.5×10^8 ohm cm^2 in the absence of KCl, confirming the interpretation that K^+ is permeable. Changes in photopotential can be brought about by other cations; Rb^+ and Cs^+ work more effectively than K^+, but Li^+ does not. The order $Cs^+>Rb^+>K^+>Li$ presumably reflects the relative permeabilities of these cations in $\underline{H}$. $\underline{halobium}$ membranes. When valinomycin is present the order of cation specificity is $Rb^+>K^+>Cs^+>Na^+ \simeq Li^+$ which is consistent with the ionophore specificity of valinomycin.

$\underline{Low\text{-}salt\ Disruption\ of\ Filter\ Membrane\ Associated\ Envelope\ Vesicles}$. When the salt concentration is lowered, the membrane of $\underline{H}$. $\underline{halobium}$ is disrupted. Membrane associated envelope vesicles are protected from disruption, possibly because of firmly bound Ca^{2+}. However, when the salt concentration is lowered before adding the vesicles to the inner compartment, the photopotential is very much reduced and the shape of the signal becomes biphasic, indicating that the orientation of the membanes had randomized. Replacing the salt solution with distilled water results in negative photopotential, indicating that when the vesicles are completely disrupted, the endogenous lipids form oppositely oriented vesicles analogous to those made from soybean lipids and bacteriorhodopsin[14]. Envelopes, suspended in distilled water, and subsequently added to the filter membrane give no photoresponse. However, if liposomes are added to the inner compartment they seem to take up the purple membrane component liberated from the envelopes, and photopotentials of negative sign and increasing magnitude are developed.

Thus, it is apparent that lipid-impregnated filter membranes are useful in studying the electrical properties of natural membrane vesicles. Our results indicate that the electrical signals measured across the two aqueous compartments separated by the lipid-impregnated filter reflect primarily the properties of the vesicles. Although the precise nature of the association between the vesicles and the planar membranes is not yet known, it may be assumed that the envelope vesicles have retained their integrity and communicate electrically with both compartments.

$\underline{Summary}$. Lipid-impregnated millipore filter membranes are stable in 4 M salt solutions. In the presence of high salt and $CaCl_2$, envelope vesicles of $\underline{Halobacterium\ halobium}$ become associated with the filter membrane. As more and more vesicles become bound, progressively larger photopotentials are developed, between 10-30 mV. Addition of small amounts of permeable ions, such as K^+ or $TPMP^+$, to the solutions bathing both compartments on either side of the planar membrane results in collapse of the photopotentials, providing a direct method for observing Na^+-proton exchange for K^+ and other cations.

III. ASSAY OF IONOPHORE ACTIVITY OF THE OLIGOMYCIN-BINDING SUBUNIT 9 OF THE
       MITOCHONDRIAL ATPase

One of the links between the formation of proton gradients by electron transport
and the synthesis of ATP has been postulated to involve the proteolipid components
of the oligomycin-sensitive ATPase[22,23]. In particular, recent studies point to
the importance of the subunit 9 polypeptide, the chloroform:methanol soluble protein component of F_0, or the membrane sector of the ATP synthetase (hydrolysis)
complex [24]. This polypeptide has been reported to have a molecular weight between
7000-8000 for yeast, Neurospora, and beef heart mitochondria. The amino acid
sequence of the Neurospora protein has been found by Sebald et al.[25-27] to be quite
homologous with the yeast proteolipid and approximately five amino acids longer.
Since this protein binds DCCD and oligomycin, both of which are reported to block
proton translocation, we have undertaken the characterization of subunit 9 mediated
proton translocation by assaying its activity in lipid impregnated millipore filters
and liposomes containing bacteriorhodopsin.

Proteoliposomes. Proteoliposomes containing either bacteriorhodopsin in the
form of purple membrane fragments and/or subunit 9 preparations were made by soni-
cating lipids in the presence of these substances as previously described (cf. ref.
8,24 and Figure 3, lower trace, for details). Preparations of subunit 9 were
obtained from wild type and OR4 oligomycin-resistant mutants, Sacchromyces cere-
vesiae [24] according to a procedure slightly modified from Tzagoloff [28]. Liposomes
containing subunit 9 were prepared by mixing a 1:1 volume ratio of subunit 9
dissolved in (2/1) chloroform:methanol solution (0.12 mg protein/ml) and lipid
stock solution (15 mg lipid/ml), then sonicated at 4°C for 3 min at high frequency.
Subunit 9 containing liposomes prepared by this method had a cloudy white appearance.

Assay of Ionophore Activity. Using a location of subunit 9 and bacteriorhodopsin
in opposite compartments as the basic test system, the effectiveness of subunit 9
preparations in both reducing the steady state photopotentials generated by bacteriorhodopsin and increasing membrane conductance were investigated. Summarized in
Fig. 4, 0.3 ml of subunit 9 liposomes (600 µg protein/ml) were added into the outer
compartment together with 100 mM $CaCl_2$. This figure illustrates the photopotentials
generated by bacteriorhodopsin, their collapse in the presence of subunit 9 liposome preparations, and the nearly complete restoration of the photopotentials
brought about by adding oligomycin. The maximum effect is achieved with a final
concentration of less than 10 µg of subunit 9 added per ml. It may be assumed that
only added protein taken up by the planar membrane was effective, because rinsing
the compartment after the subunit 9 induced increase in photopotential is complete,
does not modify the results. Maximum recovery of photopotential was reached with
oligomycin concentration near 1×10^{-6} M (Fig. 5). This corresponds to about 1
mole of oligomycin/mole of subunit 9 (molecular weight = 7500) taken up by the

planar membrane. The precise amount of subunit 9 taken up by the membrane in each experiment with bacteriorhodopsin is unknown, hence the molar ratio cannot be accurately determined, but it appears that the number of oligomycin molecules added for 50% reversal is much less than the number of subunit 9 molecules in the membrane, suggesting a cooperative interaction.

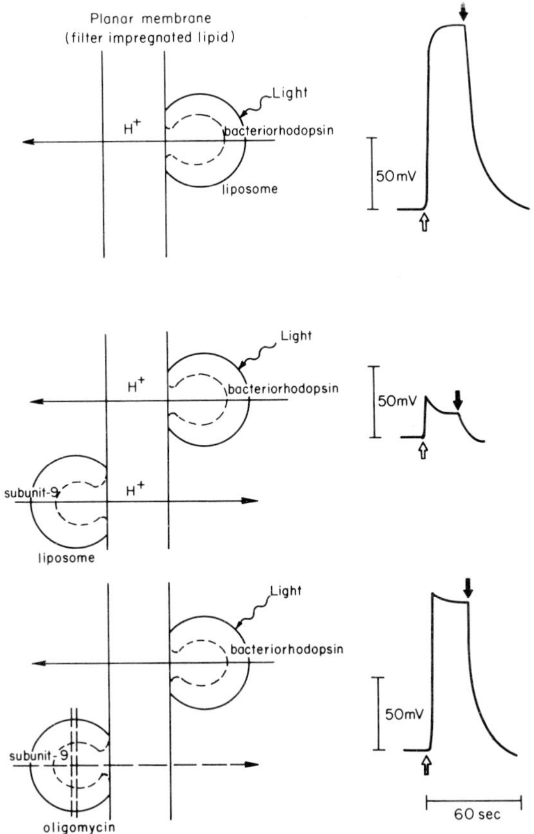

Fig. 4. Photopotential development by bacteriorhodopsin liposomes, collapse by wild-type yeast mitochondrial subunit 9 liposomes, and recovery after oligomycin addition.

These effects of oligomycin on the ionophoric properties of the proteolipid subunit of ATPase are consistent with previous studies in which a reductive alkylation procedure showed a specific interaction between subunit 9 and oligomycin[29] by adding oligomycin to the intact ATPase enzyme complex, reducing the reaction products with ^{3}H-borohydride and showing the protein-bound label was associated exclusively with subunit 9. It is important to note that alteration of this subunit by preparation of the polypeptide from oligomycin-resistant (OR4) mutants or by treatment with DCCD blocked the oligomycin-dependent labeling of subunit 9.

Correspondingly these preparations were capable of manifesting ionophore activity, but photopotentials were not recovered upon oligomycin addition. Thus, the results of binding studies and studies of ionophoric activity are qualitatively consistent.

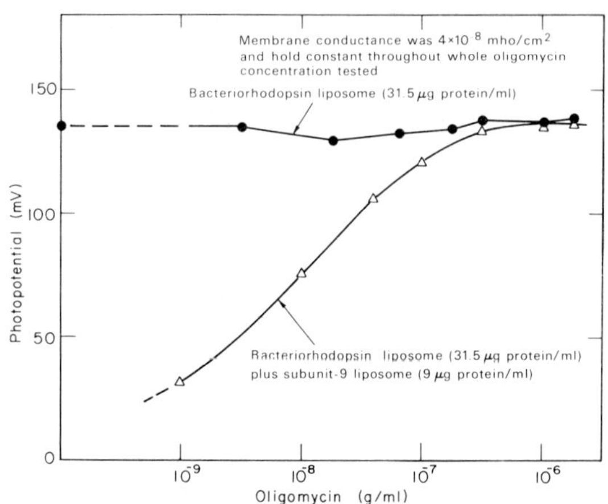

Fig. 5. Concentration dependence of oligomycin for inhibition of ionophore
activity of wild-type yeast mitochondrial subunit 9 preparations.

Summary. The hydrophobic protein subunit 9 of the mitochondrial ATPase from wild type and OR4 oligomycin-resistant mutants act as ionophores, collapsing photopotentials generated by the proton pump of bacteriorhodopsin. Only with the wild type subunit 9, which binds oligomycin, is the ionophore effect reversed by oligomycin. These results suggest that the hydrophobic subunit 9 polypeptide is the component linking ATP synthesis (hydrolysis) with proton translocation in mitochondria.

REFERENCES

1. Hirata, H., Altendorf, K. and Harold, F.M. (1973) Proc. Nat Acad. Sci. USA 70, 1804.

2. Altendorf, K., Hirata, H. and Harold, F.M. (1975) J. Biol. Chem. 250, 1405.

3. Schuldiner, S. and Kaback, H.R. (1975) Biochemistry 14, 5451.

4. Waggoner, A. (1976) J. Memb. Biol. 27, 317.

5. Drachev, L.A., Jasaitis, A.A., Kaulen, A.D., Kondrashin, A.D., Liberman, E.A., Nemecek, L.B., Ostroumov, S.A., Semenov, A. and Skulachev, V.P. (1974) Nature 249, 321.

6. Shieh, P. and Packer, L. (1976) Biochem. Biophys. Res. Comm. 71, 603.

7.   Skulachev, V.P. (1976) FEBS Letters 64, 23.

8.   Packer, L., Konishi, T. and Shieh, P. (1977) IN "Living Systems as Energy
     Converters", (R. Buvet, M.J. Allen and J.-P. Massue, eds.) p. 119, North
     Holland Publ. Co., Amsterdam.

9.   Shieh, P.K., Lanyi, J.K, and Packer, L. (1977) IN "Methods in Enzymology, (S.
     Fleischer and L. Packer, eds.), Academic Press, New York, in press.

10. Tobias, J.M., Agin, D.P. and Pawlowski, R. (1962) J. Gen. Physiol. 45, 989.

11. Tien, H.T. (1975) IN "Bilayer Lipid Membranes: Theory and Practice", p.
     Marcel Dekker, Inc., New York.

12. MacDonald, R.E. and Lanyi, J.K. (1975) Biochemistry 14, 2882.

13. Lanyi, J.K. and MacDonald, R.E. (1977) IN "Methods in Enzymology", (S. Fleischer
     and L. Packer, eds.) Academic Press, New York, in press.

14. Racker, E. and Stoeckenius, W. (1974) J. Biol Chem. 249, 662.

15. Lozier, R.H., Niederberger, W., Bogmoloni, R.A., Hwang, S.B. and Stoeckenius,
     W. (1976) Biochim. Biophys. Acta 440, 545.

16. Renthal, R. and Lanyi, J.K. (1976) Biochemistry 15, 2136.

17. Lanyi, J.K. and MacDonald, R.E. (1976) Biochemistry 15 4608.

18. Eisenbach, M., Cooper, S., Garty, H., Johnstone, R.M., Rottenberg, H. and
     Caplan, S.R. (1977) Biochim. Biophys. Acta, 465, 599.

19. Lanyi, J.K. (1977) In "Membrane Proteins in Energy Transduction", (R.A.
     Capaldi, ed.) Marcel Dekker, New York, in press.

20. Wagner, G. and Oesterhelt, D. (1976) Ber. Deutsch. Bot. Ges. 89, 289.

21. Garty, H. and Caplan, S.R. (1977) Biochim. Biophys. Acta 459, 532.

22. Racker, E. (1976) In "A New Look at Mechanisms in Bioenergetics", pp. 128-135,
     Academic Press, New York.

23. Kagawa, Y., Ohno, K., Yoshida, M., Takeuchi, Y. and Sone, N. (1977) Fed. Proc.
     36, 1815-1818.

24. Criddle, R.S., Packer, L. and Shieh, P. (1977) Proc. Nat. Acad. Sci. (US), in
     press.

25. Sebald, W. and Jackl, G. (1975) In "Electron Transfer Chains and Oxidative
     Phosphorylation" (E. Quagliariello, et al, eds.) p. 193, North-Holland,
     Amsterdam.

26. Sebald, W., Graf, Th. and Wild, G. (1976) In "Genetics and Biogenesis of
     Chloroplasts and Mitochondria" (Th. Bücher, et al., eds.) North-Holland,
     Amsterdam.

27. Sebald, W., Sebald-Althaus, M., Lukins, H.B. and Wachter, E. (1977), in press.

28. Tzagoloff, A., Akai, A. and Foury, F. (1976) FEBS Letters 65, 391-395.

29. Enns, R. and Criddle, R.S. (1977) Arch. Biochem. Biophys., in press.

QUINONES AND NON-HEME IRON PROTEINS

Bioenergetics of Membranes. L. Packer et al. ed.

INTRODUCTION

UBIQUINONES: RECONSTITUTION

A. AZZI

C.N.R. Unit for the Study of Physiology of Mitochondria and Institute
of General Pathology, University of Padova, Italy.

Twenty years ago two research groups arrived independently and
simultaneously to the determination of the chemical structure of ubi
quinone 10[1,2]. It is fortunate that this session coincides with the
twentieth anniversary of such an important event.

The role of ubiquinone as an obligatory electron carrier in the
electron transfer chian of mitochondria is no longer a matter of de-
bate, after the rapid kinetiks studies of Klingenberg and Kröger[3] in
dicating the fast reactivity of ubiquinone. Ubiquinone location however, in the sequence of electron carriers, is still open to discussion. The general consensus of opinions has been that ubiquinone is
reduced by the flavin-iron-sulphur dehydrogenases and oxidized by cytochrome b, as indicated by kinetic and reconstitution studies, which
are indeed too numerous to be listed here.

A second interaction site of ubiquinone, on the other hand, was
postulated by Mitchell[4] on the oxygen side of cytochrome b, Mitchell's
view, was not strongly supported by experimental evidence, but it in-
vited new research initiatives, in particular in its more recent for-
mulation, i.e. the Q cycle[5]. Such a concept is hard to describe in
few words, but essentially it implyes that QH_2 is the species which
reduces cytochrome c_1, the $QH^{\cdot}$ radical being formed and a proton released outside the mitocondria. The $QH^{\cdot}$ species reduces cytochrome
b_{566}, and a second proton is release outside the mitochondria. Oxidized ubiquinone migrates across the mitochondrial membrane and re-
ceives one electron from the dehydrogenase and one proton is taken up
from the matrix space. The second electron is transferred from cytochrome b_{562} (reduced by b_{566}) and a second proton taken up from the
matrix space. The formed QH_2 species migrates across the mitochondrial membrane and the cycle is completed.

This elaborated scheme of reactions maintains the original concept of a ubiquinone species which reacts on the oxygen side of cyto
chrome b. One of the papers to be presented in this session will spe
cifically and elegantly try to resolve the question of two ubiquinone reactive sites in bacterial electron transfer chians.

The Q cycle has also other important implications, which have

been indicated by Mitchell[5]. The components of the cycle must occur
at sufficiently high concentration and must have sufficient diffusio-
nal mobility along specific pathways. In particular it seems probable
that QH· have to be stabilized by complexation to specific membrane
proteins in order to have them at kinetically compentent levels and
at the same time to prevent their dismutation. The studies of Dr.
Ohnishi presented in this Symposium, point to this direction. It appears also necessary that ubiquinone acts as mobile species, crossing
the membrane by transverse diffusion or by appropriate flip-flop movements. Some of the following papers will deal with this problem,
utilizing both model and natural membrane systems.

New and stimulating research is thus being carried out in numerous laboratories on the function of ubiquinone in electron transport
chains, as suggested by the newly developed concepts in this field.I
believe that, if not in this session, in the near future we shall have an answer to the problems outlined above.

1. Wolf,D.E.,Hoffman,C.H., Trenner,N.R., Arison,B.H., Shunk,C.H.,
   Linn, B.O., McPherson,J.F., and Folkers,K. (1959) J.Am.Chem.Soc.,
   80, 4752.
2. Morton,R.A., Gloor,U., Schindler,O., Wilson,G.M., Choparddit-Jean,
   L.H., Hemming,F.W., Isler,O., Leat,W.M.F., Pennock,J.F., Rüegg,R.,
   Schweiter,U., and Wiss,O. (1958), Helv.Chim.Acta, 41, 2343.
3. Klingenberg,M. and Kröger,A. (1970). In Electron Transport and E-
   nergy Conservation (J.M. Tager, S.Papa, E.Quagliariello and E.C.
   Slater, eds) pp. 135-143. Bari: Adriatica Editrice.
4. Mitchell,P. (1966). Chemiosmotic Coupling in Oxidative and Photosynthetic Phosporylation. Bodmin: Glynn Research.
5. Mitchell,P. (1976), J. Theor.Biol. 62, 327-367.

FUNCTIONAL COMPARTMENTATION OF THE MITOCHONDRIAL QUINONE DURING SIMULTANEOUS OXIDATION OF TWO SUBSTRATES

Menachem Gutman
Biochemistry, George S. Wise Center of Life Sciences
Tel-Aviv University
Israel

ABSTRACT

The rate of succinate oxidation by submitochondrial particles was measured as a function of active succinate dehydrogenase. The rate of CoQ reduction by this substrate (Vred S) and the rate of $CoQH_2$ oxidation (Vox) when combined in the equation of Kroger and Klingenberg,[3] accurately predicts the observed rate (Vobs) of succinoxidase. Vobs = Vox.Vred/(Vox + Vred). The same equation was used to calculate the rate of CoQ reduction by NADH (Vred N).

When two substrates are oxidized simultaneously one oxidase inhibits the other so that the rates of respiration are not additive. We investigated the possibility that this nonadditivity is the direct consequence of the additivity of the rates of CoQ reduction by the two substrates, where ΣVred substitutes Vred in the above equation. Quantitive analysis of experimental results indicates that this is not the case. Failure to predict measured rates and systematic deviation from predicted results signify that the above equation is applicable only for oxidation of a single substrate, being inadequate for the physiogical situation where some substrates are utilized simultaneously. This implies that during oxidation of more than one substrate the redox level of the quinone varies between different domains. The redox potential at each domain is determined by the activity of the nearest dehydrogenase and the rate of $CoQH_2$ oxidation by the cytochrome system.

At the level of single domain the relation between Vred, Vox and Vobs are as for single substrate oxidation, only that another flux should be considered, the spillover reaction between adjacent domains of different redox level. The rate of the spillover is slow relative to the flux within each domain, yet it causes the recipient domain to assume a more reduced steady state, this reduction slows the reoxidation of the respective dehydrogenase and is the cause for the mutual inhibition between oxidases.

INTRODUCTION

The stoichiometric relationship between carriers in the dehydrogenase-CoQ[1]* region of the respiratory chain markedly deviates from the 1:1 ratio dominating the cytochrome section of the chain. The CoQ -cyt $b_{(561+560)}$ ratio is 6:1 and the

---

*Abbreviations: CoQ - ubiquinone 10; ADPR - adenosine diphospho-ribose; DQ - duroquinone

166

NADHDH/CoQ and SDH/CoQ values are 1:100 and 1:30 respectively. The vast excess of CoQ allows one molecule of dehydrogenase to supply electrons to many cytochrome segments of the respiratory system, an arrangement which compensates the scarcity of the dehydrogenases by their high turnover capacity.

In this study we wish to investigate those properties of the respiratory system which are imposed by the stoichiometric excess of the quinone, whatever is the mechanism of the electron transport at this level, linear arrangement, or the quinone cycle[2].

Kroger and Klingenberg[1,3] described the mitochondrial quinone as a homogeneous carrier, the flux through the quinone (Vobs) is a function of two partial reactions: the rate of CoQ reduction by the dehydrogenase (Vred) and the rate of QH_2 oxidation by the cytochromes (Vox). The correlation between them is given by equation 1:

$$1)\quad Vobs = \frac{Vred.Vox}{Vred+Vox}$$

This equation is true for a steady state flux through any homogeneous carrier located between electron donor and acceptor. Kroger and Klingenberg varified this expression by observing that the rate of respiration was determined only by the ratio $CoQH_2/CoQ_{active}$ and not by the nature of the substrate (succinate or NADH). From this they concluded that CoQ behaves as a homogenous pool and the active quinone interacts with any dehydrogenase or cytochrome b molecule in the membrane. Furthermore, by assuming that Vox > Vred, they also accounted for the sigmoidal nature of inhibition by antimycin. Presently there are other explanations for this sigmoidicity[4,5] and at least in one case antimycin inhibits in a sigmoidal mode a reaction for which Vred > Vox (Fig. 8 in Ref. 6). Thus, if the model should be taken as whole, its other conclusion should be reexamined too. If CoQ is so effective in spreading electrons emerging from various dehydrogenases, then each dehydrogenase contributes to the steady state redox level of all active quinone. As the rate of CoQ reduction decreases with increased steady state reduction of the quinone[3] it implies that whenever two dehydrogenases are simultaneously reducing the quinone pool one dehydrogenase should inhibit the other. This is the well known phenomenon of mutual inhibition[7-11] where two oxidases inhibit each other. The characteristic properties of mutual inhibition are as follows:

1) The rate of respiration on two substrates is less than the sum of the rates measured for each substrate alone[7-11].

2) The lack of additivity is due to selective inhibition of one oxidase while the other remains essentially unaffected[7,10].

3) The inhibition is neither at a level of dehydrogenase nor a result of competition for a common carrier at the cytochrome level[9,11].

4) The dehydrogenase with the higher activity usually inhibits the oxidation of other substrates. Inhibition of the dominating dehydrogenase or activation of

both NADH oxidase and succinoxidase activities are high[7]. Under these conditions inhibition of succinoxidase by NADH depends on the fraction of active succinate dehydrogenase. On the other hand succinoxidase causes only a marginal inhibition of NADH oxidase. In 0.18M sucrose 50 mM Tris acetate 5 mM $MgSO_4$ pH 7.4 (buffer used routinely for reverse electron transport) the absence of phosphate induces a slower rate of respiration (though that FCCP is present) and a noted inhibition of NADH oxidase by succinate takes place. The mechanism leading to the difference between the two buffers will be discussed elsewhere (Gutman to be published). In this study we compared the extent of the mutual inhibition as function of activation of succinate dehydrogenase in these two systems.

Correlation between partial rates and overall electron flux. To correlated the partial fluxes Vox and Vred to the overall fluxes, Vobs, we had to determine independently the magnitude of the partial fluxes. Vox was taken as the maximal rate of DQH_2 oxidation[3] corrected for its autooxidation. Vred for succinate dehydrogenase was taken as Vred(S) = % active SDH x Vmax (PMS). It was proposed that in membranal preparations PMS is reduced by the quinone reducing site with the same turnover number[15].

. The accuracy of equation 1 in predicting the rate of Vobs was tested by varying the magnitude of Vred(S) by controlled activation of succinate dehydrogenase and comparing the observed rate of respiration with the calculated one. As seen in Figure 1, except for a single point which deviates markedly (probably due to activation during the polaragraphic assay),the observed rate nearly coincide with the predicted ones. Thus equation 1 is suitable for correlating the partial rates and the overall flux. Accordingly we used Vobs (N) and Vox to calculate Vred (N). The NADH oxidase activity in phosphate buffer was 1.72 µmol/min mg with the corresponding Vred (N) = 4.1 µmol/mg. In sucrose tris buffer NADH oxidase activity was 0.505 ± 0.05 µmol/min mg and Vred (N) =1.1 µmol/min mg.

Simultaneous oxidation of two substrates: If the quinone pool is homogeneous then when two dehydrogenase are acting simultaneous the total rate of the reduction will be the sum of the two partial rates. Thus the nonadditivity of rates of respiration can be the consequence of the additivity of rates of CoQ reduction. Defining ΣVred = Vred(N) + Vred(S), then the rate of respiration on both substrates Vobs (N+S) will be smaller than the sum Vobs(S) + Vobs(N).

2) $\text{Vobs(S+N)} = \dfrac{\text{Vox.}\Sigma\text{Vred}}{\text{Vox+}\Sigma\text{Vred}} < \dfrac{\text{Vobs.Vred(S)}}{\text{Vobs+Vred(S)}} + \dfrac{\text{Vobs.Vred(N)}}{\text{Vobs+Vred(N)}} = \text{Vobs(S)} + \text{Vobs(N)}$

Assuming this explanation to be correct, we can calculate ΣVred from Vobs(S+N), exactly as was done for Vred(N). This assumption is investigated in Figure 2. In Figure 2A we calculated Vobs(N+S) as it increases upon increasing Vred(S). When Vred(S) = 0 (succinate is present but not oxidized) the measured and predicted values coincides but upon increasing Vred(S), the predicted value is consistantly smaller than the measured one.

others will shift the dominancy from one oxidase to the other.

In this study we investigated the mechanism of mutual inhibition and examined whether it can be explained by the terms of the homogeneous pool of Kroger and Klingenberg[1,3]. As will be documented, this model is inadequate for quantitating the fluxes when two substrates are oxidized. As alternative we propose another model which accounts both for our observations as well as for others.

MATERIALS AND METHODS

Phosphorylating submitochondrial particles (ETP$_H$) were prepared according to Hansen and Smith[12] from beef heart mitochondria (Ringler et al[13]). NADH oxidase was measured either polarographically (Clark oxygen electrode) or at 340 nm (Gilford 240). Both instruments were thermostated (30°) by the same thermostated circulating bath (Hakke T52) at maximal water flow. Succinoxidase was measured polarographically. NADH oxidase in presence of succinate was measured at 340 nm, the rate of NADH (0.5 mM) oxidation was recorded, then succinate (1M) was added (final concentration 10 mM). The changes in rate was corrected for dilution. Succinoxidase in presence of NADH was obtained by substructing the rate of NADH oxidase in presence of succinate (measured at 340 nm) from the rate of respiration in presence of succinate (10 mM) plus NADH (0.5 mM)(measured by oxygen electrode). Durohydroquinone oxidase was measured polarographically at Vmax. The extrapolated value is taken as Vox[3]. The rates were corrected for the autooxidation of DQH$_2$. Succinate dehydrogenase was measured at Vmax (PMS)[11].

Succinate dehydrogenase in ETP$_H$ was activated to controlled level of suspending the particles in 0.18M sucrose, 50 mM sucrose, 50 mM MES pH 6.2 and increasing concentration (0-100 mM) of NaBr[14]. After 30 min at 30°, the ETP$_H$ were spun down and stored at 10 mg/ml 0°C, 0.18M sucrose 50 mM Tris acetate, 5 mM MgSO$_4$ pH 7.4 and the fraction of active enzyme was determined. Full activation was measured after activation by succinate (20 mM). The product of the active-fraction times Vmax(PMS) is taken as Vred(S).

Samples of ETP$_H$ activated to various levels with respect to succinate dehydrogenase was assayed for the following activities: NADH oxidase; succinoxidase; NADH oxidase in presence of succinate; succinoxidase in presence of NADH and rate of respiration with both substrates. Each activity was measured in 80 mM K Phosphate, 50 μM EDTA pH 7.4 (phosphate buffer) or in 0.18M sucrose, 50 mM Tris acetate 5 mM MgSO$_4$ 0.11 μM FCCP (STM buffer). Protein was determined by Buiret.

RESULTS

Determination of experimental conditions: The mutual inhibition between NADH oxidase and succinoxidase is easily measured with submitochondrial particles, where both dehydrogenases are located on the outer surface of the vesicles. Our previous studies were measured in 80 mM K phosphate, 50 μM EDTA pH 7.4, a medium in which

In figure 2B we calculated ΣVred from Vobs(S+N). As expected ΣVred increases with Vred(S), but when Vred(S) is subtracted from ΣVred, we still observe a variable value which should represent Vred(N). This assumption is correct for Vred(S) = 0 where ΣVred = Vred(N) but when Vred(S) increases, the expected value of Vred (N) increases too. This increase must be artificial as it is measured only in presence of succinate. In absence of succinate Vred(N) = 4.1 µmole/min mg whatever is the level of activation of succinate dehydrogenases.

The variable nature of Vred(N), equated with ΣVred-Vred(S) is further emphasized in Figure 3, representing results measured in sucrose Tris buffer. In this system activation of succinate

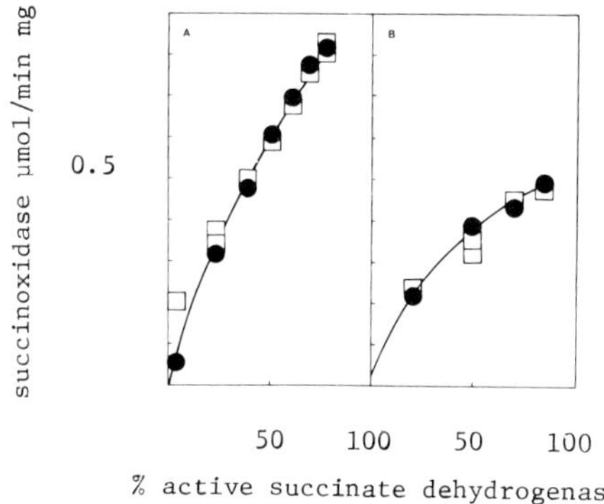

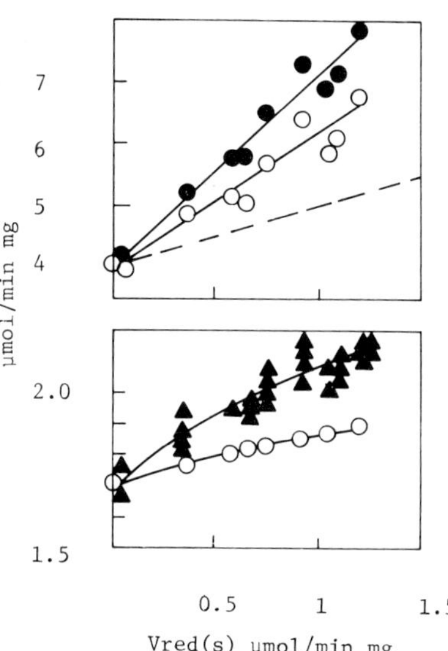

Fig. 1. Comparison between measured and predicted rates of succinate oxidation by ETP$_H$. The succinoxidase activity was measured for the same batch of ETP$_H$ either in (A) 80 mM K phosphate, 50 µM EDTA pH 7.4 or (B) in 0.18M sucrose, 50 mM Tris acetate 5 mM MgSO$_4$, 0.11 µM FCCP pH 7.4. ETP$_H$ were activated to the indicated levels by NaBr at pH 6.2 (see methods). Spun down and resuspended in 0.25M sucrose 50 mM Tris acetate 5 mM MgSO4 pH 7.4 at 10 mg protein/ml. Alliquots were taken and assayed at 30° for succinoxidase activity (□)(10 mM succinate). The rates were calculated from the initial rates before activation insitu took place. The predicted values (•) were calculated from equation 1 using for Vox the Vmax for oxidation of DQH$_2$ corrected for autooxidation, for A Vox = 2.93 µmol/min mg for B 0.91 µmol/min mg. Vred was taken as the product of % activation times Vmax for PMS reduction (30°). A) 100%=1.5 µmol/min mg for B) 100% = 1.4 µmol/min mg.

Fig. 2. Comparison between measured and predicted rates during simultaneous oxidation of NADH and succinate. ETP$_H$ were activated to the indicated levels of Vred(S) and rate of respiration was measured in 80 mM K phosphate, 50 µM EDTA pH 7.4 30° in presence of NADH 0.5 mM plus succinate 10 mM. For calculation of predicted rates Vox=2.93; Vred(N)=4.1 and Vred(S)=1.5 µmol/min mg. A. (▲) observed rate of respiration. (o) the rate calculated according to equation 1 substituting ΣVred=Vred(N) + Vred(S) for Vred. B. (•) Calculation of ΣVred from the mean rate of respiration with both substrate. (o) ΣVred-Vred(S) obtained by subtracting the corresponding Vred(S). (--) the sum of Vred(N) and Vred(S) using Vred(N)=4.1 µmol/min mg and Vred(S) as indicated on the abcissa. Note change in magnitude of ordinate between A and B.

170

dehydrogenase increases the sensitivity of NADH oxidase to mutual inhibition. At about 20% activation, addition of succinate causes only a marginal inhibition of NADH oxidase, but it increases to ∿40% inhibition at higher levels of activation. Surprisingly this decrease in rate of NADH oxidation is not reflected in ΣVred-Vred(S). This value, presumably identical with Vred(N) is identical to Vred(N)

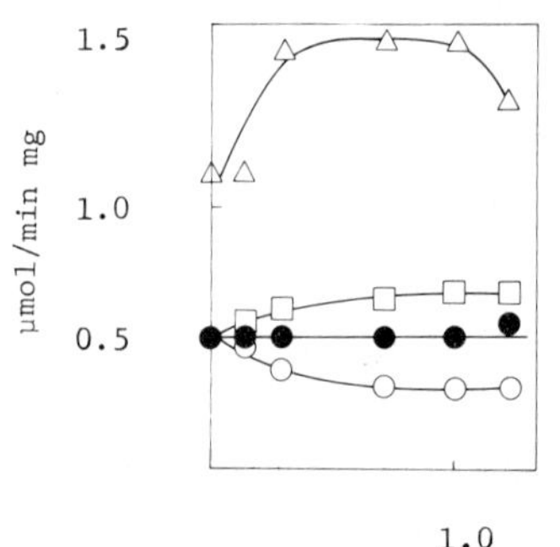

(measured in absence of succinate) at low values of Vred(S), and becomes bigger than Vred(N) when Vred(S) is high though that under these very conditions NADH oxidase is 40% inhibited.

The fact that Vred(N) calculates by this method fails to reflect 40% inhibition of NADH oxidase indicates that equation 1 is applicable only for oxidation of a single substrate but fails when two of them are oxidized simultaneously. Thus we applied equation 1 for the rate of oxidation of each substrate, and calculated Vred(N) from the

Fig. 3. The effect of increasing Vred(S) on rate of oxidation of NADH. ETP$_H$ were activated as before to the indicated levels of Vred(S). Respiration was measured in 0.18M sucrose. 50 mM Tris acetate, 5 mM MgSO$_4$, 0.11 μM FCCP pH 7.4 at 30° in presence of 0.5 mM NADH, 10 mM succinate or both. (●) NADH oxidase in absence of succinate; (o) NADH oxidase in presence of succinate, (■) respiration on NADH plus succinate, Δ ΣVred – Vred(S) calculated by equation 1 using rates of respiration on both substrates and Vox = 0.91 μmol/min mg.

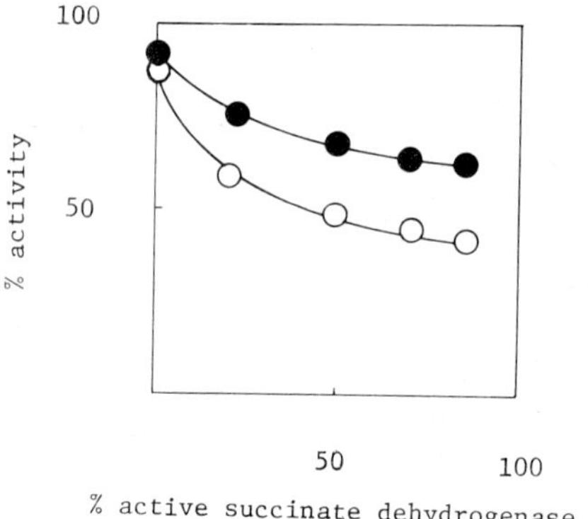
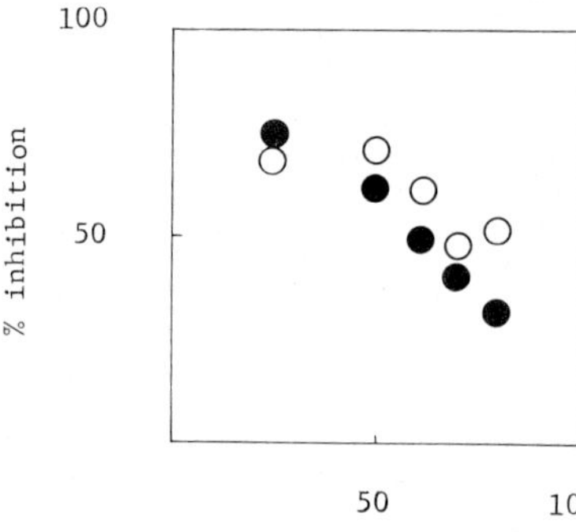

Fig. 4. The effect of activation of succinate dehydrogenase on rate of NADH oxidation by ETP$_H$ in sucrose Tris buffer. (●) NADH oxidase. (o) Vred(N) calculated from rate of NADH oxidation according to the domain model. The data was taken from Figure 3.

Fig. 5. The effect of activation of succinate dehydrogenase on the rate of succinate oxidation by ETP$_H$ in phosphate buffer in the presence of NADH. Data taken from the experiment presented in Figure 1A and 2. (●) inhibition of succinoxidase by NADH. (o) inhibition of Vred(S) calculated by equation 1 according to the domain model.

rate of NADH oxidation measured at 340 nm in presence of succinate. These values
are given in Figure 4. This treatment yields reasonable results; the inhibition
of NADH oxidase is reflected by corresponding decrease in Vred(N). In a similar
way we calculated the rate Vred(S) for oxidation of succinate, in phosphate
buffer, in the presence of NADH. The results are given in Fig. 5. In this system where NADH oxidase inhibits the oxidation of succinate, we observe a decline
of Vred(S) in parallel to the decrease of succinoxidase activity.

DISCUSSION

Kroger and Klingenberg[3] demonstrated that the rate of respiration in submitochondrial particles is related with the partial rates Vox and Vred according to
equation 1. This was concluded from the linear dependence of respiration rate on
the steady state reduction of CoQ when either Vox or Vred were varied. In Figure
1 (A and B) we varied the magnitude of Vred(S) and confirmed the accuracy of
equation 1 in two assay systems differing in their Vox. As there is no reason
to assume that NADH oxidase differs from succinoxidase we calculated by the same
equation the value of Vred(N). Kroger and Klingenberg combined equation 1 and the
assumption that Vox > Vred to explain the sigmoidal inhibition by Antimycin[3]. Yet
sigmoidal inhibition may result from the cooperative binding of the inhibitor (in
presence of substrate[4]). Furthermore, Von Jagow and Bohrer found that succinate -
duroquinone reductase of submitochondrial particles is inhibited sigmoidaly by
Antimycin though that Vred(S) >> Vobs[6]. Apparently the sigmoidal inhibition is
not of kinetic origin, a conclusion which removes the limitation Vred < Vox and
allows to accept our observations Vred > Vox as true solution for equation 1. The
high value of Vred(N) also explains how ADPR inhibits $\sim$75% of NADH-$K_3Fe(CN)_6$ reductase but only 30% of NADH oxidase activity. On this ground we regard equation 1
as an accurate expression for correlating the rate of oxidation of a single substrate with the partial rates Vred and Vox.

The purpose of this study was to investigate the mechanism of mutual inhibition
between two oxidases. As demonstrated in figures 2 and 3, the homogeneous pool
model as expressed in equation 2, fails to account for the experimental results.
Vred(S) and Vred(N) are not additive terms, their sum is smaller than ΣVred as
calculated from equation 2 (Fig. 2B). Similarly the value ΣVred-Vred(S) supposedly identical with Vred(N) is not constant. It nearly doubles upon full activation of succinate dehydrogenase. The results given in Figure 2 were measured in
phosphate buffer, where inhibition of NADH oxidase by succinoxidase is marginal.
Yet even in sucrose Tris buffer (Figure 3), where the mutual inhibition of NADH
oxidase approaches 40% the nonadditivity of Vred and Vred(S) is evident. At low
levels of activation of succinate dehydrogenase ΣVred-Vred(S) increases, as was
noted in phosphate buffer. Only at higher levels of activation, where the mutual
inhibition of NADH oxidase is maximal, ΣVred-Vred(S)assumes a constant value

which is higher than the initial one. This apparent variation in the activity of
the NADH dehydrogenase is not an indirect effect of the conformation changes of
succinate dehydrogenases, as it can be detected only when the two substrates are
simultaneously oxidized. We have to conclude that the apparent variations of
ΣVred-Vred(S), is a systematic error resulting from the erroneous assumptions
that Vred(S) and Vred(N) are additive. As alternative model we propose that equation 1 is valid, but only for calculating the flux within distinct
domains of the quinone, i.e. regions small enough to ensure that their
ox reduction level is determined primarily by the activity of the nearest dehydrogenase. A model which implies a compartmentation of the quinone into domains,
poses a question about the nature of the barriers between them. Compartmentation
does not necessarily mean that the proposed domain are physically separated. For
example the preferential reduction of mitochondrial NAD by the glutamate dehydrogenase was explained by terms of compartmentation of the NADP and NAD within the
matrix space. Recently[16] it was demonstrated that the compartmentation was of
kinetic origin, reflecting the high activity of the energy linked transhydrogenase. In analogy we attribute the different redox potential of the domains to the
fact that the redox reaction between the domains is slower than the turnover
within the domains. The relatively slow electron flux between the domains might
result from the higher proximity of the dehydrogenases to the b type cytochromes
as compared to the distance between different dehydrogenases. Alternatively, the
concentrations of CoQ at its specific binding site might be higher than its concentration in the lipid layer of the mitochondria. Evidence supporting a relative
slow flux between domains can be found in the studies of Strooband and Kabak[17].
It was demonstrated that NADH oxidation can support a rapid amino acid accumulation and build-up of $\Delta\Psi$ in _E. coli_ vesicles if they are supplemented by CoQ_1. The
fact that CoQ_1 stimulates uptake many fold more than stimulation of respiration
indicate that it does not accelerate a rate limiting step in electron flux to
oxygen but allows efficient redox equilibrium between carriers of similar potential, carriers that are compartemented by a kinetic barrier. Gutman and Silman[7]
used CoQ depleated submitochondrial particles and measured the extent of the
mutual inhibition between NADH and succinate oxidase as a function of reincorporated CoQ. At low level of reincorporated quinone the mutual inhibition was
practically nill but with higher levels of added CoQ the interaction between the
domains was enhanced with the concomitant appearance of mutual inhibition.

According to our model a difference in redox levels between adjacent domains
drives a spillover between them. If only one substrate is oxidized, no mutual
inhibition can be measured, as the spillover from the reduced domain is relatively
slow, no reduction of the recipient domain will ensue. This accounts for the fact
that Kroger and Klingenberg[1] who used either succinate or NADH observed a homogeneous quinone pool. In a similar way inhibition of $CoQH_2$ oxidation by KCN or

anaerobiasis will allow the relative slow spillover to reduce of all of the active quinone. This accounts for the observation[18] that anaerobically all active quinone is reduced either by succinate or by NADH.

When two substrates are oxidized the incrament of electron flux from the recipient domain to oxygen (dVobs) will be smaller than the rate of spillover (Vs). This is evident by differentiating Vobs in equation 1 with respect to Vred. $dVobs = (1 + \frac{Vred}{Vox})^{-2} \, dVred$. As the rate of spillover is faster than the incrament of Vobs it will increase the reduction level of the recipient domain. The increase of the ratio of QH_2/Qox is linear with the incrament of Vred $d(\frac{CoQH2}{CoQox}) = \frac{dVred}{Vox}$. The increase of the $CoQH_2/CoQox$ ratio slows oxidation of the dehydrogenase of the recipient domain either by product inhibition ($CoQH_2$), oxidant (CoQox) depleation or lowering the apparent redox potential of the domain. As a matter of fact Kroger and Klingenberg demonstrated that the rate of oxidation of the dehydrogenase falls upon increasing the $CoQH_2/CoQox$ ratio[3]. The driving force of the spillover is the difference in $CoQH_2$ concentrations between the domains (if spillover reflects a diffusion of $CoQH_2$ molecules) or the difference between the apparent redox potentials of the domains (if the spillover proceeds as redox reaction between the domain). In either case the smaller is the driving force, the slower will be the spillover. As a result the mutual inhibition will decrease too. This accounts for the fact that activation of succinate dehydrogenase lowers the sensitivity of succinoxidase towards mutual inhibition. The higher is the number of active molecules of succinate dehydrogenase reducing a certain domain (increased Vred) the higher will be the ratio $CoQH_2/CoQox$ in this domain. This will diminish the spillover's driving force, lowering the suceptibility of succinoxidase to mutual inhibition. In phosphate buffer Vred calculated for NADH dehydrogenase 4.1 µmol/min mg (Figure 3) is much higher than that of succinate dehydrogenase (1.5 µmol/min mg). The NADH dehydrogenase domains will be more reduced than succinate dehydrogenase one and as expected NADH oxidase inhibits succinoxidase. In sucrose buffer Vred for NADH and succinate are 1.1 and 1.4 µmol/mg respectively which accounts for inhibition of NADH oxidase by succinate. In both cases we observe the effect of the activation. The higher is the active fraction of succinate dehydrogenase the lesser will be its inhibition NADH oxidase (in phosphate buffer, Figure 5), or more will it inhibit NADH oxidase (in sucrose Tris buffer, Figure 4).

The domain model accounts for our previous[7] observations that ADPR (competitive inhibitor for NADH) releases succinoxidase from spillover inhibition by NADH. The partial oxidation of the NADH dehydrogenase domain by decreasing Vred(N), lowers the driving force for the spillover thus releasing succinoxidase from the mutual inhibition. Finally it was noted that at even in phosphate buffer succinoxidase can inhibit NADH oxidase given that NADH concentration (5-15 µM) is low enough to slow NADH oxidase[19]. Under such conditions Vred(N) is small enough so that its domain is more oxidized than that of succinate dehydrogenase with concomitant reversal of the direction of the spillover. Ringler, Singer and Lusty[11] noted that oxidation of succinate and choline by rat liver mitochondria is additive. But, mutual inhibition could be imposed by partial inhibition with N_3^-. Examination of their results indicate that none of this substrates was rapidly oxidized 33 and 13.6 nmoles/min mg for succinate and choline respectively. This may be a case where Vox >> Vred so that the spillover reaction is unobserved due to the high oxidation state of the respective domains. When the terminal oxidase was partially inhibited by N_3^- so that Vox became smaller than Vred, a mutual inhibition was noted and, as expected, the more active dehydrogenase (succinate dehydrogenase) inhibited choline oxidase by 62% while its own activity was only partially affected (32%).

The model we propose is not in contradiction with the observation made with single electron donor. As a matter of fact domains can not be seen with a single substrate (but no respiratory system _in vivo_ is oxidizing just one substrate). When two substrates (or more) are oxidized the Quinone around each type of dehydrogenase assumes a different redox potential so that domains are formed. These domains are not seperated by physical barriers so that they interact with each other. The result of this interaction is observed as mutual inhibition between oxidase activities.

REFERENCES

1. Kroger, A. and Klingenberg, M. (1970) In: Vitamins and Hormones (Harries, P.S. et al, Eds) Vol. 28, 539-537.

2. Mitchell, P. (1976) J. Theor. Biol. 62, 327-367.

3. Kroger, A. and Klingenberg, M. (1973) Eur. J. Biochem. 34, 358-368.

4. Berden, J.A. and Slater, E.C. (1972) Biochim. Biophys. Acta. 256, 199-215.

5. Grimmelikhuijzen, C.J.P, Marrs, C.A.A, and Slater, E.C. (1975) Biochim. Biophys. Acta. 375, 533-548.

6. Von Jagow, G. and Bohrer, C. (1975) Biochim. Biophys. Acta. 387, 409-424.

7. Gutman, M. and Silman, N. (1972) FEBS Letters 26, 207-210.

8. Wu, C-Y, and Tsou, C.-L. (1955) Sci. Sinica Peking. 4, 137.

9. Davis, E.J, Blair, P.V. and Mehoney, A.J. (1969) Biochim. Biophys. Acta. 172, 574-577.

10. Ringler, R.L. and Singer, T.P. (1959) J. Biol. Chem. 234, 2211-2217.

11. Kimura, T, Singer, T.P. and Lusty, C.J. (1960) Biochim. Biophys. Acta. 44, 284-297.

12. Hansen, M. and Smith, A.L. (1964) Biochim. Biophys. Acta. 81, 214-222.

13. Ringler, R.L, Minakami, S. and Singer, T.P. (1963) J. Biol. Chem. 238, 810.

14. Gutman, M. (1976) Biochemistry 15, 1342-1348.

15. Ackrell, B.A.C, Kearney, E.B, Mowery, P.C, Singer, T.P, Beinert, H,
Vinogradov, A.D. and White, G.A. (1976) In: Iron and Copper Proteins, Yasu-
nobu, K. et al Ed. 161-181 Plenum Press New York.

16. Hoek, J.B, Ernster, L, De Haan, J.E. and Tager, J.M. (1974) Biochim. Biophys.
Acta. 333, 546-559.

17. Stroobant, P. and Kaback, H.R. (1975) Proc. Nat. Acad. Sci. US. 72, 3970-3974.

18. Knook, D.L. and Planta, R.J. (1973) Arch. Microbiol. 93, 13-22.

19. Gutman, M. and Silman, N. Unpublished results.

THE PERMEABILITY OF QUINONES THROUGH MEMBRANES

G. Hauska
University of Regensburg, FB Biology, Botany,
84 Regensburg, Universitätsstraße 31, GFR

INTRODUCTION

Substituted benzoquinones are known to function in the electron
transport chains of respiration and photosynthesis since the late
1950s (see 1 for summarizing articles). This has been established
by the observation of absorption changes and by extraction and
reconstitution for both, ubiquinone in the respiratory chain (2,3,see
1), and plastoquinone for plant photosynthesis (4,5, see 1,6,7).
Specifying this function of quinones, proton translocation associated with electron transport has been attributed to a vectorial mode
of action following the concepts of the chemiosmotic theory (8,9).
This is formulated in Fig.1 for the combined action of plastoquinone and reaction center 1 of the electron transport in chloroplasts as one example. However, this scheme represents only one of
at least three different, feasible mechanisms of H^+-translocation
during electron transport (see 10). Another possibility results
from oxidation of a hydrogen carrying substrate on one, and reduction of an electron acceptor on the opposite surface of a membrane,
as conceived first by Lundegardh (11), and realized f.i. in photosynthetic electron flow in chloroplasts by the oxidation of water
on the inner and reduction of CO_2 on the outer surface of the membrane (see 12). The third mechanism involves the action of a macromolecular proton pump without protic redox reactions, as it is probably represented by bacteriorhodopsin in the purple membrane of
halobacteria (13).

The vectorial redox reaction of quinones as shown in Fig.1 might
be complicated by the split into two half reactions via the semiquinone radical, which have very different redox potentials. Therefore the quinone system might function twice along the redox potential cascade of electron transport chains as reasoned by Mitchell
in his "Q-cycle" (14), and as supported by recent extraction and
reconstitution studies of ubiquinone in chromatophores of photosynthetic bacteria (15).

A redox compound requires two features to function like Q in
Fig.1 - it must be reasonable lipid soluble, and it must be a
carrier of protons and electrons in the reduced form. Such compounds

178

might constitute artificial H$^+$-translocation when introduced into
photosynthetic or respiratory systems in which the function of
endogenous benzoquinones has been eliminated. Indeed, this has been
established by our own studies with artificial quinoid systems,
mainly with chloroplasts, which led to the concept of "artificial
energy conservation" (16,17,18; see 12,19 for reviews).

The studies of vectorial electron transport, conserving energy
in form of an electrochemical proton gradient, gained new stimulation by the model system of Hinkle (20,21), which has been investigated also by others (22-24). In this system the reduction of ferricyanide trapped in liposomes by an external, lipid insoluble reductant, like ascorbate, is mediated by a lipid soluble redox system
(see Fig.2).

Plastoquinone and ubiquinone posess long isoprenoid side chains
(see 1), which might direct their reactions in membranes. It has
been suggested that these side chains, which in their stretched
conformation are long enough to span the entire lipid bilayer, an-
chor the quinones by hydrophobic interaction and keep the redox
moiety from permeating through the membrane (25). Accordingly another component in addition to the quinone is required to bring
about proton translocation. We have employed Hinkle's liposome
system to test this argument. First results have been published (26).

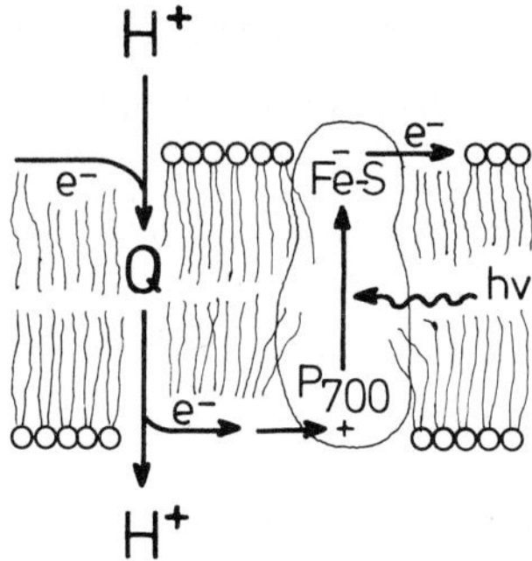

Fig. 1. Scheme for the combined vectorial function of quinoid
compounds and reaction center 1 of chloroplasts, constituting
a chemiosmotic redox loop. Q is reduced by electrons coming from
either photosystem 2 or from external ascorbate; electrons from
reaction center 1 are delivered to the NADP$^+$-reducing system or
to other final acceptors.

MATERIALS AND METHODS

Liposomes were prepared from soy bean lecithin (Sigma) as described previously (26). Quinones were incorporated by dissolving them together with lecithin in chloroform and evaporating the mixture to dryness under a stream of N_2 before adding 0.2M ferricyanide buffered at pH 8.0 with 20 mM Tricine-NaOH. The sonication mixture contained in 5 ml of ferricyanide solution 200 mg lecithin and 0.1-10 µmole of the quinone. Sonication was performed in a small beaker with a Branson sonicator using the microtip for 30 min at full output. By cooling the beaker in ice water the temperature was maintained below 25°C. Subsequently liposomes were separated from external ferricyanide on a small column of Sephadex G-50 equilibrated with 0.3M NaCl, 50 mM KCl and 20 mM Tricine-NaOH, pH 8.0. Lipid content of the liposome fractions was estimated by comparing the turbidity with that of not fractionated liposomes.

Reduction of ferricyanide at a low time resolution was measured as described (23,26) in a Zeiss-sepctrophotometer at 420 nm. Additions during measurements were performed via a plunger, furnished with a back-pushing spring, on top of the cuvette accessible for a microsyringe from outside the sample compartment. The minimal mixing time achieved was about 1 sec. Higher time resolution was obtained by employing a stop-flow apparatus, Durrum Model 110, modified by Dr. P. Bartholmes. The set up was equipped with an Osram-Xe-lamp, XBO 75 W, a monochromator from Schoeffel, GM 100-1, 1180 grates/mm, and a thermostating water bath. The optical path of the cuvette was 2 cm. The signal was fed to a Datalab transient recorder DL905 connected to a Tectronix 7623A oscilloscope and a recorder. A solution in one syringe containing the liposomes was mixed with an equal volume of a dithionite solution from the other syringe. The mixing time was 2.5 msec. The exact reaction mixture in the cuvettes are given in the legends to the figures.

Trimethyl-p-benzoquinone (TMQ) and plastoquinone (PQ) were kindly provided by Dr. A. Trebst, Bochum. The series of ubiquinones and a second lot of plastoquinone was a generous gift of Dr. Gloor and Dr. Weber, Hoffmann-La Roche, Basel. The compounds were tested for purity by TLC.

RESULTS AND DISCUSSION

<u>Our test system</u>: As shown in Fig.2, dithionite was used as
external reductant in our test system. Ascorbate did not give fast
reduction of ferricyanide trapped in quinone-containing liposomes,
presumably because its redox potential is not low enough to form
an appreciable amount of semiquinone, which is required for a rapid
turnover (27). However, dithionite itself is not totally impermeable
and can reduce internal ferricyanide in the absence of quinones, in
a concentration dependent manner (dashed arrow in Fig.2). This side
reaction is higher at lower pH, suggesting that undissociated $H_2S_2O_4$
is the permeating species. Accordingly H^+-translocation by dithionite
itself has been observed in our system (26).

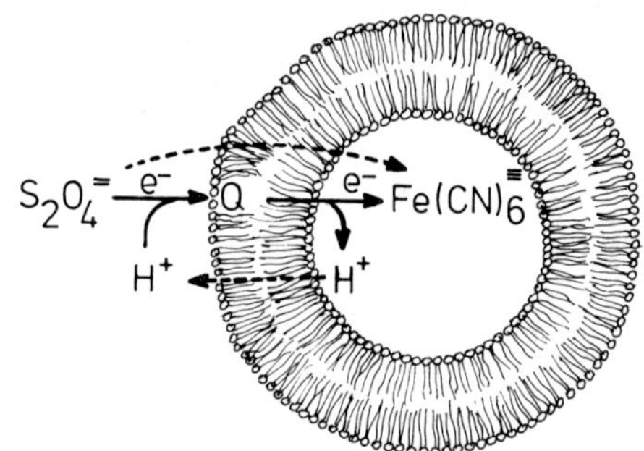

Fig. 2. Liposomal model system to test the permeability of quinones
after Hinkle (20,21).

<u>The dependence of the reaction rate on the isoprenoid side chain
of the quinone</u>: Because of the low time resolution with the Zeiss-
spectrophotometer, only at relatively low dithionite concentration
was it possible to observe catalysis of the reaction by incorporated quinones, as shown in Fig.3A. The course of the ferricyanide reduction exhibited a fast transient, which we can not
explain satisfactorily at present (26). It is best seen in the
control trace of Fig.3A, and does not reflect energization during
the course of the reaction, since uncoupling concentrations of
valinomycin plus nigericin were present. It might be due to some
externally bound ferricyanide.

As observed before (26) the benzoquinones UQ4, UQ9 and PQ with
long isoprenoid side chains were much more efficient catalysts than
their short chain homologues UQ1 and TMQ. UQ7 and UQ10 reacted
like UQ9 (not shown). The endpoint of each reduction was tested by
the addition of DAD.

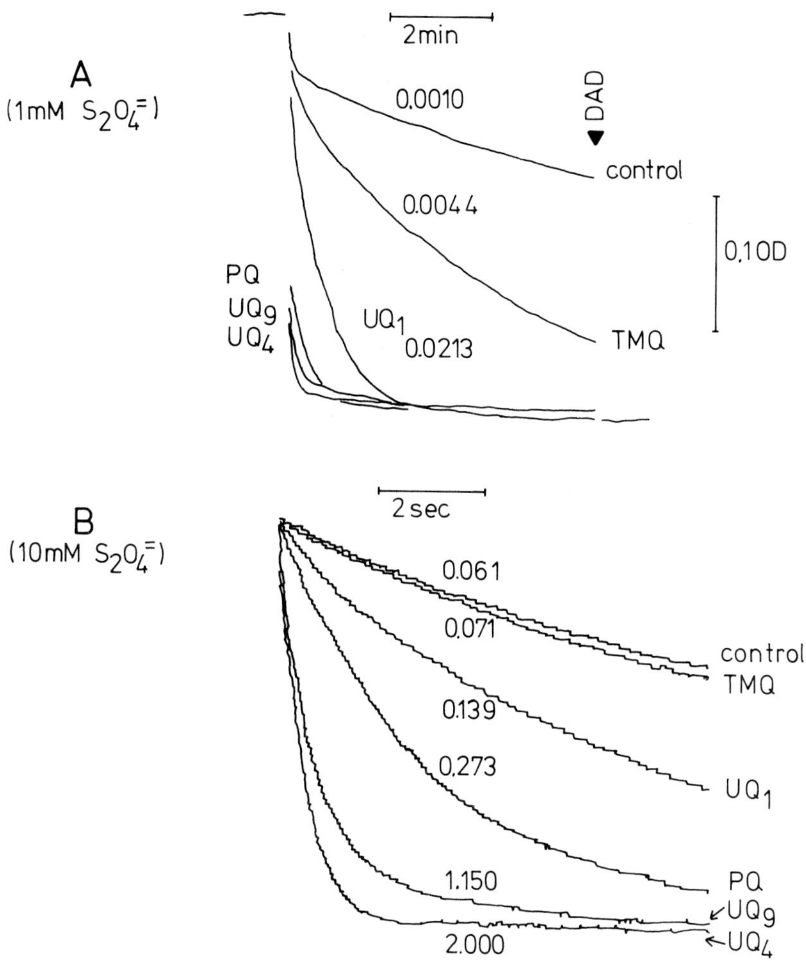

Fig. 3. Reduction of internal ferricyanide by external dithionite in quinone-containing liposomes. A) measurement in the Zeiss-spectro-photometer. B) measurement with the stop-flow apparatus. The abbreviations are: TMQ for trimethyl-p-benzoquinone, PQ for plastoquinone. UQ1, UQ4 and UQ9 for ubiquinones with 1,4 and 9 isoprene units in the side chain, respectively, DAD for 2,3,5,6-tetramethyl-p-phenylenediamine. The numbers denote pseudofirst order rate constants. The reaction mixture in the cuvette for both measurements was 0.2M NaCl, 50 mM KCl, 20mM Tricine-NaOH, pH 8.0 and 1 μg/ml of each valinomycin and nigericin. The reaction mixture was flushed with N_2 to remove most of O_2. For the measurements in A and B the amount of liposomes present corresponded to 4 and 2 mg/ml lecithin, respectively. Except for the control, 7 nmoles quinone/mg of lecithin was included. The final concentration of dithionite added was 1 mM for A and 10 mM for B, as indicated. DAD was added to a concentration of 1 μM at the end of the measurement in A.

Dithionite at 1mM concentration is not saturating, but higher concentrations made the control reaction so fast that catalysis by the quinones was difficult to observe in the Zeiss-spectrophotometer. Therefore we changed to a stop-flow apparatus for higher time resolution. The result is given in Fig.3B., for 10 mM dithionite - a concentration which is still not fully saturating (28). Under these conditions TMQ is hardly more efficient than dithionite alone. The course of the reaction in presence of the benzoquinones with long isoprenoid side chains is now resolved. The efficiency of catalysis increases in the series UQ1 - PQ - UQ9 - UQ4. Obviously the side chain facilitates the translocation of reducing equivalents. By transformation of the traces into a semilogarithmic scale pseudo-1st order rate constants were obtained, which are given as the numbers in Fig.3 (the excess of external dithionite over internal ferricyanide was 6.7 in Fig.3A and 133 in Fig. 3B).

<u>The turnover of plastoquinone and ubiquinone in our model system in comparison to their turnover in photosynthetic and respiratory electron transport, respectively</u>: The amount of quinone present in the liposomes used in the experiments for Fig.3, was 7 nmoles/mg lecithin. This comes close to the amount of plastoquinone present in chloroplasts (6,7), and also to the amount of ubiquinone present in mitochondria (29). The ratio of incorporated quinone to trapped ferricyanide was about 40, the quinones thus acting as catalysts with actual turnover. The turnover number (TON) of the quinone can be estimated from the pseudo-1st order rate constant k (sec^{-1}) multiplied by this ratio. If we do this, we arrive at 11 sec^{-1} for TON_{PQ}, and at 46 sec^{-1} for TON_{UQ9}, which are values for single electrons.

A reasonable rate of the Hill reaction in chloroplasts under uncoupled conditions would be 400 µequivalents/mg chlorophyll and hr. The amount of PQ present in chloroplasts is about 0.1 per chlorophyl on a molar basis (30,6,7). From these two values a turnover for PQ in vivo of 1.1 sec^{-1} can be derived. The turnover might actually be 3.3 sec^{-1}, if one consideres that only one third of the PQ is involved in electron transport (31,6,7).

A good rate of succinate oxidation in mitochondria would be 0.1 µmole O_2 consumed/mg protein and min. The content of mitochondria in ubiquinone is 1 - 4 nmoles/mg protein (29). A similar calculation as above gives a turnover of 1.7 - 6.7 sec^{-1} for UQ in vivo.

In both cases the turnover in our system exceeds the value in vivo, showing that our model system might be physiologically relevant.

<u>Effect of dissipating the electrochemical H^+-potential formed during the reaction</u>: Reduction of internal ferricyanide by external dithionite is stimulated by valinomycin plus nigericin in the presence of potassium, with or without catalysis of quinones (26). In the presence of quinones nigericin alone gave the same stimulation as the combination with valinomycin, while valinomycin alone had only a small stimulatory effect (experiments not shown here). This agrees with Hinkle's result studying unsubstituted benzoquinone, and suggests that the translocation of electrons and protons via the quinone proceeds predominantly in an electroneutral way.

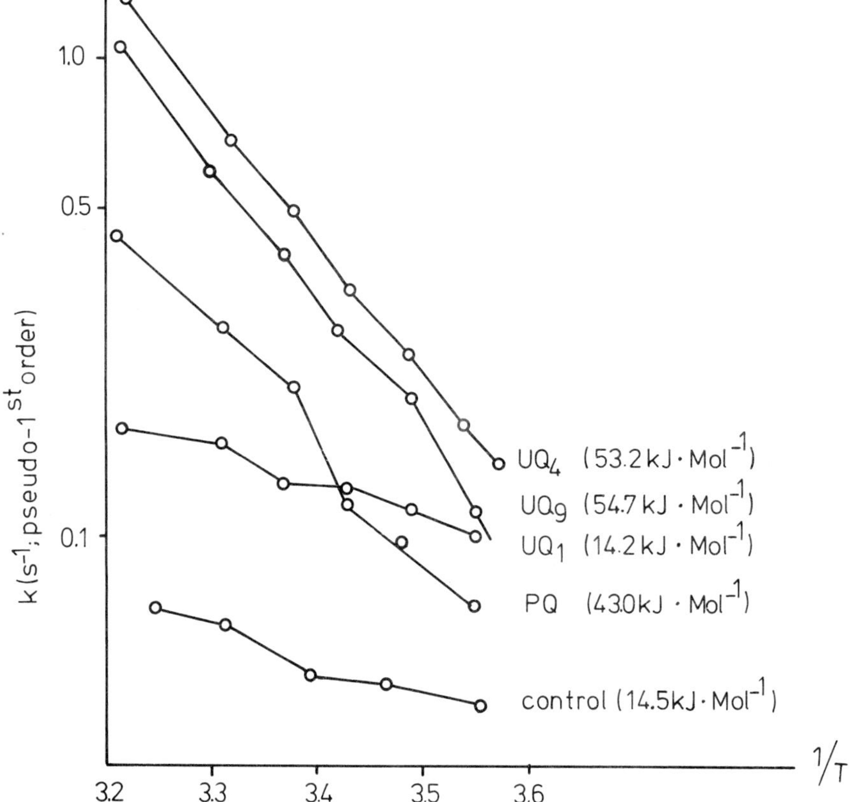

Fig. 4. Arrhenius plots of the reaction. The symbols are the same as in Fig.3. The assay was performed in the stop-flow apparatus as described under METHODS and in the legend to Fig.3B. The values in brackets give an estimate of the activation energies, when a linear T-dependence is assumed between highest and lowest T.

<u>T-dependence of the reaction</u>: Fig.4 shows Arrhenius-plots of the reaction in our system under conditions which resemble those for Fig.3, besides the variation of T. The case with TMQ is not included in the figure because it practically yielded the same values as the control. A clear difference in T-dependence of the control reaction and the reaction catalyzed by UQ1 to the one of the reaction with PQ, UQ4 and UQ9 can be observed, the latter giving the higher activation energies. Since we do not know yet what step in the reaction sequence from dithionite to internal ferricyanide is rate limiting, it is not possible to infer to the reaction mechanism from this result. It points out, however, that the isoprenoid side chain might have a profound influence on the reaction mechanism.

<u>Dependence of the reaction on the amount of quinone incorporated</u>: We have raised the content of the quinones up to 140 nmoles/mg lecithin, which corresponds to a ratio of quinone/lipid of about 10, on a molar basis. This preparation still was vesicular with trapped ferricyanide, but the leak of ferricyanide, as well as the leak of protons during ferricyanide reduction, measured by the fluorescence changes of 9-amino-acridine (26), was much increased (results not shown here). The rate of ferricyanide reduction by external dithionite was very fast in this preparation, but was superimposed by absorption changes reflecting reduction of the quinone.

At lower quinone content of the liposomes another interesting differential effect between long and short chain compounds has been observed. PQ and UQ4-10 exhibited reduction kinetics of trapped ferricyanide of at least two phases, while UQ1 and TMQ yielded monophasic kinetics besides the rapid initial transient which is also found in the control and has been mentioned above. Semilogarythmic plots for the reaction with UQ9 compared to the one with UQ1 in Fig.5 show this phenomenon. A reasonable explanation for it might be that benzoquinones with long isoprenoid side chains occupy the lipid vesicles in a heterogeneous manner. At low quinone concentration some vesicles receive most of the quinone and the rest only very little. It is possible that this phenomenon reflects the tendency of domaine formation within the lipid matrix which is governed by the isoprenoid side chain, as depicted in Fig.6. If this were so, domaine formation would be the reason for the differences in reactivity between benzoquinones with long and short side chains.

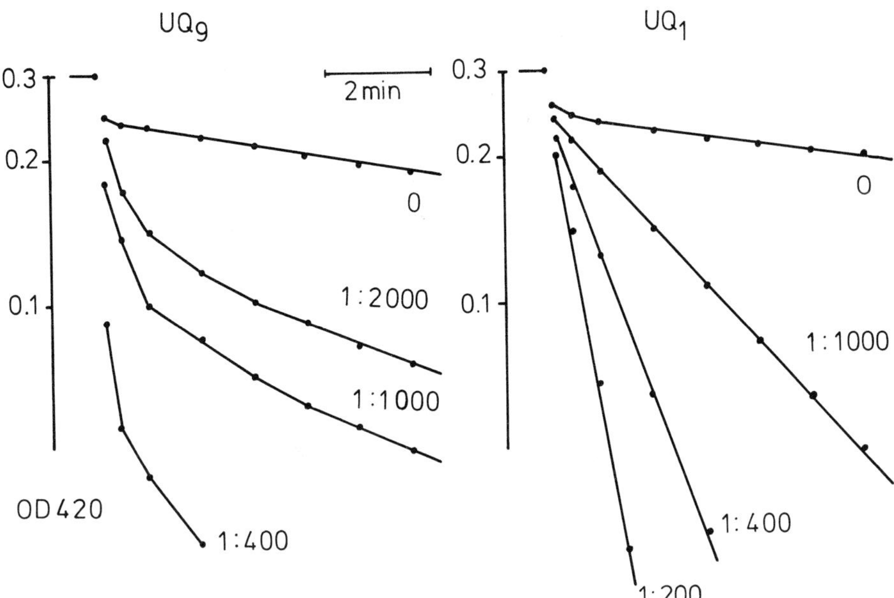

Fig. 5. Dependence of the reaction on the amount of incorporated quinone. Ferricyanide reduction was monitored in a Zeiss-spectro-photometer as described under METHODS and in the legend for Fig.3A. The recorded traces have been transformed into semilogarithmic scale. The ratios denote the estimated quinone/lipid ratio on a molar basis in the liposome preparation, which correspond to 0.7 (1:2000), 1.4 (1:1000) and 3.5 nmoles/mg lecithin (1:400).

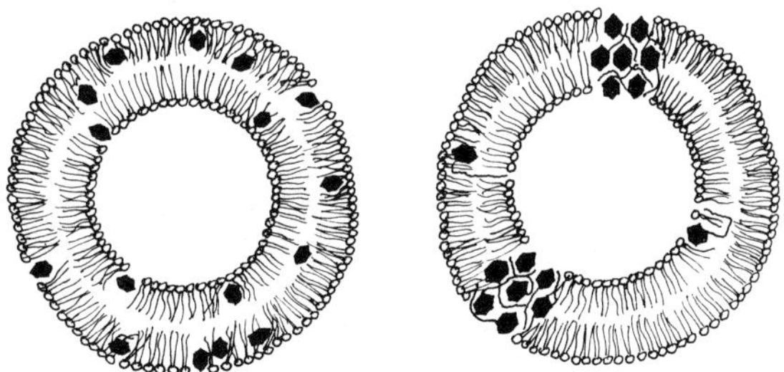

Fig. 6. Model for the occupation of liposomes by quinones with short and long side chains.

CONCLUSION AND PROSPECT

The results presented here demonstrate that plasto- and ubiquinone with long isoprenoid side chains are efficient translocators
of electrons and protons in our liposome system, in contrast to their
homologues lacking the long side chain. It is consequently reasonable to assume that they act as proton translocators per se in
vectorial electron transport of photosynthesis and respiration,
representing one mechanistic possibility to couple electron flow
to proton transport through membranes.

The influence of the isoprenoid side chain on quinone reactivity
will be further studied. Preliminary spectroscopic measurements
suggest that the semiquinone radical is transiently formed in
appreciable amounts, and that its transient concentration is higher
in the case of quinones with long isoprenoid side chains. Whether
this might result from clustering in the suggested quinone domaines,
or from an intramolecular effect of the side chain, possibly via
its isolated double bonds, remains to be studied. In the later
context it should be noted that the side chain is flexible enough
to completely wrap the quinone moiety.

ABBREVIATIONS

PQ: plastoquinone; UQ1-10: ubiquinone with 1 to 10 isoprene units
in the side chain; TMQ: trimethyl-p-benzoquinone; DAD: 2,3,5,6-
tetramethyl-p-phenylenediamine.

ACKNOWLEDGEMENTS

I am indebted to Dr. U. Gloor and Dr. F. Weber, Hofmann-La Roche,
Basel, for the gift of the ubiquinones and of plastoquinone, to
Dr. A. Trebst, Bochum, for the gift of trimethylquinone and plasto-
quinone and for continuous exchange of ideas, to Dr. R. Jaenicke
and Dr. P. Bartholmes for providing access to the stop flow appara-
tus, to Dr. P. Bartholmes for his guide to run the apparatus, to
Miss M. Gantner for perfect technical assistence and to the Deutsche
Forschungsgemeinschaft for general support.

REFERENCES

1) Hatefi, Y. et al., Biochim. Biophys. Acta 31(1959)490-501

2) Crane, F. L., Biochemistry 1(1962)510-517

3) Morton, R. A., 1965 Biochemistry of Quinones, Acad. Press

4) Klingenberg, M. et al., Nature 194(1962)379-380

5) Bishop, N. I., Proc. Nat. Acad. Sci. US 45(1959)1696-1702

6) Amesz, J., Biochim. Biophys. Acta 301(1973)35-51

7) Amesz, J. (1977) in Encyclopedia of Planta Physiology, New
Series, (Pirson, A. and Zimmermann, M. H. eds.) Vol.5, p.
238-245, Springer-Verlag, Berlin-Heidelberg-New York.

8) Mitchell, P., Nature 191(1961)144-148

9) Mitchell, P., Biol. Rev. Cambridge Phil. Soc. 41(1966)445-502

10) Papa, S., Biochim. Biophys. Acta 456(1976)39-84

11) Lundegardh, H., Arch. Bot. 32A(1945)12,1

12) Hauska, G. and Trebst, A. (1977) in Current Topics in Bioenergetics (Sanadi, D. R. ed.) Vol 7, p 151-220, Acad. Press

13) Oesterhelt, D., Angew. Chem., Int. Ed. 15(1976)17-24

14) Mitchell, P., FEBS-lett. 59(1975)137-139

15) Baccarini-Melandri, A. and Melandri, B. A., FEBS-lett., submitted

16) Hauska, G. et al., Biochim. Biophys. Acta 305(1973)632-641

17) Hauska, G. et al., Z. Naturforsch. 30c(1975)37-45

18) Hauska, G. et al., FEBS-lett. 73(1977)257-262

19) Trebst, A., Ann. Rev. Plant Physiol. 25(1974)423-458

20) Hinkle, P., Biochem. Biophys. Res. Comm. 41(1970)1375-1381

21) Hinkle, P., Fed. Proc. 32(1973)1988-1992

22) Deamer, D. W. et al., Biochim. Biophys. Acta 274(1972)323-335

23) Hauska, G. and Prince, R. C., FEBS-lett. 41(1974)35-39

24) Hill, R. et al., New Phytologist 77(1976)1-9

25) Robertson, R.N. and Boardman, N. K., FEBS-lett. 60(1975)1-6

26) Hauska, G., FEBS-lett., submitted

27) Diebler, H. et al., Z. Naturforsch. 16b(1961)629-637

28) Hauska, G., in preparation

29) Pumphrey, A. M. and Redfearn, E. R., Biochem. J. 76(1960)61-64

30) Crane, F. L. (1968) in: Biological Oxidations (Singer, T.P. ed.)
New York: Interscience, p 533-580

31) Stiehl, H. H. and Witt, H. T., Z. Naturforsch. 24b(1969)1588-
1598

Bioenergetics of Membranes. L. Packer et al. ed.

INTERACTIONS AND MOBILITY OF UBIQUINONE IN THE INNER MITOCHONDRIAL MEMBRANE

G. Lenaz, S. Mascarello, L. Landi, L. Cabrini, P. Pasquali,
G. Parenti-Castelli, A.M. Sechi and E. Bertoli

Istituto di Chimica Biologica, University of Bologna and
Istituto di Biochimica, University of Ancona (Italy)

INTRODUCTION

We have approached the problem of the role of ubiquinone (UQ) in the mitochondrial respiratory chain by testing the different UQ homologs as electron carriers. We have previously observed that NADH oxidation in pentane-extracted mitochondria is only restored by the UQ homologs having relatively long isoprenoid side chains, while all homologs have high activity in restoring succinate oxidation (1-3). Moreover we have found that the low UQ homologs act as competitive inhibitors of NADH oxidation but not of succinate oxidation in non-extracted mitochondrial membranes (3-5).

These effects might be due to either of the following reasons.

(a) The low homologs are reduced by NADH less efficiently than the higher homologs, due to a polarity or steric hindrance of their interaction with the specific site in NADH dehydrogenase.

(b) The low homologs are reduced by NADH, but their reoxidation by the following redox acceptor in the chain is impaired. This possibility involves the puzzling consideration that reoxidation of the quinone is impaired only when it has been previously reduced by NADH but not by succinate.

UBIQUINONE-3 AS ELECTRON ACCEPTOR

Ubiquinone, and in particular UQ-1, has been used as an electron acceptor for Site I of oxidative phosphorylation (6), and UQ-3 can be used as well (3). We have compared the rate of reduction of UQ-1 and UQ-3 by NADH and succinate in submitochondrial particles (SMP): although UQ-1 is a better acceptor than UQ-3 from both NADH and succinate (Table 1), the rate of reduction of exogenous UQ-3 is high

from both substrates; actually the rate of UQ-3 reduction is higher
from NADH than from succinate as electron donor.

TABLE 1

NADH-UQ REDUCTASE AND SUCCINATE UQ-REDUCTASE IN SMP

Acceptor	Oxidation rate (nmoles·min^{-1} mg^{-1})	
	NADH	Succinate
UQ-1 (300 µM)	624	303
UQ-3 (300 µM)	241	162

In order to exclude a possible interference of the endogenous
ubiquinone, we have repeated a similar experiment in pentane-extracted mitochondria. Fig. 1 shows that the rate of UQ-3 reduction is

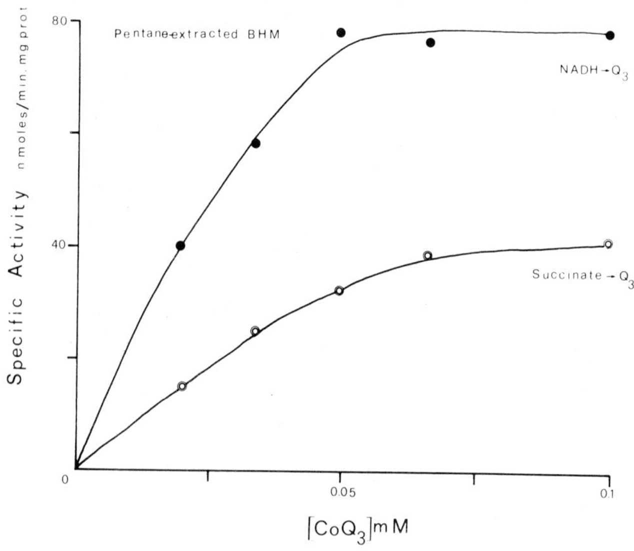

Fig. 1. NADH UQ-3 reductase and succinate UQ-3 reductase in mito-
chondria-extracted with pentane according to Szarkowska (19). NADH
UQ reductase activity was assayed at room temperature by following
the decrease in absorbance of NADH at 340 nm; the activity was rotenone sensitive. Succinate UQ reductase was assayed by following
the decrease in absorbance at 275 nm due to UQ reduction using an
extinction coefficient Δ_{ox-red} of 8.8×10^{-3} M^{-1} cm^{-1}. The assay
system contained in a total volume of 3 ml: sucrose, 750 µmoles; K
phosphate, pH 6.8, 10 µmoles; antimycin A, 1 µg; protein, 0.3 mg;
NADH, 0.4 µmoles or succinate, 50 µmoles.

higher with NADH than with succinate as the electron donor, in agreement with the data in SMP. An apparent K_M of 43 µM has been calculated for UQ-3 in NADH-UQ reductase and of 74 µM for UQ-3 in succin-
ate-UQ reductase.

It must therefore be concluded that UQ-3 can be reduced by NADH,
and therefore UQ-3 reduction cannot be the limiting factor for explaining the low NADH oxidase activity in presence of UQ-3. These
experiments allow to conclude that the second hypothesis postulated
in the Introduction is correct.

The fact that UQ-3 reduced by NADH cannot be reoxidized by the
cytochrome chain, whereas UQ-3 reduced by succinate can, is not of
obvious explanation and may involve a sidedness of UQ reduction and
reoxidation in the membrane.

TRANSMEMBRANE MOBILITY OF UBIQUINONE HOMOLOGS

We have started from the working hypothesis that UQ, when reduced
by NADH, must cross the inner mitochondrial membrane to become reoxidized, whereas UQ, when reduced by succinate, does not have to
cross the membrane; the hypothesis also requires that the low UQ
homologs, once reduced, are unable to cross the lipid bilayer, whereas the long-chain homologs, such as the natural UQ-10, can cross the
bilayer easily. The decreased water solubility of ubiquinones having increasingly long chains are in accordance with this postulation.

In order to test the validity of this interpretation, we have
investigated whether reduced ubiquinones having different lengths of
their isoprenoid side chains have also different rates of transmembrane diffusion across lipid bilayers.

We have taken advantage of non-enzymatic reduction of cytochrome
c by reduced quinones to investigate whether cytochrome c trapped
inside phospholipid vesicles can be reduced by ubiquinols added to
the external medium. The experiments have been accomplished at high
pH (9.5) in order to keep the higher quinones in solution.

Fig. 2 shows the rates of cytochrome c reduction by different
ubiquinols. When the quinols are added in absence of phospholipids,
cytochrome c is reduced at the same rate by all ubiquinols tested;

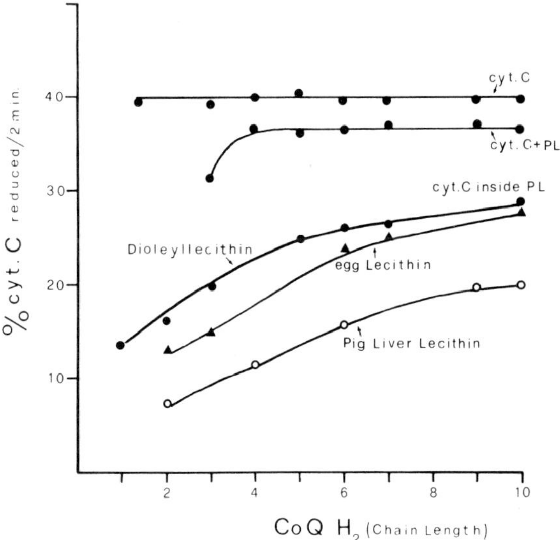

Fig. 2. Cytochrome $\underline{c}$ reduction by ubiquinols having different isoprenoid side chains. Ubiquinones were reduced according to Rieske
(20). Cytochrome $\underline{c}$ was trapped inside phospholipid vesicles according to Kimelberg and Papahadjopoulos (7). Reduction of cytochrome $\underline{c}$
was followed at 516 nm using an extinction coefficient Δ_{red-ox} of
40.3×10^{-3} $M^{-1}cm^{-1}$. The assay system contained in a total amount of
1 ml: Tris acetate 20 mM, pH 9.5, and EDTA 1 mM; phospholipids when
added were 0.3-1 μmole per assay. The final concentration of cytochrome $\underline{c}$ was 9.9 μM. The reaction was started by addition of ubiquinols, at a final concentration of 70 μM.

when the assays are accomplished in presence of phospholipids, there
are no significant changes, except for a somewhat lower rate of cytochrome $\underline{c}$ reduction by ubiquinol-3. When, however, the acceptor cytochrome $\underline{c}$ is trapped inside the phospholipid vesicles, its reduction
rate becomes strongly dependent on the length of the isoprenoid side
chain of the ubiquinol, and there is a progressive increase in rate
with increasing chain length. The half-times of cytochrome $\underline{c}$ reduction (Fig. 3) clearly express the same phenomenon: long chain quinols
have half times in the range of few seconds, contrary to short chain
quinols that have half-times of minutes.

Since cytochrome $\underline{c}$ is impermeable to lipid vesicles (7) as well
as to the inner mitochondrial membrane (8), the rate of its reduction in our system may only depend on the rate of diffusion of the

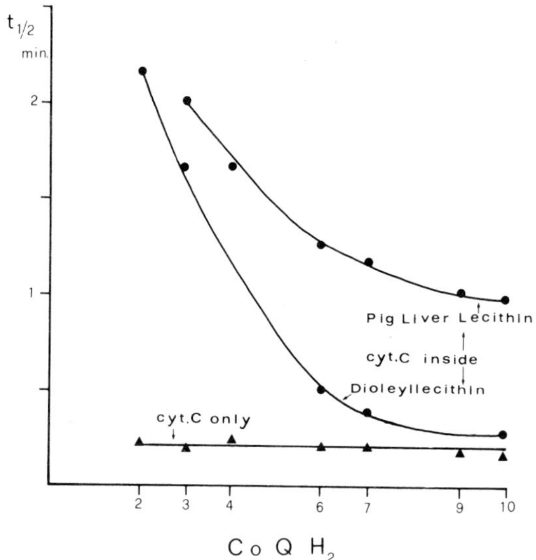

Fig. 3

Half-times of cyto-
chrome c reduction
by ubiquinols having
different isoprenoid
side chains.

quinols across the lipid bilayer, since the rate of non-enzymatic reduction of cytochrome c by ubiquinols is not limiting under our experimental conditions (cf. Fig. 2).

The low rate of transversal diffusion of the low ubiquinols may be due to the rigid disposition of these molecules in the lipid bilayer; the length of UQ-3 is very close to the width of a half-bilayer, and it is conceivable that the quinone, specially in its reduced form, becomes inserted as an amphipathic molecule together with the phospholipids in one lipid monolayer. Amphipathic molecules such as phospholipids have high rates of lateral diffusion (9) when in a fluid state, but have very little tendency to dissolve in the lipids and to cross the bilayer; the rate of "flip-flop" of phospholipids is extremely low (10) and it is likely that UQ-3 has also little tendency to flip-flop. Higher quinones, on the other hand, are more hydrophobic even in their reduced forms, and might well dissolve in the core of the lipid bilayer and diffuse across the membrane. In agreement with this interpretation is our finding that low ubiquinones, particularly when reduced, immobilize phospholipid bilayers (11, 12); an ordering effect of reduced quinones on lipids has also been described by others (13, 14).

194

EFFECT OF TEMPERATURE

If our interpretation is correct, there should be a marked effect of temperature on the kinetic properties of UQ-3 in mitochondrial membranes. In particular, if UQ-3 reduced by NADH on one side of the membrane must be reoxidized on the opposite side, reoxidation should be favored at higher temperatures, since the transversal mobility of the low quinol across the bilayer would increase by enhancing the motion of the surrounding lipid molecules. Fig. 4 shows that the inhibition of NADH oxidation by UQ-3 is progressively weaker by increasing the temperature from 10° to 37°C. It may be concluded that the increase in mobility allows the quinol to function as an efficient electron carrier across the inner mitochondrial membrane, so that it no longer competes with the natural quinone (UQ-10) for reoxidation.

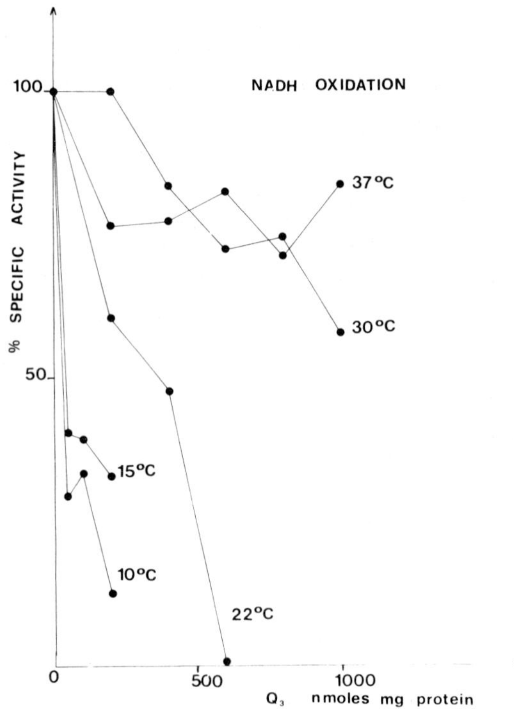

Fig. 4. Effect of temperature on the inhibition of NADH oxidation by UQ-3. NADH oxidation was assayed with an oxygen electrode (ref.3).

UBIQUINOL OXIDASE IN MITOCHONDRIA AND SUBMITOCHONDRIAL PARTICLES

Although the experiments described so far are strongly in favor of the necessity that UQ, once reduced by NADH, must cross the membrane to be reoxidized, they do not allow to conclude which is the direction of movement of ubiquinol in the inner membrane. To approach the solution of this problem, we have investigated electron transfer using exogenous ubiquinols as substrates of ubiquinol oxidase in both mitochondria and submitochondrial particles (SMP), which have an opposite polarity or sidedness: as is well known (15), SMP are inside-out and face the medium with the side corresponding to the inner or "matrix" side (M-side) of intact mitochondria. The initial velocities of ubiquinol (UQ-3, UQ-4, and UQ-5) oxidation in mitochondria and SMP are plotted in Fig. 5. The ubiquinol oxidase activities are 70-75% antimycin-sensitive. The graph shows that ubiquinol-3,

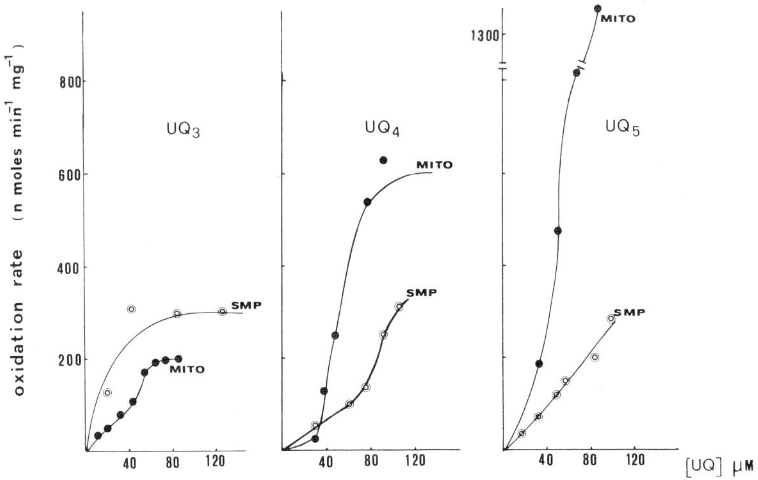

Fig. 5. Ubiquinol oxidase activity in mitochondria and SMP using ubiquinols 3 to 5 as substrates. Ubiquinol oxidase was assayed at room temperature by following the increase in absorption at 275 nm upon ubiquinol oxidation (cf. Fig. 1). The assay system contained in a total volume of 3 ml: sucrose 600 µmoles; KCl, 90 µmoles; Hepes, pH 6.8, 15 µmoles; Na malonate, 15 µmoles; rotenone, 2 µg; antimycin A, when added, 3 µg; mitochondria, 0.4 mg protein, or SMP, 0.345 mg protein. The reaction was started by addition of ubiquinols. Mitochondria for these experiments were kept frozen at -80°C.

that is more polar and water-soluble than the others, is oxidized at higher rate in SMP than in mitochondria; on the other hand, ubiquinol-4 and ubiquinol-5 (that have much lower water solubility) are oxidized at increasingly higher rates in mitochondria, with a progression resembling the increase in rate of transversal diffusion across phospholipid vesicles shown in Fig. 2.

The rates of ubiquinol-4 and ubiquinol-5 oxidation in SMP are somewhat lower than that of ubiquinol-3 and also lower than the corresponding rates in mitochondria.

The results are favorable to the existence of a site for ubiquinol oxidation which is more accessible to hydrophilic quinols from the matrix side, and to hydrophobic quinols from the outer side (C-side). A plausible interpretation is that the site of oxidation is located in Complex III near the M-side of the membrane, and is reached more easily by hydrophilic quinols (UQ-3) added to the aqueous medium at the M-side, or by hydrophobic quinols (UQ-4 and UQ-5) added to the C-side and moving through the lipid bilayer. The direction of movement of ubiquinol in the membrane would appear therefore to be from the C-side to the M-side.

This interpretation does not explain why the maximal rates of oxidation found are not those of ubiquinols, irrespective of their chain length, in SMP; it can be however postulated that the site is reached more easily when a suitable (hydrophobic) ubiquinol crosses the membrane than when any ubiquinol is added directly from the water medium near the oxidation site; perhaps the higher protein content of the inner monolayer of the inner membrane (16) prevents fast solubilization of ubiquinols in the lipids to reach the site of oxidation in an effective way.

Although this interpretation is satisfactory to explain the much higher rate of oxidation of ubiquinol-3 in SMP than in mitochondria, other interpretations are also possible; for example the site of oxidation may be a hydrophobic site near the C-side of the membrane, but ubiquinol-3 is capable of reducing a non-physiological component near the M-side; this interpretation, although more in line with current views of the sidedness of electron transfer, such as the Q-cycle

of Mitchell (17), is more difficult to accept, also in view of the
antimycin sensitivity of ubiquinol-3 oxidation in SMP. Fig. 6 reports the possible relationship of ubiquinone with the respiratory
chain in the inner membrane according to our first hypothesis.

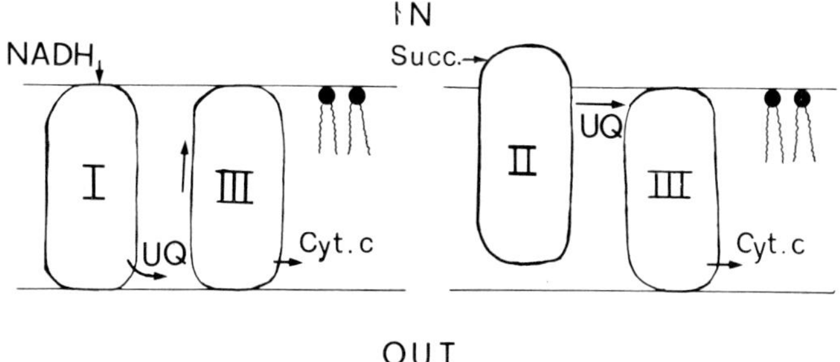

Fig. 6. Scheme of the possible relationships of ubiquinone with the
respiratory chain in the inner mitochondrial membrane.

Our explanation requires that reduced UQ, by moving hydrogen
atoms inwards, may translocate protons _inside_ the mitochondrion,
which is opposite to the postulation of the chemio-osmotic hypothesis (17); since electron transfer is known to move protons _to the_
exterior of the mitochondrion (18), in order to carry on our hypothesis we must conclude that UQ does not contribute to net proton translocation in a positive way.

ACKNOWLEDGEMENTS

This work has been supported in part by a grant from the CNR,
Rome, Italy. Samples of ubiquinone homologs were kind gifts of
Hoffmann-La Roche, Basel, Switzerland.

REFERENCES

1. Lenaz, G., Daves, D.G. and Folkers, K. (1968) Arch. Biochem.
    Biophys. 123, 539-550.
2. Lenaz, G., Castelli, A., Littarru, G.P., Bertoli, E. and Folkers,
    K. (1971) Arch. Biochem. Biophys. 142, 407-416.
3. Lenaz, G., Pasquali, P., Bertoli, E., Parenti-Castelli, G. and
    Folkers, K. (1975) Arch. Biochem. Biophys. 169, 217-226.

4.   Lenaz, G., Pasquali, P. and Bertoli, E. (1975) in "Electron Transfer Chains and Oxidative Phosphorylation" (E. Quagliariello, S. Papa, F. Palmieri, E.C. Slater and N. Siliprandi eds.), North-Holland, Amsterdam, pp. 251-256.

5.   Castelli, A., Bertoli, E., Littarru, G.P., Lenaz, G. and Folkers K. (1971) Biochem. Biophys. Res. Commun. 42, 806-812.

6.   Schatz, G. and Racker, E. (1966) J. Biol. Chem. 241, 1429-1438.

7.   Kimelberg, H.K. and Papahadjopoulos, D. (1971) J. Biol. Chem. 246, 1142-1148.

8.   Lenaz, G. and MacLennan, D.H. (1966) J. Biol. Chem. 241, 5260-5265.

9.   Lenaz, G. (1977) in "Membrane Proteins and their Interaction with Lipids" (R.A. Capaldi ed.), Marcel Dekker, New York, pp. 131-232.

10. Kornberg, R.D. and McConnell, H.M. (1971) Proc. Natl. Acad. Sci. USA 68, 2284-2288.

11. Spisni, A., Masotti, L., Lenaz, G., Bertoli, E., Pedulli, G. and Zannoni, C. (1977) Biochem. Biophys. Res. Commun., submitted.

12. Spisni, A., Lenaz, G. and Masotti, L. (1977) Symposium "Membrane Bioenergetics", Spetsai, 10-15 July 1977.

13. Cain, J., Santillan, G. and Blasie, J.K. (1972) in "Membrane Research" (C.F. Fox ed.), Academic Press, New York, pp. 3-14.

14. Chance, B., Erecinska, M. and Radda, G. (1975) Eur. J. Biochem. 54, 521-529.

15. Lee, C.P. and Ernster, L. (1966) Symposium "Regulation of Metabolic Processes in Mitochondria", BBA Library, Vol. 7, pp. 218-234.

16. Packer, L., Mehard, C.W., Meissner, G., Zahler, W.L. and Fleischer, S. (1974) Biochim. Biophys. Acta 363, 159-181.

17. Mitchell, P. (1971) in "Electron Transfer Chains and Oxidative Phosphorylation" (E. Quagliariello, S. Papa, F. Palmieri, E.C. Slater and N. Siliprandi eds.), North-Holland, Amsterdam, pp. 305-316.

18. Papa, S. (1976) Biochim. Biophys. Acta 456, 39-84.

19. Szarkowska, L. (1966) Arch. Biochem. Biophys. 113, 519-525.

20. Rieske, J.S. (1967) Methods Enzymol. 10, 239-245.

THE ROLE OF UBIQUINONE-10 IN THE b-c_2 SEGMENT OF THE CYCLIC PHOTOSYNTHETIC CHAIN OF BACTERIAL CHROMATOPHORES

Assunta Baccarini Melandri and Bruno A. Melandri
Institute of Botany - University of Bologna
via Irnerio 42 - Bologna
Italy

SUMMARY

Extraction of Ubiquinone-10 (UQ-10) with isooctane from chromatophores of Rhodopseudomonas capsulata, strain AIa pho$^+$, leads to drastic reduction of photophosphorylation, which is fully reconstituted upon reincorporation of pure UQ-10. Light induced redox changes of cytochromes $\underline{b}$ and $\underline{c}$ observed with native, extracted and reconstituted vesicles support the involvment of UQ-10 in the b-c_2 segment of the photosynthetic electron transport chain.

INTRODUCTION

In chromatophores from this facultative photosynthetic bacterium ATP synthesis is coupled to light dependent cyclic electron flow. It is generally accepted[1] that UQ acts as a secondary electron acceptor on the reducing side of cytochrome $\underline{b}$ (E'_o = +60 mV). Electrons are then transferred to cytochrome c_2 (E'_o = +340 mV) which then reduces the photochemical reaction center, P870. The presence of an additional component located between cyt.$\underline{b}$ and cyt.c_2 was suggested by Evans and Crofts[2] on kinetic and thermodynamic grounds. Crofts et al[3] proposed recently that it is a redox couple of quinone (QH$^{\cdot}$/QH$_2$) functional in an open version of the Q cycle[4].
Evidence supporting the involvment of quinone also on the oxidizing side of cyt.$\underline{b}$ is presented herein.

MATERIALS AND METHODS

A carotenoid less mutant of Rps.capsulata, strain AIa pho$^+$, was used and chromatophores were prepared as previously described[5]. Extraction of UQ-10 from lyophilized chromatophores and reconstitution of extracted vesicles with UQ-10 (Sigma, St.Louis) is detailed elsewhere[6]. Photophosphorylation activity was measured according to the procedure described in ref.5.

Redox changes of cytochrome were monitored with a dual wavelenght spectrophotometer operated at 200 cycles per second; the RC constant was 2.5 msec. Optical changes were stored in a transient recorder (Datalab DL 905, Mitcham; U.K.) and reproduced potentiometrically.

RESULTS AND DISCUSSION

When lyophilized chromatophores are repetitively extracted with isooctane, the amount of UQ-10 released varies from 0.8 to 1.4 µMoles per µMole of Bacteriochlorophyll. The pool of UQ-10 is 15-25 times greater than the total dithionite reducible cytochrome $\underline{b}$. Based on a photosynthetic unit equal to 100 Môles of Bacterioclorophyll per re-action center, the ratio of about 100 molecules of UQ-10 per reaction center can be determined. This value is in agreement with recent ob-servations by Bowyer and Crofts (personal comunication).

Repetitive extractions of UQ-10 reduce drastically the rate of photo-phosphorylation. Reincorporation of pure UQ-10 into the extracted particles restores ATP synthesis to the initial level of lyophilized chromatophores (Table I).

TABLE I

EFFECT OF UQ-DEPLETION AND UQ-RECONSTITUTION ON PHOTOPHOSPHORYLATION IN CHROMATOPHORES FROM $\underline{RPS.CAPSULATA}$, STRAIN AIA PHO$^+$

Additions	lyophilized	extracted	reconstitued
none	296	14	286
Antimycin, 4 µM	0	0	0
Antimycin, 4 µM and PMS, 100 uM	273	n.d.	432

It is important to note that Antimycin A, an inhibitor of electron flow between $cyt.b_{60}$ and $cyt.c_{340}$, inhibits completely photophospho-rylation also in reconstituted particles. Thus,in these latter vesi-cles, the native light dependent cyclic electron flow has been re-established. The light induced redox changes of cytochrome $\underline{b}$ and of

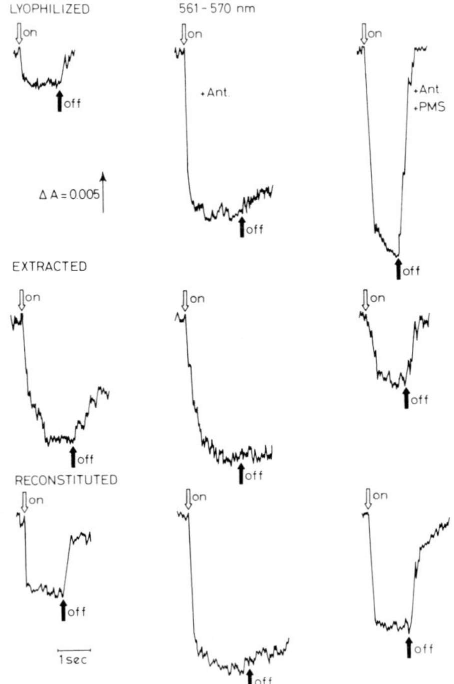

Fig.1. Light induced redox changes of cytochrome b in lyophilized, extracted and reconstituted chromatophores.

The assay mixtures contained, in a final volume of 2 ml: glycylglycine, pH 8.0, 128 µMoles; $MgCl_2$, 20 µMoles; Succinate, 0.4 µMole; KCl, 100 µMoles and membranes corresponding to 130 µg. of Bacteriochlorophyll (lyophilized chromatophores)or 100 µg of bacteriochlorophyll (extracted or reconstituted particles).
The assays were performed under strict anaerobic conditions.

cytochrome c are shown in Figs. 1 and 2.
The most important difference in the redox changes of cyt.b observed upon extraction of Ubiquinone are: 1) The rate of cyt.b photoreduction is markedly decreased, as expected from the established role of UQ-10 on the reducing side of cyt.b; 2) The rate of reoxidation of cyt.b in the dark is even more significantly decreased (the half time of reoxidation is about 10 times greater than in lyophilized particles; 3) Addition of Antimycin A to extracted particles does not induce any substantial increase in the light dependent redox steady state level of cyt.b, that is, levels comparable to those of lyophilized particles are noted even in absence of the inhibitor. This finding is opposite to that expected were UQ-10 acting only on the redu-

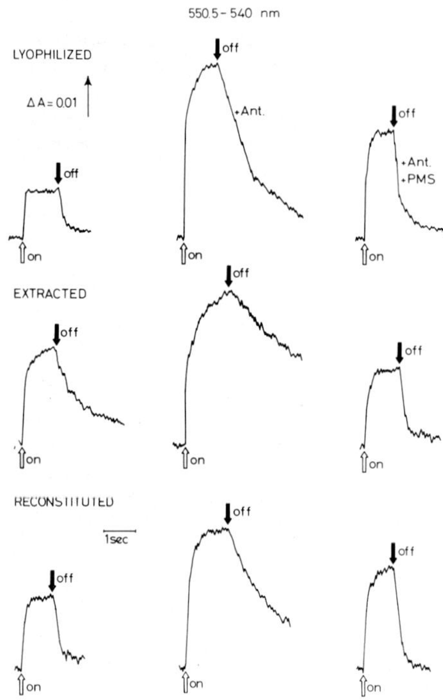

Fig.2. Light induced redox changes of cytochrome c in lyophilized, extracted and reconstituted chromatophores.

Conditions as in Fig.1

cing side of cyt.b.

PMS (100 μM) catalyzes a faster reoxidation of cyt.b in Antimycin inhibited chromatophores. However, both in lyophilized and in reconstituted particles, the steady state extent of cyt.b photoreduction is not lower in comparison to that observed with Antimycin alone. The rate of the cycle reconstituted by addition of PMS is comparable to that of the natural system. However, a difference is noted with the extracted vesicles, where PMS lowers the steady state extent of cyt.b photoreduction. Under these conditions, the effect of extraction of UQ-10 is observed rather specifically to be on the reducing side of cyt.b.

Similar effects due to extraction of UQ-10 can be observed also in

the light induced redox changes of cyt.$\underline{c}$ (Fig.2). In lyophilized particles, cyt.$\underline{c}$ undergoes a fast oxidation upon illumination which is
rapidly reversed in the dark. Antimycin A induces a large crossover
effect resulting in a many fold increase of the steady state extent
of cyt.$\underline{c}$ photooxidation. However, in this case Antimycin A decreases,
but does not block, the rate of reduction of this redox carrier.
The extraction of UQ from chromatophores leads: 1) To a decrease in
the rate of reduction of cyt.$\underline{c}$ in the dark, in agreement with the role of UQ-10 on the reducing side of cyt.$\underline{c}$; and 2) To an increase in
the steady state extent of cyt.$\underline{c}$ photooxidation which is only moderately increased by addition of Antimycin A.
PMS restores fast rates of reoxidation of cyt.$\underline{c}$ in Antimycin inhibited chromatophores. In all the three types of particles the extent
of cyt.$\underline{c}$ photooxidation is decreased by addition of PMS indicating
that this compound is an excellent electron donor to cyt.$\underline{c}$, in agreement with previously reported data. These observations of the effect
of PMS in bypassing the inhibition by Antimycin A are in agreement
with the proposal by Trebst[8] on the mechanism of PMS-mediated electron transport in bacterial photosynthesis.
The data presented (see also Ref.6) require an involvment of the UQ
pool more complex than that of merely a secondary acceptor on the reducing side of cyt.$\underline{b}$. This conclusion is supported by results already available in the literature[9,10]. The complexity of the cytochrome b-
-Ubiquinone interaction has been recently rationalized by P. Mitchell[11] in the hypothetical scheme of the Q cycle. From the results
reported in this paper it can only be proposed that redox couples of
UQ are probably operating both on the reducing and oxidizing sides
of cyt.$\underline{b}$, but no mechanistic hypothesis can be offered at this time.
Our data support the scheme proposed by Crofts et al.[3] in which an
open up version of the Q cycle was suggested. According to this sche-
me, two redox couples of UQ ($UQ/UQH^{\bullet}$ and $UQH^{\bullet}/UQH_2$) operate, respectively, on the reducing and oxidizing sides of cyt.$\underline{b}$. In this connection the lack of a complete inhibition by Antimycin A of cyt.$\underline{c}$ reduction can be interpreted as a slow dismutation reaction of UQH radicals, accumulated before the site of inhibition, to produce UQH_2,

the reducing species for cyt.c_2 (analogous concepts are offered in Gutman's paper in this volume).

However before drawing any definitive conclusions, a refined kinetic analysis by single flash spectrophotometry of the photoreactions of cyt.b and cyt.c in extracted and reconstituted particles will be required.

REFERENCES

1. Parson, W.W. (1974) Ann.Rev.Microbiol. 28, 41-59.

2. Evans, E.H. and Crofts, A.R. (1974) Biochim.Biophys.Acta 357, 89-102.

3. Crofts, A.R., Crowther, D. and Tierney, G.V. (1975) in "Electron Transfer Chains and Oxidative Phosphorylation"(Quagliarello, E., Papa, S., Palmieri, F., Slater, E.C. and Siliprandi, N. eds.) pp. 233-250, North Holland, Amsterdam.

4. Mitchell, P. (1975) FEBS Lett. 56, 1-6.

5. Baccarini Melandri, A. and Melandri, B.A. in "Methods in Enzymology" (San Pietro, ed.) vol.24, pp.96-103, Academic Press, New York and London.

6. Baccarini Melandri, A. and Melandri, B.A. (1977), FEBS Lett.(in press).

7. Dutton, P.L. and Baltscheffsky, M. (1972) Biochim. Biophys. Acta 267, 172-178.

8. Trebst, A. (1976) Z.Naturforsch. 31 c, 152-156

9. Cox, G.B. and Gibson, F. (1974) Biochim. Biophys. Acta 346,1-25.

10. Baltscheffsky, M. (1975) in "Proceedings of the III International Congress on Photosynthesis" (Avron, M. ed.) Vol.I pp.799-806, Elsevier, Amsterdam.

11. Mitchell, P. (1976) J. Theor. Biol. 62, 327-367.

INTRODUCTION

RECENTLY DISCOVERED COMPONENTS OF THE MITOCHONDRIAL ELECTRON TRANSFER SYSTEM

Helmut Beinert
Institute for Enzyme Research, University of
Wisconsin, Madison 53706

It is becoming common knowledge in these days that, in terms of species, the cytochromes as components of the mitochondrial electron transfer system are out-numbered by other electron carriers, which largely belong to the family of iron-sulfur (Fe-S) proteins. The first rapid phase of detection and characterization of these new electron carriers is over and has given way to the slower and more laborious phase of filling in essential detail and identifying those species that have so far defied such attempts. I intend to summarize here briefly what we may safely consider as established in this field and at the same time point out aspects which still need further exploration and clarification.

Even the total number of Fe-S centers in the most thoroughly studied Fe-S flavo-proteins of the electron transfer system, namely NADH and succinate dehydrogenases, is not beyond doubt. Thus, the idea that seven independent Fe-S centers are present in the former[1] is not accepted by other workers in the field[2,3] although observations leading to this suggestion have been reproduced. Also, the rather low concentration of centers 1a and 1b, as indicated by simulations of Albracht *et al.*[2], raises serious questions of stoichiometry. Similar work by the same authors has led to new values for the peak positions of centers 3 and 4, as compared to earlier work[4]. Accepting these new values, however, introduces difficulties and inconsistencies in the behaviour of the low and high field peaks of these centers in titrations. No solution to this dilemma has yet been proposed.

Work reported at this meeting by Albracht now casts doubt on the existence of center 2 of succinate dehydrogenase, at least in particulate preparations of the enzyme, such as succinate cytochrome <u>c</u> reductase or submitochondrial particles. Unfortunately, Complex II was not investigated by this author. Whereas Beinert *et al.*[5] found up to 0.5 equivalents of one Fe-S center (as compared to center 1 of succinate dehydrogenase) in soluble preparations and Complex II by direct double integration of experimental spectra, Albracht's simulations on the mentioned part-icle preparations gave no evidence that more than one Fe-S center is represented in the resonances observed in the reduced preparations. Changes in EPR signal shape and saturation with microwave power on addition of dithionite after succin-ate, as reported by others[5,6] were, however, observed. The role of center 2 had previously been questioned because of its low oxidation-reduction potential (it is not reduced by succinate!) and the lack of stoichiometry[5]. The role of the Hipip-type center in succinate dehydrogenase is, however, generally accepted.

An abstract submitted to this meeting by Ohnishi *et al.* presents interesting detail on the interactions of Fe-S centers 1 and 2 of succinate dehydrogenase. This work and the observations on changes in signal shape and saturation behaviour leave no doubt that changes do occur in the Fe-S centers of the enzyme when dithionite is added, as compared to the state reached by reduction with succinate, but whether these changes are of any biological significance remains unclear. In view of the increasing number of reports on EPR signals indicating interaction between paramagnetic centers it may be well to remember at this point that there are no biological implications in the term interaction as used here. As exciting as these manifestations may be for the physicist, in a biochemical context they mean no more than "close vicinity", which, in favorable cases, can be expressed in definitive numbers.

The low field peak of a Fe-S center which is not part of either NADH or succinate dehydrogenase is readily seen in mitochondrial preparations (stronger in those of rat and pigeon than those of beef and guinea pig heart). A corresponding Fe-S center was postulated which was initially named "center 5" and more recently "$\underline{bc}_2$" [7,8]. This Fe-S center belongs to a previously unknown Fe-S flavoprotein, that is involved in the β-oxidation pathway of fatty acids[9]. It had been reported that this Fe-S flavoprotein is reduced within a few msec, when mixed with electron transferring flavoprotein (ETF) that had been reduced by butyryl CoA in the presence of the corresponding acyl CoA dehydrogenase (also a flavoprotein)[9]. It has now been shown that re-oxidation of the reduced Fe-S flavoprotein by Q-1 is almost equally rapid. In addition, rapid re-oxidation of reduced ETF by Q-1 in the presence of catalytic amounts of the Fe-S flavoprotein could be demonstrated spectrophotometrically. Reduced ETF by itself reacts but poorly with Q-1 or oxygen.

Although little doubt exists about the association and function of the so-called "Rieske"-Fe-S center in the cytochrome $\underline{bc}_1$ complex, it may be well to point out, that stoichiometry of this center with cytochrome $\underline{c}_1$ has not been established by EPR spectroscopy, to my knowledge. In a collaborative attempt with Dr. T.E. King we failed to accomplish this, finding significantly less than stoichiometric amounts with respect to $\underline{c}_1$.

Little new information is available on the soluble Hipip-type protein of mitochondria. The solubilized protein is very autoxidizable and relatively unstable, which makes conclusive experimentation laborious and difficult. The lack of progress concerning this protein is disappointing for several reasons. First, mitochondria do contain a fair amount of this protein; secondly, it is available in highly purified (PAC-SDS electrophoresis) soluble form, which is not the case with the Hipip-type protein of succinate dehydrogenase; and thirdly, it is still the

only protein of this type (other than the originally found bacterial Hipip's),
certainly from animal tissues, for which it has been shown by purification and
analysis that it is indeed a Fe-S protein.

The Fe-S protein discovered by Backström *et al.*[10] is the subject of a report
at this meeting by Backström *et al.* It appears to be an integral protein of the
outer membrane and there are indications that it is related to the rotenone insensitive NADH-cytochrome c reductase activity of this membrane.

As far as I am aware no definitve knowledge is available on the location and
orientation of any of the Fe-S centers of the mitochrondrial inner membrane with
respect to that membrane. Thus construction of electron or proton transfer loops
involving these centers remains hypothetical.

Also, little new can be said in a definitive way on the function of Fe-S proteins in the electron transfer system. There are too many such centers and their
properties differ sufficiently that their description as "electron transfer agents"
seems certainly too general. One small group (centers 1a and 2 of NADH dehydrogenase) has been implicated in energy coupling. Whatever evidence has been produced in favor of this view has been discussed recently[8,11]. It seems, therefore,
redundant to repeat this here. A second role that is fairly obvious is that of
accomplishing the translation from two-electron to one-electron transfer as it is
required at the juncture of the substrate-dehydrogenase systems and the cytochrome
system proper. It may turn out to be a valid generalization that for placing
electrons into the cytochrom system via Q a Fe-S flavoprotein is required. At
least the three main dehydrogenase systems of heart mitochondria, namely those for
NADH, succinate and fatty acids, follow this rule. We need, of course, more examples before we could consider this statistically significant. Some experiments
and thoughts bearing on this possible function of Fe-S proteins are communicated
at another symposium of this publication series[12].

REFERENCES

1. Ohnishi, T. (1975) Biochim. Biophys. Acta, 387, 475-490.
2. Albracht, S.P.J. and Doijewaard, G. (1975) in Electron Transfer Chains and
    Oxidative Phosphorylation, Quagliarello, E. *et al.*, eds., North-Holland
    Publishing Co., Amsterdam, pp. 49-54.
3. Beinert, H. and Ruzicka, F.J. (1975) in Electron Transfer Chains and Oxidative
    Phosphorylation, Quagliarello, E. *et al.*, eds., North-Holland Publishing Co.,
    Amsterdam, pp. 37-42.
4. Orme-Johnson, N.R., Hansen, R.E. and Beinert, H. (1974) J. Biol. Chem., 249,
    1922-1927.
5. Beinert, H., Ackrell, B.A.C., Kearney, E.B. and Singer, T.P. (1975) Eur. J.
    Biochem., 54, 185-194.

6.   Ohnishi, T., Salerno, J.C., Winter, D.B., Lim, J., Yu, C.A., Yu, L. and
     King, T.E. (1976) J. Biol. Chem., 251, 2094-2104.

7.   Ohnishi, T., Wilson, D.F., Asakura, T. and Chance, B. (1972) Biochem. Biophys.
     Res. Commun., 46, 1631-1638.

8.   Ohnishi, T. (1976) Eur. J. Biochem., 64. 91-103.

9.   Ruzicka, F.J. and Beinert, H. (1975) Biochem. Biophys. Res. Commun., 66, 622-
     631.

10. Backström, D., Hoffström, I., Gustafson, I. and Ehrenberg, A. (1973) Biochem.
     Biophys. Res. Commun., 53, 596-602.

11. DeVault, D. (1976) J. Theor. Biol., 62, 115-139.

12. Beinert, H. (1977) in Structure and Function of Energy Transducing Membranes,
     Tager, J.M. *et al.*, eds., Elsevier/North-Holland, Amsterdam, in press.

EPR STUDIES OF CENTER S-3 AND SPIN-SPIN INTERACTION OF UBISEMIQUINONE PAIR

Tomoko Ohnishi, J.C. Salerno, Haywood Blum, J.S. Leigh, and W.J. Ingledew*
Dept. of Biochemistry and Biophysics, University of Pennsylvania, Phila.,PA 19174
U.S.A.

The EPR spectra of Center S-3 accompany partially overlapping signals at
g=2.04, 1.99 and 1.96, when they are observed in mitochondrial membrane preparations at intermediate redox states but not in the fully reduced or oxidized states
(1). This indicates that these signals are associated with free radical states
of the n=2 electron transfer components such as flavin or ubiquinone. Two important findings were reported by Ruzicka and Beinert (2, 3): i These multiple
signals arise from a spin-spin interaction because their field position is independent of the frequency of the applied microwaves; ii Ubisemiquinone (QH·) is
one of the interacting species because these signals disappear upon depletion of
ubiquinone (UQ) from the mitochondrial membrane and reappear upon replenishment
of UQ. These investigators initially proposed that the accompanying signals may
arise from a spin-spin interaction between the UQ· and Center S-3 spins, because
of concurrent appearance of their signals and because their short relaxation
times were typical of transition metals. The same group subsequently performed
computer simulation studies, in collaborative efforts with Sands' group, of the
EPR spectrum of Complex II which was trapped kinetically at an intermediate redox
state (4). Although a somewhat better fit was obtained for spectra simulated as
an SQ-SQ spin-spin interaction overlapped with non-interacting S-3 signals rather
than a spin-spin interaction between SQ and S-3, they assigned nearly equal
probability to another SQ (either QH· or flH·) or to Center S-3 as an interacting
partner of QH· (4). Ingledew et al, on the other hand, raised questions about
the direct spin-spin interaction between QH· and Center S-3 based on a spin concentration analysis of Center S-3 and the accompanying signals using mitochondrial
systems poised at varying redox potentials (5, 6). They also proposed QH· rather
than flH· as the interacting partner, from the measured midpoint potentials.

In the work presented here, potentiometric titration of Center S-3 and a spin-
coupled ubisemiquinone pair was more quantitatively analyzed and the existence of
a stable ubisemiquinone pair per succinate dehydrogenase was shown. It was also
demonstrated that the planes of quinone-rings as well as the line joining them
are perpendicular to the plane of the oriented multilayers of mitochondria.

In order to resolve the overlapped signals of Center S-3 and the spin-coupled
ubisemiquinone pair, contributing in different proportions (see figure 1 spectra
A-D), EPR spectra of Center S-3 and a dipolar coupled ubisemiquinone pair (QH·QH·)

---

*Present address: Dept. of Biochemistry, Medical Science Institute, University
of Dundee, Dundee DD1 4HN, Scotland, U.K.

were simulated as presented in figure 2A.

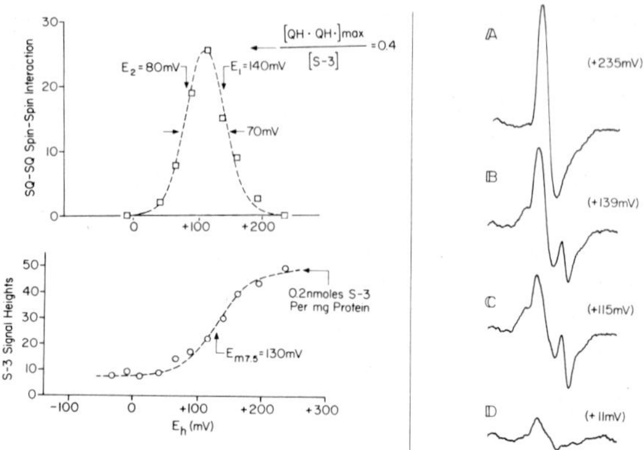

Figure 1. Potentiometric titration of EPR signals from dipolar coupled ubisemiquinone pair and Center S-3. Submitochondrial particles (SMP) (18.5 mg prot/ml)
were prepared by sonication of mitochondria in 0.25 M sucrose, 0.1 mM EDTA and
50 mM HEPES buffer at pH 7.4. Potentiometric titration was performed as described
previously in the presence of 1.4 naphthoquinone disulfonate, pyocyanine, Indigo
tetrasulfonate, Duroquinone, 1,4 naphthoquinone (the last two were disolved in
dimethyl sulfoxide) at concentrations between 20 and 50 μM. EPR conditions were:
microwave frequency, 9.18 GHz; modulation frequency, 100 KHz; modulation amplitude,
8 gauss; microwave power, 1 mW; time constant, 0.3 sec; scan rate, 250 gauss/min.;
sample temperature, 12°K.

Simulated S-3 and QH·QH· spectra were added in varying ratio to produce a
series of spectra, exemplified by spectra B and C in figure 2. Comparison of this
series with experimental spectra taken at various redox potentials allowed the
ratio of QH·QH· and S-3, and the percentage of maximum of each to be calculated.

The peak to peak height of the "g=1.99" signal and the maximum height above
base line near g=2.02 were used to calculate the ratios since the signals at
these g-values change dramatically with the proportion of QH·QH· and S-3. A good
fit was obtained between simulated spectra and experimental data in that portion
of the spectrum utilized.

Potentiometric titration of Center S-3 and spin-spin coupled ubisemiquinones
was presented in figure 1. Center S-3 shows an n=1 titration curve with a midpoint potential of +130 mV (figure 1, left bottom). Potentiometric titration of
EPR signals arising from the spin-spin interaction of the ubisemiquinone pair
gives a peak around +110 mV with about 70 mV width at half peak height. The
maximum observed spin concentration of the dipolar coupled ubisemiquinone pair
corresponds approximately to 40% of the Center S-3 concentration. The theoretical
curve for this titration and the apparent midpoint potential for Q/QH· and QH·/QH$_2$
redox couples were obtained as described below.

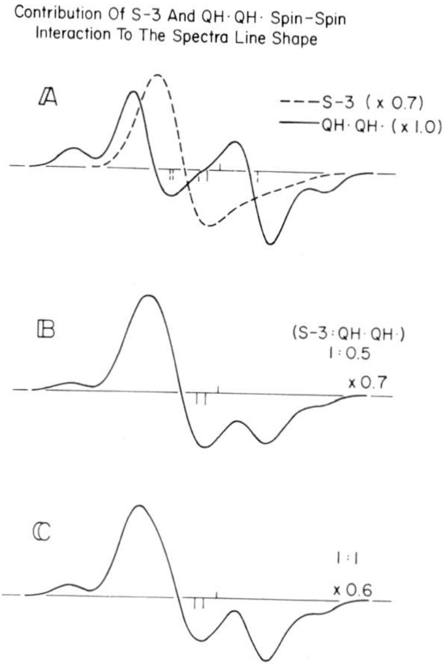

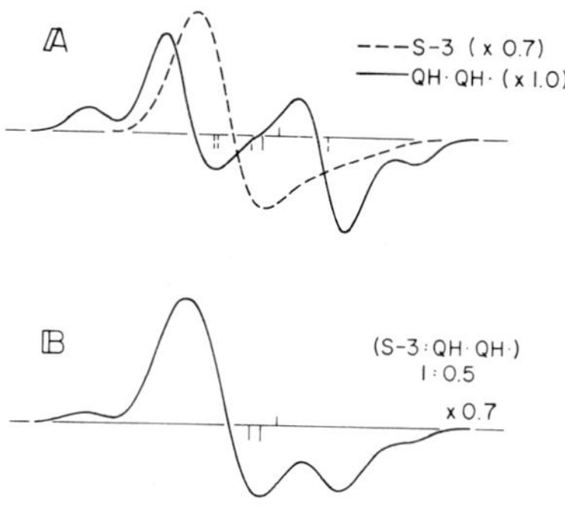

Figure 2. Computer simulation of individual spectra of Center S-3, dipolar coupled ubisemiquinone pair (A) and examples of spectra contributed from both species at varying proportions (B and C). EPR spectra were simulated by numerical integration over the angles θ and ϕ, using a PDP-10 computer. The intensity factor gp was used to correct for transition probability (7). Gaussian line shape functions were used for all simulations. Spectra of Center S-3 and dipolar coupled semiquinone pairs were simulated using the parameters of Ruzicka et al (4); small variations in simulation parameters were introduced to compensate for slight changes in line shape in different preparations.

QH· (or Q⁻·) Free Radical

$$Q \underset{E_1}{\rightleftharpoons} QH\cdot \underset{E_2}{\rightleftharpoons} QH_2$$

$$Eh = E_1 + 60\ \log_{10} [Q]/[QH\cdot]$$

$$Eh = E_2 + 60\ \log_{10} [QH\cdot]/[QH_2]$$

$$[QH\cdot] = [Q]\ 10^{(E_1 - Eh)/60} = [Q]\ \text{Exp. }a$$

$$[QH_2] = [QH\cdot]\ 10^{(E_2 - Eh)/60} = [QH\cdot]\text{Exp. }b = [Q]\ \text{Exp. }a\ \text{Exp. }b$$

$$N = [Q]\ (1 + \text{Exp. }a + \text{Exp. }a\ \text{Exp. }b)$$

$$[QH\cdot] = \frac{N\ \text{Exp. }a}{1 + \text{Exp. }a + \text{Exp. }a\ \text{Exp }b}$$

Scheme I

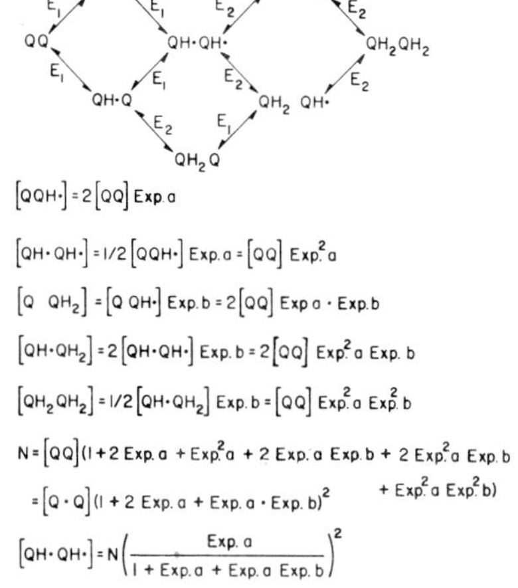

$$[QQH\cdot] = 2[QQ]\ \text{Exp. }a$$

$$[QH\cdot QH\cdot] = 1/2[QQH\cdot]\ \text{Exp. }a = [QQ]\ \text{Exp.}^2 a$$

$$[Q\ QH_2] = [Q\ QH\cdot]\ \text{Exp. }b = 2[QQ]\ \text{Exp }a \cdot \text{Exp. }b$$

$$[QH\cdot QH_2] = 2[QH\cdot QH\cdot]\ \text{Exp. }b = 2[QQ]\ \text{Exp.}^2 a\ \text{Exp. }b$$

$$[QH_2 QH_2] = 1/2[QH\cdot QH_2]\ \text{Exp. }b = [QQ]\ \text{Exp.}^2 a\ \text{Exp.}^2 b$$

$$N = [QQ](1 + 2\,\text{Exp. }a + \text{Exp.}^2 a + 2\,\text{Exp. }a\ \text{Exp. }b + 2\,\text{Exp.}^2 a\ \text{Exp. }b + \text{Exp.}^2 a\ \text{Exp.}^2 b)$$

$$= [Q\cdot Q](1 + 2\,\text{Exp. }a + \text{Exp. }a \cdot \text{Exp. }b)^2$$

$$[QH\cdot QH\cdot] = N\left(\frac{\text{Exp. }a}{1 + \text{Exp. }a + \text{Exp. }a\ \text{Exp. }b}\right)^2$$

Scheme II

For independent (non-interacting) ubiquinone, which accepts its first and second electron with midpoint potentials of E_1 and E_2 (Scheme I), the potentiometric titration curve of QH· formation is expressed as Nexpa/(1+expa+expa·expb), as detailed above.

Now assuming that a pair of ubiquinones shows a weak dipole-dipole interaction when both are in the semiquinone form (QH·QH·), but that the redox state of each quinone does not affect the redox parameters E_1 and E_2 of the other, all distinct redox species can be described as shown in Scheme II. The concentrations of individual redox species can be obtained in a similar way as in Scheme I. The (QH·QH·) concentration is obtained as the square of the independent (QH·) concentration. The theoretical redox behavior of (QH·) and (QH·QH·) with different relative midpoint potentials is plotted as shown in figure 3. The experimental results shown in figure 1 (left top) gives the best fit to the (QH·QH·) curve of case (2), which was drawn as a dotted curve in figure 1.

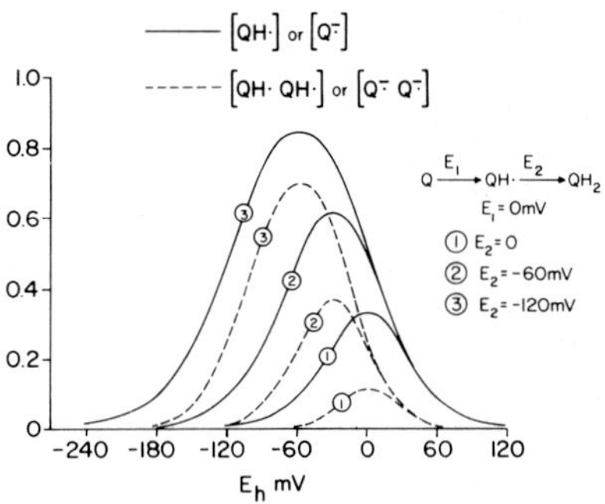

Figure 3. Theoretical redox behavior of semiquinone (QH· or Q⁻) and semiquinone-semiquinone (QH·QH· or Q⁻Q⁻) species.

As shown in figure 4, the simultaneous titration of the g=2.00 free radical signal (QH·) gave a broader curve with a peak around +80 mV. The 30 mV positive shift of signals from (QH·QH·) may arise from the constraint imposed by dipolar peak interaction of spin-coupled ubisemiquinone(s) with paramagnetic Center S-3, proposed previously by Ingledew et al (6). Approximate E_1 and E_2 values at pH 7.5 for the ubiquinone pair functioning in succinate dehydrogenase-cytochrome b region are thus estimated as +110 and +50 mV, respectively. These results show that ubiquinone functioning in this segment of the respiratory chain forms stable semiquinone; a free quinone would be expected to have an unstable free radical in the physiological pH range. Pool size of this specific ubiquinone seems to be one pair per succinate dehydrogenase molecule; thus the maximum observable QH·QH· concentration corresponds to less than 2% of the total Q pool in beef heart SMP (cf. ref. 8).

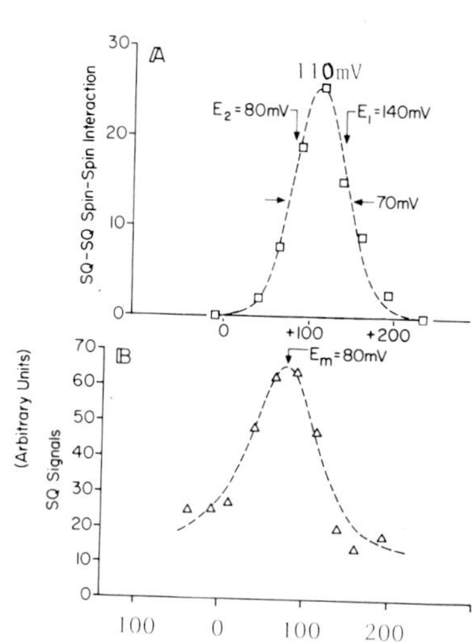

Figure 4. Potentiometric titration of EPR signals from dipolar coupled ubisemiquinone pair and g=2 free radical signal of ubisemiquinone. The same set of the potentiometric titration was used as shown in figure 1. EPR conditions used for measuring g=2.00 signal were: microwave power, 0.5 mW; sample temperature, 77°K. Other conditions were same as in figure 1.

Ruzicka et al (1) simulated the spectrum of the coupled semiquinones in Complex II (particulate succinate UQ reductase) using g_z=g_y=2.0066 (in the plane of the quinone), g_x=2.0041, and r=7.71 Å, and $\vec{r}$ in the yz plane; coupled ubisemiquinones mutually oriented edge to edge, but not face to face.

The orientation of the heme planes in cytochrome oxidase has recently been determined using oriented multilayers made by centrifugation and partial drying of purified cytochrome oxidase vesicles, electron transport particles (9) and mitochondria (10). In order to examine the orientation of these coupled ubiquinones relative to the mitochondrial membrane, their spin-spin interaction spectra were examined in oriented multilayered beef heart mitochondria.

The dipolar splitting at a particular orientation is expressed as

$$\Delta H = \frac{3\mu_A \mu_B}{4r^3} (1-3 \cos^2\psi) = D (1-3 \cos^2\psi)$$

where μ_A and μ_B are the magnetic moments of the interacting species, r is the distance between them, and ψ the angle between the line joining them and the applied magnetic field. The unoriented spectrum of a dipolar coupled pair of isotropic spins consists of two pairs of lines as shown in figure 2A (solid curve). The inner pair is displaced by the dipolar strength D to each side of the original line; the outer pair is displaced by $\pm$2D (4).

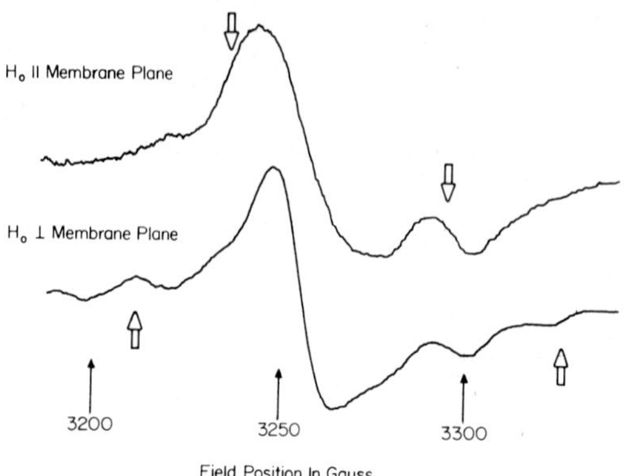

Figure 5. EPR spectra of oriented multilayers prepared from beef heart mitochondria. EPR conditions were the same as in figure 1 except following parameters: microwave power, 5 mW; modulation amplitude, 10 gauss. Arrows show the
calculated positions of the inner pair (upper arrows) and outer pair (lower
arrows). (Top) Magnetic field perpendicular to the multilayer. (Bottom) Magnetic field parallel to the plane of the multilayer.

Figure 5 (bottom) shows the spectrum of the oriented multilayers with the
field perpendicular to the membrane plane. The outer pair of lines is clearly
visible despite some overlap from Center S-3 signals.

Figure 5 (top) shows the spectrum of the same sample with the field parallel
to the membrane plane. The outer lines have disappeared and been replaced by
the inner pair of lines. The low field member of the inner pair is heavily
overlapped by the signals of Center S-3 (11, 12), and a small amount of signal
from Rieske's center. The spectrum of these two iron sulfur centers is also
dependent on th orientation of the multilayer in the magnetic field.

These results allow us to deduce the quinone orientation using $(1-3 \cos^2 \psi) = -2$,
so ψ must equal zero. Therefore the dipolar axis is perpendicular to the membrane plane (see figure 6). The positions of the signals closely corresponds
to the predictions of simulations (arrows) confirming that the dipolar axis
lies in the plane of the quinone rings as implied by simulation of the spectra
in Complex II (4). It follows that the quinone rings are also perpendicular to
the membrane plane. There is, of course, no ordering in the plane of the multilayer so rotation in the membrane plane cannot be detected. All directions within the quinone ring plane are magnetically equivalent except for the dipolar
coupling, so no information about the orientation of specific directions within
this plane can be extracted.

The EPR spectra observed both in suspension and especially in oriented multilayers indicate a degree of order which is not expected for free quinone. Re-

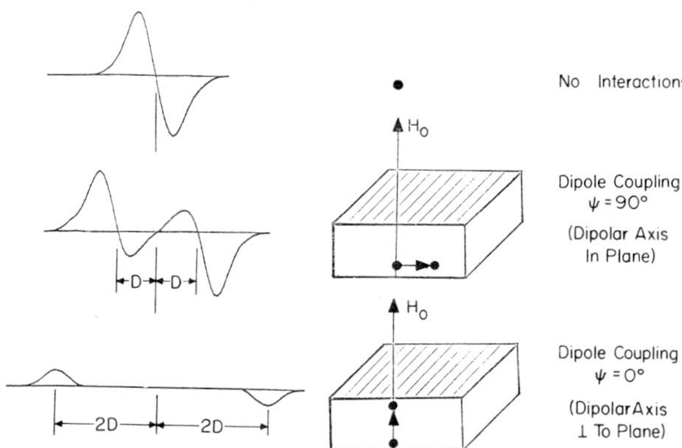

Figure 6. Schematic presentation of EPR spectra for non-interacting semiquinone spin-coupled semiquinone pair with dipolar axis in the plane, or dipolar axis perpendicular to the plane.

cently Yu et al (13) have described a hydrophobic quinone binding protein isolated from succinate cytochrome c reductase. A protein of this sort capable of binding a quinone pair would provide an obvious explanation for the high degree of order necessary to produce the observed spectra.

REFERENCES

1. Orme-Johnson, N.R., Hansen, R.E., and Beinert, H. (1974) J. Biol. Chem. 249, 1928-1939.

2. Beinert, H. (1974) in VIth International Conference on Magnetic Resonance in Biological Systems, Kanderstag.

3. Ruzicka, F.J. and Beinert, H. (1975) Fed. Proc. 34, 579.

4. Ruzicka, F.J., Beinert, H., Schefler, K.L., Dunham, W.R. and Sands, R.H. (1975) Proc. Nat. Acad. Sci. USA 72, 2886-2890.

5. Ingledew, W.J. and Ohnishi, T. (1975) FEBS Letts. 54, 167-174.

6. Ingledew, W.J., Salerno, J.C. and Ohnishi, T. (1974) Arch. Biochem. Biophys. 177, 176-184.

7. Aasa, R. and Vangaard, T. (1975) J. Mag. Resonance 19, 001-008.

8. Bäckström, D., Norling, B., Ehrenberg, A. and Ernster, L. (1970) Biochim. Biophys. Acta 197, 108-111.

9. Leigh, J.S. and Harmon, H.J. (1977) Biophysical Journal 17, 251a.

10. Erecinska, M., Blasie, J.K., and Wilson, D.F. (1977) FEBS Letts. 76, 235-239.

11. Beinert, H., Ackrell, B.A.C., Kearney, E.B., and Singer, T.P. (1975) Eur.

J. Biochem. $\underline{54}$, 185-194.

12. Ohnishi, T., Lim, J., Winter, D.B. and King, T.E. (1976) J. Biol. Chem. $\underline{251}$, 2105-2109.

13. Yu, C.A., Yu, L. and King, T.E. (1977) Fed. Proc. $\underline{36}$, 815.

Bioenergetics of Membranes. L. Packer et al. ed.

E P R DETERMINATION OF THE QUANTITATIVE RELATIONSHIPS BETWEEN THE ELECTRON TRANSPORT COMPONENTS OF THE PHOTOSYSTEM 1 REACTION CENTRE

M.C.W. EVANS and P. HEATHCOTE
Dept of Botany & Microbiology,
University College, London.
and
D.L. WILLIAMS-SMITH
Dept of Physics, Guy's Hospital
Medical School, London, U.K.

INTRODUCTION

Photosystem 1 of oxygen evolving organisms is involved in the photochemical generation of the low potential reductant required for growth. The properties of the reaction centre of photosystem 1 are less well defined than those of reaction centres isolated from purple photosynthetic bacteria. The best functional photosystem 1 preparations although free of electron transport proteins outside the reaction centre, (such as cytochromes or plastocyanin) and quinones, have a relatively high ratio of total chlorophyll to reaction centre chlorophyll (P700) of about 30:1. The initial photochemically induced electron transport reactions in photosystem 1 occur even in the solid state at cryogenic temperatures (1) and can best be observed by electron paramagnetic resonance (e.p.r.) spectrometry. The initial events involve the photooxidation of P700 and the transfer of an electron to an acceptor complex. This complex contains two, four-iron, iron-sulphur centres (2, 3, 4,), and a chemically unidentified component, X (5, 6). The components of this complex can only be certainly identified by e.p.r. at the present time. Earlier experiments suggest that X is probably the primary electron acceptor and that the bound iron-sulphur centres A and B are secondary acceptors. The photoreduction of X is reversible at low temperatures (10K) while photoreduction of the iron-sulphur centres is irreversible. The photoreduction of X and reversible photooxidation of P700 are only seen when both iron-sulphur centres are reduced. We have now improved our understanding of the photosystem 1 reaction centre by determining the quantitative relationships of the four components of the reaction centre, and determining more precisely the relationships between the redox state of the iron-sulphur centres and reversibility of P700 photooxidation.

MATERIALS AND METHODS

The experiments were carried out using photosystem 1 particles of
two types. 1) Prepared using Triton X-100 (6 and in preparation),
these particles have a P700: Chlorophyll ratio of around 1:40 and
are free of cytochromes. 2) Prepared using the French press with
P700: Chlorophyll around 1:200. E.p.r. samples were prepared and
redox titrations done by standard procedures. E.p.r. spectrometry
was done using Varian E4 or E9 spectrometers fitted with Oxford
Instruments ESR 9 cryostat for helium temperature work, or a liquid
nitrogen finger dewar. When samples were measured at two different
temperatures a 1mM copper-EDTA sample was used as a standard for
temperature calibration. Spectral simulations were computed by the
method of Leigh (7) as modified by Aasa & Vannegard (8). Full
details of the experimental procedures will be published elsewhere.

RESULTS

The organic free radical signal, signal 1, observed on illumination of chloroplast preparations has been shown to be quantitatively
related to P700 photooxidation at room temperature (9, 10). A
quantitative relationship between the photooxidation of signal 1 and
reduction of iron-sulphur centre A has been shown at low temperature
(11). However, it was apparent both from our work (12) and that of
Bearden and Malkin (13, 14) that large variations in the size of the
light induced signal 1 were observed when samples were prepared under
different conditions of pH and in the presence or absence of
reducing agents, and measurements were made at low temperatures
(below 30K) and high microwave power. There was no parallel
variation in the size of the light induced iron-sulphur centre A
signals (Fig. 1). The variations might be due to the presence of
multiple components in signal 1, or variations in the saturation
characteristics of signal 1 under different preparation conditions.

Fig. 2 shows microwave power saturation curves for signal 1 in
samples oxidised with ferricyanide, reduced with ascorbate, or
equilibrated in the dark without oxidants or reductants before
freezing. It is clear that there are considerable differences in
saturation characteristics depending on the preparation conditions,
these differences will lead to the large variations in signal size
observed when measurements are made under saturating microwave power
conditions at low temperature. The results shown in Fig. 2 were
obtained at 77K as we found that signal 1 showed saturation effects
at the lowest obtainable microwave powers (100 nanowatts) at 20K.

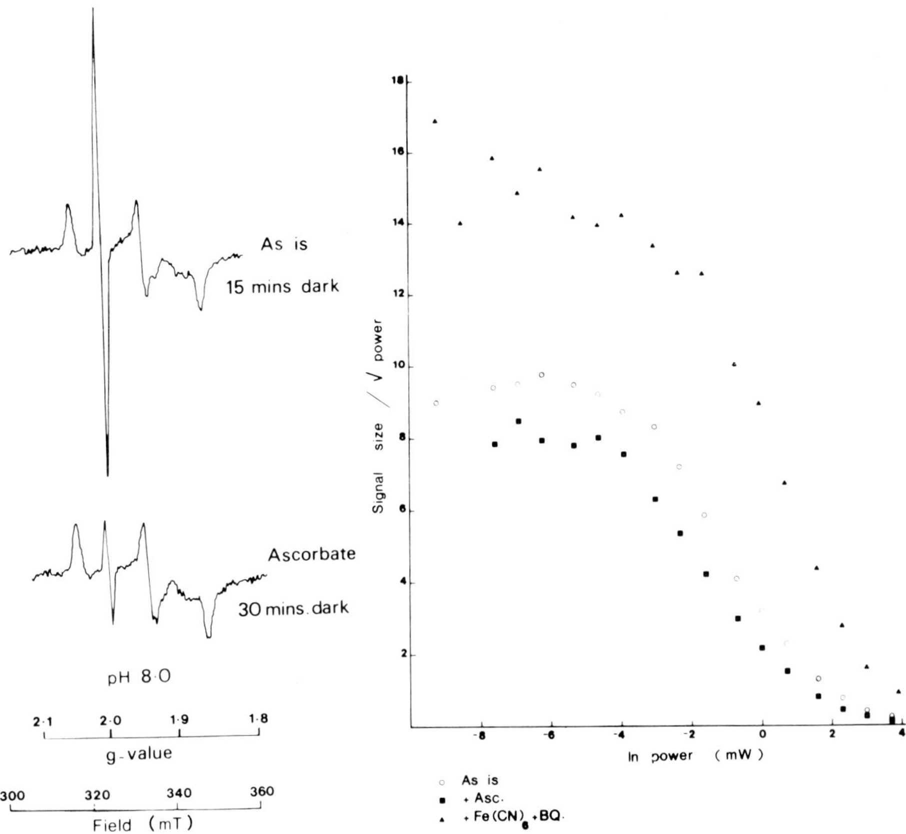

Fig. 1 (Left)
Light minus dark difference spectra of photosystem 1 particles
prepared with Triton X-100.
The samples contained particles (0.5 mg. chlorophyll/ml) in O.1M
Tris HCl buffer pH 8.0, 0.2M NaCl. Sodium ascorbate (10mM) was
added as indicated and the samples frozen after dark equilibration.
E.p.r. spectra was recorded in the dark and after illumination for
30 secs at 15K and the difference spectrum obtained by subtraction.
Instrument settings were Frequency 9.06 GHz. Microwave power 20 mW.
Modulation Amplitude 1mT. Scan Rate 500mT/Min. Gain 1000.

Fig. 2 (Right)
Saturation curves for signal 1 in photosystem 1 particles under
different redox conditions
Photosystem 1 particles prepared using Triton X-100 0.9 mg
chlorophyll/ml in O.1M Tris-HCl pH 8.0 0.2M NaCl were frozen in the
dark after a) 2 mins in the dark b) reduction by sodium
ascorbate (10mM) for 30 mins c) Oxidation by potassium ferricyanide
(10mM) and benzoquinone (100μM) for 30 mins. After illumination
for 2 mins at 77K the peak to peak signal height of Signal 1 was
determined under varying microwave power conditions at 77K with the
following instrument settings:- Frequency 9.165 GHz Modulation
amplitude 0.1mT Scan rate 12.5mT/min.

To overcome the difficulties which these variations have caused in the interpretation of experiments on the primary events in photosystem 1 we have made quantitative determinations of the relative amounts of P700 (signal 1) and Centre A produced on low temperature illumination of samples prepared with and without reductants, and of the light induced reversible P700 and X. To make these determinations we prepared simulations of the e.p.r. spectra of P700, iron-sulphur Centre A (Fig. 3) and of X (Fig. 4) which allows the relative amounts of the components to be calculated on the basis of measurements of signal heights determined under non-saturating conditions. P700 was measured as the light induced signal 1 at 77K and $5\mu W$, centre A as the g=1.96 signal (or the g=1.86 signal) at 20K and 5mW and X as the g=1.76 signal at 10K and $20\mu W$. Table 1 summarises the results of these measurements. The ratio of P700: Centre A is approximately 1:1 in agreement with the results of Bearden & Malkin (11) confirming that an electron can be transferred from P700 to centre A on illumination at low temperature. This ratio was unaffected by sample preparation conditions although there was a large apparent variation in signal 1 under the conditions used to measure centre A. The ratio of the light induced reversible P700: X was 1:1.5 and the results showed more variability than the centre A measurements. This is due to the experimental difficulties in measuring reversible, light induced, signals in two different cryostats at 10K and 77K, when it is necessary to attempt to obtain identical illumination conditions in each case. We consider that these measurements are good evidence for a quantitative electron transfer from P700 to X.

We have also used the simulations to calculate the relative amounts of centre A and X in the preparations. Centre A was measured as the signal induced by illumination at low temperature of a sample prepared with ascorbate, and X as the fully reduced signal in similar samples frozen under illumination in the presence of dithionite and methyl violegen. The ratio of 0.75:1 again indicates a quantitative role for X in photosystem 1. We have been unable to simulate the spectrum of Centre B because it is only seen in the presence of Centre A and shows strong electron-spin electron-dpin interactions with A. We have therefore estimated the amounts of Centre A and B by double integrating the spectrum of Centre A obtained on illumination at low temperature (after subtraction of Signal 1) and of A + B in an identical sample frozen under illumination in the presence of dithionite (again after subtraction of the small free radical signals in the spectrum). The ratio of

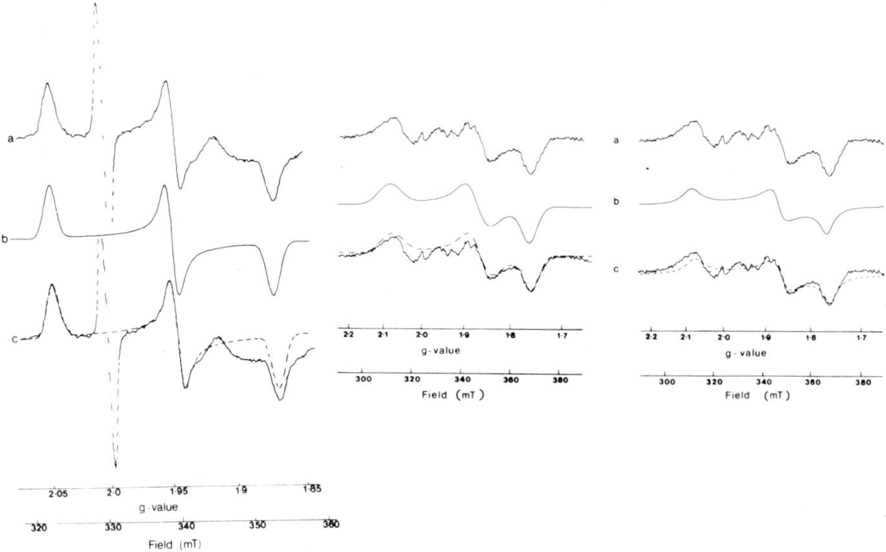

Fig. 3 (Left)
Comparison of the experimental and simulated spectra of Centre A

Fig. 4 (Right)
Comparison of the experimental and simulated spectra (gaussion and Lorentzian) of X.
a) Experimentally obtained spectrum (for Centre A this also shows the P700 radical)
b) Simulated spectra
c) Superimposition of a and b

TABLE I

The measured ratios of the components
of the photosystem 1 reaction centre

COMPONENTS		NUMBER OF DETERMINATIONS	AVERAGE RATIO	RANGE
P700: CENTRE A	1)*	7	1.02	0.8 - 1.4
	2)	8	1.07	0.9 - 1.2
P700: X	1)	8	1.54	0.8 - 2.0
CENTRE A: X	1)	7	0.78	0.6 - 1.2
CENTRE A: A+B	1)	8	1.90	1.8 - 2.0
	2)	8	2.02	1.9 - 2.1

*1 = Particles prepared with Triton X-100.

 2 = Particles prepared with the French press.

A: A + B of 1:1.95 indicates the presence of A and B in equivalent amounts. Oxidation of photosystem 1 particles with ferricyanide results in a large change in saturation characteristics of signal 1 (Fig. 2) and the observation of a very large free radical signal at low temperatures. Quantitative comparison of this signal with that of Centre A induced by illumination of matching reduced samples shows a ratio of 2.96:1. Supporting our previous suggestion that ferricyanide oxidises other components as well as P700 in these preparations (12)1

We have previously shown that P700 photooxidation becomes reversible at low temperatures only after reduction of both iron-sulphur Centres A and B before freezing. These experiments do not entirely exclude the possibility that, other components are also involved. We have therefore determined the redox potential at which P700 photooxidation becomes reversible and compared this with the reduction of the iron-sulphur centres. Fig. 5 shows the results of a titration in which samples were poised in the dark and the e.p.r. spectrum recorded. The sample was then illuminated in the e.p.r. spectrometer and the extent of reversibility of P700 photooxidation measured. At potentials more oxidised than -500mV P700 oxidation is irreversible and electrons are transferred to Centre B. The titration curves for the reduction of Centre B and increasing reversibility of P700 are essentially parallel with E_m=-585mV. This experiment shows clearly the interrelationships of the redox states of P700 and the iron-sulphur centres and that there are unlikely to be any unknown electron carriers between X and Centre B.

We have been unable to chemically reduce X which presumably has a redox potential below -650mV.

These results demonstrating the quantitative and oxidation reduction potential relationships of the components of the photosystem 1 reaction centre confirm our model of the function of this centre. X is the earliest identified acceptor of electrons from P700 which may subsequently be transferred to the iron-sulphur centres A and B. These carriers may be thought to form an electron transport chain between P700 and ferredoxin

P700 → X → Centre B → Centre A → Ferredoxin

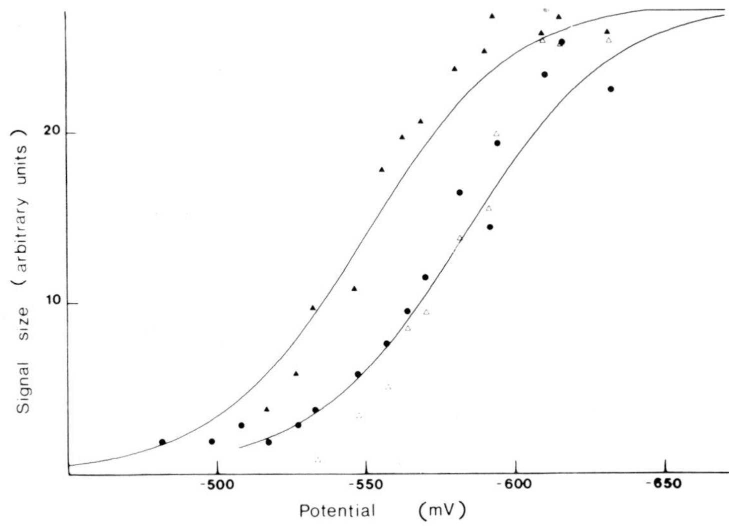

Fig. 5
Oxidation-reduction potential titration of photosystem 1 particles
Photosystem 1 particles prepared with Triton X-100 were suspended
in 0.1M glycine-KOH buffer pH 10.0 0.2M NaCl with the following
redox mediators:- Methyl Viologen, Triquat and Tetraquat all at
a concentration of 20μM. The sample was reduced to -450mV for 30
mins and then titrated by standard procedures using sodium
dithionite (2%) in 0.1 M Tris-HCl pH 9.0 as reducing agent. The
titration was done under a green safelight and the samples frozen
in complete darkness. The e.p.r. spectra were recorded at 15K
in the dark to determine the extent of reduction of the iron-
sulphur centres and then during and after illumination to determine
the size of the light induced reversible P700 signal. The
components were measured as the signal height of the g=1.96 signal
for Centre A, the g=1.89 signal for Centre B, the g=2.00 signal
(light on minus light off) for reversible P700. The signal
sizes were normalised for this figure.

▲ -   Centre A   ● -   Centre B

△ -   Reversible P700

The observations made by Malkin and Bearden (13 & 14) which

suggest that only an insignificant part of P700 photooxidation

becomes reversible is shown to be an artefact due to changes in

saturation characteristics of P700, and not to real variations in

the amount of P700 oxidised.

REFERENCES

1. Commoner, B., Heise, J. and Townsend, J. (1956) Proc. Natl. Acad. Sci. U.S.A. $\underline{42}$, 710-718.

2. Malkin, R., Bearden, A.J. (1971) Proc. Natl. Acad. Sci. U.S.A. $\underline{68}$, 16-19.

3. Evans, M.C.W., Telfer, A. and Lord, A.V. (1972) Biochim. Biophys. Acta $\underline{267}$, 530-537.

4. Evans, M.C.W. and Cammack, R. (1975) Biochem. Biophys. Res. Commun. $\underline{63}$, 187-193.

5. Evans, M.C.W., Sihra, C.K., Bolton, J.R. and Cammack, R. (1975) Nature (London) $\underline{256}$, 668-670.

6. Evans, M.C.W., Sihra, C.K., and Cammack, R. (1976) Biochem. J. $\underline{158}$, 71-77.

7. Leigh, J.S. (1970) J. Chem. Phys. $\underline{52}$, 2608-2612.

8. Aasa, R. and Vänngard, T. (1975) J. Magn. Reson. $\underline{19}$, 308-315.

9. Warden, J.T. and Bolton, J.R. (1973) J. Am. Chem. Soc. $\underline{95}$, 6435-6436

10. Baker, R.A. and Weaver, E.C. (1973) Photochem. Photobiol. $\underline{18}$, 237-241.

11. Bearden, A.J. and Malkin, R. (1972) Biochim. Biophys. Acta $\underline{283}$, 456-468.

12. Evans, M.C.W., Sihra, C.K. and Slabas, A.R. (1977) Biochem. J. $\underline{162}$, 75-85.

13. Bearden, A.J. and Malkin, R. (1976) Biochim. Biophys. Acta $\underline{430}$, 538-547.

14. Malkin, R. and Bearden, A.J. (1976) FEBS letts. $\underline{69}$, 216-220.

ACKNOWLEDGEMENTS

This work was supported in part by grants from the U.K. Science Research Council, the Commission of the European Communities, the Royal Society and the University of London Research Fund. We would like to thank Professor S.J. Wyard and Dr. R. Cammack for advice and encouragement and Dr. L.J. Dunne and R.D. Kaye for assistance with the computing.

PHOTOSYNTHESIS

ELECTROCHROMISM IN CHLOROPLASTS AND IN CHROMATOPHORES OF THE PURPLE BACTERIUM RHODOPSEUDOMONAS SPHAEROIDES

B.G. de Grooth and J. Amesz
Department of Biophysics, Huygens Laboratory,
University of Leiden, The Netherlands

INTRODUCTION

Measurements of absorbance changes upon illumination of intact cells and sub-cellular preparations of photosynthetic organisms have provided important information about the mechanism of photosynthesis. Many of these absorbance changes directly reflect the oxidation-reduction reactions of primary reactants and intermediates of photosynthetic electron transport. In addition to this, there is a second class of absorbance changes that can be ascribed to spectral changes of photosynthetic pigments and that do not directly reflect the oxidation or reduction of a specific compound.

Mainly on basis of measurements of Junge, Witt and coworkers[1-3] with spinach chloroplasts it has been generally accepted that these absorbance changes are due to electrochromic shifts of absorption bands which are caused by an electric field due to a potential difference over the thylakoid membrane. The extents of such shifts are proportional to the permanent or induced electrical dipole of the pigment molecules[2]. The membrane potentials are thought to result from vectorial electron transport, the electron acceptors of systems 1 and 2 of photosynthesis being located near the outer surface of the thylakoid, the electron donors P700 and P680 closer to the inner surface. A similar explanation was proposed for changes of pigment absorption observed[4,5] in intact cells and chromatophores of purple bacteria. With chromatophores of Rhodopseudomonas sphaeroides it was shown that the same absorbance changes as induced by illumination could also be observed if a membrane potential was generated by salt addition in the dark[6]. For a recent review of electrochromic absorbance changes in photosynthetic material we refer to ref. 7.

Although the above-mentioned explanation of the pigment changes is supported by various evidence, a direct demonstration based on analysis of the absorbance changes in vivo was lacking so far. In fact, results obtained with Rps. sphaeroides appeared to be incompatible with the field-effect hypothesis[8-10] and suggested a discrete, rather than a gradual shift of the absorption bands of the carotenoid in this bacterium. Recently we observed that the pigment shift could be strongly stimulated by secondary electron transport at sub-zero temperatures, and this enabled to measure the kinetics and spectra with enhanced precision. These results, and a quantitative analysis of the absorbance changes in chromatophores of Rps.

Abbreviation: PMS, N-methylphenazonium methosulphate.

sphaeroides are reported in this paper. More detailed accounts of these experiments are presented elsewhere[11,12].

RESULTS

<u>Chloroplasts</u>: Light-induced absorbance changes at 702 and at 515 nm were measured in chloroplasts cooled to -35 to -50 °C. The kinetics of P700 and of the electrochromic band shift of carotenoid at 515 nm were the same, if the experiments were done under conditions where photochemistry of system 2 was inhibited[11]. The absorbance changes at 515 nm were strongly stimulated if an electron donor and acceptor were added before cooling. The rate of secondary electron transport, measured either directly, or indirectly from the rate of absorbance change at 515 nm, was at least an order of magnitude lower than at room temperature. Nevertheless, continued illumination allowed the generation of large membrane potentials (up to 400 - 500 mV), due to sustained electron transport by the reaction centers of system 1, which were continuously regenerated by the added donor and acceptor. The absorbance changes reversed only slowly in the dark. Apparently the permeability of the thylakoid membrane to ions and electron carriers was strongly decreased by lowering the temperature. The permeability could be enhanced by addition of gramicidin D. The absorption difference spectrum (Fig. 1) was similar to that observed earlier at room temperature[13] and showed band shifts of chlorophyll b and chlorophyll a in addition to those of carotenoid.

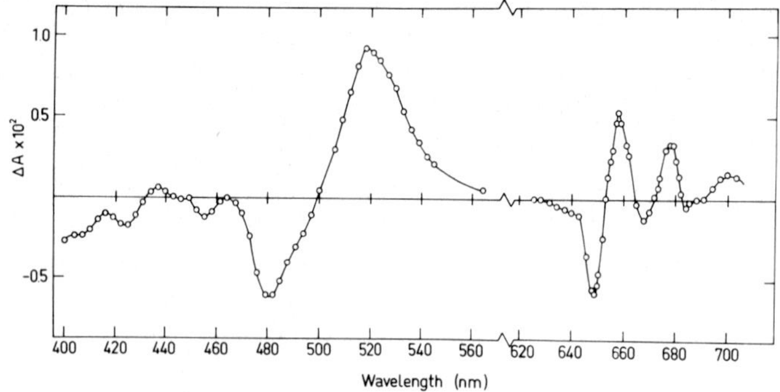

Fig. 1. Absorption difference spectrum of electrochromic absorbance changes in spinach chloroplasts upon illumination at -35 °C. The chloroplasts were suspended in a tricine-sucrose buffer, pH 7.8, to which glycol (55 % v/v) had been added before cooling. Additions: 0.17 mM PMS, 3.3 mM ascorbate. Chlorophyll concentration: 0.17 mM. Optical pathlength: 1.3 mm.

<u>Purple bacteria</u>: Strongly stimulated absorbance changes due to electrochromic band shifts of photosynthetic pigments could be obtained in bacterial vesicles in a similar way as with chloroplasts. When chromatophores of Rps. sphaeroides were cooled down in a liquid medium in the presence of partially reduced PMS, illumination at -30 to -35 °C produced absorbance changes in the carotenoid and bacteriochlorophyll region of the spectrum that were ten to fifteen times larger than occurred in the absence of additions.

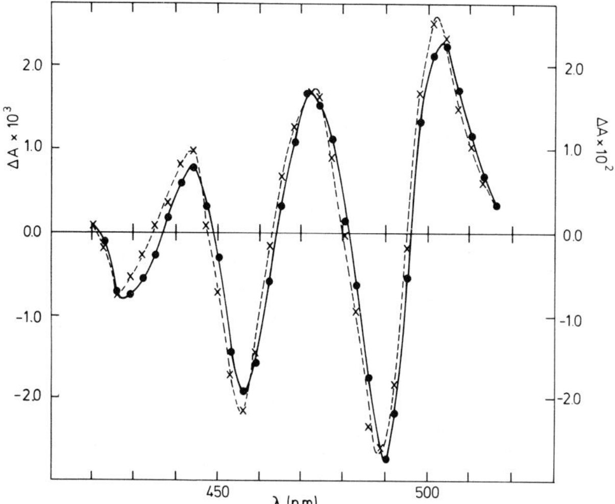

Fig. 2. Absorption difference spectra of chromatophores of the G1C mutant of Rps. sphaeroides suspended in phosphate buffer, pH 7.5, containing 50 % (v/v) glycol. Solid circles, right hand scale, conditions as for Fig. 1. Crosses, left hand scale, spectrum obtained without additions, after subtraction of the absorbance changes obtained with 50 μM gramicidin D. Bacteriochlorophyll concentration: 70 μM.

Apart from their amplitude, the difference spectra obtained with and without additions were very similar, but a careful comparison of the two spectra in the region 420 - 550 nm showed that the spectrum of the stimulated absorbance changes was located at about 1 mm longer wavelengths than the other one (Fig. 2). This indicated that the extent of the band shifts indeed increased with increasing membrane potential. In agreement with this, the absorbance changes at the zero points of the spectrum showed an increase followed by a decrease of absorbance during illumination (or vice versa), and the opposite behavior upon darkening. The kinetics near the maxima or minima showed an increase or decrease to a steady-state value, followed by a gradual decay upon turning off the light (Fig. 3).

Analysis of the kinetics of these absorbance changes enabled to calculate the extent of the band shifts and to reconstruct the shape and amplitude of the absorption spectrum of the carotenoid that shows electrochromism upon illumination. The results of such an analysis for the G1C mutant of Rps. sphaeroides are shown in Fig. 4. The major carotenoid in this mutant is neurosporene, which comprises 96 % of the total carotenoid content of this bacterium[10]. Nevertheless, as shown by Fig. 4, the analysis indicates that there are at least two pools of carotenoid. The major pool has absorption maxima at 432, 458 and 490 nm and does not show electrochromism, a second, "active" pool comprises about 25 % of the total carotenoid and absorbs at about 4 nm longer wavelengths. The latter carotenoid showed a band shift of 0.3 nm upon a single charge separation at each reaction center and

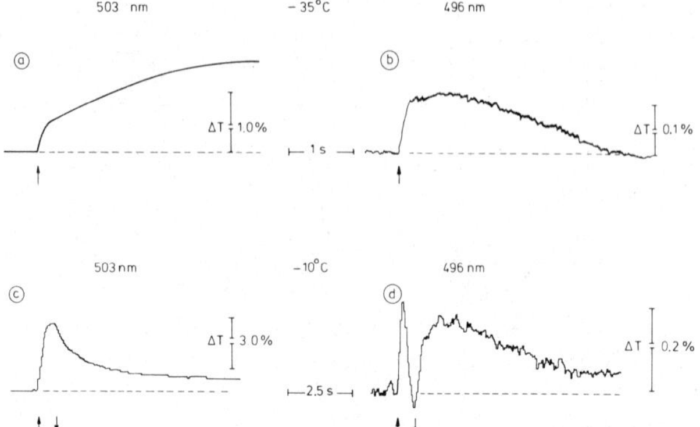

Fig. 3. Light-induced absorbance changes in the presence of PMS and ascorbate. Recordings a and b: -35 °C, c and d: -10 °C. Changes that were not sensitive to gramicidin were subtracted. Bacteriochlorophyll concentration: a and b, 23 µM, c and d, 70 µM.

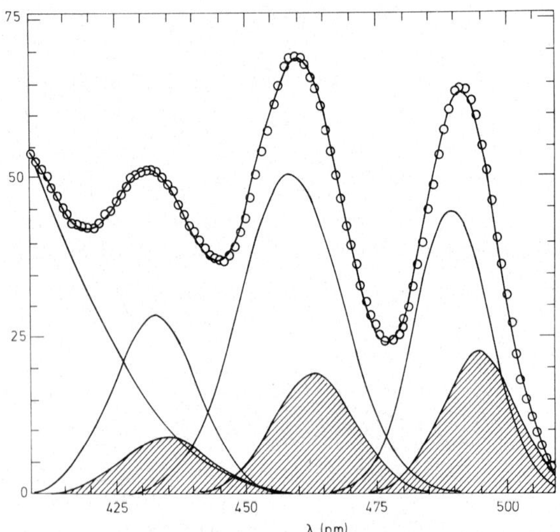

Fig. 4. Analysis of the absorption spectrum of the G1C mutant at -35 °C. The hatched area gives the absorption spectrum of the carotenoid that shows electrochromic band shifts. This spectrum was calculated from the difference spectrum of Fig. 2 (circles) on basis of the calculated band shift of 0.3 nm (see ref. 12 for details of the calculation). For sake of convenience the spectrum is resolved into three Gaussian bands. The remaining carotenoid pool is given by the other three Gaussians. Solid line: overall absorption spectrum; circles: sum of the Gaussian components.

a 8 - 12 times larger shift upon stimulation by secondary electron transport.
Similar results were obtained with the wild strain of Rps. sphaeroides, where an
absorbance shift for a single charge separation was observed which amounted to
0.25 mm for the carotenoid (both at low and room temperature) and 0.07 mm for
bacteriochlorophyll (B850). Fig. 5 shows the difference spectrum of the wild strain
in the presence of partially reduced PMS.

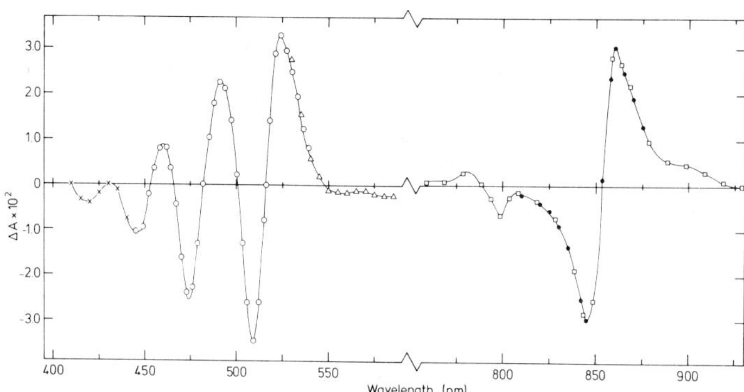

Fig. 5. Absorption difference spectrum of chromatophores of the wild strain of
Rps. sphaeroides. Conditions as for Fig. 2. Bacteriochlorophyll concentration:
80 µM.

DISCUSSION

Although the two pools of carotenoid that appear to exist in the G1C mutant of
Rps. sphaeroides have clearly different absorption spectra, one has to conclude
that both pools consist of neurosporene, since this is virtually the only carotenoid present in this bacterium[10]. Results obtained with the wild strain similarly
suggested the existence of two pools of spheroidene with different absorption
spectra[12]. In addition to spheroidene, the wild strain contains a smaller, but
variable amount of spheroidenone[14,15], but the absorption spectrum of this pigment
in organic solvent suggests that it absorbs at significantly longer wavelength.

The simplest way to explain these results is by the assumption of a local
electric field to which only part of the carotenoid molecules are subjected. Such
a field was proposed by Reich and Schmidt[2] to explain the size and linearity with
membrane potential of the absorbance changes of carotenoids (like neurosporene)
that do not possess a permanent electric dipole. The strength of the field was
estimated to be about $2 \cdot 10^{-6}$ V/cm. Such a field would be of roughly the right
magnitude to explain the difference in location of bands of the two pools, as can
be estimated on basis of the polarizability of a carotenoid such as neurosporene
or spheroidene. The "unnaturally" high long wavelength band in the spectrum of
Fig. 4 may indicate that the shift of this band is actually larger than for the

other two bands of the carotenoid. The shape of the difference spectra obtained by us indicates that an electrochromic broadening of the bands, as proposed by Conjeaud and Michel-Villaz[16] does not occur to any significant extent.

In conclusion it may be stated that our results are in good agreement with the concept that the absorbance changes of pigments as observed in photosynthetic material are caused by electrochromic shifts of the absorption bands, caused by a delocalized potential across the photosynthetic membrane. In earlier work, it was not possible to obtain absorbance changes of the large amplitude reported here, so that shifts of the zero points and transients of absorbance near these wavelengths could not be observed with reasonable certainty. On basis of the, at that time reasonable, assumption that the band shifts were due to a carotenoid with the same absorption spectrum as the overall absorption spectrum in the region around 500 nm, it was concluded that the absorbance changes were caused by large and fixed band shifts (about 8 - 10 nm) of a variable number of carotenoid molecules[8-10]. The present results do not agree with this conclusion.

REFERENCES

1. Junge, W. and Witt, H.T. (1969) Z. Naturforsch. 23b, 244-254.
2. Reich, R. and Schmidt, S. (1972) Ber. Bunsenges. Phys. Chem. 76, 589-598.
3. Witt, H.T. (1971) Q. Rev. Biophys. 4, 365-477.
4. Smith, L. and Ramirez, J. (1960) J. Biol. Chem. 235, 218-225.
5. Clayton, R.K. (1963) Proc. Natl. Acad. Sci. U.S. 50, 583-587.
6. Jackson, J.B. and Crofts, A.R. (1969) FEBS Lett. 4, 185-189.
7. Amesz, J. (1977) Progr. in Botany 39, in the press.
8. Amesz, J. and Vredenberg, W.J. (1966) in Currents in Photosynthesis (Thomas, J.B. and Goedheer, J.C., eds.), Donker Rotterdam, pp. 75-83.
9. Amesz, J., 't Mannetje, A.H. and de Grooth, B.G. (1973) in Abstr. Symp. Prokaryotic Photosynth. Org. Freiburg, pp. 34-35.
10. Crofts, A.R., Prince, R.C., Holmes, N.G. and Crowther, D. (1974) in Proc. 3rd Int. Congr. Photosynth. (Avron, M., ed.), pp. 1131-1146, Elsevier Scientific Publ. Co., Amsterdam.
11. Amesz, J. and de Grooth, B.G. (1976) Biochim. Biophys. Acta 440, 301-313.
12. De Grooth, B.G. and Amesz, J. (1977) Biochim. Biophys. Acta, in the press.
13. Emrich, H.M., Junge, W. and Witt, H.T. (1969) Z. Naturforsch. 24b, 1144-1146.
14. Goodwin, T.W., Land, D.G. and Osman, H.G. (1955) Biochem. J. 59, 491-496.
15. Shneour, E.A. (1962) Biochim. Biophys. Acta 62, 534-540.
16. Conjeaud, H. and Michel-Villaz, M. (1976) J. Theoret. Biol. 62, 1-16.

Supported by the Netherlands Foundation for Chemical Research (SON), financed by the Netherlands Organization for the Advancement of Pure Research (ZWO).

FUNCTION AND ORGANIZATION OF INDIVIDUAL POLYPEPTIDES
IN CHLOROPLAST PHOTOSYSTEM I REACTION CENTER

Nathan Nelson and Bat-El Notsani
Department of Biology
Technion-Israel Institute of Technology
Haifa, Israel

SUMMARY

Purified photosystem I reaction center consists of six different polypeptides
that were designated as subunits I, II, III, IV, V and VI in the order of decreasing molecular weights. There are two copies of subunit I, and together they
contain the P_{700} pigment along with about 40 chlorophyll a molecules. Subunit I
contains reactive groups that cause its high tendency to polymerize. Incubation
at 100°C for 1 to 5 min in the presence of sodium dodecyl sulfate brings about the
appearance of dimers, tetramers and polymers of subunit I. When photosystem I
reaction center was heated to 100°C prior to the addition of sodium dodecyl sulfate, subunit I fished out specifically subunits III and V leaving in the gel only
subunits II, IV and VI. This observation suggested to us that subunits III and V
are the nearest neighbors of subunit I. Subunit III functions in the oxidizing
side and subunit V in the reducing side of photosystem I. In photosystem I reaction center that was prepared from trypsin-treated chloroplasts, subunit I was
split into two polypeptides, I' and I" with molecular weights of 34,000 and
26,000, respectively. It was found that the reactive groups of subunit I are
located on polypeptide I". A model for the subunit structure and function of
photosystem I reaction center is proposed.

INTRODUCTION

Partial reactions of light-induced electron transport and phosphorylation are
catalyzed by specific complexes in the chloroplast membrane. The primary reactions occur in photochemical reaction centers where the light-energy is harvested
and transformed into electrochemical energy. The chloroplast electron transport
system contains two such reaction centers that are functioning in photosystem I
and photosystem II. The specific organization of reaction centers in the thylakoid membrane plays an important role in energy transduction. Therefore,detailed
understanding of the fine organization of pigments, polypeptides and reactive
groups within the reaction centers might elucidate their mechanism of action.

Biochemical studies of photosystems I and II involve the separation and purification of reaction centers while the activity of a specific partial reaction is
monitored. Using such an approach we were able to isolate a chloroplast photosystem I reaction center, active in NADP photoreduction when ascorbate serves as

electron donor and the system is supplemented with purified plastocyanin, ferredoxin and ferredoxin-NADP-reductase[1,2]. The purified reaction center consists of six polypeptides that were designated as subunits I, II, III, IV, V and VI in the order of decreasing molecular weights of 70,000, 25,000, 20,000, 18,000, 16,000 and about 8,000 respectively. Subunit I is the heart of photosystem I reaction center. The P_{700} pigment and, probably, the "very primary" electron acceptor are situated in this polypeptide. About 40 chlorophyll a molecules serve as an integral light-harvesting antenna for each P_{700} within two polypeptides of subunit I. Since in purified subunit I light can be collected to bring about the oxidation of P_{700}, this subunit was termed P_{700} reaction center.

In this paper we shall discuss further studies on the structure, function and arrangement of photosystem I reaction center and its individual polypeptides in the thylakoid membrane.

MATERIALS AND METHODS

<u>Materials</u>: Digitonin, Triton X-100, NADP, Tricine, Tris and 2(n-morpholino) ethane sulfonic acid (MES) were obtained from Sigma. N-methylphenazonium-3-sulfonate (PMS-S) was a generous gift from Dr. G. Hauska. Sodium dodecyl sulfate (SDS), acrylamide and N,N'-methylene-bis-acrylamide were obtained from Bio-Rad Laboratories.

<u>Preparations</u>: Photosystem I reaction center and P_{700} reaction center were prepared from Swiss chard chloroplasts as previously described[1,2,3]. The final preparations of these reaction centers were rather dilute, therefore, it was convenient to concentrate them as follows : about 20 ml of purified reaction center containing 0.05 to 0.08 mg chlorophyll per ml were applied on a DEAE-cellulose column (0.8x10 cm) which was equilibrated with a solution containing 50 mM Tris-Cl (pH 8) and 0.2% Triton X-100. The column was washed with 5 ml of the same buffer, and the reaction center was eluted with a solution containing 50 mM Tris-Cl (pH 8), 0.2% Triton X-100 and 300 mM NaCl. Fractions of 1 ml were collected and the reaction center thus obtained contained 0.2 to 1.0 mg chlorophyll per ml. The recovery of NADP photoreduction and P_{700} photooxidation activities was 80 to 100%. The preparations were stored at -70°C for several months.

Plastocyanin[4], Euglena cytochrome 552 [5], ferredoxin[4] and ferredoxin-NADP-reductase[6] were prepared according to published procedures.

Photosystem I reaction center was prepared from trypsin-treated chloroplasts as follows : About 230 ml of chloroplast suspension in 10 mM Tricine (pH 8), 10 mM NaCl and 400 mM sucrose at chlorophyll concentration of 1 mg per ml were incubated at room temperature for 85 min with 30 ml of trypsin solution containing 5 mg protein per ml in 3 mM H_2SO_4. Then 30 ml of trypsin-inhibitor solution (7 mg per ml) were added and the suspension was centrifuged at 20,000xg for 10 min. The pellet was suspended in 230 ml of solution containing 10 mM Tricine (pH 8), 10 mM NaCl, 5 mM $MgCl_2$ and 400 mM sucrose. The photosystem I reaction center was then prepared as previously described[2,3]. Control photosystem I reaction center was prepared similarly from untreated chloroplasts. The trypsin treatment caused about 60% inhibition of NADP photoreduction activity in the presence of ascorbate, plastocyanin and 0.1% Triton X-100.

Analytical methods: A detailed description of the analytical methods that were used in this study was given previously[1,2,3]. Chlorophyll concentration[7], protein concentration[8] and SDS gel electrophoresis[9] were performed by published procedures. Unless stated otherwise, samples for SDS gel electrophoresis were incubated at room temperature (about 23°C) prior to application for 2 hr in a solution containing 50 mM Tris-Cl (pH 8), 2% SDS, 2% mercaptoethanol and about 10% sucrose. The gels were fixed, stained with coomassie blue, destained and scanned at 600 nm as previously described[2,3].

RESULTS

Interactions among subunits in photosystem I reaction center: Figure 1 shows an SDS gel of purified photosystem I reaction center. Six polypeptide bands designated as subunits I, II, III, IV, V and VI are seen on the gel. Based on the relative intensity of the stain on the gel, a relative ratio of 2:1:1:1:1 was determined for five of the subunits. The relative amount of subunit VI, which was rather diffuse on the gels, was not measured. Quite often it was observed that another protein band appeared on the gel in the position of about 140,000 daltons. We have noticed that the conditions under which the reaction center is incubated with SDS influence the appearance of this band. Longer incubation time and elevated temperatures caused an increase in the amount of the high molecular weight band with parallel decrease in the amount of subunit I on the gel. Figure 2A depicts an SDS gel electrophoresis pattern of control photosystem I reaction center that was incubated with SDS and mercaptoethanol for 2 hours at room temperature. A small band in the position of about 140,000 daltons is seen. Upon incubation of the same preparation at 100°C for 5 min in the presence of SDS and mercaptoethanol most of subunit I disappeared and new bands in the position corresponding to dimers and polymers of this subunit appeared (Fig. 2B). The position and amounts of all other subunits did not alter. The very same phenomenon was observed with purified P_{700} reaction center (subunit I). Figure 3 shows that upon heating the P_{700} reaction center polymerization of the polypeptide takes place. In this preparation small amounts of polymers are present in the control samples. These experiments suggest that subunit I contains reactive groups that cause polymerization of this subunit upon elevation of the temperature under conditions of full dissociation of the individual polypeptides in photosystem I reaction center.

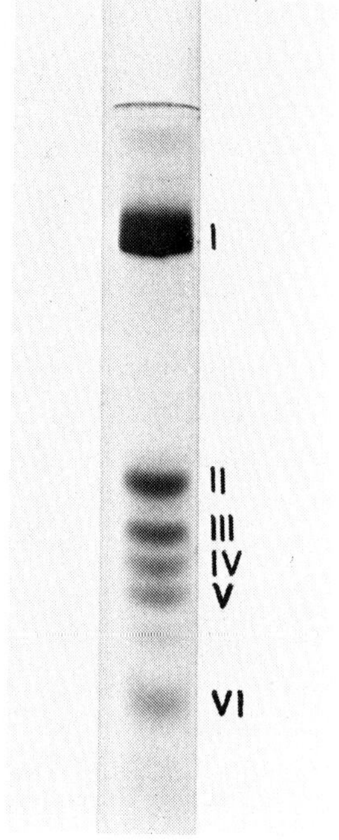

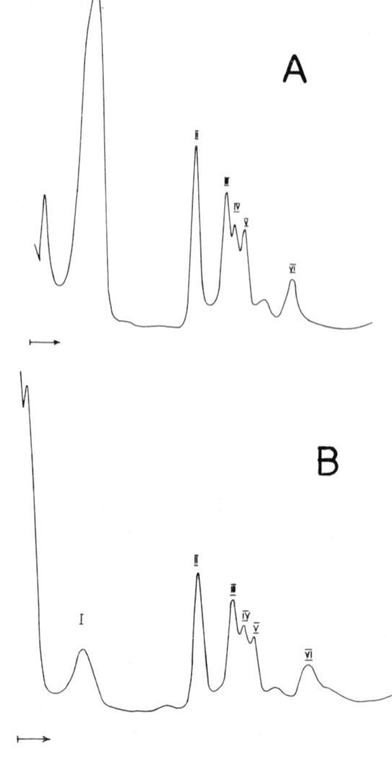

Fig. 1. The SDS gel of purified photosystem I reaction center. The sample (50 µl) containing about 100 µg protein was treated as described in the methods.

Fig. 2. The SDS gel pattern of photosystem I reaction center and heat treated reaction center in the presence of SDS. Samples containing about 80 µg protein were incubated at room temperature for 2 h in the presence of 2% SDS and 2% mercaptoethanol. A) Control. B) Temp. 100°C for 5 min during incubation.

It was interesting to determine whether the reactive groups can interact only among themselves, and thereby polymerize only subunit I, or if they will interact with other polypeptides under different conditions. If the latter is possible, a unique opportunity to study the subunit's interactions without application of external bifunctional reagents exists. Figure 4 clearly shows that upon temperature elevation prior to the SDS treatment, subunits I, III and V disappeared from the gel, while the position and amounts of subunits II, IV and VI were not changed. Incubation of photosystem I reaction center at 100°C for 5 min caused complete polymerization of subunit I with subunits III and V and the aggregates could no longer penetrate the gel. Incubation for shorter periods, or at temperatures lower than 100°C, revealed gradual polymerization of the same subunits and the appearance of penetrable polymers.

Proteolytic enzymes were employed in the study of membrane organization. The susceptability of a specific polypeptide might suggest its location in the

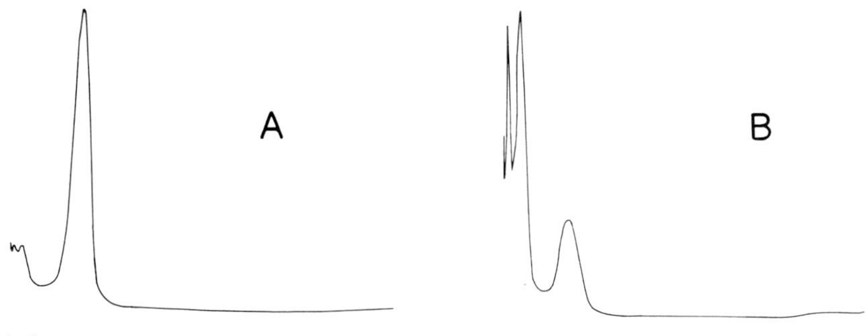

Fig. 3. The SDS gel pattern of P700 reaction center and the heat treated reaction center in the presence of SDS. Samples of P700 reaction center containing about 40 µg protein were treated as described in Fig. 2.

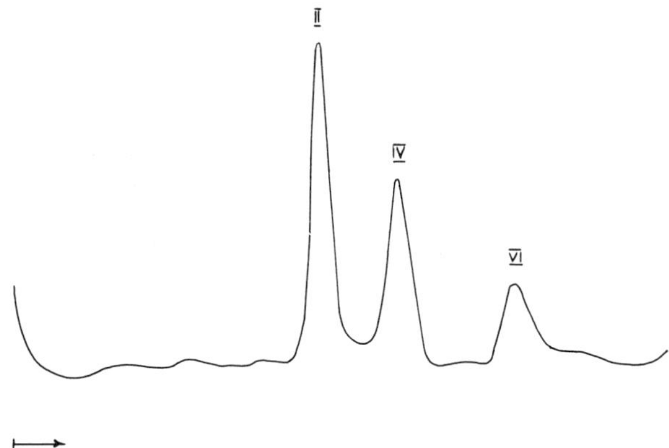

Fig. 4. The SDS gel pattern of heat treated photosystem I reaction center. A sample of photosystem I reaction center containing about 80 µg protein was incubated at 100°C for 5 min. Subsequently 2% SDS and 2% mercaptoethanol were added and further incubation for 2 hours at room temperature took place prior to the electrophoresis.

membrane. Figure 5 shows that subunits I, III, IV and V were susceptible to trypsin when it was mixed with a soluble form of the reaction center. Subunits II and VI seemed to be resistant to trypsin treatment. When the photosystem I reaction center was purified from chloroplasts treated with trypsin, only subunit I was altered. Figure 6 (top) depicts the SDS gel pattern of such a reaction center. It can be seen that the relative amount of subunit I in the reaction center decreased and two new polypeptides appeared. It seems that subunit I was cleaved by the trypsin treatment to polypeptides I' and I". The molecular weight of the newly-formed polypeptides was estimated by calibrated SDS gel electrophoresis to

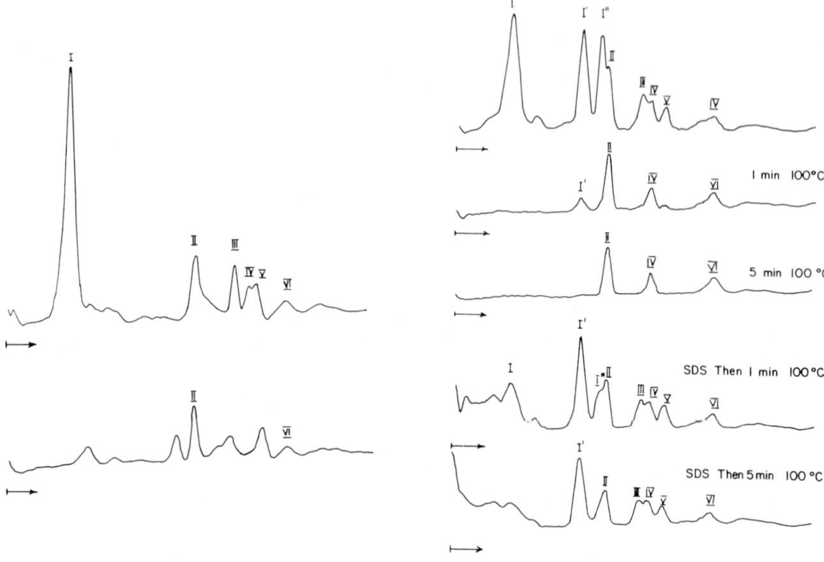

Fig. 5. Effect of trypsin treatment on the polypeptide composition of soluble photosystem I reaction center. The active fraction of the reaction center after DEAE-cellulose column was treated with trypsin at room temp. About 1.2 equivalents of trypsin-inhibitor were added and then the sample was centrifuged on sucrose gradient overnight[3]. Sample of the lower green band (30 μg protein) was incubated with SDS and electrophoresed as described in the methods.

Fig. 6. The SDS gel pattern of photosystem I reaction center that was prepared from trypsin treated chloroplasts and the effect of heat treatment on its polypeptide composition. Photosystem I reaction center was prepared from trypsin treated chloroplasts as described in the methods. Samples containing about 40 μg of protein were treated as specified in the figure.

be 34,000 for subunit I' and 26,000 for subunit I". Therefore, the trypsin treatment might digest, out of subunit I (M.W. 70,000), a small polypeptide of about 10,000 M.W. that protruded from the chloroplast membrane. The possible location of the above-mentioned reactive groups on the new polypeptides was determined by heating the reaction center prior to electrophoresis (Fig. 6). Incubation for 1 to 5 min at 100°C prior to the SDS treatment caused the disappearance of bands I, I', I", III and V from the gels. Incubation for 5 min at 100°C after the SDS treatment caused the disappearance of bands I and II" leaving bands I', II, III, IV, V and VI on the gels. Therefore, it seems that the reactive groups are

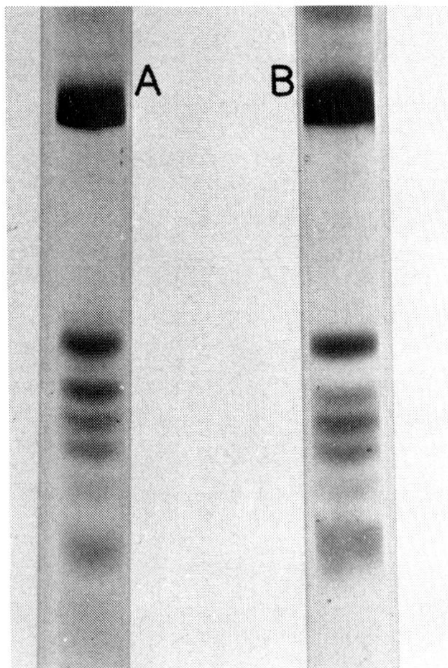

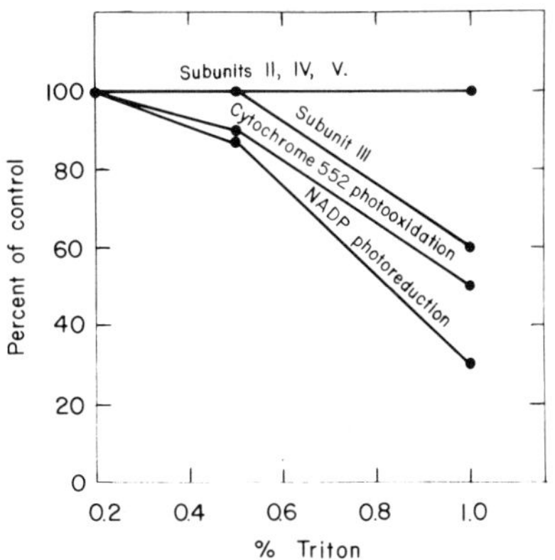

Fig. 8. Effect of Triton X-100 concentration in sucrose gradients on the photochemical activities and the polypeptide composition of the resulting reaction centers. The assay procedures are described in the methods and in ref. 3.

Fig. 7 (left). The SDS gels of photosystem I reaction centers prepared under various Triton X-100 concentrations. A) Photosystem I reaction center containing about 70 μg protein that was prepared in the presence of 0.2% Triton in the sucrose gradient. B) Reaction center containing about 70 μg protein that was prepared in the presence of 1% Triton in the sucrose gradient.

located on polypeptide I" which is part of subunit I of photosystem I reaction center.

Depletion of subunit III from photosystem I reaction center : Treatment with a high concentration of Triton X-100 proved to be effective in successful isolation of photosystem I reaction center[1,2]. However, inclusion of only 1% Triton during the DEAE-cellulose column chromatography or in the sucrose gradients caused the loss of NADP photoreduction activity of the resulting preparations. Figure 7 shows SDS gels of control reaction center that was obtained from sucrose gradient containing 0.2% Triton X-100, and a reaction center that was obtained from sucrose gradient containing 1% Triton X-100. It can be seen that subunit III is partially depleted, while the amounts of all other subunits remained unaltered. Figure 8 shows that NADP photoreduction activity and cytochrome 552 photooxidation activity were progressively inhibited by increasing Triton X-100 concentrations in the sucrose gradients. These inhibitions are concomitant with the depletion of subunit III from photosystem I reaction center. Full depletion of subunit III can be obtained by the application of the purified photosystem I reaction center on a DEAE-cellulose column and washing for several hours with a buffer containing 0.5 to 1.0% Triton X-100[3]. The eluted reaction center completely lost its subunit III and its plastocyanin dependent photoreduction activity. However, in this preparation the NADP-photoreduction activity that was mediated by PMS-S was similar to that obtained with control photosystem I reaction center[3].

P_{700} reaction center and primary electron acceptor: Treatment of the purified photosystem I reaction center with 0.5% SDS, followed by sucrose density gradient centrifugation, separated subunit I from the other polypeptides[1,2]. Since this single polypeptide preparation was active in light-induced P_{700} oxidation, it was named P_{700} reaction center. Table 1 shows the amino acid composition of the purified P_{700} reaction center. It presents a close similarity to the amino acid composition of Thornber's chlorophyll protein complex I [10]. It seems that this complex which has no detectable activity of P_{700} is denatured P_{700} reaction center. On the other hand, Thornber's P_{700}-chlorophyll protein from higher plants, was active in photobleaching of P_{700}. However, this complex contained considerable amounts of cytochromes which are not an integral part of photosystem I reaction centers.

TABLE 1

AMINO ACID COMPOSITION OF P_{700} REACTION CENTER

Amino acid	Amount per 100 μmoles of amino acids	Residues per molecule (M.W. 70,000)
Lysine	2.13	14
Histidine	3.42	22
Arginine	3.63	23
Cysteic acid	2.46	16
Aspartic acid	7.65	49
Threonine	5.83	37
Serine	7.02	45
Glutamic acid	7.73	49
Proline	4.29	27
Glycine	10.12	65
Alanine	9.83	63
Valine	6.32	41
Methionine	1.35	9
Isoleucine	6.26	40
Leucine	11.81	76
Tyrosine	3.56	23
Phenylalanine	6.58	42

Photosystem I reaction center is highly enriched in bound ferredoxin[11]. Fig. 9 shows the light-induced EPR spectrum of the purified photosystem I reaction center at 18°K. In order to obtain a similar EPR spectrum with isolated chloroplasts, a chlorophyll concentration of about 5 mg per ml was required. This represents an about 8 fold enrichment in bound ferredoxin which is in line with the increase in the NADP photoreduction activity during the purification of the reaction center[2]. Mild SDS treatments selectively released subunits IV, V and VI with parallel loss of NADP photoreduction activities and the EPR signal of bound ferredoxin[3,11]. P_{700} reaction center (purified subunit I) is completely free of detectable EPR signal of bound ferredoxin.

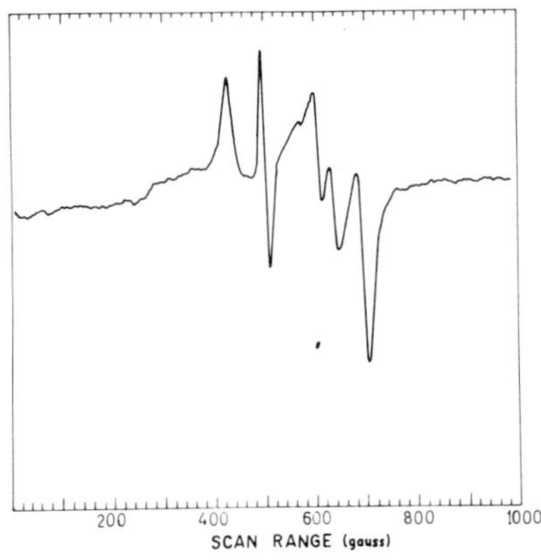

Fig. 9. The EPR spectrum of reduced photosystem I reaction center at 18°K. The spectrum was recorded by a Varian EU spectrometer under an instrument setting of : frequency 0.25 GHz, power 20 mW, modulation amplitude 10 G, scan rate 500 G/min and gain 1250. Chlorophyll concentration was about 0.6 mg per ml.

Kinetic studies that were performed in collaboration with Dr. W. Junge indicated the possible existence of a primary electron acceptor in P_{700} reaction center. By measuring light-induced absorbance changes of P_{700} and the pH indicator cresol red, electron transport from the photosynthetic electron acceptor to oxygen was indicated. By adjusting the PMS concentration, the half life of oxidized P_{700} in the presence of ascorbate was set at 50 m sec. In the presence of methyl viologen, which catalyzes a very rapid electron transfer from the primary acceptor to O_2, the half time for the pH increase was about 2 m sec. This corresponds to the rate of spontaneous dismutation of O_2^-. In the absence of methyl viologen the half time for the pH increase was about 50 m sec. This represents the rate in which the primary electron acceptor undergoes autooxidation. The above-mentioned phenomena were similar in both photosystem I reaction center and P_{700} reaction center. These observations may suggest that the "very primary" electron acceptor is situated in subunit I (P_{700} reaction center) and it is not a conventionally bound ferredoxin.

DISCUSSION

Several techniques were employed in order to gain some knowledge on the structure of photosystem I reaction center within the chloroplast membrane. An antibody that was raised against photosystem I reaction center was found to be specifically directed against subunit I [1]. The antibody inhibited NADP photoreduction by isolated chloroplasts [1,2]. Therefore, it seems that the antibody interacted with the part of subunit I that protrudes from the external side of the thylakoid membrane. This assumption is substantiated by the observation that trypsin treatment of isolated chloroplasts digested a small polypeptide out of subunit I (Fig.6). Since purified subunit I is active in light-induced P_{700} oxidation, it was concluded that the P_{700} pigment is a constituent of this polypeptide. It was measured that two subunits I are present per each P_{700}, both in photosystem I reaction center and in P_{700} reaction center. This together with the subunit stoichio-

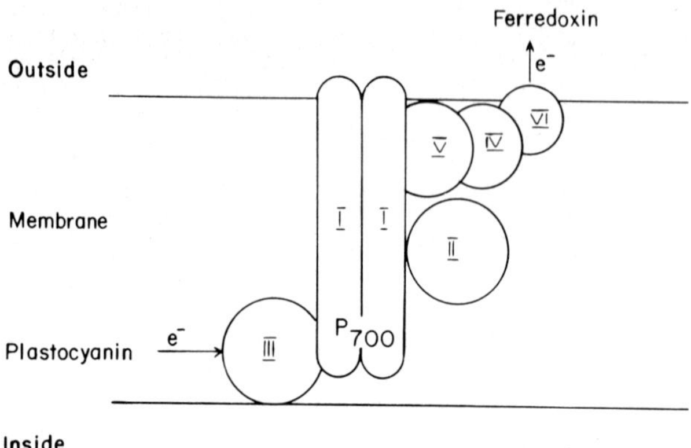

Fig. 10. A proposed model for the subunit structure and function of photosystem I reaction center.

metry of 2:1:1:1:1 which was determined for subunits I, II, III, IV and V respectively[3], suggest that a cooperation between two copies of subunit I is required to form the P_{700} pigment. However, the possibility that P_{700} reaction center (subunit I) is composed of two non-identical polypeptides having similar molecular weights was not eliminated. If this is the case then the P_{700} might be located in one of the polypeptides while the second one functions in its vicinity. The fact that heat treatment, even after SDS treatment, caused polymerization of all the subunits I is in favor with the first assumption.

Purified P_{700} reaction center contains about 40 chlorophyll a molecules per one P_{700} pigment. Studies on depletion of chlorophyll a from both photosystem I reaction center and P_{700} r.center, and light saturation experiments indicated that most of these 40 chlorophyll a molecules are capable of light harvesting for P_{700}. Further depletion of chlorophyll a can be obtained, but the treatment caused disorder in the chlorophyll organization concomitant with a marked decrease in the effectiveness of the light harvesting[3]. Therefore, it seems that about 20 chlorophyll a molecules are an integral part of each subunit I polypeptide.

It was suggested that the P_{700} pigment is situated on the internal side of the chloroplast membrane[12,13,14, 15]. Since P_{700} is located in subunit I, and a specific antibody interacted with the same polypeptide on the external side of the chloroplast membrane, it seems that subunit I is cross-membranal and it might constitute the electron channel from P_{700} to the primary electron acceptor of photosystem I. The question of the identity of the primary electron acceptor is disputed. Malkin and Bearden[16,17] suggested that a ferredoxin-type bound non heme iron protein is the primary electron acceptor. McIntosh et al[18] and Evans et al[19,20] observed a reversible light-induced EPR signal at g = 1.76 at very low temperatures, high instrument power and under conditions in which the bound ferredoxin was fully reduced. They concluded that the primary electron acceptor is not the conventional bound ferredoxin and the function of the latter is to transfer electrons from the primary electron acceptor of photosystem I to soluble ferredoxin. The kinetic studies that indicated the existence of primary electron acceptor in P_{700} reaction center, in which the signal of bound ferredoxin is absent, support this assumption. Up to three different bound ferredoxins were identified in chloroplast photosystem I preparations[16,17,21,22]. Since mild SDS treatments to photosystem I reaction center, selectively liberated subunits IV, V and VI with a parallel loss of the EPR signal of bound ferredoxin, it might be assumed that these polypeptides are in fact the bound ferredoxins.

The regularity of the donor site of photosystem I reaction center was examined by testing the activity of light-induced cytochrome 552 oxidation. Treatment of photosystem I reaction center with high Triton X-100 concentration specifically liberated subunit III of the reaction center. Parallel loss of subunit III, cytochrome 552 photooxidation activity and plastocyanin dependent NADP photoreduction activity was observed (Fig. 8). In photosystem I reaction center in which subunit III was totally depleted, nearly complete loss of the above photochemical reactions was recorded[3]. However, in the presence of PMS-S similar rates of NADP photoreduction was observed in both subunit III and control photosystem I reaction centers. This might suggest that subunit III functions as a mediator of electron transport from plastocyanin to P_{700}^{+}, and this polypeptide is probably located on the internal side of the chloroplast membrane.

Figure 10 depicts a proposed model for the subunit structure and function of photosystem I reaction center. Subunit V was placed in contact with subunit I on the reducing side of the reaction center. This is based on the heat treatment experiments in which subunit I specifically fished out subunits III and V (Fig.4). It seems that the existence of reactive groups upon subunit I was useful for our present study, but misled others who failed to demonstrate, in SDS gels of chloroplast membranes, polypeptides heavier than the α subunit of CF_1.

Further studies are required to elucidate the nature of the reactive groups, the location and structure of the primary electron acceptor, the nature of P_{700} and its environment, and the structure and role of each subunit in photosystem I reaction center.

ACKNOWLEDGEMENT

This work was supported by a grant from the Advancement of Mankind Foundation.

REFERENCES

1. Nelson, N. and Bengis, C. (1975) in: Proceedings of the Third International Congress on Photosynthesis (Avron, M., ed) Vol. I. pp. 609-620, Elsevier, Amsterdam.

2. Bengis, C. and Nelson, N. (1975) J. Biol. Chem. 250, 2783-2788.

3. Bengis, C. and Nelson, N. (1977) J. Biol. Chem., in press.

4. Yocum, C.F., Nelson, N. and Racker, E. (1975) Preparative Biochemistry 5, 305-317.

5. Perini, F., Kamen, M.D. and Schiff, J.A. (1964) Biochim. Biophys. Acta 88, 74-90.

6. Nelson, N. and Neumann, J. (1969) J. Biol. Chem. 244, 1926-1931.

7. Arnon, D.I. (1949) Plant Physiol. 24, 1-15.

8. Lowry, O.H., Rosebrough, N.J., Farr, A.L. and Randall, R.J. (1951) J. Biol. Chem. 193, 265-275.

9. Weber, K. and Osborn, M. (1969) J. Biol. Chem. 244, 4406-4412.

10. Thornber, J.P. (1975) Ann. Rev. Plant Physiol. 26, 127-158.

11. Nelson, N., Bengis, C., Silver, B.L., Getz, D. and Evans, M.C.W. (1975) FEBS Lett. 58, 363-365.

12. Hauska, G.A. (1972) FEBS Lett. 28, 217-220.

13. Nelson, N., Nelson, H. and Racker, E. (1972) Photochem. Photobiol. 16, 481-489.

14. Trebst, A. (1974) Ann. Rev. Plant Physiol. 25, 423-458.

15. Junge, W. (1975) in: Proceedings of the Third International Congress on Photosynthesis (Avron, M., ed) Vol. I. pp. 273-286, Elsevier, Amsterdam.

16. Malkin, R. and Bearden, A.J. (1971) Proc. Natl. Acad. Sci. U.S. 68, 16-19.

17. Bearden, A.J. and Malkin, R. (1976) Biochim. Biophys. Acta 430, 538-547.

18. McIntosh, A.R., Chu, M. and Bolton, J.R. (1975) Biochim. Biophys. Acta 376, 308-314.

19. Evans, M.C.W., Sihra, C.K., Bolton, J.R. and Cammack, R. (1975) Nature 256, 668-670.

20. Evans, E.H., Cammack, R. and Evans, M.C.W. (1976) Biochem. Biophys. Res. Commun. 68, 1212-1218.

21. Evans, M.C.W., Telfer, A. and Lord, A.V. (1972) Biochim. Biophys. Acta 267, 530-537.

22. Ke, B. (1975) in: Proceedings of the Third International Congress on Photosynthesis (Avron, M., ed) Vol. I. pp. 373-382, Elsevier, Amsterdam.

CATION-INDUCED MICROSTRUCTURAL CHANGES IN CHLOROPLAST MEMBRANES: EFFECTS ON PHOTOSYSTEM II ACTIVITY

Salil Bose, John J. Burke, and Charles J. Arntzen
USDA/ARS, Department of Botany, University of Illinois
Urbana, Illinois 61801 USA

ABSTRACT

A method is described which uses non-ionic detergents for the isolation and purification of a light-harvesting chlorophyll-protein complex (LHC) from pea chloroplasts. Antibodies against the LHC were prepared and their specificity was established. The antibody inhibited Mg^{++}-induced decrease of photosystem I (PS I) quantum yield, thus providing evidence for the requirement of the LHC for Mg^{++}-induced energy redistribution between the photosystems. In control chloroplasts, Mg^{++} increased the photosystem II (PS II) quantum yield by 50-100% (significantly larger than the 10-30% decrease of PS I quantum yield). The anti-LHC antibody only partially inhibited the Mg^{++}-induced increase of PS II quantum yield. The antibody insensitive portion of the Mg^{++}-effect on PS II photochemistry was ascribed to a separate effect of Mg^{++} -- one that is distinct from energy redistribution effects (i.e., not "spillover" regulation).

The steady-state flash yield of oxygen evolution in low-salt chloroplasts approximately doubled upon Mg^{++} addition. This indicates that an effect of Mg^{++} was to increase the concentration of active PS II reaction centers. The flash yield of hydroxylamine oxidation also increased after Mg^{++} addition, thus ascribing the Mg^{++}-effect as one which is close to the reaction center II chlorophyll. With saturating intensities of continuous illumination, the rate of ferricyanide reduction increased slightly ($\sim$20%) by Mg^{++}, while the rate of ferricyanide reduction mediated through silicomolybdate in presence of DCMU was approximately doubled by Mg^{++}. These latter data suggest that the rate constant of the rate limiting step of the overall electron transport was not significantly altered by Mg^{++}.

INTRODUCTION

A light-harvesting chlorophyll-protein complex (LHC) is believed to be structurally associated with the photosystem II (PS II) reaction center complex within chloroplast membranes [1,2]. In membrane fractionation studies using detergents and low cation concentration, the LHC was separated from PS II complexes and isolated as a photochemically inactive, structurally distinct component having a low chlorophyll $\underline{a}/\underline{b}$ ratio[3,4]. LHC has been shown to bind reversibly to the PS II complex through cation mediated interactions[5].

It has been generally accepted that cations, Mg^{++} in particular, regulate the distribution of excitation energy between the photosystems (see reference 6 for review). A model of the photosynthetic apparatus has been proposed which incorporates LHC as a functional component that can transfer energy to the photosystems and can participate in energy transfer from PS II to PS I[7]. The role of Mg^{++} and the participation of LHC in regulation of energy distribution have been suggested on the basis of studies with developing pea chloroplasts and mutants of barley and soybean by showing that the LHC is required for the cation-induced regulation of excitation energy distribution between the photosystems [8,9,10]. This paper will provide further evidence supporting this hypothesis and will, in addition, provide evidence for a direct effect of cations on the photosystem II reaction center complex.

MATERIALS AND METHODS

Chloroplast preparation: Chloroplasts were prepared from pea seedlings as described elsewhere[9]. Control chloroplasts were from plants grown under a 16 hr light-8 hr dark cycle. Partially developed (Intermittent Light; ImL) chloroplasts were isolated from seedlings grown in an intermittent light regime described previously[9].

Isolation and purification of LHC: For preparation of the light-harvesting complex, pelleted control chloroplasts were thoroughly dispersed in 35 ml of 5 mM EDTA (pH 7.8) containing 0.1 M Sorbitol. The pH of the resuspended sample was then adjusted to 6.0 with 1.0 N HCl, and the sample was centrifuged at 10,000 G for 10 min. The resulting pellet was dispersed in 0.1 M Sorbitol and then centrifuged again at 10,000 G for 10 min. This washed pellet was resuspended in 0.5% Triton X-100 (in deionized H_2O) to a chlorophyll concentration of 0.5 mg/ml. The mixture was incubated with stirring for 30 min at $25°C$, and centrifuged at 41,000 G for 30 min. The small pellet which formed was discarded and 8.0 ml of the supernatant were loaded on linear 0.1 → 1.0 M sucrose gradients containing 0.5 % Triton X-100. These gradients were centrifuged for 15 hr in a Beckman SW-27 rotor at 100,000 G. The portion of sucrose gradient showing high chlorophyll fluorescence was removed. (The area of very high fluorescence on the gradient tube was detected by illuminating the tube from behind with the beam of a microscope illuminator lamp.) To this fraction $MgCl_2$ and KCl were added to a final concentration of 10 mM and 100 mM, respectively. This sample was stirred briefly, and then 15 ml of the sample was loaded over 5 ml of 0.5 M sucrose in a fixed-angle rotor tube and centrifuged for 10 min at 9000 G to yield an LHC preparation.

Gel electrophoresis: Lipid extracted proteins were solubilized according to the procedure of Hoober[11] and run for 16 hrs at 12 ma on 10 → 20% polyacrylamide gradient slab gels overlaid with a 5% polyacrylamide stacking gel. The 10-20% polyacrylamide was stabilized prior to polymerization by a 5-15% sucrose gradient within the gel. Gel buffers were prepared as described by Studier[12] and pyronin y was utilized as the tracking dye. The gels were stained according to Fairbanks[13] and dried as described by Maizel[14].

Preparation of anti-LHC antibody: The purified LHC (3 to 5 mg protein) was resuspended in 1.0 ml of distilled water, emulsified with an equal volume of Freund's complete adjuvant, and 1.5 ml injected intrascapularly into rabbits.

Three weeks following the primary immunization, blood was removed from the marginal ear vein and the antisera clarified. Normal rabbit serum (NRS) was obtained from the rabbits prior to immunization. Booster injections of LHC were made 5 weeks following the primary immunization. Antigenic specificity was determined by the capillary precipitin test with various antigens.

Flash yield of O_2 evolution: The amount of oxygen evolved by single flash was measured by polarographic techniques as described by Joliot and Joliot[15]. Flashes of saturating intensity and short duration ($\sim$10 μsec) were provided in series by a General Radio Stroboslave Type 1539-A and Tectronix PG 505 pulse generator. Flash yield of hydroxylamine oxidation was measured by reversing the polarity of the electrodes.

Partial photoreactions: PS I rates of electron transport from reduced DPIP to methyl viologen (MV) were measured by continuous recording of O_2 uptake with a Clark type electrode and a YS I Model 53 oxygen monitor. PS II rates of electron transport from H_2O to ferricyanide or to ferricyanide in presence of silicomolybdate and DCMU were measured either by O_2 evolution with the oxygen electrode or by pH change of an unbuffered solution with a combination pH electrode. The reaction mixture in either case was magnetically stirred and thermoregulated at 21°C. Light from a Unitron microscope illuminator was passed through a 1-75 Corning glass filter (heat absorption) and a yellow 3-69 Corning filter. Intensity was varied by neutral density glass filters.

Fluorescence measurements: Chlorophyll _a_ fluorescence induction was monitored by an Aminco J10-280 photomultiplier microphotometer with an S-20 response phototube, stored in a Nicolet 535 digital signal averager, and recorded on an x-y recorder. Fluorescence emission spectra were measured with an Aminco-Bowman spectrofluorometer.

RESULTS

Preparation of antibodies against the light-harvesting complex (α-LHC): The isolation of a highly purified light-harvesting complex was accomplished by fractionation via a non-ionic detergent in the absence of cations. The LHC had a chlorophyll a/b ratio of 1.2 and exhibited no photochemical activity. SDS-polyacrylamide slab gel electrophoretic separation of lipid-extracted material from control chloroplast membranes and the LHC preparation is shown in Figure 1. The control chloroplasts used in these experiments were washed twice with 0.75 m_M_ EDTA to remove coupling factor and ribulose diphosphate carboxylase. Analysis of the slab gel reveals three polypeptides in the LHC preparation; these were assigned molecular weights of 30, 25, and 23 Kdaltons as determined by co-electrophoresis with protein standards.

Specificity of antiserum against the LHC was determined by the capillary

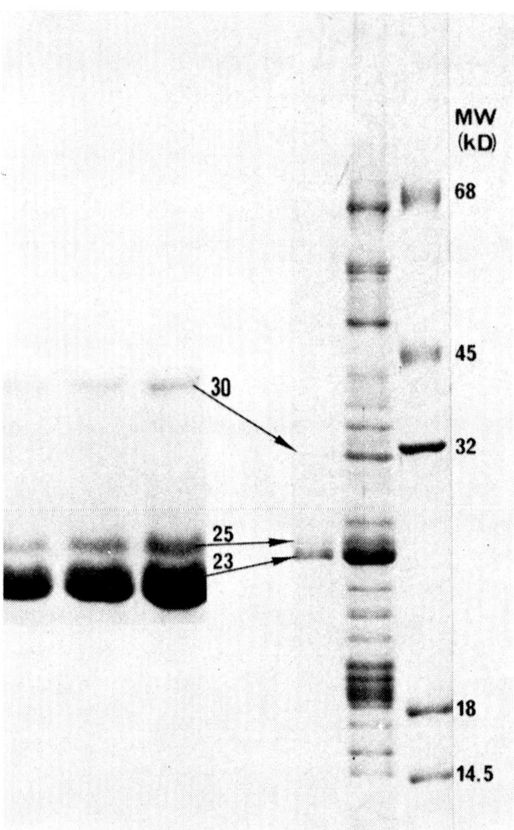

Fig. 1. SDS-polyacrylamide slab gel separation of polypeptides from the purified LHC, control chloroplast membranes, and 5 protein standards. Three polypeptides are visible in the purified LHC preparation. They exhibit molecular weights of 30, 25, and 23 Kdaltons (inset). Control chloroplast membranes were washed twice with 0.75 mM EDTA to remove coupling factor and ribulose diphosphate carboxylase.

precipitin test (Table I). Samples containing the light-harvesting complex exposed to the antiserum clearly exhibited reactivity. Class 1 chloroplasts and ImL plastids had no reaction in the presence of the α-LHC.

Involvement of the LHC in Mg^{++} regulation of "Spillover": Chlorophyll $\underline{a}$ fluorescence analyses have indicated that the LHC is required for Mg^{++}-induced regulation of energy redistribution (spillover and α-change) between photosystems I and II[9,10]. To provide further evidence for this observation, we examined the effects of an antibody elicited against a purified LHC preparation upon the Mg^{++}-induced changes of photosynthetic electron transport partial reactions and variable fluorescence of chl $\underline{a}$. Figure 2 shows that the PS I rate of electron transport in control chloroplasts in limiting light decreased by about 20% upon addition of Mg^{++}. This rate decrease is generally believed to be the result of energy redistribution in favor of PS II[6]. In the presence of the anti-LHC antibody (α-LHC), Mg^{++} had essentially no effect on the PS I rate, thus demonstrating the Mg^{++} had no effect on the regulation of energy distribution when LHC was inactivated by antibody

TABLE I

CAPILLARY PRECIPITIN TEST TO DETERMINE α-LHC SPECIFICITY

Membrane preparations were drawn into capillary tubes to approximately one-third full. The primary antiserum was then drawn up to fill another third. Identical samples were also used for NRS. The tubes were read 60 min after setting up the last tube. A(+) indicates formation of precipitant, (-) indicates lack of reaction.

Sample	α-LHC	NRS
Class 1 chloroplasts	-	-
Stacked chloroplasts	+	-
Unstacked chloroplasts	+	-
ImL chloroplasts	-	-
Purified LHC	+	-

binding. Figure 3 shows that PS II rates of electron transport, under limiting light intensity, increased following Mg^{++} to control chloroplasts. In similar samples, the extent of the Mg^{++}-stimulation of PS II was markedly lessened by the presence of the α-LHC, thus demonstating further the requirement of active LHC

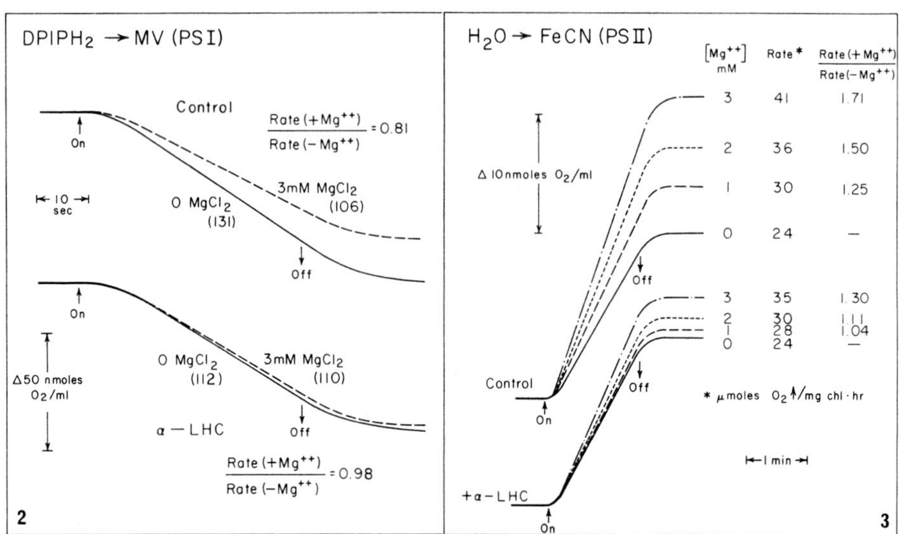

Fig. 2. Effects of α-LHC on Mg^{++}-induced decrease of PS I electron transport under limiting illumination intensities. 1.5 ml reaction mixtures contained 10 mM NaCl, 10 mM Na-Tricine, 50 mM Sorbitol, 40 μM DPIP, 1 mM Ascorbate, 5 μM DCMU and 15 μg chl. Numbers in the parenthesis indicate the rate in μeq/mg chl hr.

Fig. 3. Effects of α-LHC on the Mg^{++}-induced increase of PS II electron transport under limiting illumination intensities. 1.5 ml reaction mixtures contained 10 mM NaCl, 10 mM Na-Tricine, 50 mM Sorbitol, 0.5 mM Ferricyanide and 15 μg chl.

for Mg^{++} regulation of energy redistribution. Similar qualitative effects of
α-LHC were observed with respect to chl a fluorescence (which originates almost
entirely from PS II at room temperature). Figure 4 shows that in the absence of
Mg^{++}, α-LHC has very little effect on the magnitude of either the steady-state
level or the emission peak intensity of fluorescence, whereas in the presence of
Mg^{++}, α-LHC markedly decreased the fluorescence intensity of the steady-state and
the peak emission. Table II shows that the ratio of the variable fluorescence in
presence of Mg^{++} to that in its absence dropped from a value greater than 4 to
about 2 when α-LHC was present.

Mg^{++} activation of PS II units; non-involvement of LHC: It is important to
note that the inhibition of the Mg^{++}-effect by α-LHC on either the PS II rate of
electron transport or on chl a fluorescence was not complete whereas the α-LHC
gave virtually complete inhibition of Mg^{++}-induced PS I rate changes. Moreover,
in control chloroplasts the increase of PS II rate by Mg^{++} was significantly
higher than the decrease of PS I rate by Mg^{++}. These observations suggest that
the increase of PS II activity by Mg^{++} cannot be entirely due to energy redistribution between the photosystems. To test this idea, we examined the effects of
Mg^{++} on the flash yield of oxygen evolution under brief saturating flashes (with
a long dark period between the flashes). Under these conditions, the flash yield
is not influenced by energy redistribution and is related directly to the concentration of PS II reaction centers active in O_2 evolution. Figure 5 shows that Mg^{++}
increased the steady-state yield of control, low-salt chloroplasts by approximately

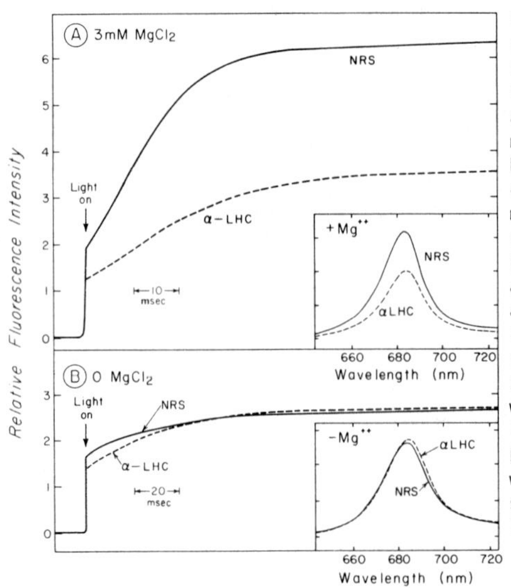

Fig. 4. Effects of antibodies on the chl a fluorescence induction and fluorescence emission (inset) charac-teristics at room temperature with and without Mg^{++}. 2.0 ml reaction mixtures contained 10 mM NaCl, 10 mM Na-Tricine, 50 mM Sorbitol and 20 µg chl. Fluorescence induction was measured with an Aminco J10-280 photomultiplier with an S-20 response phototube. The signal was stored in a Nicolet 535 digital signal averager and recorded on an x-y recorder. A Corning 2-64 (sharp-cut red) filter was mounted in front of the phototube. The actinic light was filtered through Corning 1-75 (heat absorbing) and 4-96 (broad band blue) filters. Exposure of sample was controlled by a Copal photographic shutter which gave an opening time of less than 1 msec.

TABLE II

EFFECTS OF ANTI-LHC ANTIBODY (α-LHC) ON THE Mg^{++}-INDUCED INCREASE

OF CHL A FLUORESCENCE.

Experimental conditions were as in Fig. 5. Measurement of F_O, the initial fluorescence, was taken within 2 msec after full shutter opening and F_M, the maximum total fluorescence, was measured after addition of 10 μM DCMU. NRS = normal rabbit serum. Mg^{++} concentration was 3 mM where indicated. $\Delta F = F_M - F_O$.

Addition	F_O	F_M	ΔF	$\dfrac{\Delta F (+ Mg^{++})}{\Delta F (- Mg^{++})}$
NRS (control)				
$- Mg^{++}$	42	66.5	24.5	
$+ Mg^{++}$	48.5	161.5	113	4.6
α-LHC				
$- Mg^{++}$	35.5	68.5	33	
$+ Mg^{++}$	31.5	91.5	60	1.8

twofold; this demonstates clearly that approximately half of the PS II units were photochemically inactive in the absence of Mg^{++}. Increase of flash yield by similar magnitudes was also observed after adding 100-200 mM NaCl (data not presented).

We have examined the possible involvement of the LHC on Mg^{++}-induced activation of PS II units by two different means: by using α-LHC in control chloroplasts, and by analyzing chloroplasts lacking LHC. Table III shows that Mg^{++} increased the flash yield in the presence of α-LHC to a magnitude similar to that observed in the

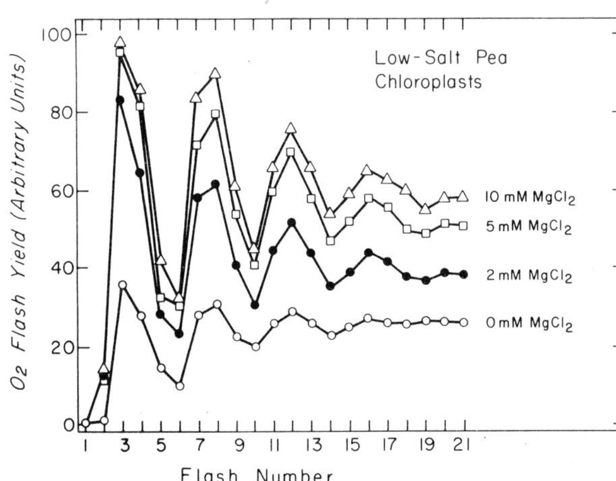

Fig. 5. Flash yield sequence observed with isolated chloroplasts in presence of Mg^{++}. 1.0 ml of each sample kept on ice contained 20 mM NaCl, 10 mM Na-Tricine, 50 mM Sorbitol, 1 mg chl, and $MgCl_2$ as indicated in the figure. An aliquot of each sample was incubated on the platinum electrode for ten minutes in the dark at room temperature before flashing with 2 flashes/sec.

TABLE III

EFFECTS OF THE ANTI-LHC ANTIBODY ON THE Mg^{++}-INDUCED INCREASE

OF STEADY-STATE FLASH YIELD OF O_2 EVOLUTION

One ml of reaction mixture kept on ice contained 20 mM NaCl, 10 mM Tricine, 50 mM Sorbitol, 0.5 mg chl, 0.2 ml antibody or normal rabbit serum, and ± 5 mM $MgCl_2$. An aliquot of each reaction mixture was incubated for ten minutes in dark at room temperature on the platinum electrode before flashing with 2 flashes/sec. Flash yields of three different samples of each mixture are shown. Values presented are arbitrary units. Multiple sample data are presented to demonstrate reproducibility of experimental sampling.

Sample	Normal Rabbit Serum				Antibody Against LHC			
Number	$- Mg^{++}$	Mean	$+ Mg^{++}$	Mean	$- Mg^{++}$	Mean	$+ Mg^{++}$	Mean
1	14		20		14		22	
2	12	13	19	20.3	13	13.6	23	21
3	13		22		14		18	

presence of normal serum. This indicates that LHC was not involved in cation-induced activation of PS II units. This suggestion was supported by the data of Table IV which show that with lmL chloroplasts, which lack the LHC[9], Mg^{++} increased the flash yield to a magnitude comparable to that with control chloro-plasts. Table IV also shows that at limiting illumination intensities, with lmL chloroplasts which did not exhibit Mg^{++}-induced energy redistribution[10], Mg^{++} increased the rate of PS II electron transport markedly (although somewhat less in comparison with the control chloroplasts).

We have attempted to further examine the site of Mg^{++} action on PS II centers by analyzing partial PS II reactions. In one set of experiments hydroxylamine was included since it has been shown to block O_2 evolution and to donate electrons close to the reaction center chlorophyll[16]. The flash-yield of hydroxylamine oxidation increased upon Mg^{++} addition by approximately twofold (Table V), indicating that the Mg^{++} action was not primarily on the water oxidizing enzymes but was more probably directly on the reaction center of PS II.

It is to be noted that the rate of electron transport to ferricyanide under continuous illumination (saturating intensity) did not increase in the presence of Mg^{++} to as large an extent as did the flash yield (Table V). This apparent contradiction can be explained by assuming that (1) the rate constant of the rate-limiting step for the overall electron transport was not altered significantly by Mg^{++}, and (2) the steady-state concentration of the reduced plastoquinone pool, which is generally believed to serve as the common reservoir of electrons for several chains[17], remained essentially unchanged by Mg^{++}. In fact, the value of the

TABLE IV

EFFECTS OF Mg^{++} ON FERRICYANIDE REDUCTION UNDER CONTINUOUS
OR FLASH ILLUMINATION

2.0 ml reaction mixtures contained 10 mM NaCl, 50 mM Sorbitol, 0.5 mM FeCN, 0.1 μM
Gramicidin and 10 μg chl. The pH was adjusted with 0.01 N NaOH to 6.8 ± 0.1.
Ben-Hayyim et al.[18] showed that at pH values of less than 7.0, FeCN was reduced
solely by PS II. pH change during illumination was limited to 0.1. Temperature
21°C. The number in parenthesis gives the ratio of values with Mg^{++}/without Mg^{++}.

| Chloroplasts | $MgCl_2$ (mM) | Rate (μeq/mg chl·hr) in continuous illumination | | Flash yield (μeq/mg chl) x 10^4 |
		Low Light	High Light	10 flashes/sec
Control	0	72	600	11.0
		(1.71)	(1.27)	(1.90)
	5	123	762	19.0
ImL	0	204	1268	15.0
		(1.52)	(1.25)	(1.67)
	5	310	1585	25.0

overall rate constant, as determined from the relation between flash yield of O_2
evolution and dark time between flashes, was found to be essentially unchanged
(data will be published elsewhere). The assumptions described above predict that
the rate of electron transport under continuous illumination of saturating intensity should increase in the presence of Mg^{++} as a consequence of activation of
PS II centers, if the plastoquinone pool and the rate limiting step is bypassed.

TABLE V

EFFECTS OF Mg^{++} ON THE FLASH YIELD OF NH_2OH OXIDATION

One ml of reaction mixture kept on ice contained 20 mM NaCl, 10 mM Tricine, 50 mM
Sorbitol, 1 mg chl, 1 mM NH_2OH, and $MgCl_2$ at different concentrations. Aliquots
of each reaction mixture were incubated for ten minutes in the dark at room temperature on the platinum electrode before flashing. Flash yields were measured with
three different aliquots for each concentration of Mg^{++}; average values are
presented.

| Mg^{++} conc. (mM) | Relative Flash Yield (2 flashes/sec) | | | |
	Sample 1	Sample 2	Sample 3	Mean
0	26	30	25	27
2	42	37	45	41.6
5	57	61	53	57
10	59	55	63	59

254

Figure 6 shows that the rate of electron transport to ferricyanide via silicomolybdate (in presence of DCMU silicomolybdate accepts electron from Q^{19}) did
indeed increase with Mg^{++} to a magnitude similar to that of flash yield. In this
experiment ImL chloroplasts were used to eliminate effects of salt-induced grana
stacking on the accessibility of silicomolybdate to the electron acceptor site
(to be discussed further elsewhere).

DISCUSSION

Involvement of the light-harvesting complex (LHC) in Mg^{++} regulation of
excitation energy distribution: A new method for the isolation of the light-
harvesting chlorophyll a/b complex using non-ionic detergents was utilized in
this study. Fractionation of cation-free chloroplast membranes yielded a structural unit containing chlorophyll a and b and polypeptides in the 23 to 30
Kdalton size classes. Antiserum prepared against the LHC reacted with the
thylakoids of chloroplasts containing the light-harvesting complex. Precipitation of ImL-chloroplast membranes failed to occur in the presence of the α-LHC.
Since ImL-plastids lack the light-harvesting complex this is an essential control
in determining antibody specificity. Class I plastids also failed to react with
α-LHC indicating that the chloroplast envelope blocked α-LHC accessibility to the
chloroplast thylakoids.

Binding of the α-LHC antibody to "low-salt" chloroplasts inhibited Mg^{++}-induced
changes in excitation energy redistribution between the two photosystems as
indicated by salt-mediated changes in PS I and PS II photochemistry and PS II
fluorescence (figs. 2,3,4). These data offer direct evidence for membrane surface
exposure of the LHC, and for an involvement of microstructural changes in the LHC

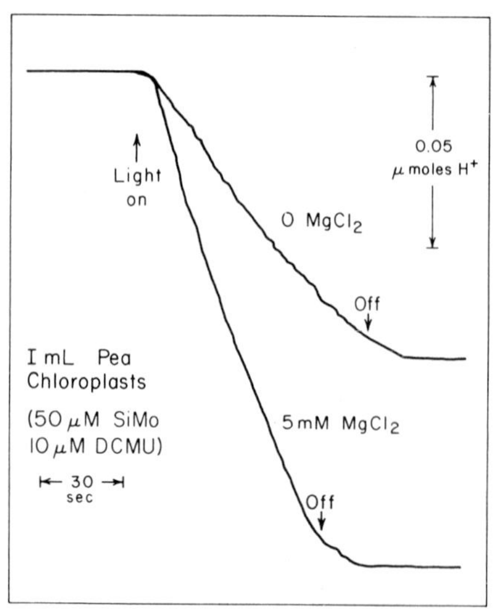

Fig. 6. Effects of Mg^{++} on silico-
molybdate-mediated electron transport
at saturating light intensity.
Reaction conditions same as in Table
IV, except that each sample contained
50 μM silicomolybdate and 10 μM DCMU.

which are induced by cations but are blocked by antibody stabilization.

Activation of PS II units - a new Mg^{++} effect: A measure of the concentration of reaction centers active in photochemistry can be obtained in experiments which use intense brief illumination and a dark time between flashes which is sufficiently long to allow for the relaxation of the excited reaction centers. The flash yield of photochemistry (i.e., of O_2 evolution or of oxidation of an artificial electron donor to the reaction center) is then proportional to the number of active units. The observations reported herein (that the flash yields of O_2 evolution and of hydroxylamine oxidation increased approximately twofold when Mg^{++} was added to low-salt chloroplasts) leads to the obvious conclusion that about half of the PS II reaction centers were inactive under low salt conditions. It should be noted that there are previous reports suggesting a cation-induced activation of PS II units[20,21,22], but none of those reports were able to provide unambiguous quantitative support for the idea (detailed discussion to be published elsewhere).

One of the immediate consequences of finding that Mg^{++} activates PS II centers is that a quantitative explanation of cation effects on PS I and PS II photochemistry can be developed. It has previously been difficult to explain these effects solely in terms of energy redistribution (spillover and α-change) between the photosystems. Butler[23] estimated from analyses of chl _a_ fluorescence at 77°K that the magnitude of increase of PS II quantum yield and concomitant decrease of PS I quantum yield is about 20% as a consequence of energy redistribution regulated by Mg^{++}. It has been demonstrated in Figure 2 and by others[24] that the PS I quantum yield does, indeed, decrease by approximately 20% in presence of cations. The PS II quantum yield, on the other hand, has been shown in Figure 3 and Table IV and by others[24] to increase by as much as 100% upon cation addition. This extra increase of PS II quantum yield, which could not be explained either by quantitative estimation of energy redistribution or from decrease of PS I quantum yield, can now be explained by the finding that Mg^{++} increases the concentration of photochemically active PS II units.

We have examined the interdependence of energy redistribution and activation of PS II units. The observation that activation occurred with chloroplasts lacking Mg^{++}-induced energy redistribution mediated by the LHC (Tables III and IV) suggests that the former is independent of the latter. However, we cannot determine if the converse is true.

The activation of PS II units may be only partially related to the dramatic increase of chl _a_ fluorescence (>200%) by cations at room temperature[9,20,25]. This is indicated by the fact that the anti-LHC antibody blocked most Mg^{++}-induced fluorescence increase (Figure 4, Table II) whereas the antibody does not inhibit the activation of PS II units (Table III). However, elucidation of the fluorescence characteristics of the inactive units would be useful in understanding the mechanism of the cation-induced fluorescence change in chloroplasts.

The mechanism of PS II unit inactivation in salt depleted chloroplasts and the reactivation of units upon Mg^{++} addition may involve membrane structural changes. The occurrence of activation with ImL chloroplasts suggests that grana-stacking and related macroscopic changes in membrane structure are not involved in the mechanism. We have found (data not presented) that glutaraldehyde prefixation of low-salt chloroplasts blocks subsequent Mg^{++} effects on the reactivation of PS II units. Thus microconformational changes involving ionic and hydrophobic interactions between membrane components may be part of this process.

Acknowledgements: This research was supported in part by NSF Grant BMS-75-03935 and NIH Predoctoral Training Grant 6M7283-1. We thank C. L. Ditto for her excellent technical assistance throughout this study.

Abbreviations: DCMU: 3-(3,4-dichlorophenyl)-1, 1 dimethyl urea (Diuron); DPIP: dichlorophenol indophenol; SDS: sodium dodecyl sulfate; EDTA: ethylene diamine-tetra-acetic acid; chl: chlorophyll.

REFERENCES

1. Thornber, J. P. (1975) Annu. Rev. Plant Physiol. 26, 127-158.
2. Anderson, J. (1975) Biochim. Biophys. Acta 416, 191-235.
3. Wessels, J. S. C., van Alphen-Van Wavern, A. and Voorn, G. (1973) Biochim. Biophys. Acta 292, 741-750.
4. Arntzen, C. J. and Ditto, C. L. (1976) Biochim. Biophys. Acta 449, 259-274.
5. Arntzen, C. J., Armond, P. A., Briantais, J-M., Burke, J. J. and Novitzky, W. P. (1976) Brookhaven Symp. Biol. No. 28, 316-337.
6. Barber, J. (1976) Topics in Photosynthesis, vol. 1, Ch. 3, North-Holland, Amsterdam.
7. Butler, W. L. and Kitajima, M. (1975) Biochim. Biophys. Acta 396, 72-85.
8. Davis, D. J., Armond, P. A., Gross, E. L. and Arntzen, C. J. (1976) Arch. Biochem. Biophys. 175, 64-70.
9. Armond, P. A., Arntzen, C. J., Briantais, J-M. and Vernotte, C. (1976) Arch. Biochem. Biophys. 175, 54-63.
10. Lieberman, J. R., Bose, S. and Arntzen, C. J. (1977) Submitted.
11. Hoober, J. K. (1970) J. Biol. Chem. 245, 4237-4334.
12. Studier, F. W. (1973) J. Mol. Biol. 79, 237-248.
13. Fairbanks, G., Steck, T. L. and Wallach, D. F. H. (1971) Biochemistry 10, 2602-2617.
14. Maizel, J. V. (1971) Methods in Virology 5, 179-246.
15. Joliot, P. and Joliot A. (1968) Biochim. Biophys. Acta 153, 625-634.
16. Joliot, A. (1977) Biochim. Biophys. Acta 460, 142-151.
17. Witt, H. T. (1971) Quart. Rev. Biophys. 4, 365-477.
18. Ben-Hayyim, G., Drechsler, Z. and Neumann, J. (1976) Plant Sci. Lett. 7, 171-178.
19. Giaquinta, R. T. and Dilley, R. A. (1975) Biochim. Biophys. Acta 387, 288-305.
20. Rurainski, H. J. and Hoch, G. E. (1971) in Proc. 2nd Int. Congr. Photosynthesis Research. Vol. 1, 133-141, W. Junk N. V., The Hague.
21. Jennings, R. C. and Forti, G. (1974) Biochim. Biophys. Acta 347, 299-310.
22. Wydrzynski, T., Gross, E. L. and Govindjee (1975) Biochim. Biophys. Acta 376, 151-161.
23. Butler, W. L. (1976) Brookhaven Symp. Biol. No. 28, 338-346.
24. Ben-Hayyim, G. and Avron, M. (1971) Photochem. Photobiol. 14, 389-396.
25. Bose, S. (1975) Ph.D. Thesis, University of Rochester.

Bioenergetics of Membranes. L. Packer et al. ed.

A COMPARATIVE STUDY OF GLUTARALDEHYDE AND DIMETHYLSUBERIMIDATE AS PROTEIN CROSSLINKING AGENTS FOR CHLOROPLAST MEMBRANES

George C. Papageorgiou and Joan Isaakidou
Nuclear Research Center Demokritos, Department of Biology,
Athens, Greece

SUMMARY

The effects of treating isolated spinach chloroplasts with the protein crosslinking agents glutaraldehyde and dimethylsuberimidate have been examined with respect to the following properties: (1) ability of chloroplasts for volume changes in response to changes of the osmotic pressure; (2) number of membrane protein amino groups available for trinitrophenylation with trinitrobenzenesulfonic acid (TNBS); (3) photosystem II and photosystem I electron transport activities ; and (4) stabilization of electron transport activities while the chloroplasts are subjected to work stress.

Both crosslinkers destroy the osmotic response of chloroplasts, but the diimidoester was found to alkylate a higher percentage of the TNBS-detectable protein amino groups (55 %)than the dialdehyde (42%). Dimethylsuberimidate inhibits electron transport through photosystem II, affecting specifically the water-splitting enzyme system, but has no detectable effect on electron transport through photosystem I. On the other hand, glutaraldehyde inhibits partially electron transport through photosystem II, and severely electron transport through photosystem I. Glutaraldehyde, which is known to assist in the preservation of the functions of resting chloroplasts, has no detectable function-stabilizing influence in the case of working chloroplasts.

INTRODUCTION

Algae, isolated higher plant chloroplasts, photosynthetic bacteria, and bacterial chromatophores have been subjected to chemical

Abbreviations: Chl, chlorophyll; DCIP, $DCIPH_2$, oxidized and reduced 2,6-dichlorophenol indophenol; DCMU, 3(3,4-dichlorophenyl)-1,1-dimethylurea; DMS, dimethylsuberimidate; DPC, diphenylcarbazide; GA, glutaraldehyde; MV, methylviologen; PDox, oxidized p-phenylenediamine; TAPS, tris (hydroxymethyl) methylaminopropane sulfonate; TNBS, 2,4,6-trinitrobenezene sulfonic acid.

modification with protein crosslinking agents with regard to two
main objectives: (1) to establish what particular photosynthetic
functions are still present in the membrane after its "immobilization" by the crosslinker; and (2) to establish whether protein crosslinking stabilizes the structure, as well as structure-dependent
functions of the photosynthetic membrane.

The term "immobilized membrane" has no precise meaning; it is
operationally defined as the state at which membrane-bounded organelles (such as thylakoids) can no longer change their volume in
response to changes in the osmotic pressure of the suspension medium.
Stabilization of membrane structure and functions is almost invari-
ably understood with reference to resting preparations, such as
those stored in darkness and at low temperature. It is assessed in
terms of initial rates, usually recorded immediately after a cold-
-stored sample is introduced into room temperature reaction medium,
and therefore it should be differentiated from the notion of stability (as used in this work) which refers to the continuous working
state of the tested preparation.

Following the work of Park et al. (1), the most commonly employed
crosslinking compound in such studies is glutaraldehyde. In aqueous
solution, glutaraldehyde condenses to a mixture of polyunsaturated
aldehydes (2), which crosslink proteins both by Schiff base formation with free $-NH_3^+$ groups of proteins, and by having their double
bonds rectified with $-NH_3^+$, SH, phenyl, and imidazolyl groups (3).
Recently, aliphatic diimidoesters have been introduced as protein
crosslinkers in studies with photosynthetic membranes (4). In contrast to glutaraldehyde, these compounds react with $-NH_3^+$ only yielding amidines that carry a positive charge in the vicinity of the
reacted protein amino group (5). Thus, they offer the double advantage of absolute chemical specificity and of membrane charge conservation.

Glutaraldehyde-treated chloroplasts are osmotically inert (6,7),
they preserve partially the ability for photoinduced transmembrane
proton transport (6,7) and for photosystem II-dependent O_2 evolution
(1,8-11), but they do not phosphorylate and they have their photosystem I reaction center inhibited (11). They also withstand prolonged cold storage (1, 11) and treatment by detergents (12), chaotropic compounds (10) and trypsin (10) better than unfixed controls.
Imidoester-treated chloroplasts, on the other hand, are capable of
photoinduced proton transport (4), but they do not photoevolve O_2
unless the crosslinking treatment was carried out at 0° C, and

even then they are quickly inactivated by continuous illumination
(L. Packer and G.C. Papageorgiou, unpublished experiments). In contrast, following treatment with DMS, the cyanobacterium <u>Anacystis</u>
<u>nidulans</u> is capable of continuous ferricyanide-dependent O_2 evolu-
tion, provided it is maintained below the solid to liquid transition
temperature of its membrane lipids (13).

In this paper, we present a systematic comparison of the properties of glutaraldehyde and DMS-treated chloroplasts. Both compounds
are capable of "immobilizing" thylakoids, but their effects on the
photoinduced electron transport across photosystem II and photosystem I are distinctly different. Glutaraldehyde is a specific inhibitor for photosystem I, whereas DMS inactivates the water-splitting enzyme system that is associated with photosystem II.

MATERIALS AND METHODS.

Chloroplasts were isolated by blending spinach leaves in cold
TAPS.KOH 50 mM, NaCl 175 mM, pH 8.3 and resuspending the fraction
which is sedimented by centrifugation between 500xg for 3 min and
3000xg for 5 min in the same buffer at a density of 1 mg Chl/ml
(stock chloroplasts). The DMS reagent was prepared fresh (13) and
it was allowed to react with chloroplasts overnight (about 18h).
Nominal only concentrations of DMS are listed in the Results section,
calculated on the basis of the added reagent. Due to the competing
hydrolysis (5), however, the actual imidoester concentrations are
smaller. Following overnight treatment with DMS at 0° C, chloroplast activities were assayed without removing the excess crosslinker and its hydrolysis products.

Grade I glutaraldehyde, a specially purified product of Sigma,
was incubated precisely for 5 min with stock chloroplasts, which
were then immediately centrifuged down at 5000xg for 5 min, and resuspended at a density of 1 mg Chl/ml. In the course of this work
we found that for reproducible preparations it was very important to
adhere strictly to the same fixation protocoll, namely glutaraldehyde
treatment for exactly 5 min at exactly 0° C.

Free amino groups in membrane proteins were determined by trinitrophenylation with TNBS (14) after extracting chloroplasts with
80 % v/v aqueous acetone. TNBS slowly penetrates into membranes (15)
and therefore it is taken to react with exposed amino groups only
(16,17). Total Chl was determined according to Arnon (18). Osmotic
volume changes were estimated macroscopically in terms of the
absorbance difference at 546 nm of identical samples equilibrated

for 10-15 min in 333 mM NaCl and in distilled H_2O.

Electron transport activities were measured by transferring stock chloroplast aliquots in sucrose 200 mM, $MgCl_2$ 3 mM, and TAPS.KOH 10 mM, pH 8.5. For ferricyanide-dependent O_2 evolution, the reaction mixture contained 30-50 µg Chl/ml and 1 mM $K_3Fe(CN)_6$. For MV-mediated O_2 uptake the reaction mixture contained 10-20 µg Chl/ml, 10 µM DCMU, 2 mM sodium ascorbate, 30 µM DCIP and 10 µM MV. O_2 evolution or uptake was measured under saturating light illumination with a concentration electrode (Rank Brothers, Cambridge, England).

RESULTS

One way to estimate the extent to which membrane proteins have been modified is to measure the number of TNBS-detectable amino groups that have been reacted with glutaraldehyde and DMS. This is illustrated in Fig. 1 where the relative number of exposed amino groups is plotted against the ratio of crosslinker to Chl (fixation ratio). Progressively fewer amino groups are available for trinitrophenylation, as the fixation ratio for both crosslinkers is increased. Evidently, DMS is more effective in alkylating a greater percentage of exposed amino groups than glutaraldehyde throughout the examined concentration range. Maximal amino group disappearance

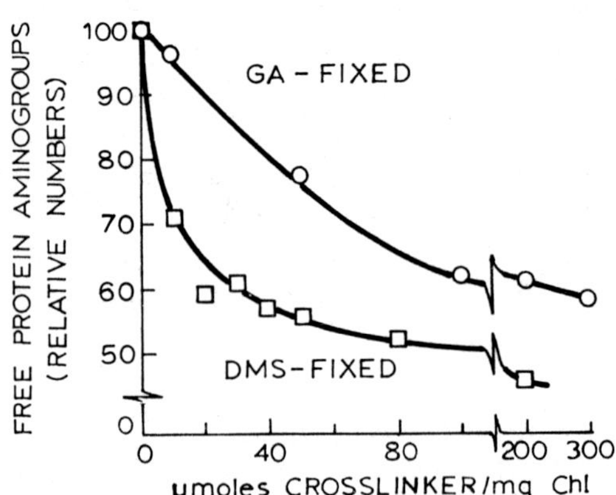

Fig. 1. TNBS-detectable amino groups in chloroplast membrane proteins as a function of the relative amount of the crosslinking reagent with which chloroplasts were treated. The TNBS assay was carried out with chloroplasts extracted with 80% (v/v) aqueous acetone.

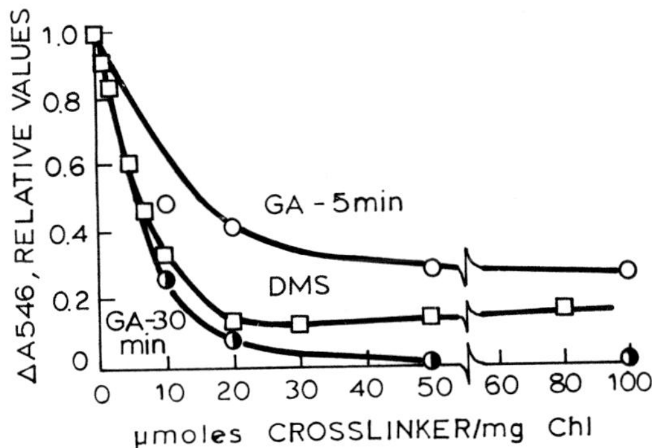

Fig. 2. Osmotic response of isolated chloroplasts as a function of the relative amount of the crosslinking reagent with which they have been treated. Volume differentials between hyperosmotic (333 mM NaCl) and hypoosmotic (distilled H_2O) media are assessed in terms of 546 nm light scattered off the $0°$-direction by the two types of chloroplast suspension. Chloroplasts were reacted at $0°$ C for 5 min and 30 min with GA, and overnight (18 h) with DMS.

is 55 % in the case of DMS and 42 % in the case of glutaraldehyde. This may be due to the fact that amidine formation upon reaction of the imidoester function with protein-NH_3^+ is irreversible (5), whereas Schiff base formation between aldehydes and primary amines is a reversible reaction (2).

On the basis of sedimentation studies with detergent extracts of glutaraldehyde-fixed spinach thylakoids Sane and Park (12) concluded that this reagent establishes both inter- and intramolecular protein linkages. We have observed in other experiments (J. Isaakidou and G.C. Papageorgiou, unpublished results) that treatment with DMS raises the molecular size of chloroplast membrane proteins above 300 kdaltons, evidencing thus the presence of intermolecular crosslinks. The establishment of intermolecular, and possibly of intramolecular, crosslinks in glutaraldehyde and DMS-reacted thylakoids is also indicated by their osmotic immobilization. Fig. 2 shows that the capacity of thylakoids for osmotic volume changes (translated into $0°$-light scattering changes) decreases with increasing fixation ratio with either crosslinker tested. In the case of DMS maximal immobilization is achieved at a nomimal ratio of 20 μmoles DMS/mg Chl. With glutaraldehyde, a multifunctional crosslinker (2,3), maximal effect depends on the duration of the fixation reaction. For 5 min incubation with this crosslinker, maximal osmotic immobilization is also

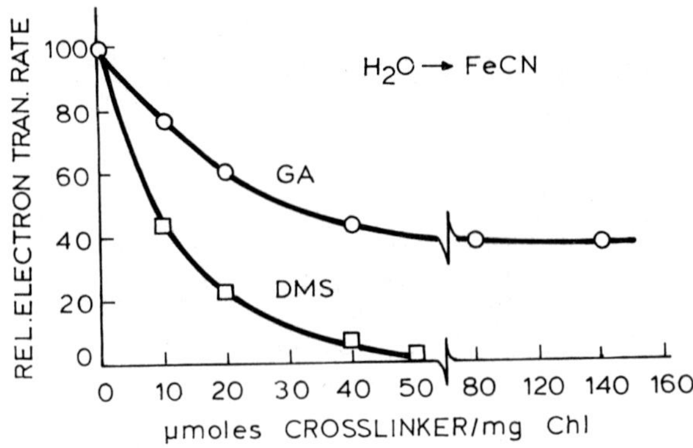

Fig. 3. Relative rates of ferricyanide-dependent O_2 evolution by isolated chloroplasts as a function of the relative amount of the crosslinking reagent with which they have been treated. Assays were carried out at 3° C with GA-fixed preparations (control rate, 375 µeq/mg Chl.h) and at 23° C with DMS-fixed preparations (control rate, 460 µeq/mg Chl.h).

realized at about 20 µmoles glutaraldehyde/mg Chl.

The effect of protein crosslinking on the rate of ferricyanide-dependent O_2 evolution (photosystem II activity) is presented in Fig. 3. Each experimental point on this figure represents a rate reading immediately after mixing a chloroplast aliquot with the re-action mixture (see Materials and Methods). Assays were carried out at 3° C for glutaraldehyde-fixed chloroplasts and at 23° C for DMS--fixed chloroplasts. A substantial fraction (ca 40 %) of photo-system II activity survives at fixation ratios that suffice to im-mobilize the thylakoids osmotically. In contrast, DMS blocks O_2 evo-lution completely at a threshold fixation ratio of 50 µmoles DMS/mg Chl. In other experiments in which we measured electron transport from H_2O to MV (i.e. through both photosystems), we observed a low level of activity (ca 20 %) persisting at fixation ratios as high as 100 µmoles DMS/mg Chl. In addition, when the artificial electron do-nor DPC was used to bypass the H_2O-splitting enzyme and donate elec-trons directly to photosystem II, the fraction of MV-mediated O_2 up-take recovered with the DMS-fixed thylakoids amounted to more than 65 % (J. Isaakidou, and G.C. Papageorgiou, unpublished results).

Recently, Hardt and Kok (11) reported that glutaraldehyde-fixed chloroplasts work better with PDox (essentially a freshly prepared mixture of p-phenylenediamine and excess ferricyanide; ref. 19) than with ferricyanide alone. They attributed this to the fact that PDox,

TABLE I

RATES OF FERRICYANIDE-DEPENDENT AND PDox-DEPENDENT OXYGEN EVOLUTION BY ISOLATED CHLOROPLASTS FOLLOWING FIXATION WITH GLUTARALDEHYDE (GA) AND DIMETHYLSUBERIMIDATE (DMS)

Type of Sample	Rate, μeq/mg Chl.h			
	H_2O	FeCN	H_2O	PDox
Control chloroplasts		362		361
Fixed chloroplasts, 50 μmoles GA/mg Chl		120		262
Control chloroplasts		305		436
Fixed chloroplasts, 10 μmoles DMS/mg Chl		135		148
Fixed chloroplasts, 50 μmoles DMS/mg Chl		0		0

as a permeant oxidant can couple directly to the electron supply of photosystem II, whereas the impermeant ferricyanide anion couples preferentially to an exposed post-photosystem I site. We confirmed these observations, but we found also that PDox does not relieve the inhibition of O_2 evolution that is introduced upon fixation of isolated chloroplasts with DMS (Table I).

Hardt and Kok (11) reported recently an 80-90 % inhibition in the photoinduced electron transport rate from $DCIPH_2$ to MV (photosystem I activity) after treating chloroplasts with 0.5 % glutaraldehyde

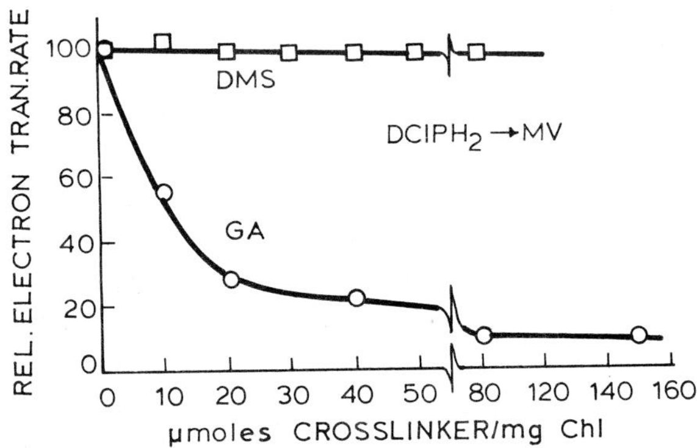

Fig. 4. Rate of MV-mediated O_2 uptake by DCMU-poisoned chloroplasts with ascorbate-reduced DCIP ($DCIPH_2$) as the electron donating substrate as a function of the relative amount of the crosslinking reagent with which chloroplasts have been treated. Assays were carried out at 3° C with GA-fixed preparations (control rate, 390 μeq/ /mg Chl.h) and at 23° C with DMS-fixed preparations (control rate, 336 μeq/mg Chl.h).

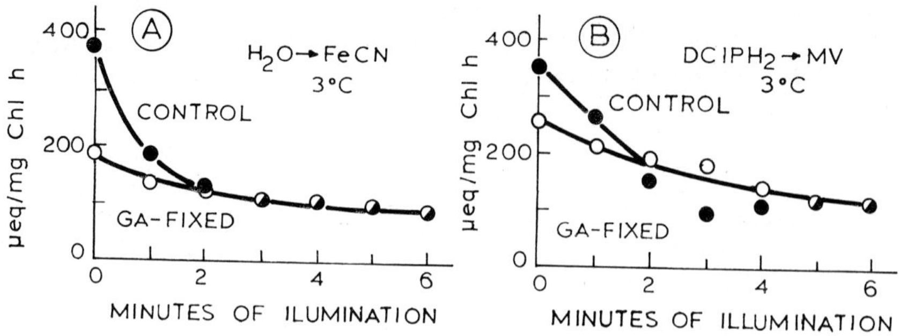

Fig. 5. Stability of photosystem II (A) and of photosystem I (B) activity of control and GA-fixed chloroplasts as a function of total time of exposure to saturating white light. The assay involved periodic illumination for 1 min, followed by 1 min dark. Fixation ratios: A, 150 μmoles GA/mg Chl; B, 10 μmoles GA/mg Chl.

(approx. 12-20 μmoles/mg Chl) for 5 min at 0° C. We have studied the effect of progressively increasing crosslinker concentrations on the same activity, under conditions of saturating illumination (Fig. 4). Glutaraldehyde is indeed inhibitory with the threshold of maximal inhibition at about 20 μmoles/mg Chl. In contrast, however photosystem I activity appears to be totally oblivious to treatment with DMS, at fixation ratios that suffice for maximal amino group disappearance (Fig. 1), membrane immobilization (Fig. 2), and photosystem II inhibition (Fig. 3).

Fixation with glutaraldehyde has a definite advantage in prolonging the ferricyanide-dependent O_2 evolution activity of chloroplasts that are stored at 0°-4° C and in darkness (1, 11). Dark storage of chloroplasts at 17° C accelerates their inactivation rate, but still the surviving activity after fixation is detectable even when the unfixed control becomes totally inert (11). Such estimates of stability are based on initial rates, recorded immediately after turning the illumination on. Membrane activity, however, involves mass transport, such as translation of electron carriers and lateral or transverse motions of protein subunits, on which the presence of inter- and intramolecular crosslinks is expected to have an effect. We asked, therefore, the question whether fixation with glutaraldehyde offers a stability advantage during the working state of chloroplasts, as it does during their resting state.

In the experiment shown in Fig. 5, control and glutaraldehyde--fixed chloroplasts were subjected to cycles of 1 min saturating illumination alternating with 1 min dark. To minimize thermal inactivation, the assays were performed at 3° C, although it should be realized that this temperature is above the solid-liquid transition range of chloroplast lipid (20). The fixation ratios are 150 μmoles glutaraldehyde/mg Chl for ferricyanide-dependent O_2 evolution, and 10 μmoles glutaraldehyde/mg Chl for MV-mediated O_2 uptake. In the second case, partial fixation was necessary in view of the inhibitory effect of glutaraldehyde on photosystem I activity (see Fig. 4). The figure clearly illustrates that fixation with this crosslinker offers no stability advantage for the two activities tested during a continuous working state of chloroplasts.

DISCUSSION

In the foregoing sections, we have shown that glutaraldehyde and DMS immobilize thylakoids against osmotic volume changes, while at the same time the number of TNBS-detectable thylakoid amino-groups is lowered.   There is ample evidence that both compounds link proteins covalently (21), but the relative importance of intermolecular and intramolecular linkages in producing "locked-in" thylakoids cannot be assessed from existing data.

Sane and Park (12) conjectured that glutaraldehyde establishes both inter- and intramolecular crosslinks, but without an apparent increase in protein size, since the sedimentation constants (S_{20}) of detergent solubilized particles from fixed thylakoids were similar, or smaller than the sedimentation constants of particles from unfixed thylakoids. Proteins solubilized from glutaraldehyde-fixed thylakoids do not enter into polyacrylamide gels, but this was attributed to the loss of protein ionizable groups upon fixation rather than to size enlargement. Similarly, detergent-solubilized proteins from DMS-reacted mitochondria (22) and chloroplasts (J. Isaakidou, and G. Papageorgiou, unpublished results) also fail to enter the gels but since protein charge is conserved in this case, the failure has been attributed to the increased size of crosslinked proteins.

It is quite plausible to assume that the monionic glutaraldehyde has access to membrane regions from which the cationic imidates are excluded. In analyzing function-dependent conformational changes in the coupling factor protein of chloroplasts, Oliver and Jagendorf (17) assumed that methyl acetimidate amidinates externally located amino groups only. Hassel and Hand (2), however, found that DMS renders insoluble up to 92% of liver tissue protein as compared to 74-90 % rendered insoluble with glutaraldehyde. A conservative conclusion is that imidates may also penetrate into membranes as neutral species, especially at the alkaline pH of the amidination reaction.

Due to the slow penetration of TNBS into biological membranes (15, 16) and the relative rapidity of the employed assay procedure (5 min; ref. 14), only exposed amino groups are trinitrophenylated. Following DMS and glutaraldehyde crosslinking, the set of nondetectable amino groups comprises those that have been alkylated, those hidden within hydrophobic barriers and thus inaccesible to TNBS, and those that became sterically shielded by the crosslinking bridges formed. In other words, it is not necessarily true that the fraction of amino groups that disappeared equals the fraction of those alkylated. This should be especially true in the case of glutaraldehyde, which establishes multiple linkages reacting with several protein sites. We have found that more amino groups disappear after treating chloroplasts with DMS overnight, than following 5 min treatment with glutaraldehyde (Fig. 1). In other experiments, in which the glutaraldehyde treatment was extended to 30 min, we found that the fraction of TNBS-detectable amino groups can be as low as 30 %. It is evident therefore that the 5 min fixation with glutaraldehyde employed by us and by Hardt and Kok (11) is not sufficient for complete reaction. Nevertheless, it suffices in obliterating osmotic volume changes in glutaraldehyde-treated thylakoids (Fig. 2).

Although both glutaraldehyde and DMS fix thylakoids, the effects of each crosslinker on the functional properties of chloroplasts are quite distinct. DMS inhibits ferricyanide-dependent O_2 evolution completely whereas with glutaraldehyde the inhibition is partial (Fig. 3). When DPC is used to bypass the water-splitting enzyme system and donate electrons directly to the reaction center of photosystem II much of the DMS inhibition is relieved. On the other hand, the lipophilic acceptor PDox does not reactivate DMS-inhibited chloroplasts as it does for glutaraldehyde-inhibited chloroplasts (ref. 11; Table I). These results outline roughly the effects of the examined crosslinkers on the function of photosystem II. DMS inhibits the water-splitting enzyme system of photosynthesis, in addition to a general inhibition due perhaps to the lower permeability of the modified membrane. Glutaraldehyde, on the other hand has no specific effect on O_2 evolution, but lowers the permeability to ionic electron acceptors, such as ferricyanide, significantly.

With regard to photosystem I activity, we have confirmed the specific inhibition by glutaraldehyde originally reported by Hardt and

Kok (11), but we also found that this activity is totally oblivious to chemical modification of the membrane by DMS.

It is well known that glutaraldehyde confers stability to chloroplasts resting in darkness both at $0°-4°$ C (1, 11) and at room temperature (11). When subjected to continuous illumination, however, glutaraldehyde-fixed chloroplasts lose activity quickly even at low temperature, becoming indistinguishable from unfixed controls (Fig. 5). We have shown earlier that photosystem II activity of DMS-fixed _Anacystis_ _nidulans_ cells is irreversibly lost at the temperature range of lipid phase transition ($20°-24°$ C; ref. 20) whereas unfixed cells can be reversibly taken to approx. $40°$ C and then brought back without reactivation. On the basis of that, we suggested that membrane phase transitions become irreversible upon establishment of cross linkages in and among the protein subunits. Since in contrast to Anacystis, spinach chloroplast lipids are in the liquid state above $0°$ C because of their high content in linolenic acid (see ref. 20), we should expect crosslinking to destabilize these membranes above $0°$ C, although transient activities can be detected at low temperature. The instability of membrane integrated functions brought about by inter-and intramolecular crosslinking should be contrasted with the stabilizing effect that intramolecular crosslinks have on the functions of monomeric or oligomeric soluble enzymes (23).

In conclusion, we may say that the bifunctional crosslinking agents glutaraldehyde and DMS exert specific and distinguishable effects on chloroplast membranes. The molecular interpretation of the properties of chemically modified membranes, however, is limited by the scarcity of direct (not conjectural) biochemical evidence, with regard to the sites affected and the location of the established crosslinking bridges.

ACKNOWLEDGEMENTS. This research was supported in part by NATO Research Grant No 855. The authors would like to thank Mrs. Thoula Lagoyanni for expert technical assistance throughout this work.

REFERENCES

1. Park, R.B., Kelly, J., Drury, S., and Sauer, K. (1966) Proceed. Natl. Acad. Sci. U.S. 55, 1056-1061.
2. Richards, R.M., and Knowles, J.R. (1968) J. Mol. Biol. 37, 231-233.
3. Habeeb, A.F.S.A., and Hiramoto, R. (1968) Arch. Biochem. Biophys. 126, 16-26.

4. Packer, L., Torres-Pereira, J., Chang, P., and Hansen, S. (1975)
    Proceed. 3rd Intern. Congress on Photosynthesis (M. Avron, edit.)
    pp. 867-872, Elsevier Scientific Publishing Company, Amsterdam.

5. Hunter, M.J., and Ludwig, M.L. (1972) Methods in Enzymology
    (Hirs, C.H.W., and Timasheff, S.N. eds.) 25, 585-596, Academic
    Press, New York.

6. Packer, L., Allen, J.M., and Starks, M. (1968) Arch. Biochem.
    Biophys. 128, 142-152.

7. West, J., and Packer, L. (1970) Bioenergetics 1, 405-412.

8. Hallier, U.W., and Park, R.B. (1969) Plant Physiol. 44, 544-546.

9. Oku, T., Sugahara, K., and Tomita, G. (1973) Plant and Cell
    Physiol. 14, 385-396.

10. Zilinskas, B., and Govindjee (1976) Z. Pflanzenphysiol. 77,
    302-314.

11. Hardt, H., and Kok, B. (1976) Biochim. Biophys. Acta 449, 125-
    -135.

12. Sane, P.V., and Park, R.B. (1970) Plant Physiol. 46, 852-854
    (1970).

13. Papageorgiou, G., (1977) Biochim. Biophys. Acta (in press).

14. Fields, R. in Methods in Enzymology (Hirs, C.H.W., and Timasheff,
    S.N., eds.) 25, 464-468, Academic Press, New York.

15. Gordesky, S.E., Marinetti, G.V., and Love, R. (1975) J. Membr.
    Biol. 20, 111-132.

16. Shimada, K., and Murata, N. (1976) Biochim. Biophys. Acta 455,
    605-620.

17. Oliver, D., and Jagendorf, A.T. (1976) J. Biol. Chem. 251,
    7168-7175.

18. Arnon, D.I. (1949) Plant Physiology, 24, 1.

19. Saha, S., Quitrakul, R., Izawa, S., and Good, N.E. (1971)
    J. Biol. Chem. 246, 3204-3209.

20. Murata, N., Troughton, J.H., and Fork, D.C. (1975) Plant Physiol.
    56, 508-517.

21. Hassel, J., and Hand, A.R. (1974) J. Histochem. and Cytochem.
    22, 223-239.

22. Tinberg, H.M., Nayudu, P.R.V., and Packer, L., (1976) Arch.
    Biochem. Biophys. 172, 734-740.

23. Kestner, A.I. (1974) Usp. Khim. 43, 1480-1511.

THYLAKOID MEMBRANE STRUCTURE : TWO-DIMENSIONAL SEPARATION OF PROTEINS BY ISOELECTRIC FOCUSING AND ELECTROPHORESIS IN SODIUM DODECYLSULPHATE

Paul-André SIEGENTHALER and Ilse NOVAK-HOFER

Laboratoire de Physiologie végétale et Biochimie, Université de Neuchâtel, 20, rue de Chantemerle, 2000 Neuchâtel, Switzerland.

INTRODUCTION

Although the mechanisms of the photosynthetic electron transport and phosphorylation reactions have been intensively studied, the components and the molecular architecture of the thylakoid membranes on which these reactions are based on are far from being elucidated (1,2). Chemical analysis reveals that the thylakoid membranes are made up of about 50% lipids and 50% proteins (1). These later compounds are essential components of thylakoid membranes. Either weakly bound to the membrane (extrinsic proteins) or firmly embedded in the lipid matrix (intrinsic proteins) they are responsible for functions such as light-driven electron transport and photophosphorylation and some of them are probably involved in the maintenance of the structure of the membrane. Although a number of extrinsic proteins have been attributed to known proteins of the chloro-plast (coupling factor, ferredoxin-NADP$^+$ reductase, ribulose-biphosphate carboxylase) the intrinsic proteins are not as well characterized because of their relative insolubility. So far the two chlorophyll-protein complexes are the only intrinsic proteins identified with certainty (1). Progress in further identification of these proteins depends on the availability of methods for their solubilization, separation and characterization.

Solubilization by the anionic detergent sodium dodecylsulphate (SDS) and electrophoresis in SDS-containing polyacrylamide gels (SDS-PAGE) have been widely used to separate membrane proteins. SDS electrophoresis of thylakoid membranes resolves up to 20 polypeptides with molecular weights ranging from 10'000 to 100'000, approximately (1,3-9,11). Although SDS electrophoresis has given much information on the intrinsic proteins of thyla-koids it suffers from some inherent drawbacks : (a) SDS

solubilizes proteins to their polypeptide subunits, thereby
destroying their native form and biological activity; (b) SDS-
polypeptides may represent aggregates or sets of polypeptides
with very similar molecular weight. A second possibility of
separating membrane proteins is by polyacrylamide gel isoelectric focusing (PAGIF). For membrane solubilization prior to
isoelectric focusing a non-ionic detergent and/or urea have to
be used, since the charged SDS molecules interfere with isoelectric focusing. We have previously shown that isoelectric focusing
in the presence of Triton X-100 separates a great number of
proteins from thylakoids (10,11). A third possibility of disclo-
sing membrane proteins lies on the combination of isoelectric
focusing with SDS electrophoresis in a two-dimensional system.
Such a system allows a separation of thylakoid membranes
polypeptides according to two independent parameters, i.e. their
charge and their molecular weight (11).

In order to assign a function to the various thylakoid
membrane polypeptides separated by electrophoresis, several
approaches have been proposed in the literature. One of these
approaches is to analyse the membrane polypeptides of mutant
strains which are either pigment-deficient or have specific
lesions in the electron transport pathway, the missing or
altered polypeptides being then correlated with the deleted
functions in the mutant (1,7,12-14). It is also possible to
follow the appearance of membrane polypeptides during the greening of leaves or algae (15-17). Serological tests have also been
proven to be a useful way for the localization and the assignement of a function to thylakoid membrane proteins (18-19).
Another approach is to fractionate the membranes by passage
through a French pressure cell, detergent or sonication into
thylakoid membrane particles enriched in either photosystem I
(PS I) or photosystem II (PS II) activities (1,4,10-11,20-24).
The protein bands obtained by PAGIF and/or SDS-PAGE are then
identified, where possible, with proteins that previously have
been isolated and characterized from chloroplasts.

In this investigation we have adopted the fractionation
approach, and compared the polypeptide profiles obtained by
one-(Triton X-100 PAGIF or SDS-PAGE) and two-dimensional
systems (Triton X-100 PAGIF versus SDS-PAGE) using fractions
of the thylakoid membranes enriched in PS I and PS II activi-

ties. Our results show that the two-dimensional method permits
a more detailed analysis of the thylakoid membrane proteins
than any unidimensional method alone i.e., to separate several
proteins which are characteristic for fractions enriched in
PS I and PS II activities, and to discriminate by their different pI polypeptides which have similar molecular weights.

MATERIAL AND METHODS

Preparation of chloroplasts and chloroplast subfractions; solubilization of samples

Spinacia oleracea var. Nobel was grown in a phytotron as
outlined before (25) and leaves were taken after 8 weeks of
growth. Chloroplasts and washed thylakoids were prepared as
previously described (10) and the preparation of photosystem
I and photosystem II fractions was carried out according to
the method of Vernon and Shaw (26). The solubilization of the
various membrane fractions before isoelectric focusing and
SDS-electrophoresis has also been described before (11).

Isoelectric focusing

The electrofocusing system for the pH range from pH 5-7
has been described elsewhere (11). The pH gradients between
pH 7 and 10 were obtained by subjecting 6% polyacrylamide gels
containing 2.5% of carrier ampholines pH 7-9 as well as 1%
Triton X-100 and 8 M urea to 5 hours of electrofocusing at 4° C.
During the experiment the current declined from 19 to 6 mA
while the voltage was rising from 70 to 490 V. On one slab gel
3 sets of 4 identical samples were separated and after electrofocusing one sample was used for the determination of the pH
gradient, two samples were stained and one sample was equilibrated in SDS-mercaptoethanol containing buffer before loading it
onto a SDS slab-gel for the second dimensional run. These procedures have been described in detail previously (11).

SDS-electrophoresis

The gel systems for uni-and two-dimensional SDS-electrophoresis are given elsewhere (11). The apparent molecular weights
(Mr) of the polypeptide chains were estimated from a calibration curve with proteins of known molecular weight (11). Mr
were found to be precise within $\pm$ 1 Kdalton (or better) with

the exception of the spots in Fig. 5 C and D which showed a
variability in the order of ± 2 Kdaltons.

RESULTS

Proteins of washed thylakoids separated by SDS-PAGE or Triton X-100 PAGIF

Polypeptides of washed thylakoids were separated by two
one-dimensional methods, sodium dodecylsulphate polyacrylamide
gel electrophoresis (SDS-PAGE) and polyacrylamide gel isoelectric focusing (PAGIF). Fig. 1 shows the densitometer tracing of
SDS-PAGE pattern obtained from water-shocked and washed chloroplasts, referred to as "washed thylakoids". The two main components of this tracing were proteins with a Mr of ca. 23-25 and
58-64 Kilodaltons (Kdaltons). These two broad peaks contained
probably several proteins, as reported also by Klein and Vernon
(22). Other proteins were distinctly observed at 100,82,70,53,
44,33 and 12 Kdaltons. Fig. 2 shows densitometer tracings of
PAGIF patterns in the pH 5-7 (Fig. 2A) and pH 7-10 regions
(Fig. 2B). In the alkaline region, two major proteins focused
at pH 9.0 and 8.95 and two minor bands appeared in a reproducible manner at pH 8.85 and 8.75. In the acidic region, around 20

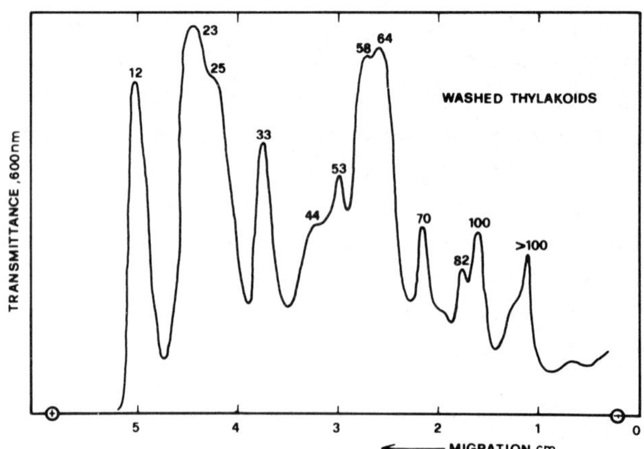

Fig. 1. Densitograms of the polypeptide patterns of washed
thylakoids (50-100 µg protein/sample) of spinach
chloroplasts. Solubilization was made with 1% SDS
and electrophoresis carried out in 9% polyacrylamide gels containing 0.1% SDS. Proteins are designated by their Mr in Kdaltons.

proteins could be detected. Some proteins focused quite distinctly at about pH 5.0, 5.35, 5.65 and 6.2. Several other proteins
appeared together as broad bands in the 5.7 to 5.8, 5.9 to 6.2
and 6.3 to 6.8 regions. An interesting feature of the above
protein patterns of washed thylakoids was the great difference
in the number of bands obtained with the Triton X-100 PAGIF and
SDS-PAGE methods. Indeed at least 20-25 bands appeared in the
Triton X-100 PAGIF pattern versus only 12 in the SDS-PAGE
pattern.

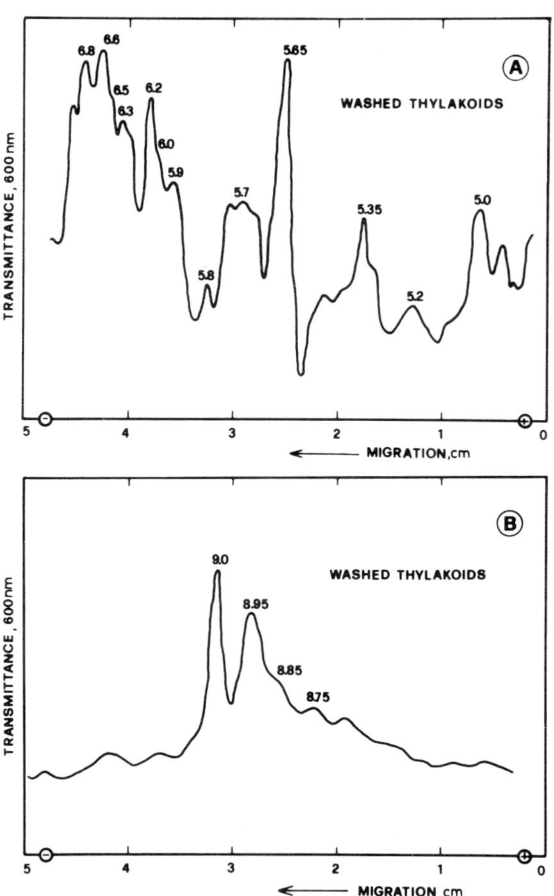

Fig. 2. Densitograms of the protein patterns of washed
thylakoids (about 100 µg protein/sample) of spinach
chloroplasts. Solubilization was obtained by 1%
Triton X-100 and the separation was carried out by
isoelectric focusing on polyacrylamide gels containing 1% Triton X-100 and 8 M urea. The pH gradient
extended from pH 5-7 (Fig. A) and from 7-10 (Fig. B).
Proteins are designated by their apparent pI.

Proteins of photosystem I and II fractions separated by SDS-PAGE or Triton X-100 PAGIF

In an attempt to identify the function of some of these membrane proteins, we have examined the composition of particles which were enriched either in PS I or in PS II activities. Both preparations were checked for their photochemical activities and found to be in agreement with the results of Vernon and Shaw (26).

Fig. 3 shows typical scannings of polyacrylamide gels following electrophoresis of SDS-solubilized proteins of PS I (3A) and PS II (3B) fractions. PS I fractions were rich in polypeptides of 95,70,60 and 14 Kdaltons. Some additional polypeptides in the 50-34 Kdalton range which were characteristic of PS I fractions were weakly stained but appeared consistently (44,42,40,35 and 34 Kdaltons). The peaks appearing in the 28-20 Kdalton range corresponded to the very strong group of polypeptides characteristic for PS II fractions (see Fig. 3B). In the 19-14 Kdaltons range, polypeptides of 19 and 17 seemed to be characteristic for PS I fractions. Major components of PS II fractions were polypeptides of 60,50,34,26,21 and 14 Kdaltons. In these experiments we used lipid containing samples, which were not drastically solubilized by 1% SDS (30 min, 4° C), in order to distinguish the two major chlorophyll-protein complexes. For complex I we found a Mr of about 95 to 100 (Fig. 3A and table I) and for complex II a broad peak from 28 to 21 Kdaltons (Fig. 3B) which appeared to comprise at least three polypeptides of 28,21 and possibly 20 Kdaltons (see Fig. 3A as a reference). Compared to the polypeptide patterns of washed thylakoids some additional polypeptides were revealed in the patterns obtained from the PS I and PS II fractions.

Fig. 4 shows typical densitometer tracings of polyacrylamide gels following isoelectric focusing of Triton X-100-solubilized proteins of PS I (4A and 4C) and PS II (4B and 4D) fractions. PS I fractions showed at least 12 proteins among them one prominent peak at pH 5.6-5.7 which only appeared strongly in gels containing urea in addition to Triton X-100. The two proteins focusing at pH 5.0 and 5.2 which appeared already in the washed thylakoids (see Fig. 2A) seemed also to be characteristic of PS I fractions. In the pH 6-7 region, more bands were present in the PS I fraction (Fig. 4A) than in the PS II fraction

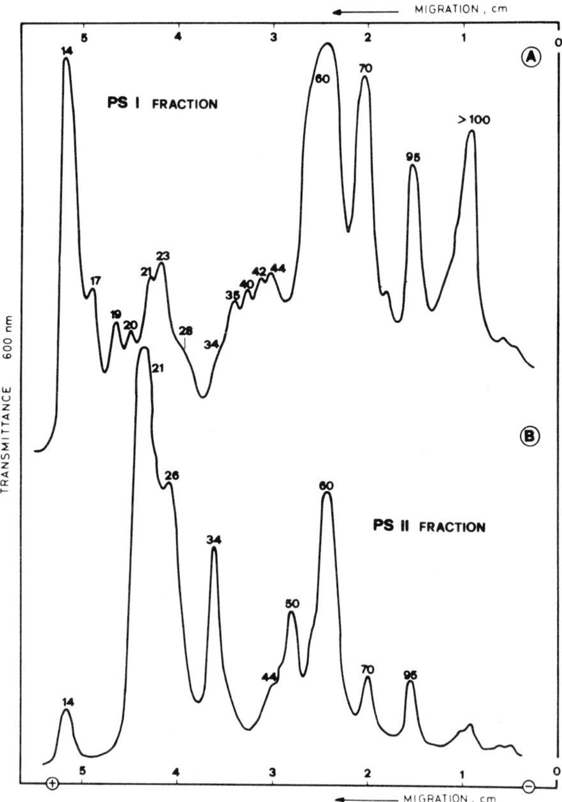

Fig. 3.   Densitograms of the polypeptide patterns of PS I (A)
and PS II (B) fractions (50-100 µg protein/sample).
Fractions were prepared according to Vernon and Shaw
(26) and were solubilized in 1% SDS. Electrophoresis
was carried out in 9% polyacrylamide gels containing
0.1% SDS. Proteins are designated by their Mr in
Kdaltons.

(Fig. 4B). Major proteins appeared at pH 5.8,5.9,6.0,6.2,6.6
and 6.8. In the pH 7-10 range, one major protein focused at
pH 8.95 which was characteristic for PS I fraction (Fig. 4C).
For PS II fractions we found less proteins than in PS I
fractions but two prominent bands at pH 5.3 and 6.3 which
were characteristic of PS II fractions (Fig. 4B). The expanded
pH gradient used in these experiments allowed to distinguish
clearly two closely spaced bands focusing at pH 6.2 (PS I) and
pH 6.3 (PS II). Apart from these major peaks, a number of minor

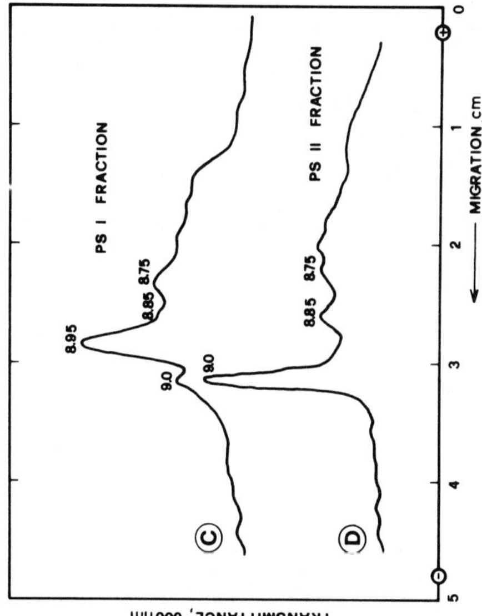
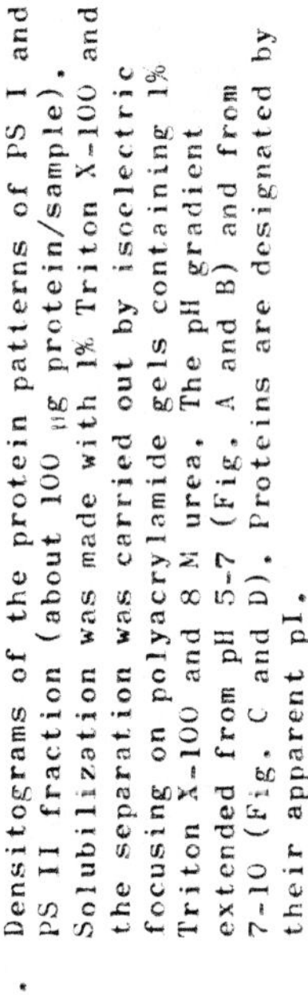

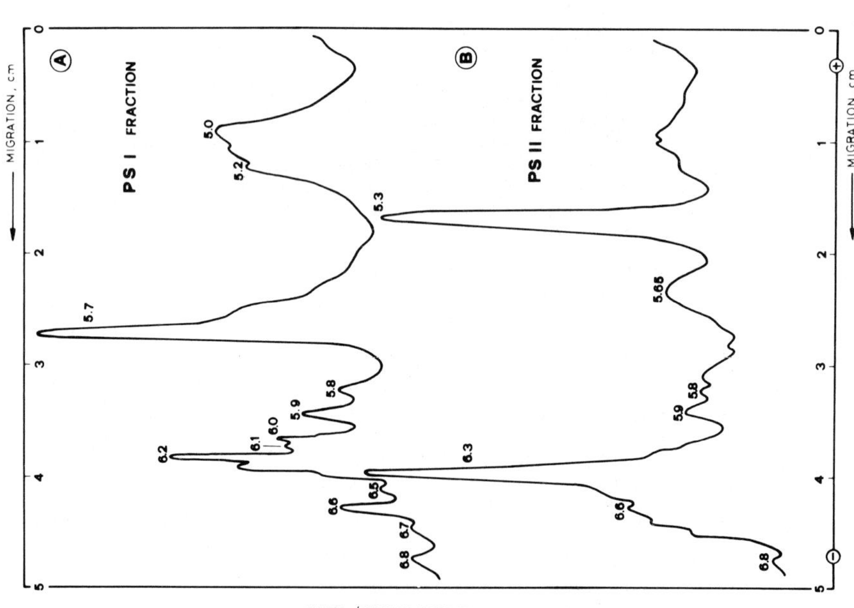

Fig. 4. Densitograms of the protein patterns of PS I and PS II fraction (about 100 µg protein/sample). Solubilization was made with 1% Triton X-100 and the separation was carried out by isoelectric focusing on polyacrylamide gels containing 1% Triton X-100 and 8 M urea. The pH gradient extended from pH 5-7 (Fig. A and B) and from 7-10 (Fig. C and D). Proteins are designated by their apparent pI.

bands appeared in a reproducible manner which were characterized
by their pI values as indicated in Fig. 4B. In the alkaline
range, one prominent band appeared at pH 9.0 which was characteristic for the PS II fraction (Fig. 4D).

Proteins of PS I and PS II fractions separated by two-dimensional electrophoresis

Fig. 5 shows typical patterns of PS I and PS II fractions
obtained by Triton X-100 PAGIF (pH 5-10) followed by SDS-PAGE
at right angle. The schematic drawings in Fig. 5A and C (PS I
fraction) and 5B and D (PS II fraction) represent typical
experiments in the pH 5-7 (Figs 5A and B) and pH 7-10 (Figs 5C
and D) region. Very strong spots appeared besides only weakly
stained spots. Sometimes, a greater amount of spots were found
but these variable spots were not included here. The comparison
of the molecular weights with those obtained by unidimensional
SDS-PAGE allows the correlation between the major isoelectric
focusing bands and the major SDS-polypeptides.

The map of the PS I fraction (Fig. 5A) shows horizontal
streakings in the high Mr range (78-60 Kdaltons) over a pH
range from about pH 4.5 to 5.9. The incorporation of urea into
the PAGIF gel reduced the diffuse streakings in the second
dimension and led to the appearance of distinct spots at about
5.6 having a Mr of about 68 and 60 Kdaltons. These spots were
the strongest of the PS I patterns; they corresponded to the
strong group of SDS-PAGE-polypeptides of 70 and 60 Kdaltons in
Fig. 3A. No polypeptides with a higher Mr than 78 Kdaltons were
found in the two-dimensional system in contrast to unidimensional
SDS-electrophoresis where a polypeptide of 95 and aggregates
higher than 100 Kdaltons were present. A series of six spots
between pH 5.8 and 6.8 having a Mr of 44,42,37 (a and a') and
33 (a and a') were also characteristic of PS I fractions.
They corresponded to polypeptides of the 50-34 kdalton range
in the unidimensional SDS-electrophoresis (Fig. 3A) where they
appeared as rather weak bands in contrast to the two-dimensional pattern where they were unexpectedly conspicuous and characteristic for PS I fractions. In the same pH range appeared
also some spots with lower Mr (25,19 (a and a') and 12 Kdaltons).
In the alkaline pH range (Fig. 5C) a spot of about 15 Kdaltons
focusing at pH 8.95 was clearly characteristic of PS I fractions.

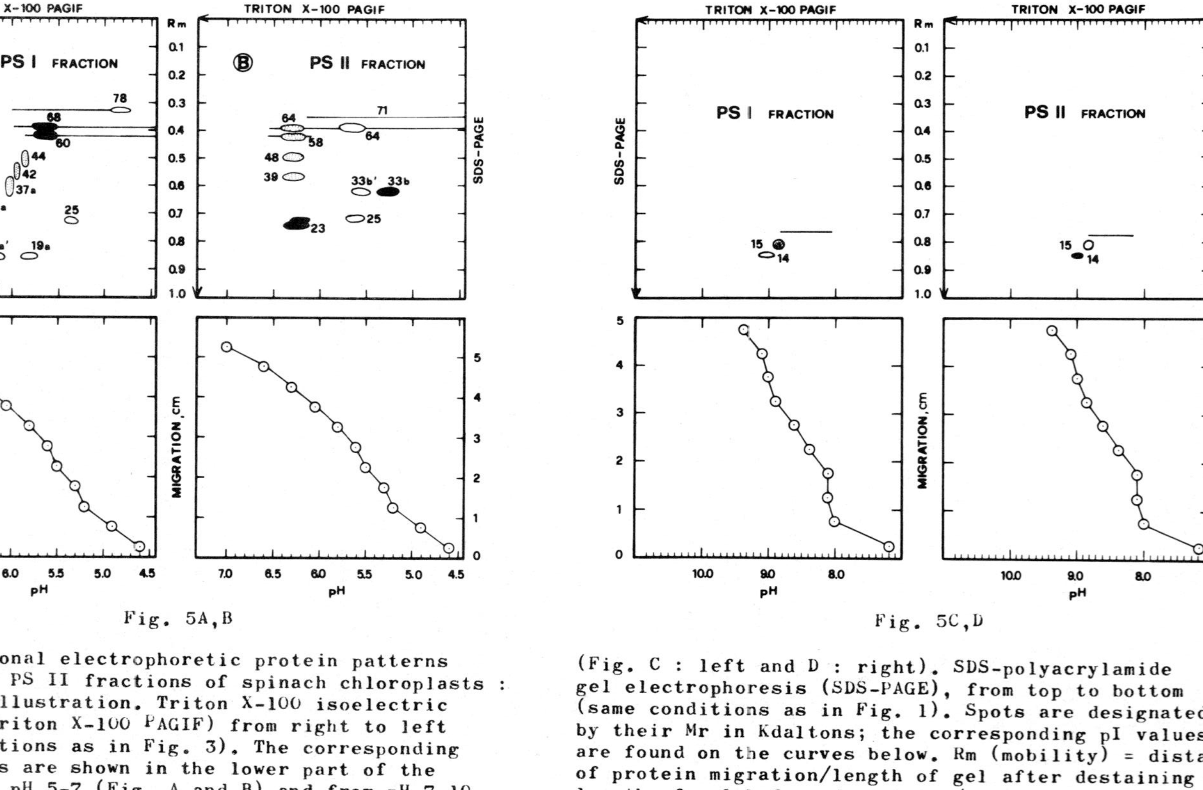

Fig. 5A,B

Fig. 5C,D

Two-dimensional electrophoretic protein patterns of PS I and PS II fractions of spinach chloroplasts : schematic illustration. Triton X-100 isoelectric focusing (Triton X-100 PAGIF) from right to left (same conditions as in Fig. 3). The corresponding pH gradients are shown in the lower part of the figure from pH 5-7 (Fig. A and B) and from pH 7-10

(Fig. C : left and D : right). SDS-polyacrylamide gel electrophoresis (SDS-PAGE), from top to bottom (same conditions as in Fig. 1). Spots are designated by their Mr in Kdaltons; the corresponding pI values are found on the curves below. Rm (mobility) = distance of protein migration/length of gel after destaining x length of gel before destaining/distance of bromophenolblue migration.

The map of the PS II fraction (Fig. 5B) shows two very strong
spots at pH 5.3 (33 Kdaltons : 33b) and 6.3 (23 Kdaltons). A
weaker spot at about pH 5.6 had a Mr of 25 Kdaltons. These three
spots corresponded to the "Group II polypeptides" (1,21) characteristic for PS II fractions observed in unidimensional SDS-
PAGE. The second characteristic feature of PS II fractions was
a series of higher Mr spots at pH 6.3 (64,58,48,39 Kdaltons).
These spots always appeared one beneath the other, above the
strong 23 Kdaltons spot (Fig. 5B). In the unidimensional SDS-
PAGE (Fig. 3B) we found peaks at 60 and 50 Kdaltons appearing
in PS II fractions, which may correspond to the 64,58 and 48
Kdaltons spots in the two-dimensional electrophoresis. In the
alkaline pH range (Fig. 5D) a spot of about 14 Kdaltons focusing at pH 9.0 belonged to PS II fractions.

DISCUSSION

The main purpose of this investigation was to separate the
polypeptides of the thylakoid membrane by polyacrylamide gel
isoelectric focusing (PAGIF), and to combine this new method
with the usual SDS-polyacrylamide gel electrophoresis (SDS-
PAGE) technique. This permitted to obtain a two-dimensional
map of polypeptides from thylakoid particles enriched in either
PS I or PS II activities (Fig. 5), where each polypeptide was
characterized within certain limits by its isoelectric point
(pI) and its molecular weight (Mr). In these experiments, one
has to remember that prior to the PAGIF separation, proteins of
the thylakoid membrane were extracted by the non ionic detergent
Triton X-100 which extracts some of the membrane proteins
without disruption of their native conformation and without
photochemical inactivation (27,28). Thus in the first PAGIF
dimension (Figs 2 and 4) this separation procedure may provide
an information on the properties of membrane proteins in their
native state. On the contrary, the disintegration of the thylakoid by the anionic detergent SDS and the separation by SDS-
PAGE in the first (Figs 1 and 3) or second dimension (Fig. 5)
result in a denaturation of the proteins and a resolution into
their subunits. It is therefore impossible to correlate directly the Triton X-100 PAGIF with the SDS-PAGE patterns. However,
it will be worth while to compare the unidimensional SDS-PAGE
with the two-dimensional fingerprints.

280

<u>Analysis of unidimensional SDS-PAGE</u>

The unidimensional SDS electrophoresis patterns of washed
thylakoids (Fig. 1) show, in agreement with previous observations (1,3-9,20-24), two characteristic groups of polypeptides,
the group I polypeptides being in the range of 50-70 Kdaltons
and the group II appearing in the range of 20-33 Kdaltons.
Subchloroplast fractions enriched in either PS I or PS II
activities (Fig. 3) differed significantly from each other. In
PS I fractions the group I polypeptides predominated, whereas
the PS II enriched fractions were more rich in the group II
polypeptides. These patterns were in many respects similar to
those obtained by Klein and Vernon (22) and other authors
(20,21,23,24). However, we did not use mercaptoethanol in
addition to SDS for solubilization. Without mercaptoethanol,
PS II patterns showed a series of higher Mr polypeptides (50-
95 Kdaltons, see Fig. 3B) which did not appear in samples
containing this sulfhydryl compound (22). On the basis of unidimensional SDS-PAGE, the 14,35,40,42,44,70 and 95 Kdaltons
polypeptides seemed to be characteristic for PS I fractions
(Fig. 3A) whereas the 21,26,34 and 50 Kdaltons appeared to
characterize PS II fractions (Fig. 3B). In these patterns the
broad 60 Kdalton band was commun to both fractions.

<u>Comparison of SDS-PAGE with the two-dimensional system</u>

The advantage of the two-dimensional map (Fig. 5) was to
reveal however that two or three polypeptides were involved
here which can be distinguished by their pI. A 60 Kdalton
polypeptide having a pI of about 5.5-5.7 was characteristic
of PS I fractions (Fig. 5A) whereas two polypeptides of 58
and/or 64 Kdaltons having a same pI of about 6.3 was characteristic of PS II fractions (Fig. 5B). The 60 (maybe also the
70/68) Kdalton polypeptide in PS I fraction might be the apo-
protein(s) of the chlorophyll-protein complex I (1,3,22,24,29),
whereas the 58 and 64 Kdalton polypeptides in fractions II
should be related to a quite different function.

Another advantage of the two-dimensional system over the
conventional SDS-PAGE was to provide informations about the
33-34 Kdaltons polypeptides which appeared with SDS-PAGE as a
stronger band in PS II than in PS I fractions (see Fig. 3A and
B). Indeed the 33 Kdalton spot of the PS I fraction was distinct

from the strong 33 Kdalton polypeptide of the PS II fraction, the two spots appearing, due to their different pI (6.2 for 33 a and 5.3 for 33 b), at a totally different position on the two-dimensional maps (Fig. 5A and B). The two weaker spots (33 a' and 33 b') may arise from an heterogeneity of charges in these polypeptides which are separated on the PAGIF gel. The 33 (a) Kdalton polypeptide of PS I fractions focusing at pH 6.2 may represent cytochrome f (9,22) whereas the 33 (b) Kdalton of PS II fraction has yet an unknown function.

<u>Analysis of unidimensional Triton X-100 PAGIF</u>

The unidimensional Triton X-100 PAGIF patterns of washed thylakoids (Fig. 2A and B) shows, at first sight, a higher resolution than the SDS-PAGE patterns. In addition to major bands denoted by their pI in Fig. 2, a number of very closely spaced bands can appear, especially in the more acidic part of the gradient, which may be due to an artifactual heterogeneity. Methodological artifacts producing multiple bands can be due to interactions of proteins with urea or other components of the system. Heterogeneity, not directly due to isoelectric focusing conditions, may arise by the interaction of proteins with different ligands such as lipids or carbohydrates. The PAGIF profiles of the two subchloroplast particles were significantly different. In PS I fractions, major proteins focused at pH 5.7, 6.2 and 8.95 whereas in PS II fractions prominent proteins occurred at pH 5.3, 6.3 and 9.0.

<u>Comparison of Triton X-100 PAGIF with the two-dimensional system</u>

The prominent peak at pH 5.7 in PS I fractions (Fig. 4A) was disclosed in the second dimension (Fig. 5A) in two conspicuous spots (60 and 68 Kdaltons) corresponding probably to the apoproteins of the chlorophyll-protein complex I. These polypeptides needed urea in addition to Triton X-100 in the PAGIF gel to focuse distinctly. The five very closely spaced bands in the 5.8 to 6.2 pH range (Fig. 4A) corresponded to five well separated spots in Fig. 5A (19,44,42,37a and 33a Kdaltons), among them probably cyt f at 33 Kdaltons (9,22) and cytochrome b_6 at 42 Kdaltons (22). In addition to the fact that cytochromes are very well extracted by Triton X-100 (27), the

TABLE I :   POLYPEPTIDES OF PS I AND PS II FRACTIONS SEPARATED BY

UNIDIMENSIONAL SDS-PAGE (extraction : 1% SDS)				TWO-DIMENSIONAL TRITON X-100 PAGIF VS SDS-PAGE (extraction : 1% Triton X-100)				POSSIBLE ASSIGNMENT TO KNOWN PROTEINS (see ref.9,22,29-31) In parenthesis, Mr in Kdaltons
PS I fraction M_r		PS II fraction M_r		PS I fraction M_r (pI)		PS II fraction M_r (pI)		
100 ± 5	S	100 ± 5	L					CP I
*82 ± 2	VL							
				78 (4.8)	S			
70 ± 2	S	*70 ± 0	L	68 (5.6)	S	71 (streaking)	L	apoprotein of CP I (?)
						64 a (6.3)	L	
64 ± 1		64 ± 1				64 a' (5.6-5.7)	VL	
≃ 60 S		≃ 60 S		60 (5.6)	S			apoprotein of CP I (?) CF$_{1\alpha}$ (59)
58 ± 0		58 ± 0				58 (6.3)	L	
*54 ± 0	L	*53 ± 1	L					CF$_{1\beta}$ (56) RUDP carboxylase (56)
*48 ± 0	L	49 ± 1	S			48 (6.3)	L	dimer of CP II
44 ± 0	VL	44 ± 2	VL	44 (5.85)	L			
42 ± 0	VL			42 (5.95)	L			cyt b$_6$ (42)
40 ± 0	VL							
*39 ± 0	VL					39 (6.3)	L	
				37a (6.05)	L			CF$_{1\delta}$ (37)
				37a' (6.7)	VL			
35 ± 0	VL							
33 ± 1	VL	33 ± 1	S	33a (6.2)	L			cyt f (33-34) in PS I
				33a' (6.7)	VL			
						33b (5.3)	S	apoprotein of CP II (?)
						33b'(5.6)	VL	

TABLE I (cont.)

POLYPEPTIDES OF PS I AND PS II FRACTIONS SEPARATED BY

UNIDIMENSIONAL SDS-PAGE (extraction : 1% SDS)				TWO-DIMENSIONAL TRITON X-100 PAGIF VS SDS-PAGE (extraction : 1% Triton X-100)				POSSIBLE ASSIGNMENT TO KNOWN PROTEINS (see ref.9,22,29-31) In parenthesis, M_r in Kdaltons
PS I fraction		PS II fraction		PS I fraction		PS II fraction		
M_r		M_r		M_r (pI)		M_r (pI)		
*28 ± 0	L	27 ± 1	S					
*25 ± 1	L	*25 ± 0	S	25 (5.35)	VL	25 (5.6)	VL	
23 ± 0	L	23 ± 0	S			23 (6.3)	S	
21 ± 0	L	21 ± 0	S					Structural protein } CF II
20 ± 0	L	20 ± 0	L					
				19a (5.8)	VL			
				19a' (6.2)	VL			
18 ± 1	L	18 ± 1	VL					$CF_1\delta$ (17.5)
*16 ± 1	L	*16 ± 1	VL					
				15 (8.9)	L	15 (8.8)	VL	
14 ± 0	S	14 ± 0	L	14 (9.05)	VL	14 (9.0)	L	Ferredoxin (14)
12 ± 0	L	12 ± 0	L	12 (6.3-6.6)	VL			$CF_1\epsilon$ (13) RUDPcarboxylase (12)

*not always visible; M_r : Molecular weights in Kilodaltons; Staining intensity : S - strong   L - light
VL - very light; CF_1 : coupling factor with the subunits α, β, γ, δ, ϵ; CP I and CP II : chlorophyll-protein complex I and II.

284

isoelectric focusing method permits to concentrate these molecules at their pI. Thus, in the second dimension system,
cytochromes f and b_6 were quite characteristic for PS I fractions
whereas in the unidimensional SDS electrophoresis, they appeared
weakly stained (Fig. 3A). The other minor bands in the 6.3-6.8
pH region were not as well characterized in the second dimension.
The unknown protein focusing at pH 8.95 (Fig. 2B and 4C) was
characteristic of PS I fractions and gave rise to a spot having
a Mr of about 15 Kdaltons (Fig. 5C).

In the PS II fractions (Fig. 4B) the two prominent proteins
focusing at pH 5.3 and 6.3 corresponded, in the two-dimensional
system, to the two strongest spots of the map at 33 (b) and 23
Kdaltons, respectively. These two polypeptides with a third,
weaker spot of 25 Kdaltons (pI : 5.6) corresponded to the "group
II" polypeptides known from unidimensional SDS-electrophoresis
(1,21) and represent most probably the apoproteins of chlorophyll-
proteins complex II (1,3,21,29). In addition to the strong 23
Kdalton spot, a series of polypeptides (64,58,48 and 39 Kdaltons)
arose in the second dimension from the strong PAGIF peak at pH
6.3 which were characteristic of the PS II fraction. The dissociation of the pI 6.3 proteins was observed upon incubation in
SDS-mercaptoethanol before the separation in the second dimension, thus suggesting that the 23 Kdalton polypeptide of this
series may exist in oligomeric forms. Finally, the protein
focusing at pH 9 (Fig. 2B and 4D) was characteristic of PS II
fractions and gave rise to a spot having a Mr of about 14
Kdaltons (Fig. 5D).

Possible functions of thylakoid membrane polypeptides

Table I summarizes all the polypeptides obtained by
unidimensional SDS-PAGE and by the two-dimensional system
for PS I and PS II subchloroplast fractions. The table shows
also a tentative assignement on a Mr basis of some of the
polypeptides to proteins or subunits of proteins which have
been previously isolated and characterized, i.e. chlorophyll-
protein complexes I and II (29), cytochrome b_6 (22) and f (9),
coupling factor (9,30) ribulosebiphosphate carboxylase (22),
ferredoxin (22) and the structural protein described by Criddle
and Park (31).

ACKNOWLEDGEMENTS

We thank the Swiss National Science Foundation for its support (Grant 3.2470.74 to P.A.S.), Mr. Philippe Sublet for his technical assistance in some parts of this work, Mrs Jana Smutny for drawing the figures and Miss Christiane Bachmann for typing the manuscript.

REFERENCES

(1) Anderson, J.M. (1975) Biochim. Biophys. Acta 416, 191-235

(2) Trebst, A. (1974) Annu. Rev. Plant Physiol., 25, 423-458

(3) Machold, O. (1975) Biochim. Biophys. Acta, 382, 494-505

(4) Nolan, W.G. and Park, R.B. (1975) Biochim. Biophys. Acta, 375, 406-421

(5) Henriques, F. and Park, R.B. (1976) Biochim. Biophys. Acta, 430, 312-320

(6) Henriques, F. and Park, R.B. (1976) Arch. Biochem. Biophys., 176, 472-478

(7) Herrmann, F.H., Schumann, B., Börner, T. and Knoth, R. (1976) Photosynthetica, 10, 164-171

(8) Lagoutte, B. and Duranton, J. (1976) Biochim. Biophys. Acta, 427, 141-152

(9) Süss, K.H. (1976) FEBS Lett., 70, 191-196

(10) Novak-Hofer, I. and Siegenthaler, P.A. (1975) FEBS Lett., 60, 47-50

(11) Novak-Hofer, I. and Siegenthaler, P.A. (1977) Biochim. Biophys. Acta (biomembranes sect.), in press

(12) Chua, N.H. and Bennoun, P. (1975) Proc. Nat. Acad. Sci., U.S.A., 72, 2175-2179

(13) Chua, N.H., Matlin, K. and Bennoun, P. (1975) J. Cell Biol., 67, 361-377

(14) Picaud, A. and Acker, S. (1975) FEBS Lett., 54, 13-17

(15) Beck, D.P. and Levine, R.P. (1974) J. Cell Biol., 63, 759-772

(16) Guignery, G., Luzzati, A. and Duranton, J. (1974) Planta (Berl.), 115, 227-243

(17) Nielsen, N.C. (1975) Eur. J. Biochem., 50, 611-623

(18) Schmid, G.H., Menke, W., Koenig, F. and Radunz, A. (1976) Z. Naturforsch., 31 c, 304-311

(19) Schmid, G.H., Radunz, A. and Menke, W. (1977) Z. Naturforsch., 32 c, 271-280

(20) Remy, R. (1971) FEBS Lett., 13, 313-317

(21) Anderson, J.M. and Levine, R.P. (1974) Biochim. Biophys. Acta, 333, 378-387

(22) Klein, S.M. and Vernon, L.P. (1974) Photochem. Photobiol., 19, 43-49

(23) Adler, K. (1974) Photosynthetica, 8, 360-367

(24) Apel, K., Bogorad, L. and Woodcock, C.L.F. (1975) Biochim. Biophys. Acta, 387, 568-579

(25) Siegenthaler, P.A. and Depéry, F. (1976) Eur. J. Biochem., 61, 573-580

(26) Vernon, L.P. and Shaw, E.R. (1971) in Methods in Enzymology (A. San Pietro ed.) vol. XXIII (A), pp. 277-289, Academic Press, New York and London

(27) Wasserman, A.R. (1974) in Methods in Enzymology (Fleischer, S. and Packer, L. eds), vol. XXXII (B),

pp. 406-422, Academic Press, New York, San Fransisco
and London

(28) Makino, S., Reynolds, J.A. and Tanford, C. (1973) J.
Biol. Chem., $\underline{248}$, 4926-4932

(29) Thornber, J.P. (1975) Annu. Rev. Plant Physiol., $\underline{26}$,
127-158

(30) Lien, S. and Racker, E. (1971) in Methods in Enzymology
(A. San Pietro, ed.) vol. XXIII, pp. 547-555, Academic
Press, New York

(31) Criddle, R.S. and Park, S. (1964) Biochem. Biophys. Res.
Commun., $\underline{17}$, 74-79

287

Bioenergetics of Membranes. L. Packer et al. ed.

INACTIVATION OF S_2 BY ALKALINE pH INSIDE   THE THYLAKOID

J-M. Briantais, C. Vernotte, J. Lavergne, C.J. Arntzen[*] and M. Picaud

Laboratoire de Photosynthèse C.N.R.S., Gif-sur-Yvette 91190 France

[*]USDA/FRS Department of Botany, University of Illinois, Urbana, Ill. 61801 USA

INTRODUCTION

The requirement of a defined spatial relationship between Photosystem II components across the membrane, in order to allow electron donation from water is suggested by inactivating treatments which can be interpreted in terms of structural
modifications. For exemple Joliot and Joliot (1) observed that hydroxylamine decreases the amplitude of the electric field induced by PS II photoact. One of the
two explanations they proposed is an hydroxylamine induced motion of chlorophyll a
II from the inside face to a position close to the outside face. Arntzen et al.(2)
showed that glutaraldehyde prefixation prevents the inactivation between water and
diphenylcarbazide donation in chloroplasts which is induced by enzymatic iodination
of the outside face of the thylakoid. This data suggests that iodination induces a
conformational change of the membrane which disconnect H_2O donation. The same conclusion is reached by Zilinskas and Govindjee (3) who observed a protection by glutaraldehyde fixation against various inactivations of the donor side of System II.

These spatial relationships are modified during the illumination. Evidence of
such structural changes comes from experiments of photoinduced inhibition of the
donor side (supposed to be on the inside face in the dark) by non permeant agents.

A photoinduced appearance of the donor side of PS II at the outside face of the
thylakoid could be responsible for the inhibition by diaminobenzene sulfonate described by Giaquinta and Dilley (4-5). An outside location of the donor side (in
the light) is also pointed out by Schmid et al.data (6) :they observed that inhibition by an antichloroplast antibody of an electron transfert site between tetramethylbenzidine and diphenylcarbazide donations is produced in the course of illumination.

Trebst and coworkers (7-8) suggested that a photoinduced conformational change
makes the water splitting enzyme labile to a high pH imposed in the inside compartment of the thylakoid.

It has been proposed that conformational changes occur as a consequence of the
energization of the coupled membrane but some changes seem induced by electron
flow. This is exemplified by the alkaline inactivation which occurs in the presence of gramicidin observed by Trebst et al.(7-8). In order to get further informations on the mechanism of the alkaline inactivation, and especially to determine
whether the accumulation of positive charges at the level of the water oxidizing
enzyme is involved we used short saturating flashes to induce alkaline inactivation.

It will be shown that the S_2 state of Photosystem II is especially sensitive to high pH inside the thylakoid.

We have attempted to determine if this inactivation is due to a structural change. Our results suggest that if a conformational change is induced by the accumulation of two positive charges on the water splitting enzyme, then it is a movement of the enzyme in the lipid phase of the membrane rather than a conformational change of the proteinic moiety of the H_2O splitting enzyme.

MATERIALS AND METHODS

Broken chloroplasts were isolated from spinach, lettuce or peas, according to a procedure previously described (2) and then resuspended in 0.4 M sorbitol, 10^{-2}M NaCl, 10^{-2}M MgCl$_2$ and 10^{-2}M Tricine pH 7.8.

Anacystis nidulans are grown autotrophically at 35°C. Green fragments of the algae are obtained through sonication (6 times 10 seconds at 10 microns) followed by a 1500 X g/10 min. centrifugation to remove unbroken cells, the supernatant is then centrifuged 10 min. at 30 000 X g, the green pellet is resuspended in the same 10^{-2}M Tricine buffer.

The flash photoinduced inactivation treatment is reported in the scheme of Fig. 1. The flash saturation has been verified. For the measurements which require a higher concentration of chloroplasts (EPR experiments) continuous illumination was used.

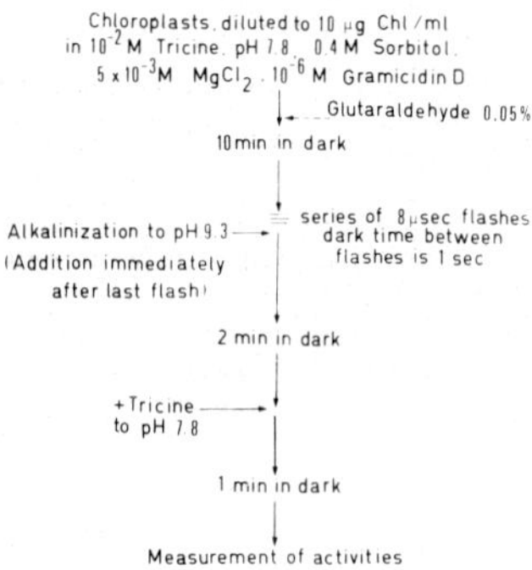

Fig. 1. Experimental procedure for the alkaline pretreatment.

The temperature dependence of the rate of the alkaline inactivation has been determined by the time of half inhibition of DPIP photoreduction rate in saturating light at pH 9.3 for each temperature. The sample contains 10 µg chlorophyll/ml 10^{-6}M gramicidin, 10^{-4}M DPIP and is buffered at pH 7.8. A five minutes incubation for each temperature is done. The glycine buffer is added in the dark to reach pH 9.3, 30 seconds later the analytic beam is turned on, the actinic beam is super imposed 15 seconds later. DPC is added to control the inactivation of the donor side.

The rate of Hill reaction is measured spectroscopically by the photoinduced changes of DPIP absorbance at 580 nm in a CARY 14 spectrophotometer with lateral illumination, in presence, when mentioned, of $5 . 10^{-4}$M diphenylcarbazide (DPC) as an artificial donor. The actinic light is red light (Corning 2-61 filter) a blue filter (Corning 4-96) protects the photomultiplier from the actinic light.

The rate of the Hill reaction with Ferricyanide as the acceptor is measured spectroscopically by the photoinduced change of absorbance at 420 nm. The same set of filters than for DPIP photoreduction is used.

Fluorescence induction at 685 nm is determined, at room temperature in presence of 10^{-5}M DCMU.

The yield of oxygen evolved at each flash (8 µs duration) of a series was measured with a rate electrode. Chloroplasts are used at a concentration of 500 µg Chl/ml.

EPR measurements. A Bruker B-ER 420 spectrometer was used. Instrument settings: 3.2 Gpp modulation, 25 mW microwave power for Signal II spectra, 16 Gpp and 100 mW for Mn^{2+}. Chloroplasts were used at a concentration of about 4 mg Chl/ml in a flat quartz cell.

RESULTS

Photoinduced inactivation in flashing light

In Figure 2 are compared a) the amplitude of the inactivation of the $H_2O \rightarrow$ DPIP photoreduction in chloroplasts which have been submitted to the pretreatment described in Fig. 1 and b) the relative concentration of S_2 measured by the oxygen yield at the flash n + 2 in a series of flashes given at pH 7.8.

The similarity of the two sequences indicates an evident correlation between the concentration of S_2 and the amplitude of the inactivation. But notice that larger inhibition than expected, according to the S_2 concentration, is obtained after two flashes.

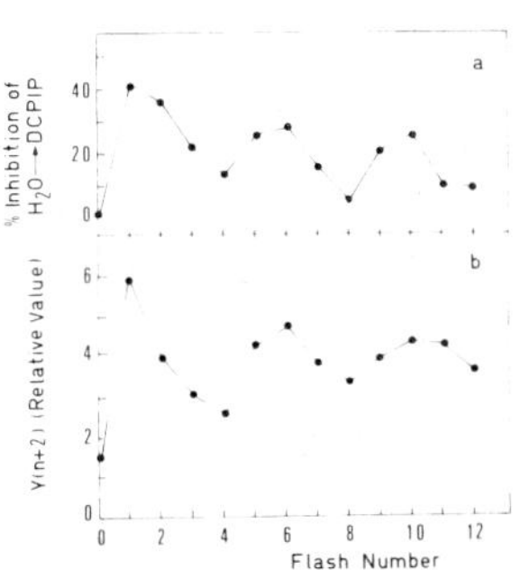

Fig. 2.a. Inhibition of the $H_2O \rightarrow$ DPIP photoreaction in saturating light as a function of the number of preillumination flashes according to the procedure of Figure 1.

b. Oxygen evolved a flash n+2 in sample at pH 7.8 without the alkaline pretreatment.

290

Table I shows that DPC restores partially DPIP photoreduction as in Trebst and col's (7-8) experiment.

When no gramicidin D is added, we observed that the preillumination by one flash does not cause any inactivation compare to the zero flash sample.

TABLE   I

RATE OF DPIP PHOTOREDUCTION

Flash number	Rate of DPIP photoreduction	
	H_2O	+ DPC
0	100[*]	100
1	35	75

[*]100 = 600 µmole DPIP reduced per hour per 1 mg Chl.

The O_2 sequence in partialy inactivated chloroplasts is not changed qualitatively but only the yield of O_2 evolved per flash is decreased by the factor of inactivation.

In Figure 3 is shown the kinetics of alkaline inactivation. It is obtained, varying the time of incubation at high pH after one flash; each inactivation is determined by comparison of the activity of the flashed sample to a suspension which has been submitted to the same time of pH 9.3 incubation but without any flash.

In Figure 4 is represented the amplitude of alkaline inactivation versus the time of darkness at pH 7.8 between the flash and the alkalinization. This kinetics is slower than the one obtained for the alkaline inactivation reaction(Fig. 3). Then the curve of Figure 4 is close to the kinetics of the disappearance of the alkaline sensitive state. The half time of 45 seconds fits the half time of S_2 deactivation which have been measured in those chloroplasts.

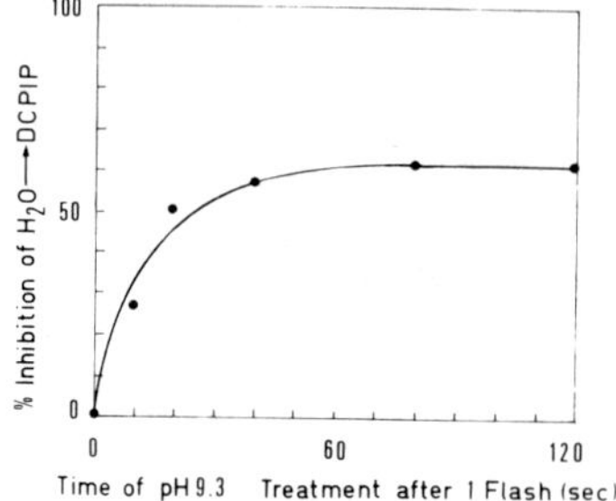

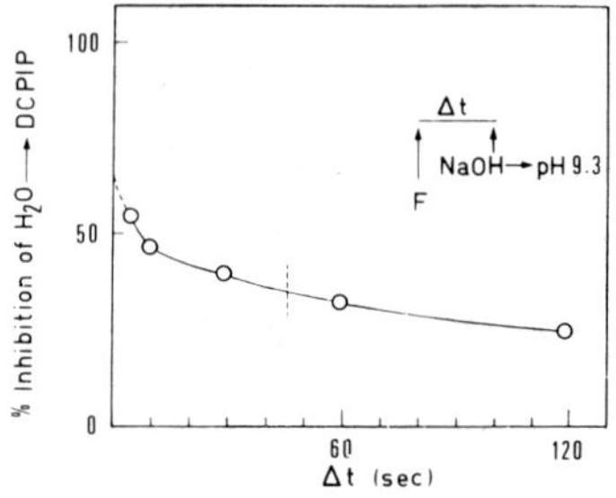

Fig. 3. Kinetics of alkaline inactivation.

Fig. 4. Dark relaxation of the alkaline pH sensitive species.

Inactivation in continuous light at pH 9.3 in presence of orthophenantroline

In Table II the inactivations of DPIP photoreduction in samples illuminated for one minute at pH 9.3 with or without o-phenantroline are compared. After pH neutralization all the samples are centrifuged, then resuspended in a medium with 0.2 M Zinc acetate in order to complex any remaining o-phenantroline. The DPIP photoreduction is assayed on these washed samples.

TABLE   II

PHOTOINACTIVATION BY pH 9.3 IN PRESENCE OF o-PHENANTROLINE

	o-phenantroline	
	0	10^{-4}M
% inhibition of $H_2O \longrightarrow$ DPIP	92	60

The value of 60 % inactivation obtained in the o-phenantroline treated sample is very close to the S_2 concentration which can be obtained by a one PS II photoact process. In contrast almost a total inactivation is reached when o-phenantroline is omitted.

Consequences of the inactivation

As previously shown, DPC addition restores the photoreduction of DPIP.

DCMU fluorescence induction : The data on fluorescence induction reported in Table III show almost no difference between the photoinactivated sample and the sample kept in the dark; then in the inactivated sample, still the Z^+Q^- formation can take place.

TABLE   III

FLUORESCENCE CHARACTERISTICS IN 10^{-5} DCMU POISONED   CHLOROPLASTS

Flash number	F_o	F_∞	$t\frac{1}{2}$ ms
	arbitrary units		
0	100	297	20
1	100	300	23.5

Manganese content and appearance of EPR Signal II fast Chloroplasts at pH 7.8 plus gramicidin D 10^{-6}M is parted into three samples. Two are alkalinized for 5 min one of these two is kept in the dark, the other one in the light, during the alkaline treatment; then they are brought back to pH 7.8. The 3 samples are spun down

and washed once in the usual medium except for low (1 mM) tricine concentration.
The activity of DPIP photoreduction is measured in the 3 samples. The concentrations of Mn^{2+} measured by the amplitude of the EPR signal are assayed in the samples before and after they have been heated at 58°C for 30 min : the difference
ΔMn^{2+} would represent complexed manganese.

In Figure 5 is the plot of the relative content ΔMn^{2+} versus the DPIP activity;
a non linear correlation exists between the deficiency of Mn^{2+} and the inactivation.

In the same figure is shown the results of a similar experiment except that no
sample has been photoinactivated, only the alkalinization duration varied (5 and
20 minutes). A similar dependence of Mn^{2+}, released by heating, versus activity is
found. In the same figure the amplitude of the light induced EPR Signal II fast
(according to Babcock and Sauer (9) nomenclature) is plotted versus the activity.
Signal II appears in inactivated samples and a non linear relationship between the
increase of Signal II fast amplitude and the decrease of activity is also obtained.

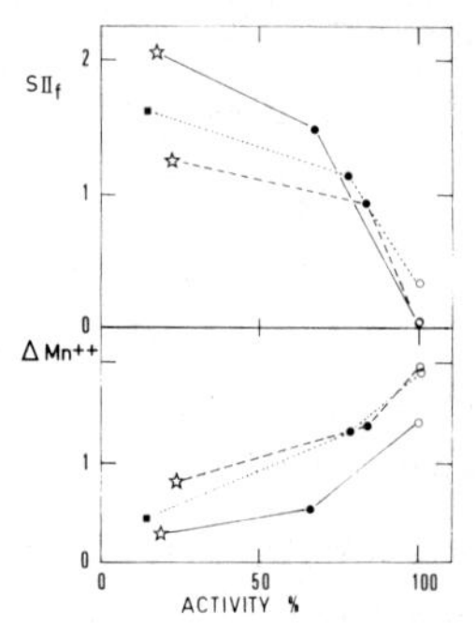

Fig. 5. ΔMn^{2+} and Signal II fast versus the $H_2O \longrightarrow$ DPIP activity.

o   control

●   5 min. at pH 9.3 in the dark

■   20 min. at pH 9.3 in the dark

✿   5 min. at pH 9.3 in the light.

The 3 sets of curves represent 3 different experiments.

<u>Relationship between the inactivation and conformational changes</u>

<u>Glutaraldehyde prefixation</u> As shown in Table IV the glutaraldehyde prefixation
does not prevent from the photoinduced inactivation.

TABLE IV

EFFECT OF GLUTARALDEHYDE PREFIXATION

% Glutaraldehyde	0		0.05 %[*]	
	0F	1F	0F	1F
$H_2O \longrightarrow$ DPIP rates	100 %	27 %	41 %	12 %
Per cent inhibition	73		71	

[*]Complete inhibition of the osmotic change in the 10 μg chlorophyll/ml
chloroplast suspension.

<u>Temperature dependence of the inactivation in green fragments of Anacystis nidu-lans and in pea chloroplasts</u> The Arrhenius plot of the rate constant of photoin-duced alkaline inactivation is presented in Figure 6. In <u>Anacystis</u> fragments two activation energies are obtained 9 and 14 Kcal. respectively above and below 16°C. In contrast the values obtained for pea chloroplast do   not show any significant break in the range of temperatures which has been investigated. The activation energy obtained for the pea chloroplasts is close to the value obtained for <u>Anacystis</u> fragments above 16°C.

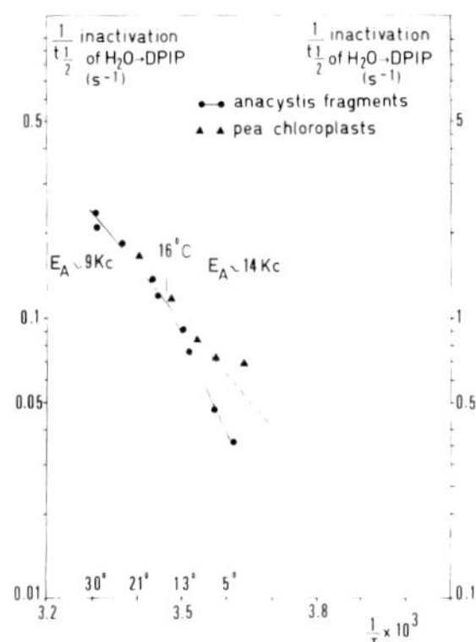

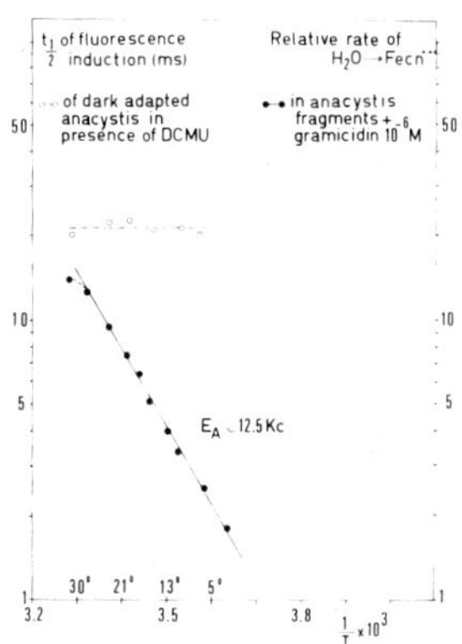

Fig. 6. Arrhenius plot of the photoin-duced inactivation of $H_2O \rightarrow$ DPIP reac-tion,($t_{\frac{1}{2}}^{1}$ is   the time require to have a rate for DPIP photoreduction = half the initial rate(see material and methods) in <u>Anacystis nidulans</u> fragments and in pea chloroplasts.

Fig. 7. Arrhenius plot of : the half rise time of variable fluorescence in the presence of DCMU o ; of the water to ferricyanide photoreaction● in <u>Anacystis nidulans</u> and in <u>Anacystis nidulans</u> fragments respec-tively.

<u>Temperature dependence of PS II activity in Anacystis nidulans</u> The rise time fluorescence induction in the presence of DCMU for dark adapted algae at various temperatures is presented in Figure 7. It is independant of the temperature. The temperature dependence of the electron transport from water to ferricyanide in the green fragments is shown by its Arrhenius' plot (same figure); only one activation energy (12.5 Kcal.) is observed.

DISCUSSION

The data presented here show that the accumulation of two positive charges on the water splitting enzyme determines its inactivation by alkaline pH inside   the thylakoid. If this inactivation occurs via a conformational changes, it is a move-ment of the entire complex through the lipid phase rather than a conformational change of the proteinic moiety.

The inactivation occurs in the dark but it is dramatically increased by PS II photoact. The inactivation in the dark could occur on S_2 also, through the equilibrium between S states or it could be due to some sensitivity of S_0 or S_1 to alkaline pH ? In the light S_2 state is the one which is sensitive to alkaline pH :this is pointed out by the correlation between S_2 concentration and the amplitude of the inactivation, it is confirmed by the 60 % inactivation obtained at pH 9.3 in continuous light in the presence of o-phenantroline. This conclusion is also consistent with the similarity between the life time of S_2 and of the alkaline sensitive species.

Several possibilities can explain the quantitative discripency between S_2 concentration and the amplitude of inactivation. A competition between S_2 deactivation (back to S_1) and the alkaline inactivation of S_2 can explain the defect of inhibition observed after one flash. Some sensitivity of S_3 to alkaline pH described previously (10) can explain the excess of inhibition observed after two flashes , or this excess can be due also to a back reaction $S_3 \longrightarrow S_2$ during alkaline incubation

The degree of inactivation seems to be dependent upon the number of inactivated centers rather than on a degree of inhibition of all centers; indeed in partially inactivated chloroplasts, the O_2 flash sequence is only modified by a factor of scale.

The sensitivity of S_2 and at a less extent of S_3 to alkaline pH has to be correlated to the specific binding of NH_3 with the oxidized states S_2 and S_3 described by Velthuys (11).

The site of inactivation seems to be the water splitting enzyme itself, indeed the inactivated centers are still able to monitor a reduction of Q and the oxidation of the secondary donor (no accumulation of P_{680}^+) according to the fluorescence characteristics in DCMU which are unchanged in the inactivated chloroplasts.The appearance of EPR Signal II fast suggests that there is a block of the reduction of a secondary donor as it is in Tris-washed chloroplasts. The release of Mn^{2+} by alkaline treatment also demonstrates the similarity between alkaline treatment and Tris washing and suggests that it is the water splitting enzyme itself which is inactivated by high pH.

We pointed out that there is no linear relationship between the inactivation of the electron flow and the manganese release. The same discrepency is obtained also for Signal II fast. Mn release and relative increase of Signal II fast are proportional. These results have been obtained several times, and are difficult to explain in a classical scheme. About the reaction of inactivation itself we can suggest the formation of an insoluble manganese compound by alkalinization or the modification of the enzyme itself.

Because the inactivation takes place only when gramicidin is present, at least for brief incubations at high pH, Trebst et al.(7-8) have proposed that a conformational change is induced in the light; a change which makes the water splitting enzyme exposed to the inside face of the thylakoid. A conformational change of the proteinic moiety of the membrane is doubtful because we observed that glutaraldehyde prefixation does not protect from the inactivation. If the conformational change is a movement of a protein accross the lipids of the membrane, this motion would not be suppressed by glutaraldehyde fixation but would depend upon the physical state of the lipids. According to Murata and Fork (12) the "solid" $\longleftrightarrow$ "smectic" transition of the lipids in the photosynthetic membranes of <u>Anacystis nidulans</u> occurs above 0°C in contrast to chloroplasts of many green plants. This is related to the higher proportion of saturated fatty acids in <u>Anacystis</u>. The transition temperature varies from $\sim$ 20 to 10°C depending on the temperature of growth. The Arrhenius' plot for the alkaline photoinactivation shows two activation energies in the case of <u>Anacystis</u> membrane with a break around 16°C; this indicates that the inactivation reaction depends upon the physical state of the lipids, it is easier in the "smectic" state than in the "solid" configuration. It is one phase of the inactivation itself which depends upon the lipid state, indeed the PS II photoact which makes S_2 is insensitive to the temperature. Then one part of the alkaline inactivation process could be a motion of the water splitting enzyme from the inside of the membrane to the inside face.

Because water to ferricyanide electron transport is independent of the physical state of the lipids, one conclude that the conformational change is not required for the functioning of the electron chain on the donor side of PS II. Nevertheless the appearance of the donor side on the inside face of the thylakoid plus its sensitivity to pH could make the donation rate sensitive to the photoinduced pH changes inside the thylakoid. This could regulate the extent of proton uptake. Van Gorkom <u>et al.</u>(13) showed that at pH < 4.5 there is an inactivation of the donor side of PS II. This regulation could explain the changes of electron flow induced by the proton gradient (14).

REFERENCES

1. Joliot, P. et Joliot, A. (1976) C.R. Acad. Sc., 283 D, 4, 393-396.

2. Arntzen, C.J., Vernotte, C., Briantais, J-M. and Armond, P. (1974) Biochim. Biophys. Acta, 368, 39-53.

3. Zilinskas, B.A. and Govindjee (1976) Z. Pflanzenphysiol., Bd 77, 302-314.

4. Giaquinta, R.T. and Dilley, R.A. (1974) Proc. 3rd Intern. Cong. Photosynth. Research (Avron M. ed.) Vol. II p. 883-895 Elsevier, Amsterdam.

5. Giaquinta, R.T. and Dilley, R.A. (1975) Biochemistry, 14, 4392-4396.

6. Schmid, G.H., Menke, W., Koenig, F. and Radunz, A. (1976) Z. Naturforsh., 31c., 304-311.

7. Harth, E., Reimer, S. and Trebst, A. (1974) FEBS Letters, 42, 165-168.

8. Reimer, S. and Trebst, A. (1975) Biochem. Physiol. Pflanzen, 168, 225-232.

9. Babcock, G.T. and Sauer, K. (1973) Biochim. Biophys. Acta, 325, 483-503.

10.   Wraight, C.A., Kraan, G.P.B. and Gerrits, N.M. (1972) Biochim. Biophys. Acta, 283, 259-267.

11.   Velthuys, B.R. (1975) Biochim. Biophys. Acta, 396, 392-401.

12.   Murata, N. and Fork, D.C. (1975) Plant Physiol., 56, 791-796.

13.   Van Gorkom, H.J., Pulles, M.P.J., Haveman, J. and Den Haan, G.A. (1976) Biochim. Biophys. Acta, 423, 217-226.

14.   Avron, M. (1971) 2nd Int. Cong. on Photosynth., Stresa (Forti ed.) Vol. II p. 861-871, Junk pub. The Hague.

Bioenergetics of Membranes. L. Packer et al. ed.

FLUORESCENCE SPECTROSCOPY OF CHLOROPHYLL-PROTEIN COMPLEXES*

Jeanette S. Brown
Carnegie Institution of Washington
Stanford, CA 94305, U.S.A.

SUMMARY

Fluorescence excitation and emission spectra (-196 C) of Photosystem I particles, P700-chlorophyll-protein complexes from spinach and several algae, and another chlorophyll-protein from Euglena are presented. P700-chlorophyll-proteins fluoresce maximally at 696 nm providing that the ratio of antenna chlorophyll to P700 is less than about 45. Apparently Triton-solubilization removes the longer wavelength "forms" of chlorophyll a which fluoresce beyond 700 nm, and if more than 45 antenna chlorophylls are still attached to the reaction center, their high-yield fluorescence near 680 nm masks the 696 nm emission. Excitation spectra for the 696 nm emission have maxima at wavelengths between 440 and 500 nm in addition to the expected chlorophyll absorption maxima near 438 and 420-425 nm. These longer wavelength, blue bands indicate efficient energy transfer between carotenoids and some absorbing "forms" of chlorophyll a in isolated P700-chlorophyll-protein complexes. A chlorophyll-protein from Euglena that does not contain P700 has an unusually narrow emission band at 687 nm which is not excited by absorption between 440 and 500 nm. A diagramatic scheme is presented to suggest possible paths of energy transfer and fluorescence emission between and from the various biological "forms" of chlorophyll a.

INTRODUCTION

Spectroscopic measurements of chlorophyll in vivo have long been studied to learn about the essential nature of the pigment which enables it to function in photosynthesis. Absorption and derivative absorption spectra of chloroplast fragments from widely diverse algae and higher plants strongly suggest that chlorophyll a in vivo exists in four major and several minor "forms", each with a characteristic absorption maximum and band-width in the red spectral region[1,2,3]. It is not known what causes the distinctive absorption of each "form", nor has any "form" been isolated from the others, but chloroplast fractions enriched in Photosystem I have a higher

---

*CIW-DPB No. 600

proportion of the longer wavelength absorbing "forms" ($>$ 680 nm), and
Photosystem II fractions have little or no absorption beyond 690 nm[4].
The "forms" have been visualized and quantified by curve analysis of
absorption spectra; each "form" is designated by its wavelength maximum, e.g. $\underline{Ca}$683.

Photosystem I particle fractions have in recent years been further
separated by the detergent Triton X-100 and hydroxylapatite chromatography into a chlorophyll $\underline{a}$-protein complex highly enriched in the
reaction-center chlorophyll, P700, (CPI*) and solubilized chlorophyll[5].
The absorption spectrum of CPI is the same as that of the original
particle fraction except for slightly less absorption near 668 relative to 678 nm[6,7], and therefore both fractions have approximately
the same proportions of the long wavelength "forms" of chlorophyll $\underline{a}$.

Fluorescence spectroscopy can be used to investigate possible
energy transfer between closely associated pigment molecules. Light
of a higher energy level (e.g. blue light) absorbed by a pigment molecule may be emitted at a lower energy level of the same pigment (as
red light), or the absorbed energy may be transferred to an adjacent,
different pigment molecule before emission. Thus fluorescence excitation and emission spectra can be used to identify individual pigments
in a mixture and to indicate the closeness of their association.
Such spectra cannot usually be used to measure the quantity of an
individual pigment except under rigidly standardized conditions
because the fluorescence yield may vary considerably. Also artifacts
caused by reabsorption of emitted light by pigments in concentrated
solutions or in particles must be avoided or accounted for.

Current studies presented in part here were undertaken to elucidate
the nature of the "forms" of antenna chlorophyll $\underline{a}$ and energy transfer
between antenna and reaction-center chlorophyll.

METHODS

Chlorophyll-protein complexes enriched in P700 were prepared according to Shiozawa et al[8] from market spinach or from unicellular algae.
The algae were grown in the appropriate autotrophic media bubbled
with air enriched with 3% CO_2. The cells were harvested and broken in
a French press; the fragments were washed with 50 mM Tris buffer, pH
7.5 to remove soluble proteins including phycocyanin in the case of
the blue-green alga, $\underline{Anabaena}$ $\underline{cylindrica}$, and finally sedimented for
storage in a freezer until used for CPI extraction with Triton X-100.

---

*CPI = P700-chlorophyll $\underline{a}$-protein complex

P700 was calculated from the light induced difference in absorption between 697 nm (ϵ = 64^{-1}cm^{-1}) and a 725 nm isosbestic point, or between 430 nm (ϵ = 45^{-1}cm^{-1}) and a 408 nm isosbestic point[9]. The absorption differences were measured with a Perkin-Elmer 356 Dual Wavelength Spectrophotometer equipped for side actinic illumination from a tungsten-iodide lamp and appropriate blue (Corning #9782) or red (Schott RG 2) filter.

Fluorescence was measured with a Perkin-Elmer MPF-3L Fluorimeter fitted with a glass Dewar compartment which allows the sample, in a 5 mm OD NMR tube, to be immersed in liquid N_2 during the measurement. Emission spectra were corrected for the spectral sensitivity of the instrument by an electronic compensator attachment previously calibrated against a standard lamp. Excitation spectra were recorded and stored in the memory of a Hewlett-Packard Computer (#2116C). They were later corrected for the emission lines of the Xenon lamp and spectral sensitivity of the monochrometer by dividing by an excitation spectrum of Rhodamine B recorded at the same slit-width. The quantum yield of concentrated Rhodamine is essentially flat over the blue spectral region, and its excitation is therefore proportional to the lamp intensity.

RESULTS AND DISCUSSION

Chlorophyll-protein complexes enriched in P700 have been isolated by Triton extraction and hydroxylapatite chromatography from a number of plant species, but by far the greatest number of fluorescence measurements have been done with spinach CPI. It has been noted[7,10] that the relative height of the 696 nm fluorescence maximum at -196 C is directly proportional to the P700 concentration of the protein complex. When chloroplasts are treated with 1% Triton, the fluorescence yield of the mixture is very high with a maximum near 670 nm from solubilized chlorophyll a. Solubilized chlorophyll b fluorescence near 650 nm can also be observed when excitation is at 470 nm. The chlorophyll-protein that binds to hydroxylapatite from the mixture and that can be eluted in 0.2 M phosphate before extensive Triton-washing has a somewhat lower fluorescence with a maximum near 683 nm at -196 C. This behavior is in contrast to the fraction I particles prepared by the French press which have their emission maximum near 735 nm (see Fig. 1,a,b) even though their P700 concentration

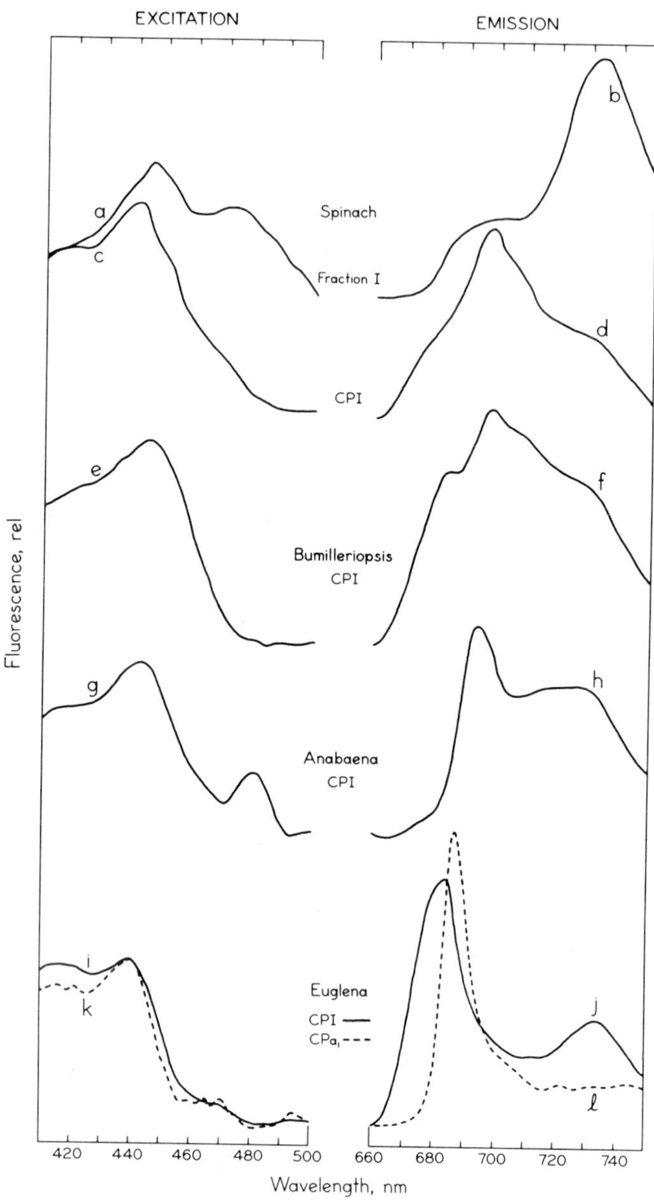

Fig. 1. Fluorescence excitation and emission spectra measured at
-196 C of a chloroplast particle fraction from spinach and chlorophyll-protein complexes from several plants. Excitation spectra
recorded with a slit-width of 5 or 6 nm for emission near the maximum of the corresponding emission spectrum. Emission spectra
recorded with a slit-width of 3 nm from excitation at 450 nm.

is similar. This change in shape of the emission spectrum following exposure to Triton may be caused by preferential solubilization of the long wavelength chlorophyll "forms", C$\underline{a}$693 and C$\underline{a}$700-705 and also by the difference in possible reabsorption artifacts between particles and solubilized protein complexes.

However, when more antenna chlorophyll is removed from the bound protein complex by Triton-washing, the subsequently eluted CPI has a chlorophyll to P700 ratio between 25 and 40 and an emission maximum at 696 nm (see Fig. 1,c,d). The total fluorescence yield is inversely proportional to the P700 concentration[10]. Some other fluorescence characteristics of spinach CPI have been described previously[10]. For example, the shift in the emission maximum from 683 to 696 nm is proportional to the decrease in temperature from 20 to -196 C and does not take place suddenly, e.g. upon freezing. Dilution of the sample to 50% in glycerol before freezing causes an increase in the proportion of 680 to 696 nm emission, and this effect is reversible when the glycerol is removed. Fifty percent glycerol also causes a decrease in quantum yield or rate of P700 oxidation at room temperature.

Two possibilities have been suggested to explain these fluorescence results: 1) The 696 nm emission may be from the chlorophyll $\underline{a}$ "form", C$\underline{a}$684, that emits at the lowest energy level after removal or increase in distance from other antenna chlorophyll "forms"[10]. 2) The unoxidized chlorophyll of an oxidized P700 dimer (chl$^{+}\cdot$chl), absorbing at about 686 nm, may fluoresce at 696 nm[10] after loss of exciton interaction[11]. Because of the very low fluorescence yield of CPI, it is necessary to use intense exciting light which presumably oxidizes P700. As yet no way has been found to observe whether the 696 nm emission maximum is still present when P700 is reduced at -196 C.

A comparison of excitation spectra in Fig. 1 between spinach Fraction I (Fig. 1,a) and CPI (Fig. 1,c) shows that absorption by chlorophyll $\underline{b}$ and carotenoids in the region between 450 and 500 nm contributes to the long wavelength emission in Fraction I (Fig. 1,b), but their contribution is much less in CPI. The shoulders near 450 and 470 nm on curve c indicate that the small amount of β-carotene in CPI may be bound closely enough to antenna chlorophyll $\underline{a}$ to transfer energy to it efficiently.

Chlorophyll $\underline{a}$ is the only photosynthetic pigment in the yellow-green alga, $\underline{Bumilleriopsis}$ $\underline{filiformis}$. The CPI from this alga with a

chlorophyll to P700 ratio of about 43 had an excitation maximum near
444 nm (Fig. 1,e) although the blue absorption maxima (not shown) were
at 420 and 435 nm. A similar difference between absorption and excitation maxima was observed in all the CPI preparations from the different plants examined. This finding indicates that "forms" of
antenna chlorophyll $\underline{a}$ absorbing between 440 and 460 nm, although
indistinguishable on absorption spectra, may function in the antenna
pigments of Photosystem I together with carotenoids. A carotenoid
excitation band at about 480 nm appears in the $\underline{Anabaena}$ $\underline{cylindrica}$
CPI preparation (Fig. 1,g) for fluorescence near 696 nm (Fig. 1,h).

Another CPI preparation from $\underline{Bumilleriopsis}$ that had a higher
ratio of chlorophyll $\underline{a}$ to P700 (60) also had a higher ratio of fluorescence near 680 relative to 696 nm than shown on curve f; the same
correlation was mentioned earlier for different spinach CPI preparations.

The $\underline{Euglena}$ $\underline{gracilis}$ CPI preparation measured for Fig. 1 had a
chlorophyll to P700 ratio of about 60, considerably higher than that
of the other plants. Therefore, it is not surprising that the emission maximum is at 683 nm (Fig. 1,j). It was noted previously[7] that
the P700-chlorophyll $\underline{a}$-protein is eluted from hydroxylapatite by 10 mM
phosphate in $\underline{Euglena}$ preparations instead of by 0.2 M phosphate as
with other plants. When 0.2 M phosphate is applied to a column after
elution of CPI by 10 mM phosphate and the usual Triton-washing,
another distinctive chlorophyll $\underline{a}$ complex termed $CP\underline{a}_1$ can be collected.
The absorption spectrum of $CP\underline{a}_1$[6] showed a distinct maximum at 683 nm
in addition to a larger 670 nm band. Curve analysis of this spectrum
showed the necessity of using a very narrow component curve at 683 nm
having less than one-half the bandwidth used for other component
curves[6]. The unusual sharpness of the 687 nm emission band (Fig. 1,ℓ)
of this preparation may be a related characteristic. Although CPa_1
may not exist as such in vivo, further study of this complex may
reveal important features of chlorophyll-protein binding. It should
also be noted that the calculated difference spectrum between excita-
tion spectra for $\underline{Euglena}$ CPI and $CP\underline{a}_1$ (Fig. 1, i,k) showed bands near
450 and 480 nm characteristic of CPI.

The diagram in Fig. 2 presents a possible scheme for energy transfer between the "forms" of antenna chlorophyll $\underline{a}$ and the reaction centers of Photosystems I and II. A similar scheme was suggested by Dr.
G. I. Garab (personal comm.). The absorption maxima of the biological
"forms" of chlorophyll $\underline{a}$ are listed on the lefthand side and fluorescence emission maxima which have been observed with different plant

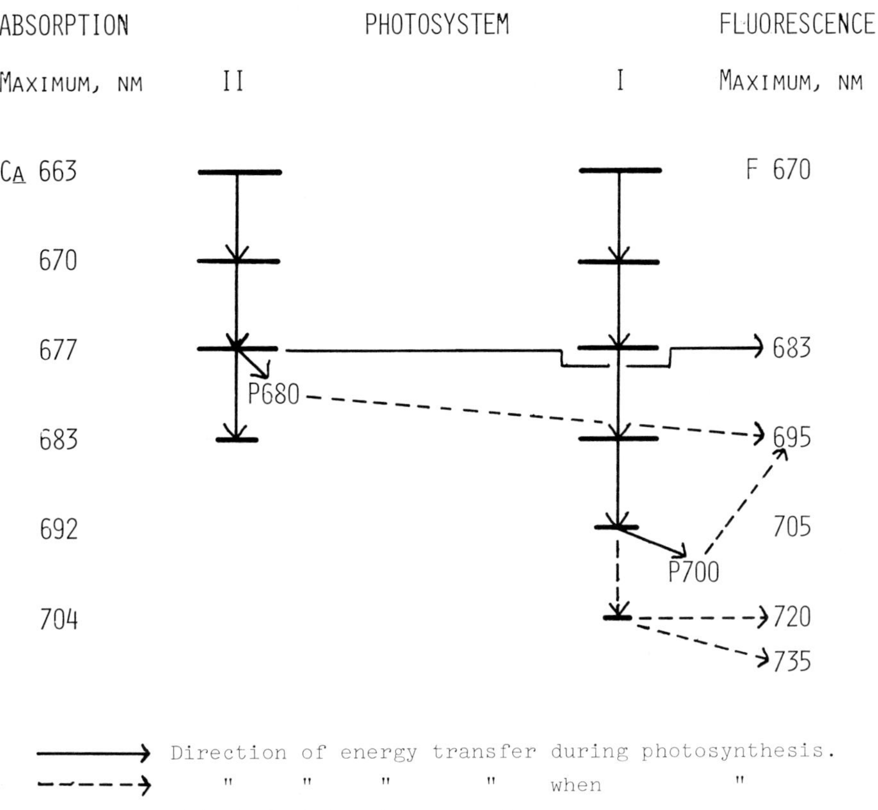

Direction of energy transfer during photosynthesis.
 "   "   "   "   when   "
blocked at low temperature.

Fig. 2. Possible pathway of energy transfer and fluorescence emission between and from the biological "forms" of chlorophyll a.

materials and under various conditions on the righthand side. Solid horizontal lines indicate the chlorophyll "forms" in each Photosystem. Solid line arrows indicate the path of energy transfer and fluorescence emission during photosynthesis under physiological conditions. The dashed arrows show the possible sources of fluorescence emission when photochemistry is blocked at low temperature. P680 and P700 are the reaction-center chlorophylls for Photosystems II and I resp.

Seely[12] showed by calculation of a planar model array how the grouping of chlorophyll "forms" absorbing at different wavelength maxima around two reaction centers could greatly increase the efficiency of energy transfer between light absorbing pigment molecules and reaction centers. Ca663 may be solubilized in a lipid environment in vivo[3]. Fluorescence at 670 nm is not observed from intact chloroplasts, but only after severe membrane damage caused by such treatments as heating or detergent solubilization. It does not seem likely that chlorophyll dissolved in lipid would function in an energy transfer chain from the light harvesting chlorophyll a-b protein[6], that has absorption maxima at 650 and 670 nm, to Ca670. However, each of the isolated chlorophyll-protein complexes also contains a 663 nm component, and perhaps this "form" exists in both a lipid and protein environment in vivo. Since light pretreatments influence the distribution of energy between the Photosystems[13,14], it may be assumed that at least some of the "forms" are physically distinct in vivo, but there may possibly be energy transfer between "forms" in the two Photosystems under particular conditions. Ca692 and Ca704 were not seen in spectra of Photosystem II particles[4].

An emission maximum between 680 and 685 nm (F683) is nearly always observed from all plant material at ambiant temperatures. When the temperature is lowered below -180 C, one or two of the other four maxima may be observed depending upon the preparation. The chapter by Govindjee et al[15] contains an extensive account of the absorption and fluorescence of chlorophyll in vivo. The dashed arrow from P700 to F696 is suggested by the work presented above in this paper.

REFERENCES

1. Brown, J. (1972) Ann. Rev. Plant Physiol. 23, 73-86.
2. Gulyayev, B. A. and F. F. Litvin (1967) Biofizika 12, 845-854.
3. French, C. S., J. S. Brown and M. C. Lawrence (1972) Plant
        Physiol. 49, 421-429.
4. Brown, J. S. and R. Gasanov (1974) Photochem. Photobiol. 19,
        139-146.
5. Brown, J. S., R. S. Alberte and J. P. Thornber (1975) in Proc.
        3rd Intern. Congr. Photosynthesis (Mordhay Avron, ed.)
        Elsevier, Amsterdam III, pp. 1951-62.
6. Brown, J. S., R. S. Alberte, J. P. Thornber and C. S. Franch
        (1974) Carnegie Inst. Year Book 73, 694-706.
7. Brown, J. (1976) Ibid 75, 460-465.
8. Shiozawa, J. A., R. S. Alberte and J. P. Thornber (1974) Arch.
        Biochem. Biophys. 165, 388-397.
9. Hiyama, T. and B. Ke (1972) Biochim. Biophys. Acta 267, 160-171.
10. Brown, J. S. (1977) Photochem. Photobiol. In press.
11. Philipson, K. D., V. L. Sato and K. Sauer (1972) Biochemistry
        11, 4591-95.
12. Seely, G. R. (1973) J. theor. Biol. 40, 189-199.
13. Bonaventura, C. and J. Myers (1969) Biochim. Biophys. Acta 189,
        366-383.
14. Murata, N. (1969) Biochim. Biophys. Acta 172, 242-251.
15. Govindjee, G. Papageorgiou and E. Rabinowitch (1973) in Practical Fluorescence by George G. Guilbault, Marcel Dekker, Inc.,
        New York, pp. 543-575.

***The author is grateful to Steven Graff for helping to grow the algae and prepare the CPIs.

Bioenergetics of Membranes. L. Packer et al. ed.

THE ROLE OF MANGANESE IN THE
OXYGEN EVOLVING MECHANISM OF PHOTOSYNTHESIS

Govindjee, T. Wydrzynski and S. B. Marks
Departments of Physiology & Biophysics and Chemistry
University of Illinois, Urbana, Illinois 61801
U.S.A.

INTRODUCTION

It is generally accepted that the ultimate source of O_2 in photosynthesis is water, although how O_2 is released from water is not known. In the past, we have made several attempts to understand the O_2 evolving side of system II but did not have much success.

First, we prepared antibodies against chloroplast suspension[1] or against an extract from frozen and thawed preparations of washed thylakoids[2]; these antibodies were tested for their specificity against the O_2 evolving side of photosynthesis. We succeeded in obtaining one preparation[2] which gave us specific inhibition of the O_2 evolving side, but the inhibition was only upto 30%. This could be taken to confirm what was already surmised before, that a protein may be somehow involved in O_2 evolution and that the O_2 evolving components may lie on the inner side of the thylakoid membrane (see Trebst[3], Fowler and Kok[4] and Babcock and Sauer[5]).

Next, we started with the hope that bicarbonate (or CO_2) may have something to do with O_2 evolution, even if in a catalytic fashion (see Metzner[6]). Initially, we thought that the site of stimulation of the Hill reaction by CO_2 was located on the water side (see Stemler and Govindjee[7-9]) but later experiments did not support[10] this conclusion. In fact, all our efforts[11] to find an effect on the water side failed. On the other hand, we were able to show that a major effect of CO_2 lies in stimulating electron flow from Q^- (Q being the "primary" acceptor of system II) to the plastoquinone (PQ) pool. The halftime of the decay of chlorophyll (Chl) _a_ fluorescence yield, after a flash, was reversibly slowed down by about five fold in CO_2-depleted chloroplasts (Jursinic _et al._[11]), and this qualitatively explained the reversible slowing down[12] of the relaxation of Sn' to Sn+1 states involved in O_2 evolution kinetics (S being the charge accumulator for oxygen evolution, see Joliot and Kok[13]). More recently, we have shown[14,15] that the major effect of CO_2 depletion is in the electron flow from R(or B), the secondary two electron acceptor, to the PQ pool. The ^{14}C binding studies of Stemler[16] also support a site of CO_2 effect on the electron acceptor side of system II. Independent biochemical studies of Khanna _et al._[17] provided further confirmation that the bicarbonate (or CO_2) effect lies somewhere between Q and the PQ pool. Thus, if CO_2 must play a direct part in O_2 evolution, it still remains to be discovered.

Attempts to isolate "oxygen evolving components" have thus far failed. However there is clear evidence that manganese is required for O_2 evolution (see review[18]). But there has been no evidence, until recently, that manganese undergoes dynamic changes during oxygen evolution. This is what we shall describe here.

Our present approach to understand the O_2 evolution mechanism became promising when we recognized that water proton longitudinal (or, spin-lattice) relaxation rate, $1/T_1$, of thylakoid membranes was strongly dependent upon the presence of bound manganese (Wydrzynski _et al._[19]). When thylakoids were depleted of their bound manganese either by alkaline TRIS or $NH_2OH/EDTA$ treatment, the $1/T_1$ was reduced to 0.4 of its original value. Furthermore, reductants such as tetraphenylboron (TPB) increased and oxidants, such as potassium ferricyanide, decreased the $1/T_1$ of thylakoids. We suggested that $1/T_1$ can monitor bound-manganese and that it exists as a mixture of oxidation states. Soon thereafter, we measured the water proton transverse (or spin-spin) relaxation rates, $1/T_2$, as a function of light-flash number, and discovered oscillations with a period of four, with maxima at the 3rd, 7th, 11th, 15th etc. flashes (Wydrzynski _et al._[20]). These maxima correspond with those in the O_2/flash as a function of flash number in a series of flashes. We, thus, suggested[20] that $1/T_2$ monitors the O_2 evolving mechanism. Significant differences in the $1/T_2$ and O_2 patterns were not explained at that time.

In this paper we present a brief review of our recent data[21,22] on proton relaxation rates (PRR) showing (1) the relationship of $1/T_1$ and $1/T_2$ to manganese concentration and O_2 evolution; (2) the frequency dependence of $1/T_1$ and $1/T_2$, which suggests that the electron spin relaxation ($1/\tau s$) of Mn[II] dominates the PRR, and (3) the relationship of $1/T_2$ and O_2 evolution patterns as a function of flash number under a variety of conditions which show that O_2 evolution can be uncoupled from the intermediate monitored by PRR. Finally, we present a hypothetical manganese model of O_2 evolution.

MATERIALS AND METHODS

Chloroplast membrane fragments were prepared by homogenizing leaves in a medium consisiting of 50 mM N-2-hydroxyethylpiperazineN-2 ethane solfonic acid (HEPES) buffer, adjusted to pH 7.5 with NaOH, 400 mM sucrose and 10 mM NaCl. 0.5% bovine serum albumin and 10 mM sodium ascorbate were included during the grinding procedure. After appropriate filteration and differential centrifugation, chloroplast fragments were given osmotic shock and then suspended in the medium described above, but adjusted to appropriate pH. For PRR measurements, thylakoid suspensions containing 2-3 mg Chl/ml were used.

O_2 evolution/flash was measured on a Joliot-type electrode. The light source was a Phase-R Model DL 2100 A dye laser (λ, 590 nm) having a pulse width of 0.6 μs at half height, terminating in about 2 μs. The dark time between flashes was 4s. The same illuminating conditions were used for the measurements of $1/T_2$ as a function of flash number. Manganese concentrations were measured with a Perkin-

Elmer Model 303 Atomic Spectrometer or by neutron activation analysis.

Proton relaxation rates were measured in a pulsed nuclear magnetic resonance spectrometer (constructed in the laboratories of Drs. Paul Schmidt and of H.S. Gutowsky). (See Fig. 1 for the basic design and Wydrzynski et al.[21] for details.)

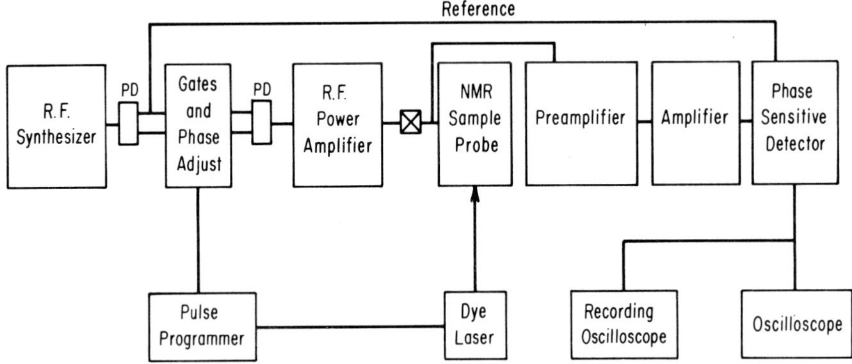

Fig. 1 Generalized Block Diagram of a Pulsed NMR Spectrometer. The rf transmitter consists of rf synthesizer, gating and phase adjust unit and rf power amplifier. The rf receiver consists of appropriate amplifiers and a phase sensitive detector. PD refers to power divider. The NMR sample probe is positioned in strong magnetic field. A dye laser (λ, 590 nm) is used for light excitation of the sample. A pulse programmer determines the rf and light pulse sequence. Signals are recorded by an oscilloscope and oscillographic recorder. (after ref. 23.)

The $1/T_1$ was measured by the inversion recovery method. The sample was placed in magnetic field, then a 180° radiofrequency (rf) pulse was given, followed by 90° rf pulses after various times (τ). The $1/T_1$ was calculated from a plot of $\ln \frac{Mo-Mz(\tau)}{2\,Mo}$ as a function of τ, where Mo is the equilibrium magnetization and $Mz(\tau)$ is the magnetization at time τ after the 180° pulse. The $1/T_2$ was measured as follows. A Carr-Purcell-Meiboom-Gill pulse sequence was given: a 90° rf pulse was followed by time τ and then a series of 2,100 180° pulses spaced 2τ apart ($\tau=$ 500 μs) were given. The $1/T_2$ was calculated from the decay of spin echo envelopes with time (τ); the slope of $\ln \frac{Mz(\tau)}{Mo}$ as a function of τ is proportional to $1/T_2$.

RESULTS AND DISCUSSION

1. <u>Proton</u> <u>Relaxation</u> <u>Rates</u> <u>Monitor</u> <u>Bound</u> <u>Manganese</u>

Using the method of Chen and Wang[24], we were able to vary the concentration of chloroplast manganese by replacement with magnesium. Fig. 2 (left) shows $1/T_1$ and $1/T_2$ as a function of manganese concentration. Both $1/T_1$ and $1/T_2$ decrease as manganese concentration is decreased till about 50% of manganese is left. The $1/T_2$ continues to decrease linearly, whereas $1/T_1$ reaches a constant value representing some background contribution. A similar pattern is observed if $1/T_1$ and $1/T_2$ are plotted as a function of O_2 (Fig. 2 (right)). A direct relationship be-

308

tween PRR, manganese content and O_2 evolution is thus established. It appears that $1/T_1$ monitors mainly the loosely bound manganese related to O_2 evolution (c.f. ref. 18), where $1/T_2$ may monitor the total manganese content of the membrane due to greater non-site specific contributions to $1/T_2$ (see Wydrzynski[23] for further details).

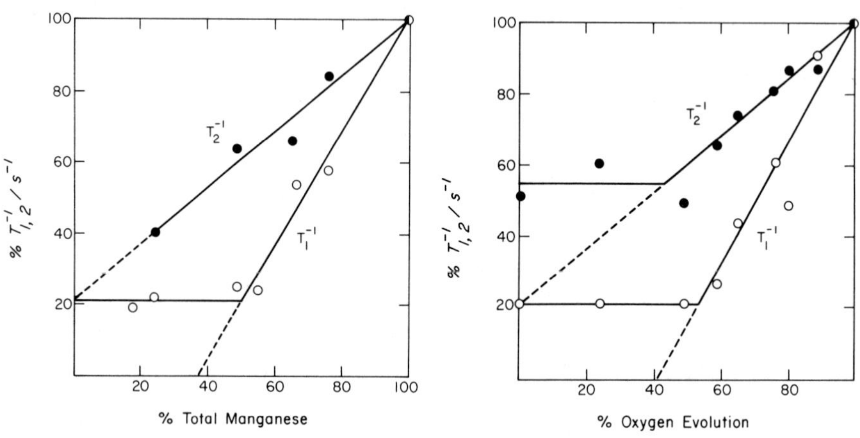

Fig. 2 (left):Plot of Percent $1/T_1$ and $1/T_2$ versus percent total Mn content. (right):Plot of Percent $1/T_1$ and $1/T_2$ versus percent O_2 activity. Mn extraction was achieved by incubating pea chloroplasts for 2 hr in dark at 4°C in HEPES buffer medium containing $MgCl_2$ at the following [Mg]/[Chl] ratios: 0, 167, 332, 667, 2,500 and 10,000. After incubation the chloroplasts were centrifuged and pellet was resuspended to 3 mg Chl/ml. Mn content was determined by neutron activation analysis. For control. Mn content was 0.62±0.03 μg Mn/mg Chl. Rate of O_2 evolution for control was 120 μmoles O_2/mg Chl-hr. (After ref. 21.)

2. _Proton_ _Relaxation_ _Rates_ _Monitor_ _Bound_ _Mn[II]_

Fig. 3 shows dependence of $1/T_1$ and $1/T_2$ as a function of NMR frequency. The $1/T_1$ shows a broad peak in the 10-20 MHz range (open circles, lower curve), whereas $1/T_2$ shows only an increase (open circles, upper curve). This characteristic frequency behavior is obtained when the electronic relaxation of bound paramagnetic ions dominate the PRR. The solid lines through the experimental points in Fig. 3 are the "best" fit theoretical curves based on Solomon-Bloembergen- Morgan analysis for relaxation in paramagnetic systems.

Both $1/T_1$ and $1/T_2$ are related to the dipolar correlation time (τ_c) by Solomon-Bloembergen equation[25]. The major molecular processes which lead to dipolar interactions include the electron spin relaxation ($1/\tau_s$), rotational motions ($1/\tau_r$) and the rate of chemical changes ($1/\tau_m$). The $1/\tau_c$ is related to these processes as follows: $\dfrac{1}{\tau_c} = \dfrac{1}{\tau_s} + \dfrac{1}{\tau_r} + \dfrac{1}{\tau_m}$. For macromolecules, τ_r is long ($\sim 10^{-6}$ s) and thus

$1/\tau_r$ is insignificant with respect to $1/\tau_s$ and $1/\tau_m$. If τ_s dominates τ_c, one expects (see ref. 25) the frequency behavior as shown in Fig. 3 since τ_s itself changes with frequency. The frequency dependence of τ_s is given by the Bloembergen and Morgan equation:

$$\frac{1}{\tau_s} = B \left(\frac{\tau_v}{1+\omega_s^2\tau_v^2} + \frac{4\tau_v}{1+4\omega_s^2\tau_v^2} \right)$$

where, τ_s is electronic Larmor frequency, B is a constant related to the zero field splitting parameters and τ_v is the correlation time for the modulation of the zero field splitting. The NMR parameter values obtained from the "best" fit curves in Fig. 3 are $\tau_v{=}20{\times}10^{-12}$, $\tau_m{=}2.2{\times}10^{-8}$ s and $B{=}0.9{\times}10^{19}(rad/s)^2$. These values compare favorably with those for Mn[II]-pyruvate kinase systems, $\tau_v{=}14{\pm}4{\times}10^{-12}$ s, $\tau_m{=}0.4{\pm}1.04{\times}10^{-8}$ s, and $B = 0.8{\pm}0.1{\times}10^{19}(rad/s)^2$ (after Navon[26]). The correlation times for Mn[III] and high spin Fe[II] and Fe[III] are 2-3 orders of magnitude different (lower) than for Mn[II]. And, since the copper in plastocyanin has no effect on PRR (see ref. 27), the PRR of chloroplasts, thus, must be monitoring only Mn[II] contributions.

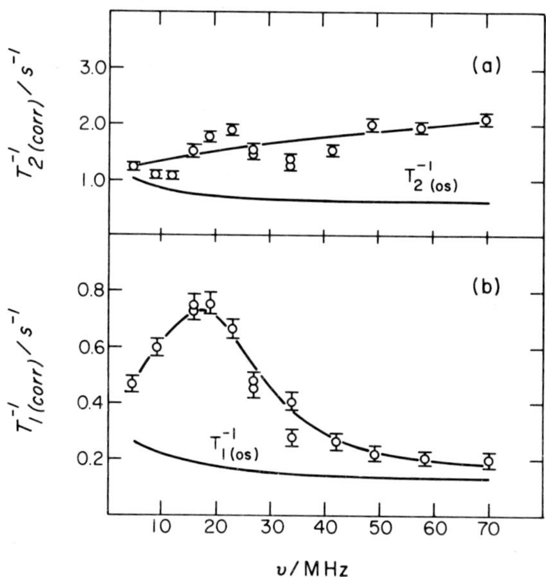

Fig. 3. "Best" Fit Theoretical Curves to the Frequency Dependence of the Relaxation Rates for Dark-Adapted Chloroplast Membranes. (a) $1/T_2$ relaxation ; (b) $1/T_1$ relaxation; $1/T_{1,2(corr)}= 1/T_{1,2(obs)}-1/T_{1,2(T-A)}$. Rates were measured at room temperature (23-25°C) and normalized to the same Mn concentration at each frequency. $1/T_{1,2(os)}$ refers to the theoretical outer sphere contribution on a translational diffusion model. Pea chloroplasts at 3 mg Chl/ml were used. (After ref. 21.)

3. The $1/T_2$ Monitors the S States in O_2 Evolution

Fig. 4 shows $1/T_2$ and O_2/flash as a function of flash number in a series of

310

flashes for three different samples of pea chloroplasts, at pH 6.7. The absence
of O_2 in the second flash indicates that there are no double hits with our laser
flashes. The main peak in O_2 or $1/T_2$ curve is at the 3rd flash, the second peak
is at the 7th flash in $1/T_2$ (curves c & e), but for O_2 yields, the 7th and 8th
flashes have about the same value. When manganese is extracted from chloroplasts
by TRIS-acetone wash method[28], both oscillations in $1/T_2$ and O_2 disappear. Addition of DCMU to chloroplasts which stops turnover of reaction center II, after
one flash, also eliminates $1/T_2$ oscillations (Fig. 5). These results suggest
that $1/T_2$ is monitoring the O_2 evolving mechanism.

Differences between $1/T_2$ and O_2 patterns are: (a) the ratio of $1/T_2$ in the
3rd to the 2nd flash is very low as compared to that for O_2; (b) a minimum in $1/T_2$
is at the 5th flash, whereas in O_2 it is at the 6th flash and (c) the dark $1/T_2$
level is significantly higher than in the first flash, whereas there is no O_2 in
the first flash as in the darkness.

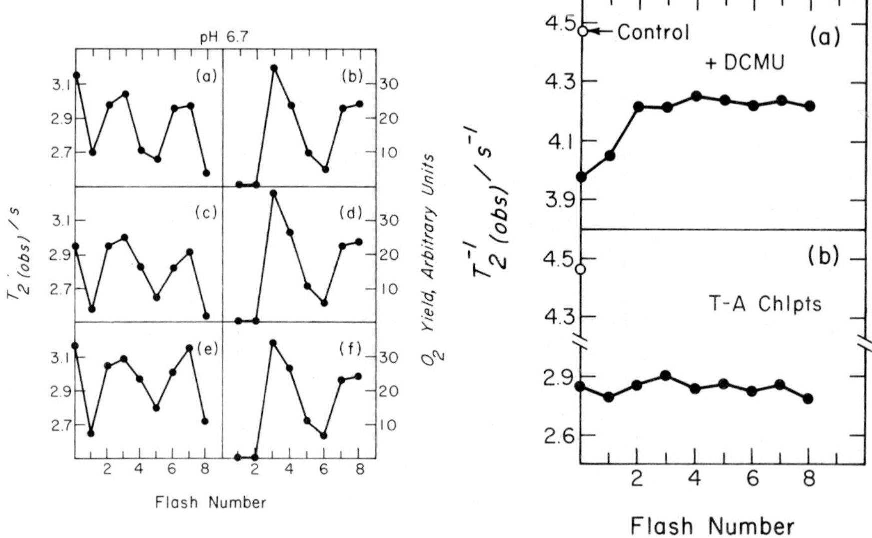

Fig. 4 (left) Observed $1/T_2$ Rates and O_2 Yield Measured as a Function of
Saturating Flashes of Light for three samples of Pea Chloroplast Membranes.
The $1/T_2$ and O_2 measured after each flash in a pulse sequence. Saturating
light flashes obtained from a pulsed dye laser (λ, 590 nm) were spaced 4s
apart. $1/T_2$ was measured at 27 MHz, 23°C. Sample contained 2 mg Chl/ml.
(After ref. 22.)

Fig. 5 (right) Effect of DCMU and Manganese Extraction on $1/T_{2(obs)}$ Flash
Pattern. (a) [DCMU]/[Chl]=0.089; (b) TRIS-acetone washed chloroplasts.
Flash procedure and conditions for pea chloroplast in Fig. 4 were used.
(After ref. 22.)

Note also that $1/T_2$ in dark is decreased by DCMU. Data reported by Wydrzynski[23] show that the dark $1/T_2$ level can also be decreased by (a) ferricyanide,

(b) fixation of chloroplasts with glutaraldehyde; and (c) removal of sucrose from the suspension medium. Furthermore, unpublished observations of Rita Khanna (in our laboratory) show that in intact cells of blue-green alga _Phormidium luridum_ the dark level is also low. These observations suggest that the significance of the high dark level of $1/T_2$ in isolated chloroplasts should not be exaggerated although it still needs to be explained.

When tetraphenylboron (TPB), hydroxylamine (NH_2OH) or carbonylcyanide m-chlorophenylhydrazone (CCCP) is added to chloroplasts, the flash pattern of $1/T_2$ is changed with the peaks appearing at the 2nd and 6th flashes instead of the 3rd and 7th flashes (Fig. 6). No O_2 is evolved in the TPB case in the first 10 flashes, but then slowly increases with succeeding flash, with NH_2OH the O_2 peaks are on the 6th and 10 flashes, and no O_2 is evolved when CCCP is present. All of the above suggest that $1/T_2$ oscillation can be uncoupled from the O_2 evolution. The differences in $1/T_2$ and O_2 patterns suggest that $1/T_2$ is monitoring more than just the final O_2 evolving step. It is easy to qualitatively understand this difference because O_2 comes off only during the last step of the following set of reactions, according to Kok _et al._ (see ref. 13):

$$S_0 \xrightarrow{h\nu} S_1 \xrightarrow{h\nu} S_2 \xrightarrow{h\nu} S_3 \xrightarrow{h\nu} S_4$$
$$2H_2O \longrightarrow O_2 + 4H^+$$

where, the subscripts refer to the number of oxidizing equivalents on the intermediate S; PRR may be monitoring the S intermediate directly. Since PRR is suggested (in section 2) to monitor Mn[II], it is logical to propose that $1/T_2$ changes induced by light flashes indicate dynamic changes in Mn[II] concentration during O_2 evolving process. The relationship is, however, complex (see section 4).

4. _The Model_

A search for a model to explain $1/T_2$ as a function of flash number began as soon as we observed[20] the flash number dependance of $1/T_2$ in spinach chloroplasts. Many differences between $1/T_2$ and O_2 had to be recognized. First, the minima were at 4th, 8th, 12th flashes in $1/T_2$ in contrast to minima at 6th, 10th etc. flashes in O_2 evolution. Second, the dark $1/T_2$ was as high as that at 3rd etc. flashes. Third, when O_2 yield decreased in going from 4th to 6th flash, $1/T_2$ increased. Fourth, overall the O_2 yield decreased with increasing flash number, but the $1/T_2$ increased. In these experiments, chloroplasts stayed at room temperature for a long time, particularly for flash numbers greater than 9, because $1/T_2$ was measured only at the end of a series of flashes. The data on peas reported here ($1/T_2$ measured after each flash in a series of flashes) and lettuce (measured as in the spinach case) show essentially the same major features upto the 8th flash as observed in spinach except that the $1/T_2$ after the first flash is almost the same as dark (lettuce) or is intermediate between that for spinach and lettuce (peas).

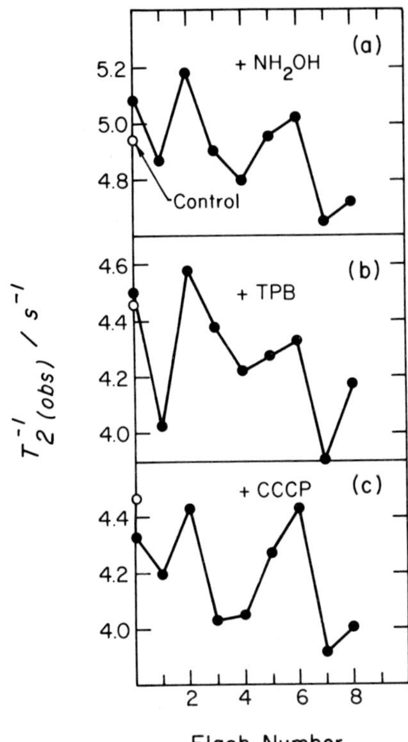

$T_{2(obs)}^{-1}$ / s^{-1}

Figure 6. $1/T_{2(obs)}$ Flash Pattern in the Presence of NH2OH, TPB and CCCP. (a) [NH2OH]/[Chl]=0.44; (b) [TPB]/[Chl]=0.016; (c) [CCCP]/[Chl]=0.089. The $1/T_2$ for the dark-adapted controls are shown with open circles. Flash procedure and conditions for pea chloroplasts in Figure 4 were used. (After ref. 22.)

Another difference was that the minimum in lettuce or peas was on the 5th flash in contrast to the 4th flash in spinach. No significant difference in O_2 flash pattern was, however, observed. Many of these differences could be accounted for by variations in the following model.

We used the basic features of Kok et al.'s model and procedure (ref. 29) to calculate the concentrations of S_0 and S_1, α (misses) and β (double hits) from O_2/flash as a function of flash number for a sample at pH 6.7. This gave us $[S_0]$ = 0.30, $[S_1]$= 0.70, α = 0.10 and β = 0. From these initial conditions, we calculated the concentration of all the S states after each flash. Then, we assigned various weighting factors to each S state and multiplied these by the concentration for that S state after each flash. These values were then summed to give the relative theoretical $1/T_2$ values after a flash and normalized to the experimental point (corrected for non-specific $1/T_2$ value in TRIS-acetone washed chloroplasts) at the 3rd flash. The weighting factors which best fit the $1/T_2$ oscillations for pea chloroplasts at pH 6.7 were 2, 1, 1, 3 for S_0, S_1, S_2, S_3 respectively, but, in addition, we had to assume that S_0 in darkness had a value of 4 to fit the high $1/T_2$ in darkness (Fig. 7). This high value of Mn[II] contribution in darkness may be due to the existence of a special dark-adapted state in isolated chloroplast described as S_{-1} by Velthuys[30].

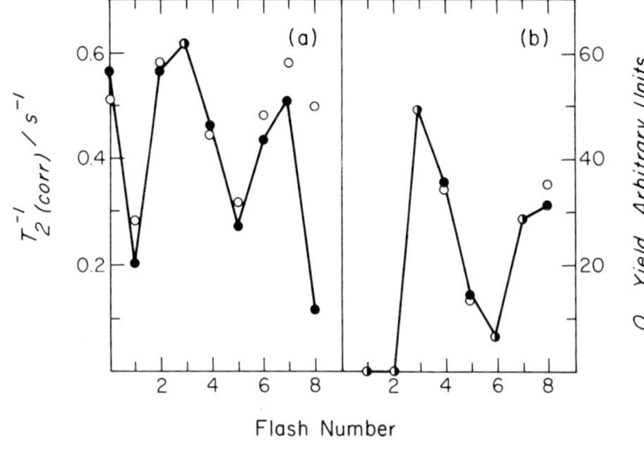

Figure 7. Theoretical Fit to the $1/T_2$ and O_2 Yield Flash Patterns at pH 6.7. (a) $1/T_2$ relaxation; (b) O_2 yield. Closed circles are experimental data points, $1/T_2$(corr); open circles are theoretical points. For the theoretical fit $[S_0]$=0.30, $[S_1]$=0.70, α=0.10, β = 0, and weighting factors for S_0, S_1, S_2, S_3 are 2, 1, 1, 3 respectively except S_0=4 in the dark (After ref. 22.)

Now, the question is: what is the significance of 2, 1, 1, 3 contribution of Mn[II] for S_0, S_1, S_2, S_3, respectively. First, it shows heterogeneity of the S states at the time $1/T_2$ measurements are made. The contribution of Mn[II] for S_1, being smaller than S_0 is in line with Kok et al.'s model that S_1 is more oxidized than S_0--S_0 having more Mn[II] character than S_1. However, the following step of S_1 to S_2 is different because at the time of $1/T_2$ measurement (200-300 μs after the flash) S_2 becomes equivalent in terms of Mn[II] contribution. This is possible if the initial oxidation of S_1 to S_2 is followed by another reaction (not recognized in Kok et al.'s model)in which H_2O is used to reduce Kok's S_2 (which is more oxidized than S_1) to our version of S_2 (Figure 8) which has the same oxidizing equivalent as S_1 in terms of Mn[II]. The next step from S_2 to S_3 has more dramatic differences. In Kok's picture, S_3 is clearly the most oxidized species among S_0--→S_3, but in our picture, S_3 attains the most reducing character at the time of $1/T_2$ measurement--its Mn[II] character is maximal except, perhaps, for the dark-adapted $S_0[S_{-1}]$. This implies reduction steps following (or simultaneously to) the oxidation steps of the S states. A possible consequence of the reduction steps is that when H_2O reduces the S state, protons may be released. This hypothesis is consistent with the recent observations of C.F. Fowler (personal communication) and of A.R. Crofts and coworkers (personal communication) which show that protons may be released in steps much earlier than shown in Kok et al.'s original scheme.

Fowler suggests that the release of H^+ is as follows: S_0--→S_1 (0.7 H^+); S_1→S_2 (0 H^+); S_2→S_3 (1.3 H^+) and S_3→S_0 (2 H^+). It is possible to produce a model

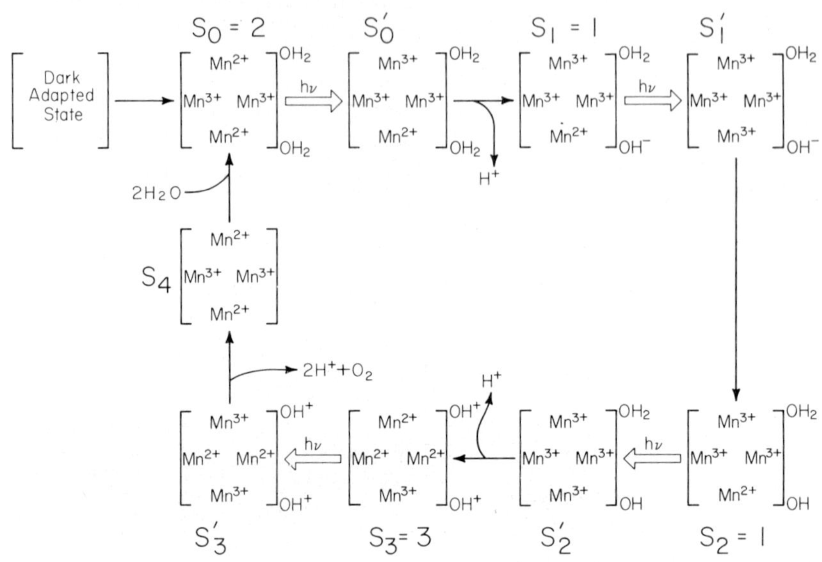

Figure 8.   Working Hypothesis for Oxygen Evolution in Photosynthesis.

which would have similar H^+ releases, if not identical, and retain the 2, 1, 1, 3 picture (Fig. 8). If we assume that Mn[II] contributions, that we have been discussing, are proportional to both the concentration of Mn[II] and the number of exchangeable protons, then 2, 1, 1, 3 pattern would imply a Mn[II] concentration of 0.5, 0.3, 0.3, 1.5 (or, 1.67, 1.0, 1.0, 5.0) for S_0, S_1, S_2 and S_3, respectively. This does not change the model significantly except that S_3 has still greater Mn[II] contributions. We have no way yet to make the distinction between the "pure" effect of Mn[II] concentration and the combined effect of Mn[II] concentration and the number of exchangeable protons.

A glaring difference between the 2, 1, 1, 3, model and the experimental data is the poor fit at the 8th flash. This needs further analysis.

Fig. 8 assumes that there are 4 Mn atoms, and that S_0 and S_1 are mixtures of different redox states of Mn (see ref. 19). A titration of $1/T_1$ as a function of [TPB] shows[31] four plateaus as if it reflects the four reducible Mn[II] in the system. The chemistry of H_2O oxidations is unknown, and, therefore, in our scheme only the charges have been balanced, the species shown are only symbolic. Data of Fig. 6 can be simulated by proposing a 0, 1, 1, 3 model. Other variations in the patterns can be generated by altering the weighting factors. We,thus, hope that the extension of the present approach would lead to a further understanding of the O_2 evolving mechanism of photosynthsis.

5. SUMMARY AND CONCLUDING REMARKS

(1) The water proton relaxation rates monitor bound Mn[II] in chloroplast membranes. (2) A mixture of manganese oxidation states exists in dark-adapted chloroplasts. (3) Bound manganese participates directly in the S intermediate, the precursor to photosynthetic O_2 evolution. (4) A special state of the S intermediate, having a high Mn[II] contribution, exists in the dark-adapted isolated chloroplasts. (5) The cycling of the various states of the S intermediate (S_n, where n = 0...4) in the light involves changes in the manganese oxidation states. (6) Water, apparently, donates electrons to the S intermediate in reactions prior to the release of O_2, at times proton relaxation rates are measured (~200-300 μs after the flash).

ACKNOWLEDGEMENTS

We thank Drs. Paul G. Schmidt and H. S. Gutowsky for collaboration in this research. Results summarized here will be published jointly with these colleagues in separate publications[21,22]. We are thankful to research grants from the National Science Foundation (to G. and H.S.G.), the National Institutes of Health (to P.G.S.) and the Office of Naval Research (to H.S.G.). T.W. was supported by a U.S. Publich Health Service Training Grant in Cellular and Molecular Biology (GM-7283-1 Sub. Proj. -604) and G. by a National Science Foundation Grant (PCM 76- 11657).

REFERENCES

1. Zilinskas, B. and Govindjee (1972) FEBS Lett. 25, 143-146.

2. Zilinskas-Braun, B. and Govindjee (1974) Plant Sci. Lett. 3, 219-227.

3. Trebst, A. (1974) Ann. Rev. Plant Physiol. 25, 423-458.

4. Fowler, F. and Kok, B. (1974) Biochim. Biophys. Acta 357; 308-318.

5. Babcock, G.T. and Sauer, K. (1975) Biochim. Biophys. Acta 396, 48-62.

6. Metzner, H. (1975) J. Theor. Biol. 51, 201-231.

7. Stemler, A. and Govindjee (1973) Plant Physiol. 52, 119-123.

8. Stemler, A. and Govindjee (1974) Plant Cell. Physiol. 15, 533-544.

9. Stemler, A. and Govindjee (1974) Photochem. Photobiol. 19, 227-232.

10. Wydrzynski, T. and Govindjee (1975) Biochim. Biophys. Acta 387, 403-405.

11. Jursinic, P., Warden, J. and Govindjee (1976) Biochim. Biophys. Acta 440, 322-330.

12. Stemler, A., Babcock, G.T. and Govindjee (1974) Proc. Natl. Acad. Sci., USA 71, 4679-4683.

13. Joliot, P. and Kok, B. (1975) in "Bioenergetics of Photosynthesis" (Govindjee, ed.) Academic Press, New York, p. 387-412.

14. Govindjee, Pulles, M.P.J., Govindjee, R., Van Gorkom, H.J. and Duysens, L.N.M. (1976) Biochim. Biophys. Acta 449, 602-605.

15. Siggel, U., Khanna, R., Renger, G. and Govindjee (1977) Biochim. Biophys. Acta, in the press.

16. Stemler, A. (1977) Biochim. Biophys. Acta, in the press.

17. Khanna, R., Govindjee and Wydrzynski, T. (1977) Biochim. Biophys. Acta, in the press.

18. Cheniae, G. (1970) Ann. Rev. Plant Physiol. $\underline{21}$, 467-498.

19. Wydrzynski, T., Zumbulyadis, N., Schmidt, P.G. and Govindjee (1975) Biochim. Biophys. Acta $\underline{408}$, 349-354.

20. Wydrzynski, T., Zumbulyadis, N., Schmidt, P.G., Gutowsky, H.S., and Govindjee (1976) Proc. Natl. Acad. Sci. USA $\underline{73}$, 1196-1198.

21. Wydrzynski, T., Marks, S.B., Schmidt, P.G., Gutowsky, H.S. and Govindjee (1977) in preparation.

22. Wydrzynski, T., Govindjee, Marks, S.B., Schmidt, P.G., and Gutowsky, H.S. (1977) in preparation.

23. Wydrzynski, T., (1977) Ph.D. thesis, University of Illinois, Urbana, Illinois.

24. Chen, K-Y, and Wang, J.H. (1974) Bioinorg. Chem. $\underline{3}$, 339-352.

25. Dwek, R.A. (1975) "Nuclear Magnetic Resonance (N.M.R.) in Biochemistry: Application to Enzyme Systems", Clarendon Press, Oxford.

26. Navon, G. (1970) Chem. Phys. Lett. $\underline{7}$, 390-394.

27. Blumberg, W.E. and Peisach, J. (1966) Biochim. Biophys. Acta $\underline{126}$, 269-273.

28. Yamashita, T. and Tomita, G. (1974) Plant Cell Physiol. $\underline{15}$, 252-266.

29. Forbush, B., Kok, B. and McGloin, M. (1971) Photochem. Photobiol. $\underline{14}$, 457-475.

30. Velthuys, B.R. (1976) Ph.D. Thesis, The State Univeristy, Leiden, The Netherlands.

31. Wydrzynski, T., Govindjee, Zumbulyadis, N., Schmidt, P.G. and Gutowsky, H.S. (1976) in "Magnetic Resonance in Colloid and Interface Science" (H.A. Resing and C.G. Wade, eds.) Am. Chem. Soc., pp. 471-482.

ELECTRON TRANSFER BETWEEN DIFFERENT PHOTOSYSTEMS I IN CHLOROPLASTS

Wolfgang Haehnel
Lehrstuhl für Biochemie der Pflanzen
Ruhr-Universität Bochum
Postfach 102148, D-4630 Bochum 1
Germany (West)

SUMMARY

The kinetics of the absorbance changes of plastoquinone and of the
ESR amplitude of P700 (chlorophyll a_I) induced by excitation with
long flashes of saturating intensity were measured in spinach chloroplasts. The influence of partial inhibition of electron transport by
2,5-dibromo-3-methyl-6-isopropyl-p-benzoquinone (DBMIB) and KCN incubation was studied.

1. The kinetics of plastoquinone provide direct evidence for a
specific inhibition of the electron transfer step from the pool of
plastoquinone to the electron carriers of Photosystem I. An estimation of the maximal number of inhibition sites of DBMIB yielded two
per light reaction.

2. The reduction kinetics of $P700^+$ showed a homogeneous time
course even at an inhibition of more than 90% of linear electron
transport by DBMIB. This result suggests a fast electron exchange
after the electron transfer from plastoquinone between at least ten
different Photosystems I.

3. At increasing inhibition with KCN the $P700^+$ reduction showed
biphasic kinetics with an increasing portion of the slow component
(half-time approx. 0.5 s). The time course of the remaining fast
component was always similar to that in uninhibited chloroplasts.
This result indicates that P700 is reduced via tightly associated
plastocyanin. A possible involvement of loosely bound plastocyanin
in the electron exchange between different Photosystems I is discussed.

Abbreviations: DBMIB, 2,5-dibromo-3-methyl-6-isopropyl-p-benzo-
quinone; DCMU, 3-(3,4-dichlorophenyl)-1,1-dimethylurea; ESR, electron spin resonance.

INTRODUCTION

Linear electron transport between the two light reactions in photosynthesis of higher plants involves a pool of approximately seven plastoquinone molecules per light reaction II[1,2]. Stiehl and Witt[1] and Siggel et al.[3] reported evidence for a cooperation of at least ten electron transport chains via a strand of combined pools of plastoquinone. This was derived from the effect of 3-(3,4-dichlorophenyl)-1,1-dimethylurea (DCMU) on the absorbance changes of plastoquinone and P700 (chlorophyll a_I). The corresponding model of linear electron transport is shown in Fig. 1. The electron carriers of

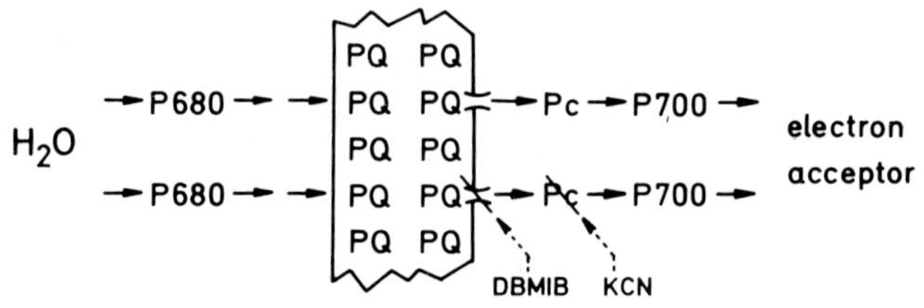

Fig. 1. Simplified model of linear electron transport with an electron exchange between (at least 10) separate chains via a strand of plastoquinone according to refs. 1 and 3. The inhibition site of DBMIB is indicated as proposed by Böhme et al.[4] and that of KCN according to the results in refs. 5-7.

Photosystem I, cytochrome f, plastocyanin and P700, with an electron capacity of three per light reaction II[1,8,9], were implied to act as separate units.

The association of P700 with a large protein-chlorophyll a complex[10,11] and the rather uniform distribution of smaller and larger particles in the stacked region of the thylakoids[10,12] seem to be consistent with an independent function of different P700.

However, we reported evidence from the absorbance changes of P700 after a short flash for an equilibration of electrons between different P700[13]. The alternatives can be investigated by an independent approach using partial inhibition of linear electron transport behind the pool of plastoquinone. If Photosystems I act as separate chains as shown in Fig. 1 a reduction would only be observed of that P700 which is connected to the uninhibited chains. Appropriate inhibitors are 2,5-dibromo-3-methyl-6-isopropyl-p-benzoquinone (DBMIB) and KCN.

The plastoquinone antagonist dibromothymoquinone (DBMIB) introduced by Trebst[14,4] inhibits probably the reoxidation of plastoquinone by the electron carriers of Photosystem I[4,15,16]. Investigations at high concentrations of DBMIB indicated a less specific
disconnection of the plastoquinone pool from light reaction II[17]
and an additional function of DBMIB as electron acceptor (for a
review see refs. 18 and 19).

Incubation of chloroplasts with KCN inhibits plastocyanin in
situ[7] with a high degree of specifity (for a review see ref. 19).
P700 as well as cytochrome f appear to be quite immune to KCN as
detected spectroscopically[6,20]. We found no influence of DBMIB on
the fast electron transfer from plastocyanin to P700$^+$ with a half-
time of 200 μs. But this reaction was completely inhibited by KCN[9].

In this communication evidence will be reported that at a specific inhibition of the electron transfer sites from plastoquinone
to Photosystem I by DBMIB an electron exchange faster than 20 ms
between at least ten P700 is possible. After KCN incubation this
electron exchange was not observed. The function of plastocyanin
will be discussed.

MATERIALS AND METHODS

Hypotonically broken and Class II chloroplasts of spinach were
isolated as described in ref. 13. KCN treatment was carried out
according to the procedure of Izawa et al.[20]. After dilution to
terminate the treatment the chlorophyll concentration was increased
by centrifugation for 3 min at 1500xg. For differential inhibition
by KCN[6] the incubation time was varied as given in the legends of
the figures.

Measurements of absorbance changes of plastoquinone were carried
out as previously described[13,9]. The reaction mixture contained
chloroplasts at a concentration of 10 μM chlorophyll, 20 mM N-tris
(hydroxymethyl)methylglycine adjusted to pH 7.2 with NaOH, 20 mM
KCl, 1 mM MgCl$_2$, 10 μM benzylviologen and 1 μM gramicidin D. Further
additions are noted.

ESR-measurements: Electron spin resonance measurements were carried out with a Varian E-9 spectrometer. Light-induced transients of Signal I, the ESR signal of $P700^+$ [21], were monitored at the peak of the low-field derivative maximum. The microwave power was 20 mW, the modulation amplitude 4.0 G and the time constant 3 ms. Signal averaging was performed using a Tracor TN-1500 digital signal analyser. Appropriate timing circuits (Systron Donner 100 A) synchronized the inition of the averager sweep and the illumination period of about 100 ms. A Compur-electronic m-1 shutter enabled illumination with a rise and a decay time shorter than 2 ms. The actinic light was filtered through 7 cm of water, a heat-reflecting filter Calflex C from Balzers and a Schott filter RG 610 (3 mm). The light was guided through a light pipe to the grid in front of the cavity. The incident light upon the flat sample cell (thickness 0.3 mm) had a saturating intensity of 1300 $W \cdot m^{-2}$.

The reaction mixture contained chloroplasts at a concentration between 1 and 2.5 mg chlorophyll per ml, 5 mM $MgCl_2$, 20 mM KCl, 20 mM N-tris(hydroxymethyl)methylglycine adjusted to pH 7.6 with NaOH and 0.2 mM methylviologen. Deviating conditions are noted. All measurements were carried out at room temperature of 20-22°C.

RESULTS

Fig. 2 shows the effect of 30 nM DBMIB on the time course of the absorbance changes of plastoquinone induced by a flash group. The amount of electrons accumulated in the plastoquinone pool during the flash group of 16 flashes exceeded the electron capacity of the electron acceptors on the Photosystem I side. This is indicated by the amplitude of the slow phase of the plastoquinone reoxidation after the flash group[1,9]. Three results should be noted: 1. The half-time of the fast reoxidation of plastoquinone was increased from 16 ms to 34 ms. 2. The amount of reduced plastoquinone was increased. 3. The amplitude of the fast reoxidation was not changed. It is obvious from this experiment that DBMIB inhibits the rate-limiting electron transfer of linear electron transport from the plastoquinone pool to the electron carriers of Photosystem I. This substantiates previous conclusions derived from more indirect lines of evidence[4,19].

The increase of the half-time by more than a factor of two indi-

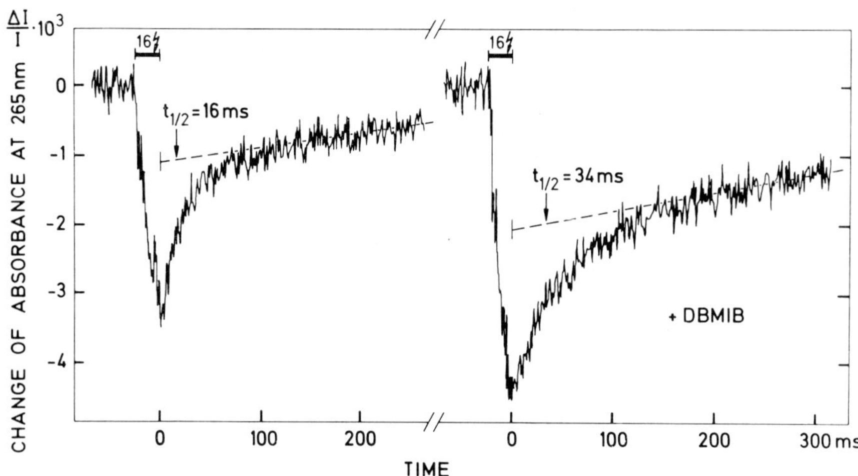

Fig. 2. Absorbance changes of plastoquinone at 265 nm as a function
of time. Every trace is the average of 32 signals induced by 16
flashes spaced at 1.6 ms with a repetition rate of 0.1 Hz. Continuous illumination with far-red light (Schott interference filter IL
720) had an intensity of $3 \cdot W \cdot m^{-2}$. Broken spinach chloroplasts were
present at a concentration of 10 µM chlorophyll. During the measurement of the right hand transient the reaction mixture additionally contained 30 nM DBMIB. The indicated half-times of 16 ms and
34 ms were obtained from semi-logarithmic plots of the fast reoxidation of plastoquinone. The origin of the time scales is at the
last flash of the flash group indicated by the bar.

cates the inhibition of more than 50 per cent of the electron
transfer sites to Photosystem I. If unspecific binding and the concentration of free DBMIB are assumed to be negligible the number of
inhibitor binding sites[22] would be one per 200 chlorophyll molecules.

The slowed down reoxidation of plastoquinone by DBMIB must diminish the electron release from plastoquinone during the dark time
t_d of 1.6 ms between the single flashes. Thus, the accumulated
amount of plastohydroquinone should increase. The resulting difference of the absorbance changes between two flashes $\Delta(\Delta I/I)_1$ can be
estimated from the amplitude of the fast reoxidation $(\Delta I/I)_{fast}$
(approx. $2.3 \cdot 10^{-3}$ in Fig. 2) and its half-time $t_{1/2}$ without and
$t_{1/2}^{i}$ with inhibitor:

$$\Delta(\Delta I/I)_1 = (\Delta I/I)_{fast} \cdot (\exp(-\ln2 \cdot t_d/t_{1/2}^{i}) - \exp(-\ln2 \cdot t_d/t_{1/2})).$$

If the initial increase of the amplitude of the fast reoxidation is
taken into account (cf. ref. 13) the total increase of the amplitude

over all 15 dark intervals summarizes to $\Delta I/I = 10^{-3}$. This is in excellent agreement with the detected increase of the amplitude from $3.4 \cdot 10^{-3}$ to $4.44 \cdot 10^{-3}$ in the presence of DBMIB. Thus, any other effect besides the inhibition of the plastoquinone reoxidation is excluded at the low concentration used in this experiment.

The third aspect concerns the equal amount of electron carriers of Photosystem I being reduced by plastohydroquinone although more than 50 per cent of the electron transfer sites were blocked. This seems to indicate a reduction of Photosystems I connected to inhibited sites via an electron exchange with Photosystems I connected to intact sites. This conclusion is not in agreement with the model shown in Fig. 1 and was additionally examined by the time course of the $P700^{+}$ reduction.

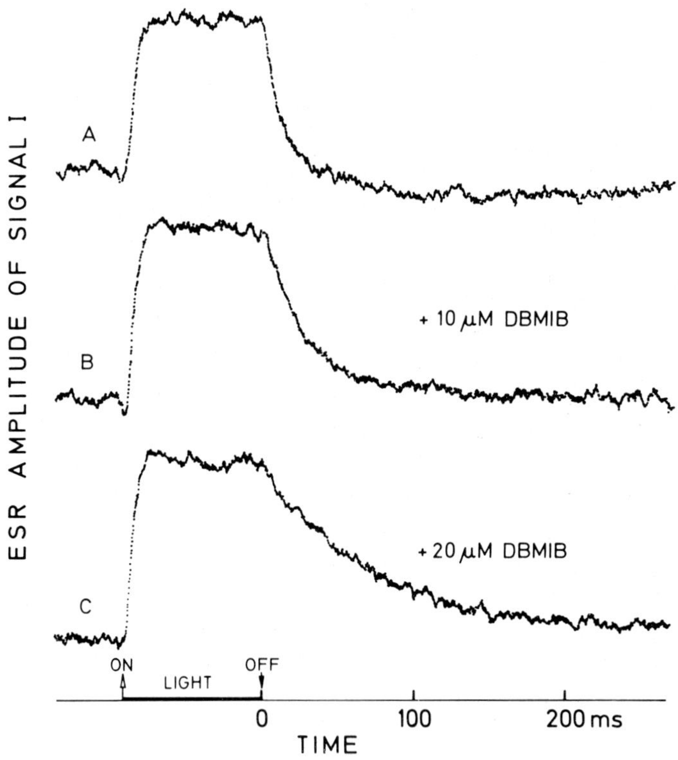

Fig. 3. Time course of the ESR amplitude of Signal I. Every trace is the average of 50 signals induced by long flashes of 92 ms with a repetition rate of 0.5 Hz. The reaction mixture contained spinach chloroplasts at a chlorophyll concentration of 2.5 mg/ml. Trace A, control; B, 10 /uM DBMIB; C, 20 /uM DBMIB.

Fig. 3 shows the time course of the ESR signal of P700 (chlorophyll a_I) in the presence of DBMIB. During the illumination with saturating light P700 and the other electron carriers of Photosystem I are oxidized. At cessation of illumination the transfer of the electrons accumulated in the plastoquinone pool via the rate-limiting step is observed (cf. Fig. 2 and refs. 1 and 9). In the presence of 10 /uM and 20 /uM DBMIB - corresponding to a molar ratio of DBMIB/chlorophyll equal to 1/270 and 1/135, respectively - the first half-time of the P700[+] reduction was increased from 9 ms to 19 ms and 48 ms, respectively. Higher concentrations caused greater half-times, e.g. at 50 /uM DBMIB a first half-time of 125 ms was observed. The P700[+] reduction showed a homogeneous time course at all concentrations of DBMIB. The shape of these transients was almost equal to that without DBMIB. This was easily seen from the fit of the curves after appropriate upset of the time scales (not shown). It is concluded that all P700 molecules equilibrate rapidly even at an inhibition of more than 90 per cent of the electron transfer sites between the plastoquinone pool and Photosystem I. This electron exchange may involve plastocyanin or occur directly between different P700.

To discriminate these possibilities we have used KCN for differential inhibition of plastocyanin[6]. Fig. 4 shows the effect on the time course of the ESR signal of P700. The reduction kinetics after the long flash were not changed by 0 min (i.e. less than 0.5 min) incubation with 30 mM KCN. At maximal inhibition after 90 min incubation of the chloroplasts[6] the total P700[+] is reduced slowly with a half-time of about 0.5 s in agreement with the previous result of Izawa et al.[20]. At partial inhibition after 20, 35 and 50 min incubation (see Fig. 5) a biphasic reduction was found with an amplitude of the fast phase of 65, 45 and 20 per cent of the total amplitude. The time course of this fast phase was almost equal to that at 0 min incubation. The time course of the slow phase resembled that at complete inhibition.

At increasing inactivation of plastocyanin an increasing part of P700[+] was obviously not supplied with electrons from plastoquinone via the normal electron transfer in contrast to the other part of P700. Thus, an electron exchange between different P700 can be excluded as well as via a direct interaction of plastocyanin with different P700[23] In the latter case remaining plastocyanin could

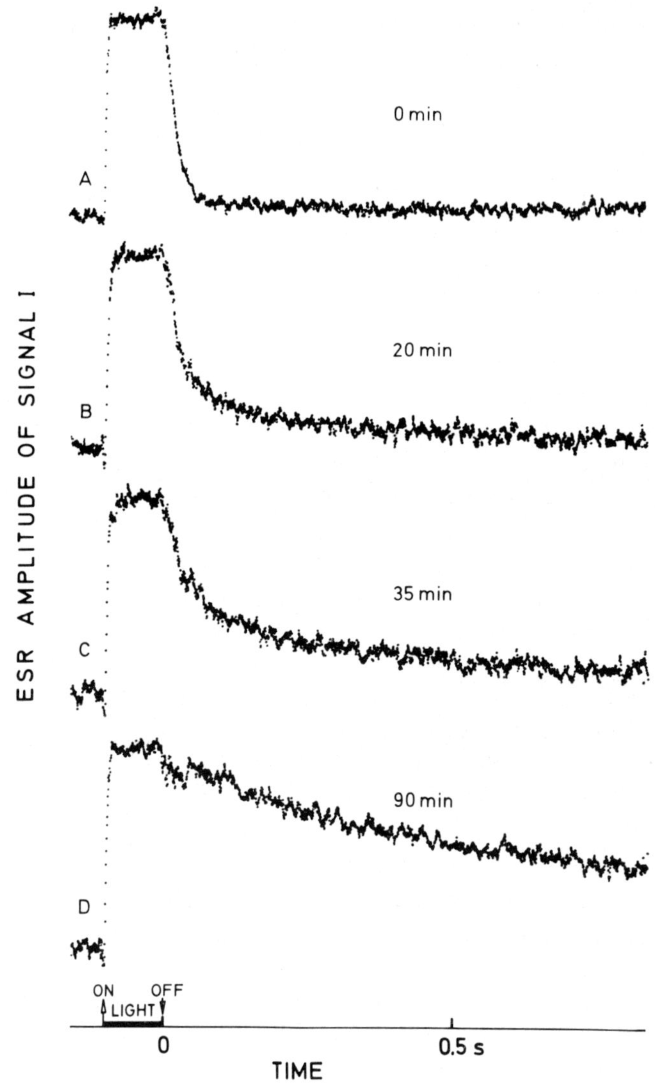

Fig. 4. Time course of the ESR amplitude of Signal I. Every trace is the average of 150 signals induced by long flashes of 105 ms with a repetition rate for traces A, B, C of 0.5 Hz and trace D of 0.2 Hz. The incubation time of the spinach chloroplasts with KCN was for trace A less than 0.5 min; B, 20 min; C, 35 min; D, 90 min. Because chloroplasts were present at varying chlorophyll concentrations between 1.3 and 1.7 mg/ml the amplitudes of the signals were normalized by an appropriate factor choosen in the signal averager.

reduce the total P700 at partial inhibition. P700 is concluded to be tightly associated with a plastocyanin molecule. The reduction as well as the electron exchange between different P700 should only be possible via this plastocyanin.

DISCUSSION

DBMIB was shown to be a very effective inhibitor even at the extreme chlorophyll concentration of 2.5 mg/ml (Fig. 3). The maximal relative concentration of inhibition sites was estimated as 1 per 200 chlorophyll molecules. The site of inhibition was directly identified as the electron transfer site from the plastoquinone pool to the electron carriers of Photosystem I. The ratio of these electron transfer sites per light reaction should therefore be equal or smaller than two.

This result excludes an electron transfer from every plastoquinone molecule of the pool to Photosystem I. The specific binding site of DBMIB could be a specialized plastoquinone molecule different from the pool or a structural site which enables the reaction between plastoquinone molecules and an electron carrier of Photosystem I.

A possible electron exchange between Photosystem I chains was investigated from the effect of differential inhibition. The two clearly distinguishable alternatives have been found and the re-sulting reduction kinetics of $P700^+$ are contrasted in Fig. 5.

The finding of the small number of inhibition sites of DBMIB per light reaction was prerequisite for the argument of an electron exchange between different P700. In case of successive inhibition of two or three electron transfer sites to separate Photosystems I a biphasic reduction of $P700^+$ could be seen only at an inhibition greater than 50 or 67 per cent of linear electron transport, respect. However, e.g. the half-time of the homogeneous reduction kinetics at 20 /uM DBMIB (Fig. 5 A) indicates an inhibition greater than 80 per cent. An electron exchange faster than 20 ms between at least 8-10 P700 is concluded.

The previous suggestion of an electron distribution via plasto-quinone pools based on the effect of DCMU on the absorbance changes of $P700^3$ disregarded a possible electron exchange between Photo-

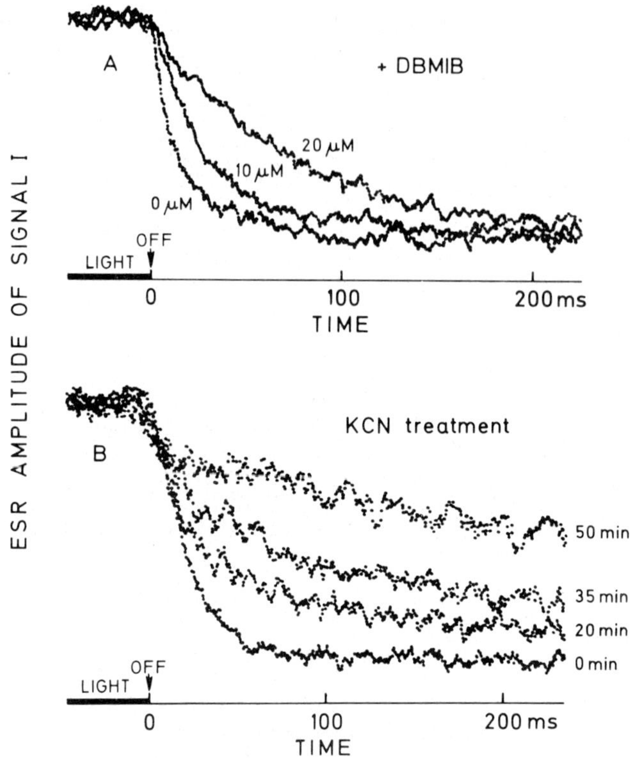

Fig. 5. Time course of P700$^+$ (chlorophyll a$_I^+$) reduction at partial inhibition of linear electron transport with DBMIB (A) and after KCN incubation (B). A: The time courses of Signal I after the flash shown in Fig. 3 were replotted. B: The time courses of Signal I after the flash shown in Fig. 4 A, B and C were plotted at an expanded time scale. In addition the time course after an incubation with KCN for 50 min is shown.

systems I. However, an additional electron exchange between the two light reactions via plastoquinone pools seems to be indicated by the absorbance changes of plastoquinone[1,3].

The tight association of P700 with a plastocyanin molecule concluded from the lack of electron exchange between P700 after KCN incubation is consistent with the finding of Ke et al.[24] of bound plastocyanin in Triton particles of Photosystem I. A large amount of loosely bound plastocyanin is solubilized by treatment with Triton X-100 (25). It may be this loosely bound plastocyanin which mediates the electron exchange between different P700 via the bound plastocyanin.

ACKNOWLEDGEMENTS

The author is indebted to Professor Dr. A. Trebst for stimulating discussions. This work was supported by Deutsche Forschungsgemeinschaft.

REFERNCES

1. Stiehl, H.H. and Witt, H.T. (1969) Z. Naturforsch. 24 b, 1588-1598

2. Amesz, J., Visser, J.W.M., van den Engh, G.H. and Dirks, M.P. (1972) Biochim. Biophys. Acta 256, 370-380

3. Siggel, U., Renger, G., Stiehl, H.H. and Rumberg, B. (1972) Biochim. Biophys. Acta 256, 328-335

4. Böhme,H., Reimer, S. and Trebst, A. (1971) Z. Naturforsch. 26 b, 341-351

5. Trebst, A. (1963) Z. Naturforsch. 13 b, 317-321

6. Ouitrakul, R. and Izawa, S. (1973) Biochim. Biophys. Acta 305, 105-118

7. Selman, B.R., Johnson, G.L., Giaquinta, R.T. and Dilley, R.A. (1975) Bioenergetics 6, 221-231

8. Marsho, T.V. and Kok, B. (1970) Biochim. Biophys. Acta 223, 240-250

9. Haehnel, W. (1977) Biochim. Biophys. Acta 459, 418-441

10. Sane, P.V., Goodchild, D.J. and Park, R.B. (1970) Biochim. Biophys. Acta 216, 162-178

11. Vernon, L.P., Shaw, E.R., Ogawa, T. and Raveed, D. (1971) Photochem. Photobiol. 14, 343-357

12. Staehelin, L.A. (1975) Biochim. Biophys. Acta 408, 1-11

13. Haehnel, W. (1973) Biochim. Biophys. Acta 305, 618-631

14. Trebst, A., Harth, E. and Draber, W. (1970) Z. Naturforsch. 25 b, 1157-1159

15. Lozier, R.H. and Butler, W.L. (1972) FEBS Lett. 26, 161-164

16. Böhme, H. and Cramer, W.A. (1972) Biochemistry 11, 1155-1160

17. Kouchkovsky, Y. de and Kouchkovsky, F. de (1974) Biochim. Biophys. Acta 368, 113-124

18. Trebst, A. (1974) Ann. Rev. Plant Physiol. 25, 423-458

19. Izawa, S. (1977) in: Encyclopedia of Plant Physiol., Vol. 5 (Trebst, A. and Avron, M., eds.) pp. 266-282, Springer, Berlin,

Heidelberg, New York

20. Izawa, S., Kraayenhof, R., Ruuge, E.K. and DeVault, D. (1973) Biochim. Biophys. Acta 314, 328-339
21. Beinert, H., Kok, B. and Hoch, G. (1962) Biochem. Biophys. Res. Commun. 7, 209-212
22. Tischer, W. and Strotmann, H. (1977) Biochim. Biophys. Acta 460, 113-125
23. Bouges-Bocquet, B. (1975) Biochim. Biophys. Acta 396, 382-391
24. Ke, B., Sugahara, K. and Shaw, E.R. (1975) Biochim. Biophys. Acta 408, 12-25
25. Sane, P.V. and Hauska, G.A. (1972) Z. Naturforsch. 27 b,932-938

Bioenergetics of Membranes. L. Packer et al. ed.

STRUCTURAL AND QUANTITATIVE ANALYSIS OF MEMBRANE-BOUND COMPONENTS OF THE PHOTO-SYSTEM I COMPLEX OF SPINACH CHLOROPLASTS BY IMMUNOLOGICAL METHODS

Herbert Böhme
Fachbereich Biologie
Universität Konstanz
775 Konstanz
W-Germany

INTRODUCTION

Immunological techniques have been successfully applied for functional and structural studies of membrane bound electron transport proteins [1-13]. By chloroplast agglutination studies a number of membrane-surface located proteins have been recognized, since the antibody is a large molecule which does not penetrate a lipo-protein bilayer and is highly specific to its antigen. Using antibodies as specific inhibitors of these surface located proteins a more detailed functional analysis of the electron transport system was possible. As shown in this study, monospecific antibodies may be also used for a quantitative analysis of electron transport components by the technique of immunoelectrophoresis in antibody containing agarose gel slabs.

MATERIALS AND METHODS

Intact chloroplasts were isolated from spinach leaves (variety Atlanta) essentially according to Jensen & Bassham [14] omitting the $NaNO_3$ and Na-isoascorbate additions. Intactness was determined by ferricyanide reduction [15] in the presence of 2 mM NH_4Cl. The supernatant solution of Table 2 was obtained by suspension and two successive washes of intact chloroplasts in a hypotonic medium containing: 5 mM N-tris (hydroxymethyl) methyl glycine/NaOH, pH 7.8, 5 mM NaCl, 1 mM $MgCl_2$ (=buffer 1). After centrifugation, the supernatant solutions were pooled and concentrated by freeze drying. The remaining shocked and washed chloroplasts were either frozen once in liquid nitrogen before immunoelectrophoresis or incubated for 5 h with 1 % Triton X-100.

Purified ferredoxin, ferredoxin-$NADP^+$ reductase and plastocyanin were prepared in good yield from the supernatant of a freeze-thawed leaf-homogenate [13]. The chromophore/protein ratios were 0.53 for

ferredoxin, 0.13 for ferredoxin-NADP$^+$ reductase and 0.8 for plastocyanin. The following extinction coefficients were used: ε 420 nm=
9.8 mM^{-1} x cm^{-1} for ferredoxin [16], ε 457 nm=10.3 mM^{-1} x cm^{-1} for
reductase [17] and ε 597 nm=4.9 mM^{-1} x cm^{-1} for plastocyanin [18].
Cytochrome f and P 700 were determined by difference spectroscopy
(Aminco, DW 2) [19] using a differential extinction coefficient of
ε 554 nm=19.7 mM^{-1} x cm^{-1} [20] and ε 700 nm=64 mM^{-1} x cm^{-1}, respectively [21]. Chlorophyll was determined in an 80 % acetone extract
according to Arnon [22].

Double diffusion tests were carried out in agarose gels (0.8 %
agarose dissolved in 0.06 M phosphate-buffer, pH 7.8). For staining
the gel slabs were washed in 0.15 M NaCl, dried and stained with
amido-black [13]. For agglutination reactions intact chloroplasts were
suspended and washed in the five-fold concentrated buffer-1 solution pH 7.8 (details see ref. 13).

Immunization of rabbits by highly purified ferredoxin, reductase and plastocyanin followed standard procedures (compare ref. 12).
The inhibitory properties of the antisera and the corresponding
test reactions as i) cytochrome-c reduction by NADPH, ii) NADP$^+$-
and cytochrome b$_6$ photoreduction, iii) ascorbate photooxidation by
photosystem I fragments, have been already described [12,19]. Purification and fractionation of the antisera was performed by repeated
$(NH_4)_2SO_4$ precipitation and gel chromatography on Sephadex G-200
(see refs. 23,24); γ G- and γ M-immunoglobulins were identified
with goat-antirabbit γ -globulin by immunoelectrophoresis. Protein
content of the antisera was determined according to Kalckar [25].

Quantitative immunoelectrophoresis in antiserum containing agarose gels according to Laurell [26] is described in detail in ref. 13.
Both the samples and the corresponding antigen standard dilutions
(10-50 pmol) were subjected to electrophoresis simultaneously,
allowing a calibration of each individual run. After 18 h electrophoresis at 120 V/4 mA, 15° C, the agarose gel slabs were dried and
stained [13].

RESULTS AND DISCUSSION

Antisera were prepared against three electron-carriers of the
photosystem I complex; namely, ferredoxin, ferredoxin-NADP$^+$ reductase and plastocyanin. The antisera contained monospecific antibodies since they produced only single precipitation arcs in a double
diffusion test with a complex chloroplast extract (fig. 1). Each
antigen (=purified electron carrier) was precipitated by its re-

spective antiserum and no cross reactions were observed (fig. 1,
compare Table 1).

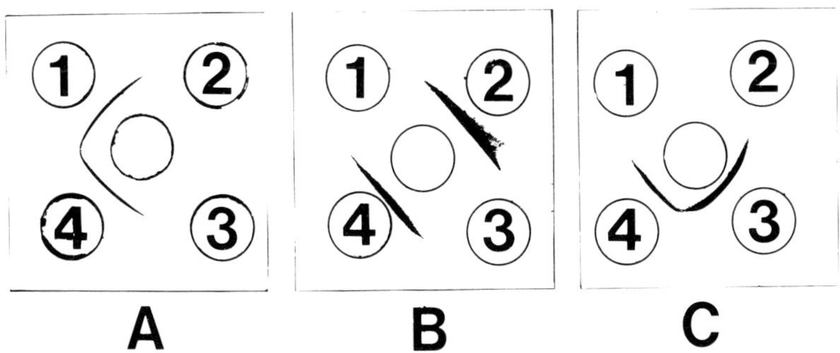

Fig. 1 DOUBLE DIFFUSION TEST FOR MONOSPECIFICITY OF ANTISERA
AGAINST (A) FERREDOXIN, (B) PLASTOCYANIN AND (C) FERREDOXIN-NADP$^+$
REDUCTASE.
The center well contained the respective antiserum; in (A) 20 μl,
(B) 7 μl, (C) 20 μl. The four outer wells contained (1) 40 pmol
ferredoxin, (2) 80 pmol plastocyanin, (3) 40 pmol reductase and
(4) spinach chloroplasts equivalent to 10 μg chlorophyll. Diffusion
period: 16 h at room temperature (see ref. 12).

Agglutination reactions were performed with spinach chloroplasts
without outer envelope. These chloroplasts are prepared according
to standard procedures showing high rates of electron transport
coupled to phosphorylation. Photosynthetic control ratios of 3 by
addition of ADP and 6 by addition of 2 mM NH_4Cl were generally obtained; the ADP/O ratio was in the range 1.2 - 1.6. It could be
shown that purified IgG-fractions of all three antisera used in
this study agglutinated chloroplasts directly and specifically.
This means that all three photosynthetic proteins, ferredoxin, reductase and plastocyanin, are localized on the outer thylakoid
membrane surface. Furthermore, after osmotic shock at least part of
these three electron carriers remains membrane bound, which was
not expected for ferredoxin (see e.g. ref. 5). It has been shown,
however, that this residual, membrane bound ferredoxin may play a
role in cytochrome b_6 photoreduction [12]. From these agglutination
experiments it can be additionally concluded that the coupling

factor does not represent a protruding structure thereby preventing direct agglutination by steric hindrance, as has been suggested [3,4]. If however, the antiserum preparation contains a substantial amount of γM-immunoglobulins direct agglutination might be observed in spite of protruding membrane structures. The pentameric γM-antibody allows a span of about 350 Å to be bridged compared to 150 Å for the monomer (= γG). Therefore the antiserum preparations were further fractionated by Sephadex-gel filtration yielding relatively pure γM and γG fractions [13].

TABLE 1

PROPERTIES OF ANTISERUM FRACTIONS AGAINST FERREDOXIN-NADP$^+$ REDUCTASE, FERREDOXIN AND PLASTOCYANIN.

Antiserum against	Agglutination	Precipitation	Inhibition of enzymatic activity
Reductase:			
γM-globulins	yes	yes	yes
γG-globulins	yes	yes	yes
Ferredoxin:			
γM-globulins	yes	no	no
γG-globulins	no	yes	yes
Plastocyanin:			
γM-globulins	yes	no	no
γG-globulins	no	yes	yes

In the case of ferredoxin-NADP$^+$ reductase antiserum both fractions, γM and γG, precipitated reductase in a double diffusion test and agglutinated chloroplasts directly. In these chloroplasts the span of one bivalent antibody is sufficient to observe agglutination. This observation argues against location of this enzyme between coupling factor molecules, which were supposed to protrude at least 80 Å [4]. Binding of either species of the reductase antibodies led to complete inhibition of the diaphorase activity of the solubilized enzyme (cytochrome c reduction by NADPH) [12] as well as to inhibition of NADP-photoreduction catalyzed by membrane bound reductase (Table 1). This could mean that the ferredoxin-NADP$^+$ reductase is surface located on the membrane with its catalytic site exposed to the matrix site of chloroplasts.

Fractionation of the ferredoxin antiserum on the other hand led to different results. Only the γM-globulins but not the γG-globulins agglutinated chloroplasts (Table 1). Dissociation of the γM-pentamer by treatment with 2-mercaptoethanol prevented agglutination completely [13]. The necessity for a pentameric antibody to observe a positive agglutination of chloroplasts might be interpreted as production of cavities by osmotic shock, whereby most of the ferredoxin is released from the thylakoid membrane. Only the γG- but not the γM-fraction of the ferredoxin antiserum precipitated ferredoxin in a double diffusion test [13] indicating the presence of only one species of antibody in the γM-fraction. Anti-ferredoxin γG-globulins also were inhibitory in the diaphorase test (cytochrome c reduction by NADPH), which requires addition of ferredoxin. Similarly $NADP^+$ and cytochrome b_6 photoreduction catalyzed by bound ferredoxin could be completely inhibited [12]. The γM-immunoglobulins which agglutinated chloroplasts had no influence on these photoreductions, nor did they influence the diaphorase test. A tentative explanation might be that an antigenic determinant of the ferredoxin molecule is occupied by the antibody which does not impair the catalytic activity.

Similar observations could be made with γG- and γM-immunoglobulines of the plastocyanin antiserum. Only γM-globulins agglutinated chloroplasts, whereas the γG-globulins did not. On the other hand only the γG-fraction formed a precipitation arc with isolated plastocyanin and was inhibitory. Ascorbate photooxidation by digitonin photosystem I fragments, which is dependent on plastocyanin addition, could be largely blocked [19]. Moreover linear electron transport (methylviologen as acceptor), whether it was coupled or uncoupled, was inhibited by the γG-fraction when the chloroplasts were preincubated with the immunoglobulin in the dark. However, inhibition of electron transport was far from complete (up to 40% inhibition) and was not observed with all chloroplast preparations. It appears that plastocyanin is less accessible to antibody than is ferredoxin or ferredoxin-$NADP^+$ reductase (compare also ref. 8). These data are in contrast to the findings of Hauska et al. [10], who concluded that plastocyanin resides at the inner membrane side, since their antiserum neither agglutinated chloroplasts nor inhibited electron transport activity unless the chloroplasts were sonicated in the presence of antiserum.

To summarize the data so far presented it appears that ferredoxin, ferredoxin-NADP$^+$ reductase and plastocyanin are peripheral rather than integral membrane proteins located on the outer membrane surface and accessible to antibodies. An explanation for the contradictory results discussed above may be found in the dynamic nature of the membrane; depending on the energy state, electron transport components become more or less surface exposed [9,19]. Moreover, different antiserum preparations might contain antibodies directed towards different antigenic determinants of the electron carrier molecule, which might or might not be exposed when bound to the membrane. With respect to agglutination reactions, this work indicates further that the relative concentration of γM- and γG-antibodies in the serum (= immune response) might determine whether positive or negative reactions are observed.

In the course of these investigations it was noted that osmotic shock of intact chloroplasts releases a soluble protein fraction containing large amounts of ferredoxin (as expected) but also certain amounts of ferredoxin-NADP$^+$ reductase and plastocyanin (compare Fig.1). In an attempt to estimate quantitatively these enzymes a simple and sensitive technique was applied: quantitative immunoelectrophoresis. The principle of the method involves an antigen (= solubilized electron carrier) moving under the influence of an electrical field in an antiserum containing gel; a stable antigen-antibody precipitate will be formed at the equivalence point, resulting in a direct relationship between final migration distance and the amount of antigen present [26]. Due to the high specificity of an antibody to its antigen and the high sensitivity of detection of antibody-antigen complexes crude chloroplast extracts can be checked directly without partial purification.

All quantitative studies were performed with chloroplasts isolated from spinach leaves, variety Atlanta. As mentioned earlier [12,19], this spinach variety contains 1000 chlorophylls per photosynthetic unit rather than 300-500, as generally reported [18,32,33]. Consistently values of 1.0 - 1.2 nmol P700 or cytochrome f per µmol chlorophyll were measured. Intact chloroplasts (85-95%) were subjected to various treatments: osmotic shock, freeze-thawing and Triton X-100 extraction. The solubilized protein fractions were then directly subjected to immunoelectrophoresis. Each antiserum containing agarose-gel slab was calibrated in the same run with the corresponding purified electron transport protein. The results are summarized in Table 2.

TABLE 2

QUANTITATIVE DETERMINATION OF FERREDOXIN, FERREDOXIN-NADP$^+$ RE-
DUCTASE AND PLASTOCYANIN BY QUANTITATIVE IMMUNOELECTROPHORESIS

	nmol electron carrier released/ μmol chlorophyll		
	Ferredoxin	Reductase	Plastocyanin
Supernatant after osmotic shock (1)	4.2	0.5	0.6
Freeze-thaw, shocked chloroplasts (2)	0.5	2.1	3.2
Triton X-100 treated shocked chloroplasts (3)	0.9	2.6	3.4
Total amount: (1)+(3)	5.1	3.1	4.0

It can be seen that the supernatant solution after osmotic shock
contains most of the ferredoxin (80%), but also reductase and plastocyanin. For the latter two electron carriers values between 0.5
and 1.0 nmol/μmol chlorophyll were routinely obtained. The remaining membrane-bound components are released by freeze-thawing and
more completely by Triton X-100 treatment of osmotically shocked
and washed chloroplasts. The ratio of ferredoxin : reductase : plastocyanin approximates 5 : 3 : 4, respectively, per photosynthetic
unit containing one P700 and one cytochrome f. A more complete quantitative analysis regarding b-cytochromes and plastoquinone content
of these chloroplasts is given in Table 1 of refs. 12 and 19.

CONCLUSION

Immunological methods may be conveniently used for structural,
functional and quantitative studies of membrane bound electron
transport systems. As shown herein and also by others [2,4,7,8], ferredoxin, ferredoxin-NADP$^+$ reductase and plastocyanin appear to be
peripherally located on the outer thylakoid membrane of higher
plant chloroplasts. From similar studies it was concluded that
P700, the reaction center I chlorophyll [11], cytochrome f [9] and

other more or less specified components of photosystem I [2,27-29]
are also accessible to antibodies. These data support the idea of
an asymmetric distribution of electron transport components across
the membrane, where the constituents of photosystem I are located
more on the exterior side and of photosystem II on the interior
side of the chloroplast membrane [30,31]. The frequently inferred
1 : 1 or 1 : 2 molar relationship of electron carriers per photosystem I unit could not be confirmed [18,32,33]. The values found
for ferredoxin, ferredoxin-NADP$^+$ reductase and plastocyanin rather
approach a 5 : 3 : 4 stoichiometry per P700 or cytochrome f. As
additional aspect this investigation shows that osmotic shock and
even more pronounced freeze-thawing of chloroplasts (in liquid
nitrogen) results in a more or less complete release of all three
electron carriers from the thylakoid membrane. This limits the
general use of osmotically shocked or frozen chloroplasts in photosynthesis research.

ACKNOWLEDGEMENT

This work was supported by the Deutsche Forschungsgemeinschaft
through Sonderforschungsbereich 138, project C_2. I wish to thank
Prof. A. San Pietro and Prof. P. Böger for reading the manuscript
and helpful suggestions.

REFERENCES

1. Keister, D.K., San Pietro, A., and Stolzenbach, F.E. (1962)
   Arch. Biochem. Biophys. 98, 235-244
2. Berzborn, R.J., Menke, W., Trebst, A., and Pistorius, E. (1966)
   Z. Naturforsch. 21b, 1057-1059
3. Berzborn, R. (1968) Z. Naturforsch. 23b, 1096-1104
4. Berzborn, R. (1969) Z. Naturforsch. 24b, 436-446
5. Hiedemann-VanWyk, D., and Kannangara, C.G. (1971) Z. Naturforsch. 26b, 46-50
6. Tel-Or, E., and Avron, M. (1974) Eur. J. Biochem. 47, 417-421
7. Schmid, G.H., and Radunz, A. (1974) Z. Naturforsch. 29c,
   384-391
8. Schmid, G.H., Radunz, A., and Menke, W. (1975) Z. Naturforsch.
   30c, 201-212
9. Schmid, G.H., Radunz, A., and Menke, W. (1977) Z. Naturforsch.
   32c, 271-280

10. Hauska, G.A., McCarty, R.E., Berzborn, R.J., and Racker, E. (1971) J. Biol. Chem. 246, 3524-3531

11. Racker, E., Hauska, G.A., Lien, S., Berzborn, R.J., and Nelson, N. (1972) 2nd Int. Congr. Photosynthesis, Stresa, (eds. G. Forti, M. Avron, A. Melandri) pp. 1097-1113

12. Böhme, H. (1977) Eur. J. Biochem. 72, 283-289

13. Böhme, H. (1977) Eur. J. Biochem., submitted

14. Jensen, R.G., and Bassham, J.A. (1966) Proc. Nat. Acad. Sci. 56, 1095-1101

15. Heber, U., and Santarius, K.A. (1970) Z. Naturforsch. 25b, 718-728

16. Buchanan, B.B., and Arnon, D.I. (1971) Methods Enzymol. 23, 413-430

17. Foust, G.P., Mayhew, S.G., and Massey, V. (1969) J. Biol. Chem. 244, 964-970

18. Katoh, S., Suga, I., Shiratori, I., and Takamiya, A. (1961) Arch. Biochem. Biophys. 94, 136-141

19. Böhme, H. (1976) Z. Naturforsch. 31c, 68-77

20. Plesnicar, M., and Bendall, D.S. (1970) Biochim. Biophys. Acta 216, 192-199

21. Hiyama, T., and Ke, B. (1977) Biochim. Biophys. Acta 267, 160-171

22. Arnon, D.I. (1949) Plant Physiol. 24, 1-5

23. Böhme, H. (1977) Eur. J. Biochem., submitted

24. Nowotny, A. (1969) Basic Exercises in Immunochemistry, Springer Verlag, Berlin, pp. 5-15

25. Kalckar, H.M. (1947) J. Biol. Chem. 167, 461-475

26. Laurell, C.B. (1966) Anal. Biochem. 15, 45-52

27. Regitz, G., Berzborn, R., and Trebst, A. (1970) Planta 91, 8-17

28. Koenig, F., Menke, W., Craubner, H., Schmid, G.H., and Radunz, A. (1972) Z. Naturforsch. 27b, 1225-1238

29. Bengis, C., and Nelson, N. (1975) J. Biol. Chem. 250, 2783-2788

30. Briantais, J.M., and Picaud, M. (1972) FEBS Lett. 20, 100-104

31. Arntzen, C.J., Dilley, R.A., and Crane, F.L. (1969) J. Cell Biol. 43, 16-31

32. Haslett, B.G., and Cammack, R. (1976) New Phytol. 76, 219-226

33. Tagawa, K., and Arnon, D.I. (1962) Nature 195, 537-543

THE USE OF TRYPSIN AS A STRUCTURALLY SELECTIVE
MODIFIER OF THE THYLAKOID MEMBRANE

Gernot Renger
Max Volmer-Institut für Physikalische Chemie und Molekular-
biologie der Technischen Universität
Straße des 17. Juni 135, D 1000 Berlin 12

INTRODUCTION

The fundamental steps of biological energy conserving and transforming orga-
nellae (chloroplasts, mitochondria) occur in reaction sequences, which chemical
components are anisotropically embedded into structurally highly organized membranes. Hence, for an understanding of the mechanism of the whole machinery a
thorough knowledge is required for both, the spatial arrangement as well as the
chemical and physical properties of the reactive species.

Based on flash spectroscopic and polarographic studies of the molecular events
of photosynthesis it was inferred, that electrons are driven from water to $NADP^+$
by two light reactions in a characteristic zig-zag-pattern through the thylakoid
membrane (for rev. s. ref. 1,2). According to this scheme the photochemical processes at the reaction centers of system I and II, respectively, lead in $\leq$20 ns
to electron transfer from inside to outside, whereas a plastoquinone pool funnels
the electrons via dark reactions in the reverse direction from the acceptor side
of system II to the donor side of system I. Hence, the reaction center complexes
are assumed to be intercalated into the thylakoid membrane with the primary
electron donors (special chlorophyll-a-complexes) near the inner side and the
corresponding primary electron acceptors (Fe-S-protein ? and X 320) at the outer
side. The plastoquinone/plastohydroquinone system provides the natural shuttle
for net electron and proton transport from outside to inside. Furthermore, the
watersplitting enzyme system Y, realizing water cleavage to protons and molecular
oxygen by oxidizing equivalents produced at chlorophyll-a$_{II}$, is supposed to be
located near the inner side, whereas the enzymes at the acceptor side of system I,
responsible for $NADP^+$-reduction coupled with proton uptake, are attached to the
outer side of the thylakoid membrane.

Therefore, a structural modification of the thylakoid membrane should give
rise to a specific change in the functional pattern of the presumed zig-zag-
electron flow.

During the last years selective modifiers became available for the investigation of the correlation between structural order and functional pattern. In this
respect two types of substances are of current interest: A) Antibodies, which
selectively bind to the corresponding component (for rev. s. ref. 3) and B)
Chemicals,which react with specific amino acid groups (for rev. s. ref. 4), such

as sulfhydryl-groups (labelled e.g. by N-ethylmaleimide), carboxyl-groups (modified e.g. by carbodiimides), tyrosine- and histidine-residues (transformed via
lactoperoxidase catalized iodination) or arginine and lysine-residues (attacked
by trypsin).

In the present paper the structural anisotropy and functional sidedness of
system II will be studied with the aid of the proteolytic enzyme trypsin.

MATERIALS AND METHODS

<u>Chloroplast preparation and assay conditions</u>:   Stripped spinach chloroplasts
have been prepared according to the method of Winget et al (5) as is described
in ref. 6. Trypsinated chloroplasts were obtained by addition of commercially
available trypsin (Boehringer) to the chloroplast suspension. The measurements
were made after an incubation time indicated in the abscissae or the legends of
the figures. As the time required for the detection of oxygen or an absorption
change was short enough, frequently no trypsin inhibitor was applied to stop the
digestion process.

<u>Measurements</u>:   Oxygen was detected with a Clark-type electrode by a method
outlined in ref. 7, the measurements of the absorption changes are described in
ref. 8.

For the detection of the pH-changes in the inner phase of the thylakoid
membrane neutral red was used as indicator and imidazole as buffer according to
the method described in ref. 9.

<u>Chloroplast suspension</u> :   If not otherwise stated, the reaction mixture
contained: chloroplasts (50 /uM chlorophyll), 10 mM KCl, 2 mM $MgCl_2$, 20 mM
Tricine-NaOH, pH = 7,0. For trypsination 50 /ug trypsin/ml suspension were applied.
The addition of electron acceptors and of other agents is given in the figures
or in their legends.

RESULTS AND DISCUSSION

As trypsin is a large water soluble enzyme which does not readilly penetrate
through the thylakoid membrane (10) it can be anticipated to attack primarilly
proteins at the outer side of the thylakoid membrane. Hence, according to the
zig-zag-scheme the acceptor sides of system I and II, respectively, are expected
to be sensitive to tryptic digestion. This was confirmed for the secondary
reactions of the acceptor side of system I leading to $NADP^+$-reduction. On the
contrary, the direct electron transfer from photosystem I to artificial electron
acceptors is not impaired (10).

Fig. 1 summarizes the effect of trypsin on the linear electron transport
mediated by external redox agents. It is seen that in the presence of p-benzoquinone or benzylviologen as electron acceptor trypsination leads to a complete
interruption of the linear electron transport chain. As absorption changes due to

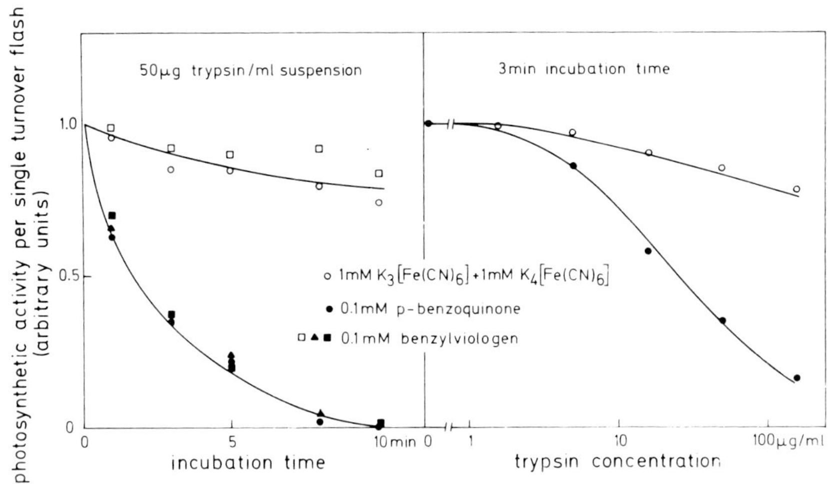

Fig. 1. Photosynthetic activity per single turnover flash as a function of the incubation time (left side) and of the concentration (right side) of trypsin in the presence of different electron acceptors.
The symbols indicate: O ● average oxygen yield, ◻ ∎ initial amplitude of the absorption change at 703 nm (◻ in the presence of 20 μM PMS and 2 mM ascorbate) ▲ initial amplitude of the electrochromic absorption change at 520 nm.
Excitation: saturating short flashes (duration t_f = 20 μs), time t_d between the flashes 250 ms.

the P 700-turnover were shown to be fully restored by addition of PMS, whereas the chlorophyll-a_{II}-reaction remained blocked even in the presence of system-II-electron donors (s. ref. 8), it was inferred that in accordance with the zig-zag-scheme trypsination blocks the linear electron transport at the reducing side of system II, too.

This interpretation was shown to be correct by the application of $K_3\left[Fe(CN)_6\right]$ as electron acceptor (11). Fig. 1 shows that under these circumstances oxygen evolution remains resistant to trypsin up to a few minutes. This result clearly indicates that the watersplitting enzyme system Y, which is known to be very sensitive to protein denaturing procedures, withstands tryptic attack much more better than the reducing side of system II. The simplest explanation for this effect provides the assumption that at pH = 7,0 system Y is not very trypsin-accessible because of its localization towards the inner side of the thylakoid membrane.

Since the electron carriers at the reducing side of system II are pure lipidic compounds (plastoquinone molecules with different properties) the localization of the inhibition by trypsin into this segment requires the postulation of a trypsin-izable proteinaceous component, which is indispensible for the functional integrity of the linear electron transport chain. This proteinaceous component could also contain the binding site for the powerful DCMU-type-inhibitors completely blocking the reoxidation of the photosystem II electron acceptor either by the plastoquinone pool or by external redox agents. Hence, in the light of this speculation the removal or serious modification of the proteinaceous component could also impose a high resistance to DCMU-blockage.

Taking this idea as a working hypothesis, one can anticipate, that trypsination should lead to a high resistance to DCMU of oxygen evolution in the presence of $K_3\left[Fe(CN)_6\right]$ as electron acceptor. In chloroplasts completely blocked by sufficiently high DCMU-concentrations oxygen evolution should firstly increase with trypsin-incubation time due to the digestion of the presumed proteinaceous component, whereas at longer time a slower decrease is expected to occur, caused by a destruction of the watersplitting enzyme system Y. On the contrary, in the absence of DCMU oxygen evolution should decrease continuously with incubation time, because only the thermal degradation of system Y (forced by the enzymic attack of trypsin), which is known to be the most labile part of system II in the absence of trypsin (for rev. s. ref. 12), should be observed.

The obtained results at pH = 7,0 depicted in fig. 2 are in correspondence with the expected pattern. In the absence of DCMU the oxygen evolving capacity continuously declines with a half life time of about 15 min. As the chloroplasts were excited with single turnover flashes, this curve represents the destruction of the number of the watersplitting enzyme systems Y. In the presence of 1 µM DCMU, completely blocking the electron transport at the reducing side of system II, however a different pattern arises. During the first minutes the average oxygen yield per flash increases with the incubation time until after about 10 min the curve reaches the same level and then declines in the same manner as the curve in the absence of DCMU. This result is fully consistent with the above mentioned

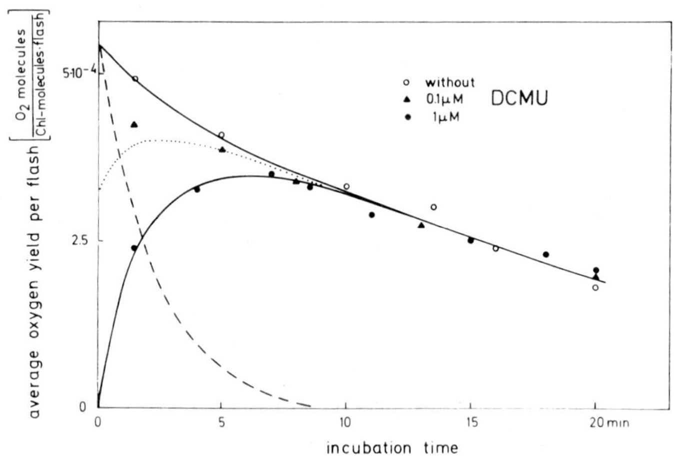

Fig. 2. Average oxygen yield per flash as a function of incubation time of trypsin in the absence and presence of DCMU.
Excitation: white flashes, t_f = 20 µs, time t_d between the flashes 250 ms.

hypothesis that trypsination modifies the proteinaceous component in such a way that the inhibitory effect of DCMU completely disappears. If one presupposes that DCMU does not markedly influence the rate of trypsination, then the difference between both curves directly reflects the digestion of this proteinaceous component. The difference curve indicated by the dotted line has a half life time of about 1,5 min. Thus, this process is about 10 times faster than the destruction of the watersplitting enzyme system Y. Accordingly, at moderate DCMU concentrations, leading to only a partial inhibition, a curve should be observed, which falls in between the range of the curves representing the extreme cases of none and of full blockage, respectively. At 0,1 /uM DCMU under our conditions about 40 % of the systems II are inhibited in the absence of trypsin. The slightly dotted curve is computed based on the assumption that the rate of modification of the proteinaceous component is invariant to DCMU concentration. The experimental data given by full triangles fit this theoretical curve fairly well. Hence, the data of fig. 2 support the thesis that trypsin attacks primarilly a proteinaceous component at the outer side of the thylakoid membrane, which is of functional importance for the electron transport on the reducing side of system II and its inhibition by DCMU.

As the enzymic activity of trypsin is strongly dependent on pH, with a pronounced maximum in the range of 8 - 9, the effect described in fig. 2 is expected to reflect this pH-characteristics of trypsin. Measurements at pH = 6,0 (data not shown) indicate that the digestion rate is about 2 - 3 times slower, in agreement with the decrease of tryptic activity. At pH = 8,0 however, an even more complicated pattern arises. The expected increase in the digestion rate becomes masked by a significant acceleration of the degradation rate of the watersplitting enzyme system Y. This might be caused by a combined effect, the enhanced lability of system Y at higher pH-values (13,14) and the higher activity of trypsin not only towards the proteinaceous component, but also to system Y. Hence, the structurally selective action of trypsin is optimal in the range of $6,0 \leq pH \leq 7,0$. This effect explains also earlier measurements made under suboptimal trypsination conditions suggesting the localization of the inhibitory site of trypsin on the donor side of system II (for rev. s. ref. 4).

The rate of photosynthetic oxygen evolution is determined by two factors:
a) the total number of functioning electron transport chains and
b) the rate limiting step of the overall electron transport from water to the terminal electron acceptor.

In the preceding section the influence of trypsination on the former factor was described, now it remains to investigate the consequences of trypsination on the overall kinetics. In normal chloroplasts the rate limiting step was shown to be the reoxidation of the plastoquinone pool (15), whereas the electron transfer from system II into the pool is much more faster (15,16). The overall rate of

344

oxygen evolution under saturating continuous light or high frequency repetitive
flashes is independent of the acceptor concentration, provided that secondary
effects (uncoupling, inhibition) or exhaustion of these agents can be excluded.
In trypsinated chloroplasts, however, the electron flow from system II into the
plastoquinone pool is assumed to be interrupted and directly deflected to
$K_3\left[Fe(CN)_6\right]$. Therefore, the oxygen evolution rate v_{O_2} is expected to become
strongly dependent on the acceptor concentration.

This is confirmed by the data depicted in fig. 3. From the double logarithmic
plot the relation

$$v_{O_2} \sim M_{O_2}(t_d = 2\ ms) = a \cdot \left[K_3\left[Fe(CN)_6\right]\right]^{0,6} + b$$

was derived, where a and b are constants and $\left[K_3\left[Fe(CN)_6\right]\right]$ represents the concentration of $K_3\left[Fe(CN)_6\right]$.

The mathematical structure of the equation implies a more complicated mechanism than a simple diffusion controlled acceptor transport to the reactive site
(17). If $K_3\left[Fe(CN)_6\right]$ directly accepts the electrons from X 320$^-$, the photoreduced
primary electron acceptor of system II, then the relaxation kinetics of the absorption change at 334 nm reflecting the reoxidation of X 320$^-$ (s. ref. 15) has
to show the same dependency on $K_3\left[Fe(CN)_6\right]$ concentration as the rate of oxygen
evolution. This was experimentally confirmed (s. ref. 17), as is shown in fig. 3.
Despite of the problems in the interpretation of the 334 nm relaxation kinetics
arising due to the existence of a special plastoquinone-connector molecule
(carrier B, s. ref. 18) between X 320 and the pool (for discussion, s. ref.19)

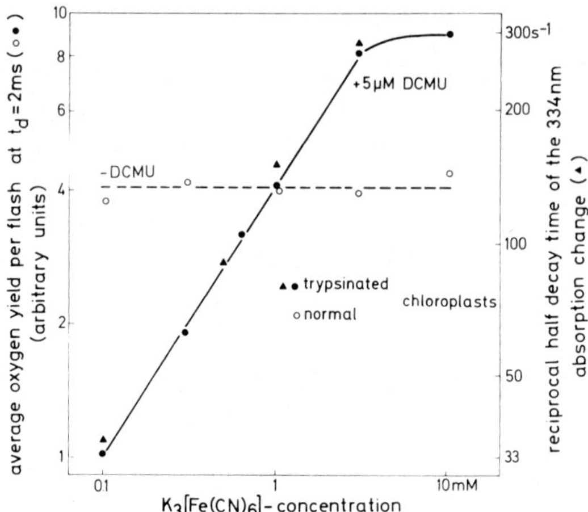

Fig. 3. Dependency on the $K_3\left[Fe(CN)_6\right]$ concentration of the average oxygen yield
per flash at t_d = 2 ms in normal and in trypsinated chloroplasts and of the reciprocal half decay time of the 334 nm absorption change in trypsinated chloroplasts measured at a 2 Hz repetition rate of the short saturating red flashes.

the data of fig. 3 are compatible with the assumption of a direct electron transfer to $K_3[Fe(CN)_6]$ in trypsinated chloroplasts, even in the presence of DCMU.

This is confirmed by latest fluorescence measurements (20) leading to the conclusion that a modification of the thylakoid membrane, either by lowering the external pH below 5 or by trypsination, makes the primary electron acceptor of system II accessible to external redox agents and that the electron transfer rate depends on the 0,75 - 0,80-th power of the $K_3[Fe(CN)_6]$ concentration, in fairly good agreement with the data presented here.

The photosynthetic electron transfer processes are coupled with protonation/ deprotonation reactions leading to net proton pumping through the thylakoid membrane (for rev. s. ref. 1,2). With respect to system II it is assumed that proton uptake occurs due to protonation of the photoreduced acceptor (probably at the special plastoquinone-connector molecule B, s. ref. 21) by protons from the external aqueous phase. Subsequently, these protons are funnelled concomitantly with the corresponding electrons via the plastoquinone pool into the inner thylakoid, where oxidation by photosystem I causes the acidification of the inner phase. Complementary to the proton transfer reactions coupled with the electron transfer at the acceptor side of system II a proton release into the inner phase of the thylakoid was shown to occur due to water oxidation catalized by system Y (9, 22, 23). Furthermore, a second site for proton uptake from the outer aqueous phase is located on the reducing side of system I (1,2). These reactions lead to the light induced generation of a proton gradient across the thylakoid membrane, which relaxes in the dark by reverse slow proton effluxes (1,2).

Kinetic studies indicate that the light induced proton uptake from the outer aqueous phase is retarded by a diffusion barrier (24). In the light of the present data one wonders whether this barrier could be established by the trypsin sensitive proteinaceous component covering up the acceptor side of system II. If this would be really the case, then 2 effects are expected to occur for the proton transport at the reducing side of system II in trypsinated chloroplasts:
A) If proton uptake occurs, its rate should be drastically increased.
B) As the electron flow to the plastoquinone pool is presumed to be interrupted and substituted by a direct electron transfer to the external electron acceptor, the protons taken up can not be transported through the thylakoid membrane. Accordingly, the protons should be released directly back into the outer aqueous phase as the reduced endogeneous acceptor becomes reoxidized by external $K_3[Fe(CN)_6]$. The kinetics of this release has to depend on the $K_3[Fe(CN)_6]$ concentration in a similar way as the electron transfer rate (s. fig. 3).

Preliminary results (Renger and Tiemann, in prep.) show, that in trypsinated chloroplasts in the presence of DCMU and $K_3[Fe(CN)_6]$ a very fast proton uptake occurs, followed by a comparitively rapid proton release, too fast to be explainable simply by an uncoupling effect of trypsin. These data are in line with the

346

above mentioned hypothesis, that the trypsin sensitive proteinaceous component
not only provides a protective shield to exogeneous redox agents, but also acts as
a barrier to proton diffusion. Furthermore, they give a hint for the artificially
induced electron transport from system II to exogeneous acceptors in trypsinated
chloroplasts to be coupled with protonation/deprotonation reactions. The kinetical
and mechanistical details of this mechanism remain to be clarified.

In contrast to the drastic modification of the kinetic pattern of the protolytic
reactions coupled with electron transport at the acceptor side of system II, the
processes at the watersplitting enzyme system Y should be only slightly influenced
by mild trypsination. According to fig. 2 a slow decline in the extent of proton
release to the inner thylakoid phase could be expected to arise, caused by the
degradation of the oxygen evolving capacity. The kinetics of proton liberation
should remain uneffected, whereas the proton efflux might be accelerated due to a
possible uncoupling effect by trypsin.

The measurements depicted in fig. 2 were kindly performed by Dr. W. Ausländer,
applying neutral red as indicator for pH-changes in the inner thylakoid space.

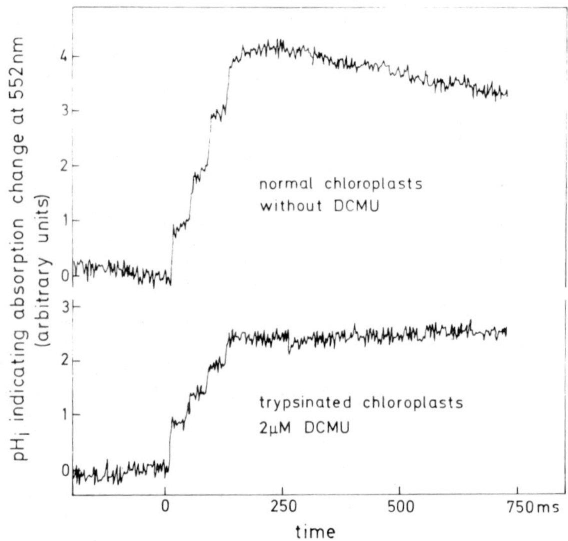

Fig. 4. pH$_i$-indicating absorption change at 552 nm in normal chloroplasts in the
absence of DCMU and in trypsinated chloroplasts in the presence of 2 μM DCMU.
Assay conditions: chloroplasts (10 μM chlorophyll), 0,3 mM K_3[Fe(CN)$_6$] + 1 μM
DBMIB as electron acceptor for system II electron transport, 10 mM KCl, 2 mM MgCL$_2$
2 mM MES-NaOH, pH = 7,2.
Trypsination: 10 μg trypsin/ml suspension at an incubation time of 6 min at room
temperature.
Excitation: flash groups containing 4 saturating short flashes (20 μs) per group
with a dark time of 40 ms between the flashes, the time between the flash groups
was 3,2 s
The signals are the differences of measurements made in the absence and presence
of imidazole

In order to eliminate absorption changes not related to pH_i-indicator response
signals obtained in the presence of the permeable buffer imidazole, completely
suppressing pH_i-changes, were subtracted. As the same results are obtained for
difference measurements made in the presence and absence of.the indicator dye
neutral red, the correction appears to be justified.

DBMIB was added for functional separation of system II electron transport, as is
described in ref. 23.

The data of fig. 4 show, that in trypsinated chloroplasts in the presence of
DCMU the extent of proton release due to excitation with repetitive flash groups,
each containing 4 flashes separated by a dark time 40 ms (repetition rate of the
groups $\sim$ 0,3 Hz), is nearly halfed, except for the first flash, which appears to
be uneffected. Surprisingly, even in normal DCMU-inhibited chloroplasts a proton
release was found in the first flash of each group under our excitation conditions,
whereas the next flashes in the group are ineffective (data not shown). Therefore,
the acidification induced by the first flash might be explainable by another me-
chanism(Renger and Ausländer, in prep.). Within the limits of the present time
resolution the kinetics of proton release appears not to be markedly influenced by
trypsination (in some preparations an accelerated relaxation is observed which
might be caused by an uncoupling effect of trypsin).

Thus, the results depicted in fig. 4 support the conclusion that under mild
tryptic digestion of the thylakoid membrane, the watersplitting enzyme system Y
remains active for both, oxygen evolution and acidification of the inner thylakoid
phase by water oxidation.

However, it could be possible that a slight proteolytic attack influences the
intrinsic mechanism, i. e. the storage life time of the intermediary oxidation
states of water cleavage generated by four sequential univalent redox steps between photooxidized chlorophyll-a_{II}^{+} and system Y (for rev. s. ref. 12).

In isolated spinach chloroplasts the life time of the oxidizing equivalents
stored in system Y can be selectively modified by a class of chemicals denoted as
ADRY-reagents (7,25). In order to check, whether trypsination might cause an ADRY-
type destabilization of the intermediary oxidation states, comparitive studies
were made in normal and in trypsinated chloroplast, respectively, in the absence
and in the presence of the most powerful ADRY-reagent 2-(3,4,5-trichloro-)anilino-
3,5-dinitrothiophene (ANT 2s, s. ref. 26). The obtained results depicted in fig. 5
show that in the absence of ANT 2s the relative oxygen yield per flash remains
practically constant with increasing times t_d between the flashes in normal as well
as in trypsinated chloroplasts for the depicted time range (at longer times above
2 - 3 s the normal decline appears). This indicates, that trypsination does not impose an ADRY-like effect on the watersplitting enzyme system Y, even in the presence of rather high DCMU-concentrations (up to 50 /uM DCMU no typical ADRY-effect
was found, data not shown).

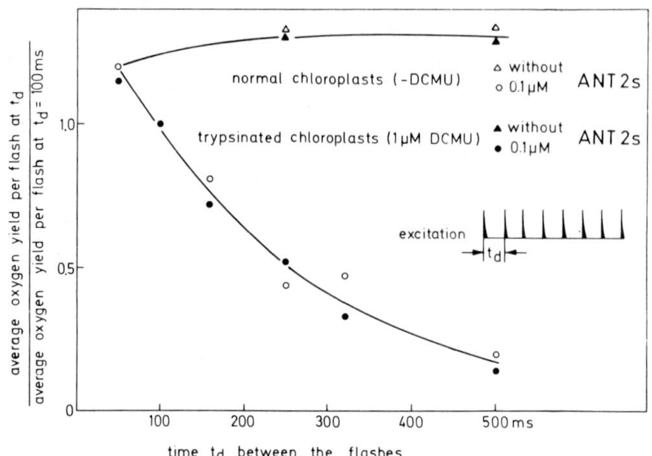

Fig. 5. Relative average oxygen yield per flash as a function of the time t_d between the flashes in the absence and presence of 0,1 /uM ANT 2s, respectively, in normal chloroplasts without DCMU and in trypsinated chloroplasts in the presence of 1 /uM DCMU.
Excitation: white short flashes (20 /us)

On the contrary , in the presence of 0,1 /uM ANT 2s a pronounced ADRY-effect is observed in normal as well as in trypsinated chloroplasts. These results indicate that trypsination does neither modify the characteristics of system Y for charge storage, nor prevents its influence by ADRY-reagents.

The failure of comparitively high DCMU-concentrations to cause a significant ADRY-effect in trypsinated chloroplasts proves the destabilization of the states S_2 and S_3 earlier detected at much lower DCMU-concentrations (27,28) not to be a true ADRY-effect. Another mechanism is responsible for this DCMU action.

CONCLUSIONS

The data presented here indicate that mild trypsination of stripped spinach chloroplasts preferentially degrades a proteinaceous component located at the acceptor side of system II, whereas the watersplitting enzyme system Y withstands much more better the tryptic attack. As system Y is known to be the most sensitive part of system II, the present findings support evidence for its localization near the inner side of the thylakoid membrane, whereas the trypsin-sensitive proteinaceous component covers up the reducing side of system II towards the outer aqueous phase.

The discovery of this new component implies the question about its functional role and its chemical nature.
Though the sensitivity to trypsin points to the importance of the alkalinic amino acids lysine and arginine (pK of the order of 9), the identification of the proteinaceous component must await further investigations.

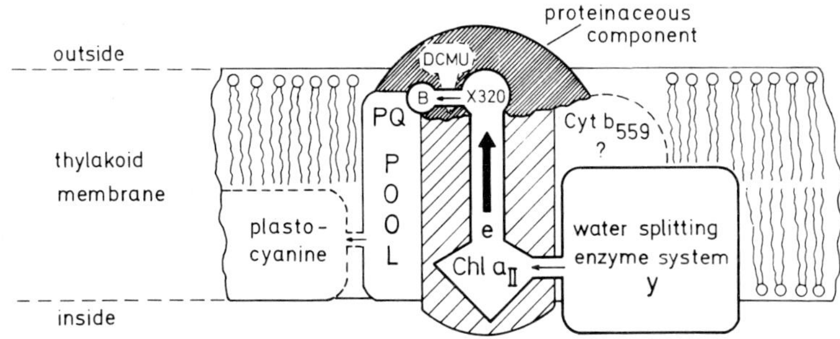

Fig. 6. Simplified scheme of system II (for explanation s. text)

With respect to its function, earlier findings (11,17) and the present data favour the assumption, that this component plays a central role as regulatory element for the electron transport at the reducing side of system II:
1) Its structural integrity is indispensible for the electron transport between systems II and I. As no experimental evidence exists for the direct participation of a proteinaceous component as electron carrier between photosystem II and the plastoquinone pool, it is assumed to act as an allosteric regulator for the reactions of the plastoquinone components X 320, B and the PQ-pool (s. fig. 6).
2) The proteinaceous component provides a diffusion barrier for protons as well as for exogeneous redox agents. Thus, it prevents dissipative electron leakage of redox equivalents at the reducing side of system II.
3) The proteinaceous component is supposed to contain the binding site for DCMU-type inhibitors. Accordingly, an allosteric mode of inhibition seems to be responsible for the action of DCMU-like poisons. This mechanism is supported by the equality of the constants for binding and inhibition, respectively (29), and by the kinetical coincidence of the activity decline of the linear electron transport from water to system I (s. fig. 1, left side) and of the removal of the DCMU-block (fig. 2,dotted curve). Despite these arguments it can not be excluded, that only the trypsin-induced accessibility of X 320 (and/or B ?) to exogeneous acceptors circumvents the DCMU-block, whereas the poison still remains bound at its inhibitory site (s. ref. 11).

Finally, it must be emphasized that the components of the thylakoid membrane do not form a fixed crystal-like lattice structure, but a rather flexible ensemble, whose structure depends on the environment. In this sense, fig. 6 represents only a very rough scheme, especially with respect to the localization of system Y, whose accessibility to agents from the outer aqueous phase might be strongly dependent on parameters like the external pH. However, the functional interaction between chlorophyll-a$_{II}$ and the intact system Y is regulated mainly by the inner phase (30).

ACKNOWLEDGEMENTS

The author gratefully acknowledges the financial support by ERP-Sondervermögen.

REFERENCES

1. Witt, H. T. (1975) in: Bioenergetics of Photosynthesis (Govindjee, ed.)
    pp. 493 - 554, Academic Press, New York
2. Junge, W. (1977) in: Encyclopedia of Plant Physiology, New Series Vol.5 (Trebst,
    A. and Avron, M. eds) pp. 59 - 93, Springer, Berlin
3. Berzborn, R. J. and Lockau, W. (1977) in: Encyclopedia (1. c. 2), pp. 283 - 296
4. Giaquinta, R. and Dilley, R. A.(1977) in: Encyclopedia (1. C. 2), pp. 297 - 303
5. Winget, G. D., Izawa, S. and Good, N. E. (1965) Biochem. Biophys. Res. Commun.
    21, 438 - 443
6. Renger, G. and Wolff, Ch. (1975) Z. Naturforsch. 30c, 161 - 171
7. Renger, G. (1972) Biochim. Biophys. Acta 256, 428 - 439
8. Renger, G., Erixon, K., Döring, G. and Wolff, Ch. (1976) Biochim. Biophys. Acta
    440, 278 - 286
9. Ausländer, W. (1977) Thesis, Technische Universität Berlin
10. Reqitz, G. and Ohad, I. (1976) J. Biol. Chem. 251, 247 - 252
11. Renger, G. (1976) Biochim. Biophys. Acta 440, 287 - 300
12. Renger, G. (1977) in: Topics in Current Chemistry (Boschke, F. L. ed.) Vol. 69,
    pp. 39 - 90, Springer, Berlin
13. Reimer, S. and Trebst, A. (1975) Biochem. Physiol. Pflanzen, 168, 225 - 232
14. Renger, G. (1969) Thesis, Technische Universität Berlin
15. Stiehl, H. H. and Witt, H. T. (1969) Z. Naturforsch. 24b, 1588 - 1598
16. Vater, J., Renger, G., Stiehl, H. H. and Witt, H. T. (1968) Naturwiss. 55,
    220 - 221
17. Renger, G. (1976) FEBS Letters 69, 225 - 230
18. Bouges-Bocquet, B. (1973) Biochim. Biophys. Acta 314, 250 - 256
19. Siggel, U., Khanna, R., Renger, G. and Govindjee,Biochim. Biophys. Acta (in
    press)
20. Itoh, S. and Nishimura, M. (1977) Biochim. Biophys. Acta 460, 381 - 392
21. Fowler, Ch. F. (1977) Biochim. Biophys. Acta 459, 351 - 363
22. Fowler, Ch. F. and Kok, B. (1974) Biochim. Biophys. Acta 357, 299 - 309
23. Ausländer, W. and Junge, W. (1975) FEBS Letters 59, 310 - 315
24. Ausländer, W. and Junge, W. (1974) Biochim. Biophys. Acta 357, 285 - 298
25. Renger, G., Bouges-Bocquet, B. and Delosme, R. (1973) Biochim. Biophys. Acta
    292, 796 - 807
26. Renger, G. (1973) Biochim. Biophys. Acta 314, 390 - 402
27. Bouges-Bocquet, B., Bennoun, P. and Taboury, J. (1973) Biochim. Biophys. Acta
    325, 247 - 254
28. Renger, G. (1973) Biochim. Biophys. Acta 314, 113 - 116
29. Tischer, W. and Strotmann, H. (1977) Biochim. Biophys. Acta 460, 113 - 125
30. Renger, G., Gläser, M. and Buchwald, H. E. (1977) Biochim. Biophys. Acta 461
    (in press)

Bioenergetics of Membranes. L. Packer et al. ed.

OPTICAL STUDIES OF PHOTOSYSTEM I PARTICLES :
EVIDENCE FOR THE PRESENCE OF MULTIPLE ELECTRON ACCEPTORS.

Kenneth SAUER[*], Suzanne ACKER, Paul MATHIS and Jasper A. VAN BEST
Département de Biologie. Centre d'Etudes Nucléaires de Saclay
B.P. n° 2 -91190- Gif-sur-Yvette. France.

SUMMARY

The absorption changes of P700 in chloroplast TSF1 particles enriched in Photosystem I exhibit fast kinetic components under reducing conditions. The addition of dithionite to an anaerobic sample at pH 10 induces a rapid (250µs) exponential decay of P700 absorption changes following laser pulse excitation. This decay is approximately 100 times faster than the recombination of $P700^+$ with $P430^-$ seen under less reducing conditions. In separate experiments where a good electron donor, neutral red, is added to the anaerobic dithionite-containing samples, a still faster (3µs) reversal of $P700^+$ is seen when a laser pulse is superimposed on red background illumination. These results are interpreted in terms of two new intermediate electron acceptors that lie between P700 and P430. Preliminary experiments with broken chloroplasts show a similar rapidly decaying component under strongly reducing conditions.

INTRODUCTION

The photoreactions of Photosystem I of chloroplasts can be monitored readily using either optical [1] or electron paramagnetic resonance [2] techniques. Such studies have established the characteristics of the primary electron donor P700, several condidates for electron acceptor species, $P430^2$ and X^3, and the secondary electron donors, cytochrome f and plastocyanin. Through the use of selective detergent treatment [4] it is possible to obtain subchloroplast particles that are greatly enriched in Photosystem I relative to Photosystem 2. In 'such preparations the secondary donors are usually not well connected to P700, and exogenous electron donors must be used to support non-cyclic electron flow. Nevertheless, such particles provide unique advantages for studying the relation between the primary donor and the acceptors of Photosystem I.

Hiyama and Ke[5] discovered the optical component P430, which appeared to have the characteristics of the primary electron acceptor of Photosystem I. It is a species with a low reduction potential that becomes reduced within 100 ns following a brief flash, and it participates readily in both cyclic and noncyclic

[*] On leave from University of California, Berkeley, USA.

352

electron flow mediated by Photosystem I. A low temperature EPR component reported at about the same time by Malkin and Bearden was also proposed as a candidate for the primary acceptor [6]. Subsequent studies showed that this EPR component, which has the features of a membrane-bound iron-sulfur protein, is probably the same as the optically detected P430[7].

More recent experiments have led to the report of intermediates between P700 and P430. In low temperature EPR studies by McIntosh ,et al [8] and by Evans, et al [9], a photoreversible component designated X is assigned to this role. X$^-$ has a distinctive EPR spectrum which does not closely resemble those of any known iron sulfur proteins [2]. It is formed in the light and decays in the dark along with P700$^+$ even at 6°K [8]. Furthermore, studies of the spin polarization of the EPR of P700$^+$ by Dismukes, et al [10] are consistent with an interaction with a species like X . Recent optical studies by Ke, et al [11] at low reduction potentials have confirmed the presence of an intermediate between P700 and P430.

We describe the results of rapid transient optical absorption spectroscopy applied to Triton-solubilized Photosystem I particles (TSF1) under reducing conditions. In the presence of a good electron donor, such as neutral red, and with background illumination the terminal electron acceptors become progressively reduced. Then a laser pulse inducing the formation of P700$^+$ is followed by the back reaction with more primary reduced electron acceptors. From studies of the rate of the back reaction of P700$^+$ we have obtained evidence for the occurence of two intermediate states between P700 and P430.

MATERIAL AND METHODS

Chloroplasts from spinach were prepared in sucrose, 0.4 M, tricine buffer, 50 mM, pH 7.6. Triton-solubilized Photosystem I particles (TSF 1) were prepared following the procedure of Vernon and Shaw [4] and were kept frozen at -20°C until needed.

Dimethyl triquat (DMT; 1,1'-trimethylene-4,4'-dimethyl-2,2'-bipyridilium bromide) was obtained as a gift that was kindly provided by Imperial Chemical Industries, Ltd, Bracknell, Berkshire, England. Neutral red (3-amino-7-dimethylamino-2-methylphenazine hydrochloride) or PMS (phenazine methosulfate) were dissolved in water at 2 mM and 1 mM, respectively.

For each experiment a TSF 1 sample was thawed and homogenized with the appropriate buffer and added reagents, except that solid dithionite when present was added later from the side arm of an evacuable cuvette with 1 cm path length. The reaction mixture placed in a second side arm was frozen and degassed three times under vacuum. Mixing with dithionite in darkness immediately preceded the experimental study. Glycine buffer, 0.2 M, pH 10 was used for all experiments that included dithionite; for studies using ascorbate, the buffer was tricine, 0.02M, pH 7.6.

Measuring beam light from a quartz tungsten-iodine incandescent lamp passed to the sample through a water filter and, typically, a RG 630 filter (Schott) when background illumination was desired. In other experiments in the red and near IR spectral region with low background illumination, either a Wratten 87 gelatin filter or an interference filter (bandwidth 3 nm at half maximum) was added before the sample. The light transmitted by the sample was focussed on the entrance slit of a Bausch and Lomb grating monochromator (500 mm) with 7 nm bandpass and, in some cases, a supplementary interference filter. Light was detected using either a silicon photodiode, PIN-10 (UDT, Santa Monica, California, USA), in the red and near IR or a photomultiplier in the blue or near UV. Signals were recorded using a Tektronix R 7912 Transient Digitizer coupled to a Didac 4000 (Intertechnique) multichannel analyzer. Flash excitation of the sample was provided at 90° to the measuring beam using either a Q-switched ruby laser (Quantel, France; λ 694 nm,10 ns duration) or a flash-lamp pumped dye laser (Electro-Photonics, Belfast; λ 620 nm, 1 μs duration).

For each set of experimental transient signals recorded, an equal number of flash artefacts with the measuring beam blocked was subtracted.

RESULTS

Initial studies were carried out to confirm the observations of Hiyama and Ke [1,5]. In the presence of ascorbate and $DCIPH_2$ as electron donor and benzyl viologen as acceptor we observed flash-induced absorption changes at 703 or 820 nm that reversed slowly, during several seconds. The spectrum of these changes between 370 and 500 nm and between 703 and 1000 nm is that of P700, reported previously [1]. In particular, we observed isosbestic wavelengths at about 408, 445 and 725 nm. In the absence of benzyl viologen, for a sample of TSF 1 particles containing ascorbate/ $DCIPH_2$ and degassed, the flash-induced absorption changes at 703 or 820 nm reversed more rapidly but in a biphasic decay pattern. Approximately 75% (depending on the wavelength) reversed with a halftime about   30 ms; the remaining portion exhibited a much slower decay. The difference spectrum in the region longer than 703 nm is the same as in the presence of benzyl viologen, but in the blue the signals were consistent with the sum of absorbance changes from P430 and P700 [1]. No faster transient components of the decay could be observed at 703 or 820 nm, to the instrument limit of about 1 μs, under either set of experimental conditions.

Samples of TSF 1 particles containing dithionite(2 $mg.ml^{-1}$) added after degassing exhibit a more rapid relaxation (250 $\pm$ 20 μs) at room temperature, as shown for 703 or 820 nm wavelengths in Fig.1. Semi-logarithmic plots show that the decay occurs almost entirely (>90%) by a single exponential. The difference spectrum in the region from 675 to 900 nm closely resembles that of P700, but it is distinctly different in the blue. By contrast with P700, the absorption changes in the presence of dithionite are essentially zero between 395 and 400 and between 445 and 455 nm, regions where P700 exhibits definite positive changes. The absorption change spectrum is not markedly different from that of P700 + P430, although the decay is nearly 100 times faster than is observed for the recombination of $P700^+$ and $P430^-$ at room temperature.

Dithionite

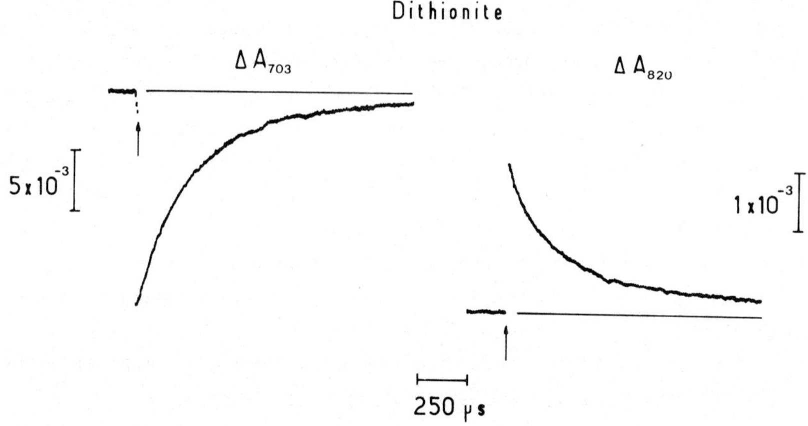

Fig.1 Absorption transients at 703 nm (left) and 820 nm (right) induced by a ruby laser flash on a sample of TSF 1 particles. Anaerobic reaction mixture contains sodium dithionite (2mg ml^{-1}), Triton (0.015%), glycine (0.2M, pH 10) and TSF 1 particles sufficient to give A_{672} nm = 1.61 cm^{-1}. At 703 nm, $t_{1/2}$ = 270 µs(96%); at 820 nm, $t_{1/2}$ = 245 µs (90%). No background illumination, ambient temperature.

   Shuvalov, et al have reported that the addition of a good electron donor like neutral red, in the presence of dithionite, leaves the system in the state P700. $P430^-$ at a time 2 ms following illumination [12]. For a similar reaction mixture we observe a slow decay of $P700^+$ following a laser flash prior to applying a strong background illumination (Fig.2, left). However, in the presence of a background of red light (λ > 630 nm) a large portion of the transient absorption change at 820 nm reverses much more rapidly (Fig.2,right). Ke, et al have shown that the low potential mediator dimethyl triquat (DMT) facilitates the reduction of P430 by dithionite. In the presence of neutral red as donor, DMT/dithionite at pH 10 to achieve a low potential (in the absence of oxygen) and under background illumination, the decay kinetics following a submicrosecond ruby laser pulse exhibits a fast (3 µs) decay component at 703 and 820 nm (Fig.3). Subsequent studies showed that apparently identical behavior obtains if DMT is omitted. While the spectrum of this absorption change resembles that of P700 in the region from 703 to 920 nm, there is a definite shift in the isobestic near 720 nm. This is seen most clearly at 720 nm, where a negative transient absorption change in the absence of background illumination becomes positive when background light is added to the same reaction mixture. Furthermore, the major components of the pronounced biphasic decay seen with background illumination exhibit different wavelength dependences throughout this spectral region.

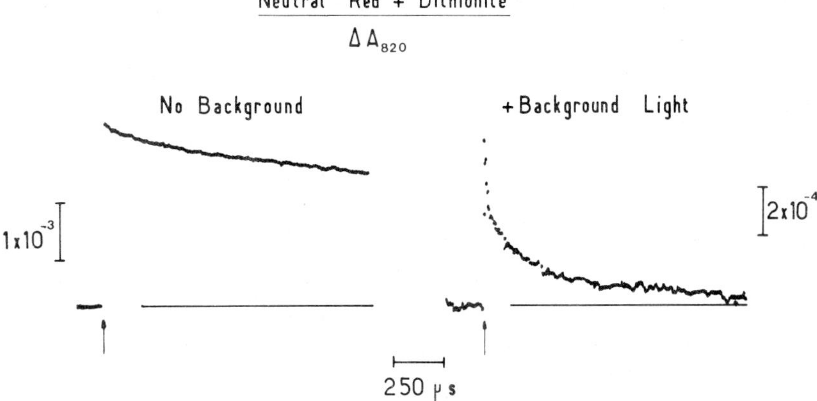

Fig.2 Absorption transients at 820 nm induced by a ruby laser flash on a TSF 1 sample without background illumination (left; Wratten 87 filter before sample) and with background illumination (right; no Wratten 87 filter). Anaerobic reaction mixture contains sodium dithionite (2 mg ml^{-1}), Triton (0.015%), neutral red (10 µM), glycine (0.2M, pH 10) and TSF 1 particles sufficient to give $A_{673\ nm}$ = 1.17 cm^{-1}. Ambient temperature.

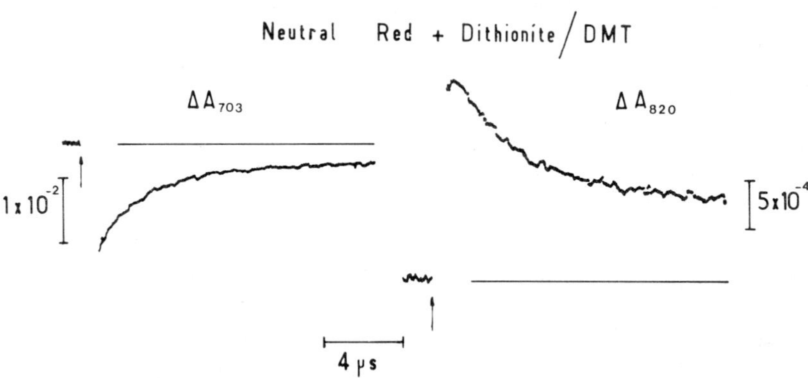

Fig.3 Absorption transients at 703 nm (left) and 820 nm (right) induced by a ruby laser flash on a sample of TSF 1 particles. Anaerobic reaction mixture contains sodium dithionite (2 mg ml^{-1}), Triton (0.015%), neutral red (10 µM), DMT (8 µM), glycine (0.2M, pH 10) and TSF1 particles sufficient to give $A_{672\ nm}$ = 1.43 cm^{-1}. At 703 nm, $t_{1/2}$ = 2.7 µs (85%); at 820 nm, $t_{1/2}$ = 2.8 µs (68%). With background illumination; ambient temperature.

When air is admitted to the anaerobic samples containing dithionite, either with or without added neutral red and DMT, the rapid transients are not immediatly quenched as would be expected for signals from triplet species. On the contrary, as the sample becomes fully oxygenated following thorough agitation under air, the decay becomes very much slower for the absorption transients in the long wavelength region.

Preliminary experiments with broken spinach chloroplasts were carried out anaerobically in the presence of dithionite and DMT. Although the signal-to-noise was poorer owing to the greater turbidity of the chloroplasts compared with TSF 1 particles, nevertheless a distinct decay component of about 4 μs was observed following ruby laser flashes. The magnitudes of the signals and the reversal in sign between 703 and 820 nm indicate that this is not a light scattering transient. In separate experiments it was demonstrated that Photosystem 2 does not contribute a signal with this relaxation time (Van Best and Mathis, to be published).

DISCUSSION

The light reactions associated with Photosystem 1 have been investigated by optical changes associated with the oxidation of P700 (bleaching at 700 and 430 nm; increase in absorption from 730 to 850 nm, 450 to 550 and 300 to 400 nm) and the reduction of P430 (broad, weaker, absorption decrease from 370 to 470 nm). The detailed studies of Ke and coworkers have established not only the wavelength dependence but also the conditions under which noncyclic electron flow, cyclic electron flow or an internal back reaction restores one or both of these components to the initial state [1]. The present research expands those studies by establishing kinetic and spectroscopic evidence for the existence of two intermediate states associated with Photosystem 1.

The major portion of our studies was carried out using Triton-solubilized Photosystem 1 particles from spinach. In this preparation, known as TSF 1, the normal electron donors and terminal acceptors are either absent or disconnected. They do contain active P700 and iron-sulfur proteins, including the membrane bound iron-sulfur protein, P430. We have investigated the properties of these particles using laser-pulse excitation under four distinct sets of conditions :

    1 Ascorbate/$DCIPH_2$ + benzylviologen provides a good electron acceptor, benzyl viologen, and a relatively inefficient electron donor, $DCIPH_2$, under mildly reducing conditions. The oxygen of air appears to serve alternatively to benzyl viologen as an agent that reoxidizes $P430^-$ following photoactivation by a brief flash. The absorption changes observed are characteristic of P700 alone, and the slow relaxation (several seconds in duration) is limited by the donation of electrons from $DCIPH_2$ to $P700^+$. The process can be summarized by the equations

$$DCIPH_2 + P700 \cdot P430 + BV \xrightarrow{\quad\quad} DCIPH_2 + P700^+ \cdot P430 + BV^-$$
$$\downarrow 0.5 - 2 \text{ s}$$
$$DCIP + P700 \cdot P430 + BV^- \tag{1}$$

The rate of reoxidation is a function of $DCIPH_2$ concentration.

2   Ascorbate/$DCIPH_2$ (anaerobic) does not provide a suitable electron acceptor for $P430^-$ and, as a consequence, the back reaction dominates in the
relaxation of $P700^+$. Under these conditions the difference spectrum is
characteristic of P700 and P430 together, with a relaxation time of about
30 ms for both components

$$P700 \cdot P430 \xrightarrow{\quad\quad} P700^+ \cdot P430^- \xrightarrow[30 \text{ ms}]{\quad\quad} P700 \cdot P430 \tag{2}$$

The rate of relaxation is independent of $DCIPH_2$ concentration. Both of
the conditions [1] and [2] were recognized and characterized by Ke[1].

3   Dithionite (anaerobic) establishes a sufficiently low reduction potential
to prevent the reoxidation and to provide for the accumulation of $P430^-$
following the first few flashes. This results in the progressive decrease
in the time of relaxation of $P700^+$ until it reaches a limiting value of
about 250 μs at room temperature. We attribute this observation to the
photoreduction of a new intermediate electron acceptor, which we designate provisionally A_2 and which subsequently undergoes a back reaction with
$P700^+$. The difference spectrum is clearly distinct from that of P700 alone, but it appears to resemble P700 + P430. The instability of the samples during long periods under the experimental conditions causes the difference spectrum to be difficult to measure with precision. The equation
representing this process is

$$P700 \cdot A_2 \cdot P430^- \xrightarrow{\quad\quad} P700^+ \cdot A_2^- \cdot P430^-$$
$$\downarrow 250 \text{μs} \tag{3}$$
$$P700 \cdot A_2 \cdot P430^-$$

It is possible that A_2 is the low potential intermediate implicated by
Ke, _et al_ [11]; another candidate is the iron-sulfur protein observed at low
potential (Center B) in the low temperature EPR studies[3,7]. This would be
consistent with the apparent similarity between the optical difference
absorption spectrum of A_2 and that of P430. The clues that we have, such
as the absence of rapid quenching upon the admission of air to the sample,
suggest that condition [3] does not involve the triplet state of chlorophyll.

4   Neutral red (or PMS) + dithionite ($\pm$ DMT) + background illumination (anaerobic) provides strongly reducing conditions and an efficient electron
donor[12,13]. A new and much faster (3 μs) decay of $P700^+$ now dominates the
decay curve. A simple explanation is provided by extending the above
reasoning to suppose that the good electron donors compete with the back
reaction of [3] so that A_2 becomes progressively reduced by the background illumination prior to the laser flashes. In this view, the 3 μs
back reaction reflects the recombination of $P700^+$ with a second new intermediate species, which we designate analogously A_1^-. This reaction is

$$\xrightarrow[\text{light}]{\text{Background}} P700 \cdot A_1 \cdot A_2^- \cdot P430^- \xrightarrow{} P700^+ A_1^- \cdot A_2^- \cdot P430^- \quad\downarrow 3\ \mu s \qquad P700 \cdot A_1 \cdot A_2^-\ P430^-$$

[4]

The identity of A_1 is completely unknown. Its rapid decay has prevented
our examining spectral regions shorter that 703 nm for evidence of its
absorption characteristics.

Shuvalov, et al [12] used a phosphoroscopic instrument to measure absorption changes immediately following (2 ms) actinic illumination under the conditions [3] and
[4] . With their technique they would not have seen the rapid transients that we
observe, but they did find evidence for the build up of reduced acceptors that
they attributed to P430. It now appears that, with dithionite and neutral red
present, their difference spectrum should represent the accumulation of A_2^- as
well. Indeed, what they observed appears to be somewhat different from that reported by Ke[1]. The data of Shuvalov (see also ref.13) show a double-peaked spectrum in the blue with a dip at 430 nm, and appreciably greater contributions near
500 nm.

If A_1 and A_2 prove to be intermediate electron acceptors between P700 and P430,
then we might expect to see evidence for their participation in the forward reactions under nonreducing conditions. We have looked for optical changes of this
nature but we have failed to find them, to the limit of detectability of our present apparatus. Hiyama and Ke[5] observed the rise time of P430 reduction to be
less than 100 ns; however, this result may need to be reinterpreted because of
the apparent similarity of the absorption changes associated with P430 and with
A_2. It may prove difficult to distinguish these two species under ambient redox
conditions until distinctive spectral features are recognized. The EPR spin polarization studies of Dismukes, et al [10] suggest that an electron acceptor different from P430 is paired with $P700^+$ on a time scale shorter than 5 μs. Because
this acceptor has EPR characteristics attributable to X^{10} and because it should

be the component closest to P700 to give the spin polarization, we tentatively associate it with the earliest component detected in our optical studies, A_1.

The presence of a rapid (4 μs) component of P700 absorption change relaxation in chloroplasts in the presence of PMS and dithionite (anaerobic) suggests that the fast component [4] is not an artefact introduced by the detergent treatment in preparing TSF 1 particles. During the normal photosynthetic electron transport the back reaction that we observe is, of course, wasteful of the absorbed energy. To prevent substantial losses by this process, the normal transfer of electrons from A_1 to A_2 and P430 should occur in less than 100 ns. It is in this difficult time range that we must explore for confirmation of their role in electron transport in the reaction centers of Photosystem 1.

ACKNOWLEDGMENTS

This research was supported, in part, by the Solar Energy Program Contract 014/76 ESF of the Commission of European Communities. One of us (KS) received generous support from the Guggenheim Memorial Foundation.

REFERENCES

1. Ke,B.(1973) Biochim. Biophys. Acta 301, 1-33
2. Bearden, A.J. and Malkin, R. (1977) Brookhaven Symposia in Biology 28, 247-265
3. Evans, M.C.W., Sihra, C.K. and Cammack, R. (1976) Biochem. J. 158, 71-77
4. Vernon, L.P. and Shaw, E.R. (1971) Methods in Enzymol. 23, 277-289
5. Hiyama, T. and Ke, B. (1971) Proc. Natl. Acad. Sci. U.S. 68, 1010-1013
6. Malkin, R. and Bearden, A.J. (1971) Proc. Natl. Acad. Sci. U.S. 68, 16-19
7. Ke, B., Hansen, R.B. and Beinert, H. (1973) Proc. Natl. Acad. Sci. U.S. 70, 2941-2945
8. Mc Intosh, A.R., Chu, M. and Bolton, J.R. (1975) Biochim. Biophys. Acta 376, 308-314
9. Evans, M.C.W., Sihra, C.K., Bolton, J.R. and Cammack, R. (1975) Nature (London) 256, 668-670
10. Dismukes, G.C., Mc Guire, A.,Blankenship, R.E. and Sauer, K. (1977) Biophys.J. submitted
11. Ke, B. Dolan, E., Sugahara, K., Hawkridge, F.M., Demeter, S. and Shaw, E.R. (1977) Plant and Cell Physiol., in press.
12. Shuvalov, V.A., Klimov, V.V. and Krasnovskii,A.A. (1976) Molec. Biol. (Engl. Transl.) 10, 261-272
13. Shuvalov, V.A. (1976) Biochim. Biophys. Acta 430, 113-121.

EFFECTS OF TRYPSIN ON CHLOROPLAST MEMBRANES

Paolo Gerola, Emanuele De Benedetti, Silvia Rizzi, Giorgio Forti and
Flavio M. Garlaschi
Istituto di Scienze Botaniche, Università di Milano
Via Giuseppe Colombo, 60, Milano
Italy

INTRODUCTION

Several methods have been devised to locate electron carriers, enzymes and other components of the photosynthetic apparatus in the chloroplast membrane. A fruitful approach has been the investigation of the effects of macromolecules, such as specific antibodies against individual components of the photosynthetic apparatus and hydrolitic enzymes, on the assumption that macromolecules should not penetrate the membrane. Their effect under carefully controlled conditions was therefore taken as an indication that the component affected by a specific macromolecular reagent is located on the external surface of the membrane. On the basis of this rationale, the effect of treatment with trypsin has been investigated in several laboratories. The following effects were established : a) trypsin activates the Ca^{++}-dependent ATPase of the coupling factor CF_1 (1); b) it inactivates photosystem II (PS II) activity (2-6); c) it develops a DCMU-resistant PS II activity, with ferricyanide as the electron acceptor (8, 6).

A controversy has arisen as to the mode of inactivation of PS II by trypsin. Selman and Bannister (3) supported the idea that trypsin inactivates the oxidizing side of PS II, at two different sites, on the basis of their observation that the photooxidation of H_2O was more sensitive than that of alternative donors such as 1,5-diphenilcarbohydrazide. Renger supported the concept that the site of inhibition is solely on the reducing side of PS II, because he observed that the O_2 yield per flash was decreased till complete suppression by trypsin with p-benzoquinone as electron acceptor, but was unaffected when ferricyanide was used (6). Simultaneously, he observed that ferricyanide reduction became DCMU resistent upon trypsin treatment. This author concluded that the digestion by trypsin of some "proteinaceous material" on the surface of the thylakoid membrane makes the primary PS II acceptor ("X320", identical to Q in other authors terminology) directly available to ferricyanide. This was proposed to be the mechanism of DCMU-resistence, and of the inhibition of electron transport from PS II to PS I (6).

We have re-investigated the mechanism and time-course of trypsin action on chloroplast membranes. Our results showed that the suppression of Mg^{++}-stimulated fluorescence is an early change induced by trypsin, by far the most sensitive to attack. This is followed, in the order, by inactivation of H_2O oxidation and uncoupling, then by inactivation of NH_2OH oxidation and by the development of DCMU resistence of ferricyanide and 2,4-dichlorophenolindophenol (DCPIP) reduction.

MATERIALS AND METHODS

Chloroplasts were prepared from freshly-harvested spinach (Spinacia oleracea) leaves. The leaves were ground for ca.3-5 seconds in a blendor with 3 volumes of cold tricine-NaOH buffer, 30 mM, pH 7.8, containing 10 mM NaCl and 0.4 M sucrose. The homogenate was quickly squeezed through 10 layers of muslin, and centrifuged 5 min. at 1000 x g. The supernatant was discarded and the green pellet resuspended in the same buffer, recentrifuged as above and finally resuspended at the concentration of 1 to 2 mg of chlorophyll per ml. All operations were carried on at $2\text{-}4^{\circ}C$.

Ferricyanide reduction was measured as the decrease of absorbance at 420 nm, in a medium containing : tricine-NaOH 30 mM, pH 8, unless otherwise stated; NaCl 10 mM; ferricyanide 0.8 mM, chloroplasts containing 20 μg of chlorophyll/ml. Other additions were as indicated in the different experiments. Illumination of saturating intensity was provided by a quartz-iodine lamp, filtered through 5 cm of water, a heath filter and a broad band red filter. DCPIP reduction was measured in the same apparatus, following the decrease of absorbance at 620 nm. O_2 uptake due to reoxidation of reduced methylviologen was estimated by means of a Clark-type O_2 electrode at $20^{\circ}C$. The medium was the same as above, with the omission of ferricyanide and the addition of methylviologen 0.5 mM and NaN_3 0.5 mM. The chloroplast suspension in the electrode cell was stirred by a magnetic bar. Illumination was provided by a 500 W tungsten lamp, filtered through 5 cm of water and a heath filter. When only PSI activity was measured, 5 μM DCMU was added and the electron donor system ascorbate (5 mM) and DCPIP 30 μM. The addition of superoxide dismutase had no effect on O_2 uptake under these conditions. Trypsin treatment was performed in the dark, at $20^{\circ}C$; the action of trypsin was stopped by the addition of 5-fold excess of soybean trypsin inhibitor.

Photophosphorylation was measured as previously described (9).

Fluorescence emitted at 685 nm was measured in a Perkin-Elmer MPF-3 fluorimeter, with exciting beam of 440 nm. The medium contained tricine-NaOH buffer, pH 8, NaCl 10 mM and 5 to 10 μg of chlorophyll. $MgCl_2$ 5 mM was added where indicated.

All reagents were analytical grade. Trypsin and trypsin inhibitor were obtained from Sigma (St. Louis, Missouri).

RESULTS

The effect of trypsin concentration (in a 4 min treatment) on the stimulation of fluorescence by Mg^{++} and on the PS II-dependent ferricyanide reduction is shown in Table 1. A very mild trypsin action, having no effect on electron transport nor on the photosynthetic control by coupling, was sufficient to suppress the effect of Mg^{++} on fluorescence. The presence of Mg^{++} during the trypsin treatment protected (data not shown), indicating that the Mg^{++}-binding site is more susceptible to trypsin attack when "uncharged " than when Mg^{++} is bound to it.This agrees with observation by Jennings (R.C.Jennings, personal communication).

TABLE 1

EFFECT OF TRYPSIN ON Mg^{++}-STIMULATED FLUORESCENCE AND ELECTRON TRANSPORT

Trypsin added, μg/ml	Mg^{++}-stimulation %	Ferricyanide reduction μequiv. mg^{-1} Chl. hr^{-1}	
		C	NH_4Cl 5 mM
0	102	220	1100
0.125	87	–	–
0.250	0	232	1140
0.500	0	288	1056
1.0	–	528	812
2.0	–	656	656

Condition : see "Methods". Trypsin treatment was for 4 min.

Figure 1 shows the effects of trypsin on photophosphorylation, the photosynthetic control and electron transport as a function of its concentration. The photosystem I (PS I)-dependent electron transport from $DCPIPH_2$ to methylviologen was not affected in the presence of uncoupler (fig.1 b), while the "basal" rate was progressively stimulated to reach the same level as the uncoupled one (fig. 1 b), indicating uncoupling by trypsin. The inhibition of photophosphorylation in the presence of phenazinemethosulfate, a PS I-dependent process, followed similar first order kinetics as the uncoupling of electron transport, suggesting that the two events are probably due to the same action of trypsin (fig.1 b). The upper part of the figure (fig.1 a) shows that uncoupled electron transport depending on both photosystems was inhibited, as previously shown (2-6). No differentiation (on the basis of trypsin concentration or the time-course at constant trypsin) could be obtained between PS II electron transport inhibition and uncoupling, when absolutely pure trypsin was used.

364

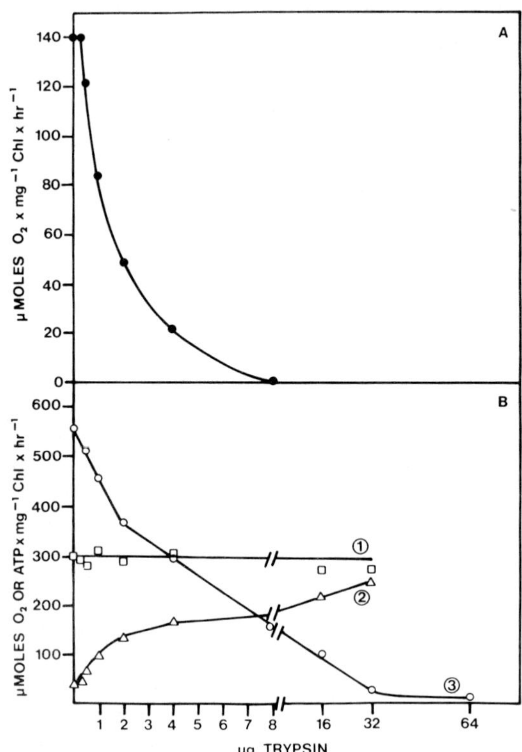

Fig.1   Inhibition and uncoupling of electron transport by trypsin treatment.
Abscissa : TRYPSIN μg . ml^{-1}. Conditions as in "Methods"; buffer of pH 7.8;
A) electron transport from H_2O to methylviologen in the presence of NH_4Cl, 5 mM.
B) electron transport from $DCPIPH_2$ to methylviologen in the presence (curve①)
and absence (curve②) of NH_4Cl 5 mM and PMS-dependent photophosphorylation
(curve③).

Figure 2 shows the time-course of PS II inhibition and the development of the resistance to DCMU inhibition of electron transport from H_2O to ferricyanide and to DCPIP.
It was observed that the DCMU-resistant electron transport is only a fraction of the initial activity (fig.2 a). If the initial uncoupled rate is considered as a control for the
trypsin-uncoupled DCMU-resistant rate, the latter is of the order of 10% of the former
(see also fig.4). This fact was not apparent in Renger's experiments (6), because his
measurements of O_2 yield per flash were done with very long dark time (250 msec) between single turnover flashes, a condition which would not detect a substancial inhibition

of a rate-limiting dark reaction. When water was the electron donor, DCPIP reduction reached complete inhibition and no DCMU resistance was developed (fig. 2 c). On the contrary, if NH_2OH is used as the electron donor, the time-course of the development of DCMU resistance of DCPIP reduction was very similar to that observed with ferricyanide (fig.2 b). The failure to observe this with water as the donor could be attributed to the fact that $DCPIPH_2$ can be reoxidized by PS II (10), thereby establishing a cycle around PS II. NH_2OH prevented this by inhibiting the reoxidation of $DCPIPH_2$ If this is true, any other method of preventing $DCPIPH_2$ electron donation to the oxidizing side of PS II should reveal the development of DCMU-resistant DCPIP reduction.

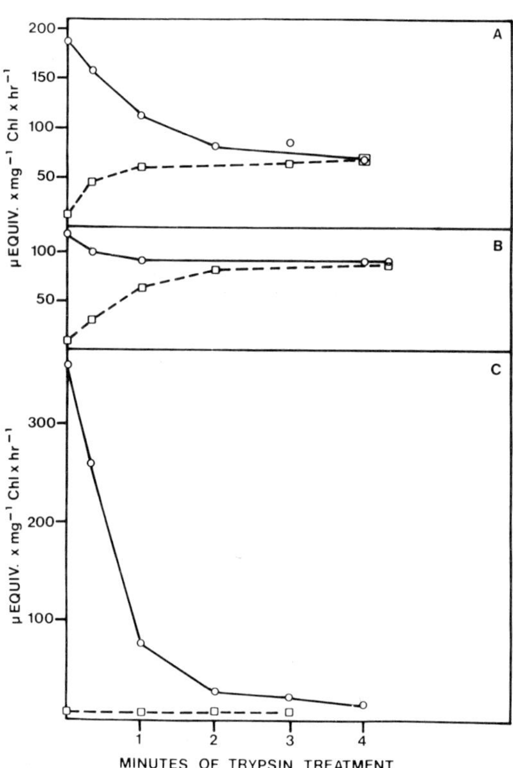

Fig.2   Inhibition and DCMU-resistence of electron transport by trypsin treatment
o———o control; □———□+ DCMU 5 µM. Conditions as in "Methods"; buffer at pH 7.8; trypsin 10 µg/ml. a) ferricyanide reduction; b) DCPIP reduction, NH_2OH added, 20 mM; c) DCPIP reduction.

The addition of cytochrome c to reoxidize DCPIPH$_2$ fulfilled this expectation. Table 2 shows that the time-course of the development of DCMU-resistant cytochrome c reduction (in the presence of DCPIP) upon trypsin treatment was no different from what was observed with ferricyanide as acceptor (compare with fig. 4).

Whether or not trypsin treatment affects the oxidizing side of PS II is a controversial point (2, 6). We have investigated the effect on H$_2$O oxidation as compared to NH$_2$OH oxidation at different pH values.

TABLE 2

EFFECT OF TRYPSIN ON DICHLOROINDOPHENOL REDUCTION

Time of Trypsin digestion	μequiv. hr^{-1} mg^{-1} Chl Control	DCMU 5 μM
0	544	7.5
30"	391	8.7
1'	355.2	14.7
2'	283.7	20.2
8'	67.9	37.3
15'	46.6	34.2

Conditions : tricine-NaOH 30 mM, pH 7.0; NaCl 10 mM; DCPIP 30 μM; gramicidin 5 μM Cyt c (oxidized) 0.1 mM; chloroplast containing 20 μg of chl. Trypsin : 10 μg. ml^{-1}. The reduction of cyt C was measured at 550 nm.

Fig. 3 shows that the inhibition by trypsin of both reactions was considerably lower at pH 7 than at pH 8, and that in both conditions the oxidation of NH$_2$OH was considerably more resistant than water oxidation. In the case of the experiment done at pH 8, it could be argued that the larger inhibition of water splitting could be attributed to the uncoupling by trypsin which would bring the pH in the internal thylakoid space close to the value of the external medium; it was shown that high internal pH specifically inhibits H$_2$O splitting (11), though this requires pH values higher than 8. However, this reasoning does not apply to the observation made at pH 7, a condition where very little uncoupling is observed (see fig. 3, insert), and anyway the pH is low enough to rule out any pH effect on water-splitting. We conclude therefore that the water-splitting reaction is a direct target of trypsin action. This is in agreement with Selman and Bannister (2).

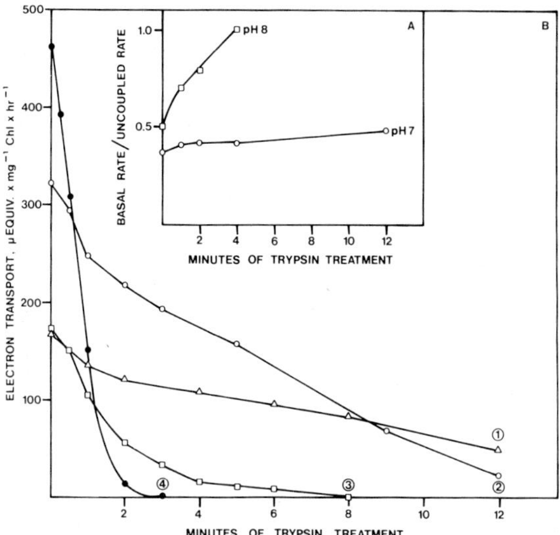

Fig. 3 Inhibition of electron transport and uncoupling by trypsin at different pH's.
Conditions as in "Methods"; gramicidin 5 μM and methylviologen 0.5 mM. Trypsin :
10 μg . ml^{-1} . Curve ①: NH_2OH 20 mM, buffer pH 7. Curve ②: buffer pH 7 ;
curve ③:NH_2OH 20 mM, buffer pH 8; curve ④: buffer pH 8.
Insert: PS I electron transport (see "Methods"); trypsin 10 μg/ml. Uncoupled rate was
measured in the presence of gramicidin, 5 μM.

A clear indication that at least two different sites of trypsin action are involved in
the inhibition of PS II electron transport was also provided by the observation that the
inhibition of ferricyanide reduction from water (in the presence of uncoupler)appeared
before any development of DCMU resistance at pH 7 (see fig.4). At pH 8 this was not
observed, because of the rapid onset of DCMU resistance at this pH (fig. 4).

CONCLUSIONS

The observations presented here demonstrated that trypsin inhibition of PS II-dependent
electron transport is the result of the action of the enzyme on at least two different
sites, besides its action on the coupling factor, which leads to uncoupling. One of them
is on the reducing side of PS II, and is responsible for the development of DCMU-resistant electron transport to ferricyanide (6, 8) and to 2-4,dichlorophenolindophenol(fig.1).
The inhibition at this site causes the interruption of electron flow from PS II to PS I (6).

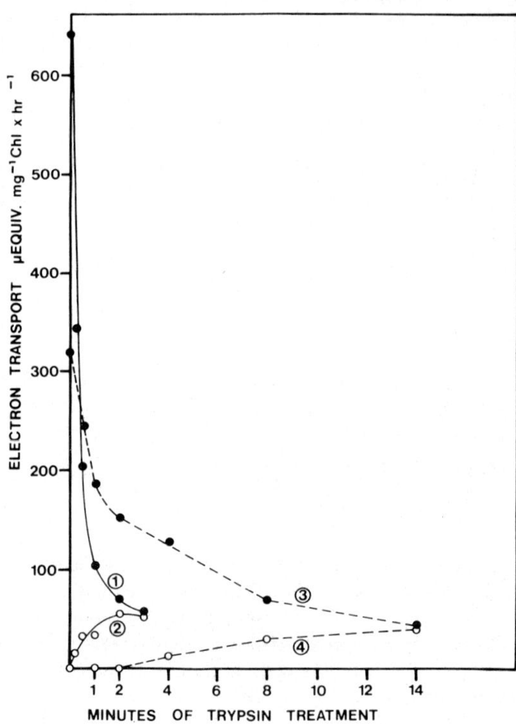

Fig. 4 Electron transport inhibition and DCMU-resistance at different pH's o————— o control; o —————o + DCMU, 5 µM. Conditions as in "Methods"; trypsin 10 µg/ml. Curves ① and ② :buffer at pH 8; curves ③ and ④: buffer at pH 7.

Another site of trypsin inhibition is located on the oxidizing side of PS II. It is indicated by the higher sensitivity of H_2O oxidation as compared to the oxidation of NH_2OH a compound known to donate electrons to PS II by-passing the site sensitive to Cl-removal (12). The inhibition of NH_2OH oxidation itself may be explained by the same action of trypsin, on the reducing side of PS II, leading to the DCMU resistance; however, the existence of another site of trypsin attack in the span from the hydroxylamine donation to PS II cannot be excluded.

The evidence now available that trypsin inactivates directly the water-splitting reaction on the thylakoid membranes indicates that this reaction involves protein(s) located on the outer surface of the membrane. This conclusion was also reached by Schmid et al.

on the basis of inhibition by the antibody against a thylakoid polypeptide (13).

Our previous observations on the suppression of Mg^{++} stimulated fluorescence by trypsin (7) have been extended now by the finding that this is the thylakoid membrane function most sensitive to trypsin attack (table 1). This indicates that a specific membrane protein, located on the outer surface of the membrane, is required for the binding of Mg^{++}, and the other divalent cations having the same effect on fluorescence.

ACKNOWLEDGMENT

This work was supported by contract No 030-76 ESI from the Commission of European Communities.

REFERENCES

1. Vambutas, V.K. and Racker, E. (1965) J. Biol.Chem. 240, 2660.

2. Mantai, K.E. (1970) Plant Physiol. 45, 563.

3. Selman, B.R. and Bannister, T.T. (1971) Biochim.Biophys. Acta 253, 428.

4. Selman, B.R., Bannister, T.T. and Dilley, R.A. (1973) Biochim. Biophys.Acta 292, 566.

5. Renger, G., Erixon, K., Döring,G. and Wolff, Ch. (1976) Biochim. Biophys. Acta 440, 278.

6. Renger, G. (1976) Biochim.Biophys. Acta 440, 287.

7. Jennings, R.C. and Forti, G. (1974) Biochim.Biophys.Acta 347, 299.

8. Regitz, G. and Ohad, I. (1974) In "Proc. III Int. Congress on Photosyn." n° 3, 1615. M. Avron Ed.- Elsevier Scient.Pub.Comp., Amsterdam, The Netherlands.

9. Forti,G. and Rosa, L. (1971) FEBS Letters 18, 55.

10. Kok, B., Malkin,S., Owens,O. and Forbush, B. (1966) In "Energy Conversion in the Photosynthetic Apparatus", Brookhaven Symposia in Biology, n° 19, 446.

11. Harth, E., Reimer,S. and Trebst, A. (1974) FEBS Letters 42, 165.

12. Izawa,S., Heath, R.L. and Hind, G. (1969) Biochim.Biophys. Acta 180, 388.

13. Schmid, G.H., Menke,W., Koenig ,F. and Radunz,A. (1976) Z. Naturforsch 31c,304.

ELECTRON TRANSPORT, COUPLING AND PHOSPHORYLATION

Bioenergetics of Membranes. L. Packer et al. ed.

INTRODUCTION

PATHWAYS OF ELECTRON TRANSPORT AND ENERGY COUPLING

Lars Ernster

Department of Biochemistry, Arrhenius Laboratory,
University of Stockholm, Stockholm, Sweden

I wish to thank the Organizers for kindly inviting me to introduce this session, the topic of which undoubtedly occupies a central position in Membrane Bioenergetics, the subject matter of this Symposium.

Although it may appear as a truism, I feel it appropriate to begin my introduction by presenting a schematic picture of the four leading hypotheses of electron transport-linked phosphorylation that have been proposed over the last three decades (Fig. 1). These are: the "chemical" hypothesis, involving covalent high-energy intermediates between the electron-transport and ATP-synthesizing systems; the "conformational" hypothesis, postulating a direct interaction between conformationally modified energy-transducing catalysts; a hypothesis involving intramembrane proton fluxes as a means of energy transduction; and the "chemiosmotic" hypothesis involving transmembrane proton gradients as the basic energy-transfer device.

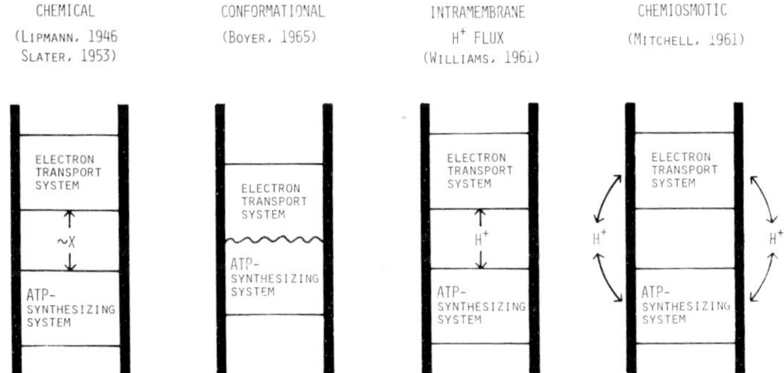

Fig. 1. Proposed mechanisms of energy transfer between membrane-associated electron-transport and ATP-synthesizing systems.

I think most workers in this field now agree that among the four hypotheses the chemiosmotic hypothesis has received most experimental support. It is generally accepted that 1. Membrane-associated energy-transducing catalysts — both electron-transport and ATP-synthesizing systems — are capable of reversibly generating a proton gradient across the membrane in which they are located. 2. Energy-transducing units located in the same membrane can interact in transferring energy through this gradient. 3. Agents that abolish the proton gradient uncouple this interaction.

Although the available evidence strongly supports the chemiosmotic hypothesis, the fundamental problem of the chemical mechanisms by which electron-transport and

ATP-synthesizing catalysts can generate and utilize a transmembrane proton gradient remains to be solved. There are in particular three questions, in my opinion, that will have to be answered before the chemiosmotic hypothesis can be wholeheartedly accepted and, in general, before the mechanism of electron transport-linked phosphorylation can be regarded as understood in a chemically satisfactory fashion: 1. What are the chemical mechanisms by which membrane-associated energy-transducing catalysts generate and utilize a transmembrane proton gradient? 2. Is the generation of a proton gradient by various energy-transducing catalysts the primary event of energy conservation or is it preceded by chemical, e.g. conformational, changes that in turn can give rise to a proton gradient? 3. Is interaction through transmembrane proton gradients the only way of transferring energy between different energy-transducing catalysts located in the same membrane, or are there instances of localized interactions between them within the membrane? I am sure that these questions will be touched upon by several speakers in this programme. Let me, however, first briefly summarize my own views on these questions.

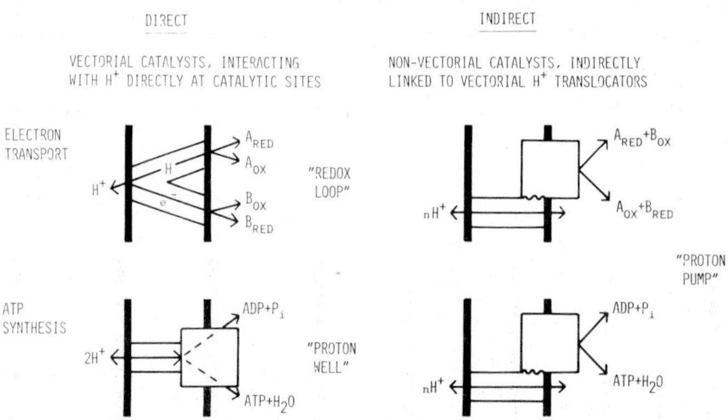

Fig. 2. Mechanisms of generation of transmembrane proton gradients by membrane-associated electron-transport and ATP-synthesizing systems.

Regarding the first question, one may consider two types of mechanisms (Fig. 2): a "direct" mechanism, involving vectorial catalysts interacting with protons directly at their catalytic sites; and an "indirect" mechanism, involving nonvectorial (or not necessarily vectorial) catalysts indirectly linked to vectorial proton translocators. According to Mitchell — and his views are shared by a good number of workers in the field — both the electron-transport and the ATP-synthesizing systems function as "direct" proton translocators, the former consisting of "redox loops" built up by alternating transmembrane hydrogen and electron carriers, and the latter consisting of an ATPase facing with its catalytic site a "proton well" across the membrane. Other investigators (including myself) favour the "indirect" type of mechanism, involving what may be termed "proton pumps", partly because of difficulties in accepting some of Mitchell's postulates, and partly because of growing experimental evidence supporting the occurrence of such pumps. To take the difficulties first, there seems to be no convincing evidence for the existence of any transmembrane hydrogen or electron carriers at coupling sites 1 and 2 of the respiratory chain, or of any transmembrane hydrogen carrier at coupling site 3. Similarly, the nicotinamide nucleotide transhydrogenase lacks transmembrane hydrogen or electron carriers. Moreover, repeated estimates indicate H^+/e^- stoichiometries significantly exceeding 1, which is incompatible with the proposed redox loops. Likewise, the ATPase model of Mitchell has been questioned on the basis of both the postulated reaction mechanism and estimates of H^+/ATP stoichiometries.

The concept of "proton pump" as depicted in Fig. 2 involves a conformational interaction between a proton translocator and the catalytic site of an energy-transducing catalyst, e.g. an electron-transport enzyme or an ATPase. The proton translocator and the catalytic site may reside in the same polypeptide (such as in bacteriorhodopsin) or in different polypeptides (such as in the proton-translocating F_0F_1-ATPase). There is now evidence for proton pumps associated with the cytochrome b-c_1 complex and with cytochrome c oxidase, and the operation of the latter seems to be geared to a spectral shift indicative of a conformational change of the oxidase. There is also kinetic evidence for an energy-linked conformational change of transhydrogenase coupled to a proton pump. In the case of ATPase there are several lines of evidence for energy-linked conformational changes, which may well be linked to proton translocation, just as the Na^+K^+- and Ca^{2+}-ATPases function as ion pumps.

We are now ready to consider my second question: Is proton translocation the primary event of energy conservation or is it preceded by an energy-conserving chemical – e.g. conformational – change of the catalyst involved? Clearly, once we accept the existence of an "indirect" type of proton pumps driven by conformationally energized catalysts, we automatically also accept the occurrence of energy-linked conformational changes as the primary events of energy conservation. The difficulty here seems to be not so much the acceptance of this concept but rather the definition and actual demonstration of an energized conformational state. Can bacteriorhodopsin be regarded to be in an "energized" conformational state between the first spectral shift following a light flash and the release of the first proton? Does the F_1-ATPase occur in an "energized" and a "de-energized" conformational state? How do these states differ chemically? How large a part of the cytochrome c oxidase molecule is involved in the conformational change indicated by the energy-dependent spectral shift? Obviously, we need more knowledge of the structure and dynamics of the energy-transducing catalysts and their membrane environments before we can answer these questions.

Finally: Are there instances of localized interactions between energy-transducing catalysts in the membrane? The answer is that we do have indications for such interactions, e.g. between ATPase and transhydrogenase, but the evidence is not yet conclusive. To further substantiate this concept it would be desirable to study these reactions in a nonvesicular membrane or a nonmembranous system. Promising approaches in this direction are being taken but, again, we need more knowledge of the catalysts involved before definite results can be expected. Meanwhile, it appears to be wise to leave the possibility open that, in addition to chemiosmotic coupling, there may occur localized interactions between energy-transducing catalysts. These may proceed by way of direct conformational interactions, intramembrane proton fluxes, and - who knows - even by covalent high-energy intermediates. The scheme in Fig. 3 (right), unifying the four hypotheses referred to earlier in my introduction, does not mean to say that all four mechanisms are likely to be operating; what it means to say is that they are not mutually exclusive and that elements of them may well coexist in the final picture.

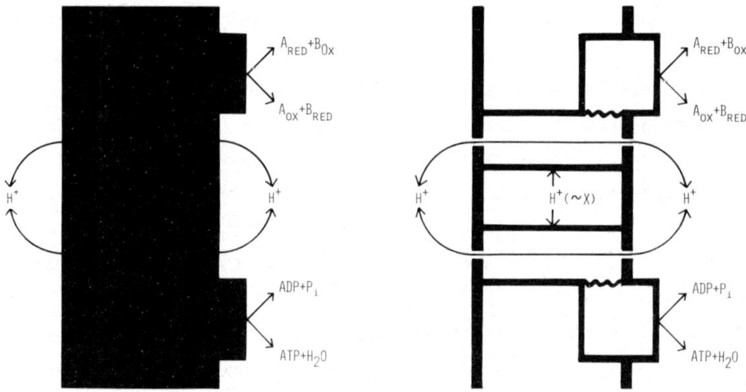

Fig. 3. A "black-and-white" aspect of membrane bioenergetics.

In his Ciba Medal Lecture that appeared about a year ago Peter Mitchell classified bioenergeticists denoting them with black and white bars according to their presumed positive or negative attitude toward the chemiosmotic rationale. I think we now should point out another "black-and-white" aspect of the problem (Fig. 3). Over the last decade many laboratories have concentrated their efforts on studying what is happening _across_ the membrane. This is understandable, since the chemiosmotic hypothesis was new and it meant a real breakthrough in membrane bioenergetics. Today we all believe in the feasibility and importance of chemiosmosis. What remains to be done is to explore the inside of the black box, to understand what is going on _within_ the membrane. Not until this is accomplished can the problem of the mechanism of electron transport and energy coupling be regarded as solved in a satisfactory fashion.

REFERENCE

Boyer, P.D., Chance, B., Ernster, L., Mitchell, P., Racker, E., and Slater, E.C. (1977). Ann. Rev. Biochem. 46, 955-1026.

Bioenergetics of Membranes. L. Packer et al. ed.

PROTON TRANSFER REACTIONS ASSOCIATED TO OXIDO-REDUCTIONS IN THE CYTOCHROME SYSTEM OF MITOCHONDRIA

S. Papa, M. Lorusso, F. Guerrieri, D. Boffoli, G. Izzo and F. Capuano
Institute of Biological Chemistry, Faculty of Medicine and Centre
for the Study of Mitochondria and Energy Metabolism,
C.N.R., University of Bari, Italy.

INTRODUCTION

The cytochrome system of mitochondria transfers reducing equivalents from dehydrogenases to oxygen and stores the energy thereby made available in a utilizable form. It is composed of 9 one-electron
functional redox centers and some fifteen polypeptides (see ref.1 for
review). These components are organized in two lipoprotein polymers,
the b-c_1 segment or complex III and cytochrome oxidase or complex IV,
connected by cytochrome c.

The cytochrome system,as a whole,and the two complexes, when present in the native membrane or inlaid in phospholipid vesicles, directly and compulsorily convert redox energy into a transmembrane thermodynamic potential difference of protons,which can be used to drive
ATP synthesis by the ATPase complex (see refs.1,2 for review).

Two general models are available to explain the redox proton pump.
The protonmotive redox-loop mechanism of Mitchell[2], which is based on
physical asimmetry and specific mobility in the membrane of the prosthetic groups of redox catalysts. The vectorial Bohr mechanism[1],
which visualizes the proton pump as due to conformational linkage
between the redox state of prosthetic groups and protolytic equilibria in the apoproteins.

According to the linear redox-loop model of Mitchell[2] the redox
centers of the cytochrome system were proposed to be arranged in the
inner mitochondrial membrane so to mediate, twice, vectorial electron
flow from the external or C side to the inner or M side: the first time by two b cytochromes, the second by the four redox centers of cytochrome oxidase. The model required, in addition, that hydrogens were
twice transferred from the M to the C side of the membrane.

Ubiquinone was originally proposed to be the last hydrogen-carrying arm in the cytochrome system[3,4]. If this were the case the aerobic oxidation of quinol had to be accompanied by the outward translocation of 1 H^+ per e^-. Measurements in our[5,6] and other laboratories[7] have, however, shown that at neutral pH's this H^+/e^- ratio approximates 2. Thus the quinone has to be involved in the first hydrogen carrying arm of the cytochrome system.

In Table I the H^+/e^- ratio for proton release from mitochondria associated to rapid electron flow from exogenous quinol to oxygen and from quinol to cytochrome c is shown. The H^+/e^- ratio was at pH 7.2, 1.8 in both cases. This ratio for proton release is comparable to that of 1.9 measured at pH 6.8 for proton uptake in sonic particles associated to oxidation of endogenous ubiquinol[6]. It should also be recalled that in reconstituted complex III phospholipid vesicles electron flow from duroquinol to cytochrome c exhibited proton release with an H^+/e^- ratio of about 2[8,9].

Table I. H^+/e^- RATIOS FOR ELECTRON FLOW FROM DUROQUINOL TO OXYGEN OR TO CYTOCHROME c IN BEEF HEART MITOCHONDRIA.

H^+ release was measured with a continuous-flow pH-meter (mixing ratio 1:60). For reductant pulses duroquinol (200 µM) oxidation was calculated by measuring oxygen consumption with a Clark electrode. The reaction mixture contained: 150 mM KCl, 5 mM K-malonate, 2 mM Glycilglycine, 0.2 mM N-ethylmaleimide, 0.5 µg/mg protein rotenone, 1 µg/mg protein oligomycin, 0.5 µg/mg protein valinomycin and beef-heart mitochondria (2 mg protein/ml). Final pH, 7.2. Temp. 25°C. For oxidant pulses, cytochrome c reduction was measured at 550-540 nm with a stopped-flow apparatus (mixing ratio 1:68). The reaction mixture contained: 150 mM KCl, 5 mM K-malonate, 1 mM KCN, 2 µg/mg protein oligomycin, 0.5 µg/mg protein rotenone, 0.5 µg/mg protein valinomycin,100 µM duroquinol and beef-heart mitochondria (1 mg protein/ml). After 2 min mitochondria were pulsed with 8 µM cytochrome c. Final pH 7.2. Temp. 25°C. H^+ release and electron flow were corrected for antimycin insensitive reactions.

	REDUCTANT PULSES $DQH_2 \longrightarrow 0$			OXIDANT PULSES $DQH_2 \longrightarrow Cyt.c$		
No.Expt.s	H^+/e^- (180 msec)	S.E.M.		No.Expt.s	H^+/e^- (45 msec)	S.E.M.
28	1.76	±0.12		12	1.79	±0.14

The fact that the same H^+/e^- stoichiometry for proton release from mitochondria is observed when electrons flow from quinols to oxygen or to cytochrome c, shows that the proton pumping activity in the cytochrome system is confined to the $b-c_1$ complex and is thus consistent with the view that cytochrome oxidase catalyzes vectorial electron flow from the C to the M side of the membrane (see refs.1-5, contrast however ref. 10).

The minimum functional unit of complex III contains one mole of cytochrome c_1,two moles of b cytochromes(b_{566} and b_{562}), one mole of Fe-S protein (g=1.90) and traces of ubiquinone[11,12]. There are additional protein components: the antimycin binding protein, a protein required for electron flow, and two core proteins[11]. Which of these components could be involved in the proton pump? A pure electron carrier cannot. However hemoproteins or metalloproteins can act as effective hydrogen carriers and be in this or in other way directly involved in proton transport if the redox state of the metal is linked to protolytic equilibria in the apoprotein[1,13]. This linkage does indeed exist in b cytochromes[14,15] and the Fe-S protein[16] of complex III as shown by the fact that their midpoint redox potential is pH dependent. The pK of the group linked to the redox state of the metal is, in the oxidized state, 6.8 for b cytochromes and 7.9 for the Fe-S protein. These pK's are shifted in the reduced state to values higher than 8.5. As the result of coupling of proton transfer to electron transfer b cytochromes and the Fe-S protein can switch from electron to effective hydrogen carriers. From the relative pK's it can be calculated (see Fig.1) the extent to which b cytochromes and the Fe-S protein change to hydrogen carriers as a function of pH. In the physiological pH range these components act both as electron and hydrogen carriers.

Fig.1 shows also that whilst the quinol, with the two pK's around 12^{17}, functions as an hydrogen carrier up to pH 11, the semiquinone, with a pK of 6^{17} changes,going from pH 5 to 8, from an hydrogen to an electron carrier. These properties of the redox carriers of complex III can be used to design experiments to test their role in active proton transport and should be taken into account by any hypothesis brought forward to explain the redox proton pump.

In this respect we have studied the pH dependence of crossover effects exerted on electron flow in the b-c segment of the respiratory

380

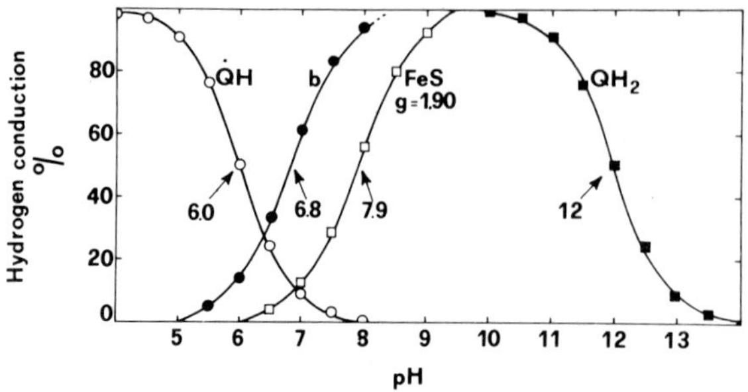

Fig.1. - Percentage of hydrogen conduction by redox carriers in the Q-b region (pK's indicated by arrows).

chain by the electrical and chemical component of the protonmotive force[18] and the pH dependence of the H^+/e^- stoichiometry of the redox proton pump of this segment[19].

Fig.2 illustrates the pH dependence of the effect of valinomycin on the steady-state redox level of b and c cytochromes in respiring mitochondria and submitochondrial particles. Fig.2-A shows that in mitochondria valinomycin caused, in the pH range 6.2 to 7.6 a large oxidation of b_{566}. Cytochrome b_{562} was also oxidized but to a much lower extent. Valinomycin caused in sonic particles a similar pattern of oxidation of b cytochromes, the effect was however considerably smaller.

These results provide evidence that collapse of membrane potential exerts a crossover effect between b_{566} and b_{562}. First it has to be noted that the valinomycin induced oxidation of b_{566} is much larger in mitochondria than in sonic particles. Measurements of the relative contribution of $\Delta\psi$ and ΔpH to the protonmotive force in state 4 have in fact revealed that whilst in mitochondria $\Delta\psi$ represents more than 2 third of the total force, in sonic particles it is the ΔpH component to predominate[21]. Second, the pH profile shows that the oxidation of b_{566} caused by collapse of the membrane potential is particularly large at lower pH's but decreases as the pH is raised. At the lower pH's b cytochromes act in part as electron carriers. As the pH is brought to more alkaline values the redox-linked protonation-deprotonation reactions of b cytochromes become more important, electron flow is progressively replaced by hydrogen flow and the electric-field

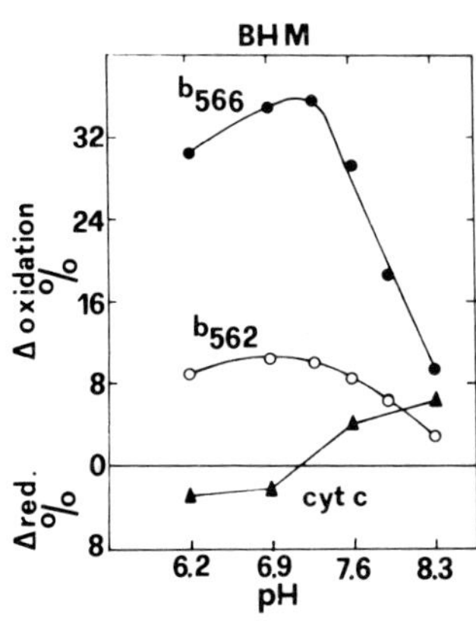

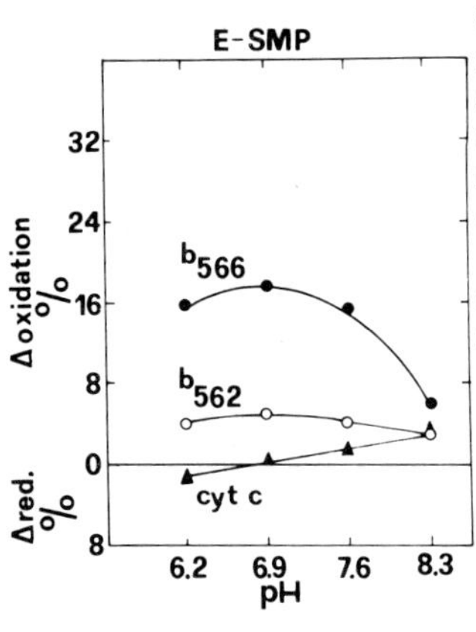

Fig.2. - pH dependence of the effect of valinomycin on the steady-state redox level of b and c cytochromes in respiring mitochondria and submitochondrial particles.

The redox transitions of b_{566} and b_{562} were calculated from the absorbance changes measured at 566/575 nm and 562/575 on the basis of the spectral contribution of the two b cytochromes at these wavelength couples[20]. The reaction mixture contained: 200 mM sucrose, 20 mM KCl, 10 mM K-succinate, 1 µg/mg protein rotenone, 2 µg/mg protein oligomycin, 0.2 µg/mg protein purified catalase and EDTA-particles or mitochondria (2 mg protein/ml). Respiration was activated by adding 5-15 µl 2% H_2O_2 to anaerobic suspension. Valinomycin was added at the concentration of 0.4 µg/mg protein to steady-state respiring particles or mitochondria. Final volume 1.5 ml. Temperature 20°C.

control of the flow of reducing equivalents disappears. These observations provide evidence that reducing equivalents are transferred vectorially from cytochrome b_{566} at the M side to b_{562} at the C side of the membrane (see Fig.5). It can however be noted that the pH dependence of the valinomycin-induced oxidation of b_{566} cannot be solely explained on these basis. The b cytochromes progressively change to effective hydrogen carriers already in the pH region between 6.2 and 7 (see Fig.1). If valinomycin acted simply by promoting electron transfer from b_{566} to b_{562}, the oxidation of b_{566} should start to decrease already going with the pH from 6.2 to 7.0. On the contrary it increased in this pH range. Evidently valinomycin in some way is also depressing cytochrome b_{566} reduction. An explanation for this seems to be offered by the pH profile of the ionization state of the semiquinone (see Fig.1).

Various experimental observations suggest that the quinol donates its electrons to two different one electron carriers of the b-c_1 complex[20,22,23]. Of the various possible mechanism it seems feasible

that the quinol-semiquinol couple, which is likely to have a relatively high midpoint redox potential[23], transfers electrons to a redox
center on the oxygen side of the antimycin inhibition site[20](Fig.5).
The semiquinone so formed, with a more negative redox potential, will
transfer reducing equivalents to cytochrome b_{566}. The membrane potential, positive outside, will favour movement of the semiquinone anion
from the hydrophobic core to the heme of b_{566}, at the C side, and its
consequent reduction (see Fig.5). This positive effect will be suppressed when $\Delta\psi$ is dissipated by valinomycin. These effects will be
more evident as the percentage of the radical anion increases with
the pH.

Fig.3 shows that the H^+/e^- ratio for proton release from mitochondria associated to electron flow from duroquinol to cytochrome c was 2
at pH 6.2 decreased to about 1.5 at pH 8.1 but then sharply increased
above 4 as the pH was raised to 9[19].

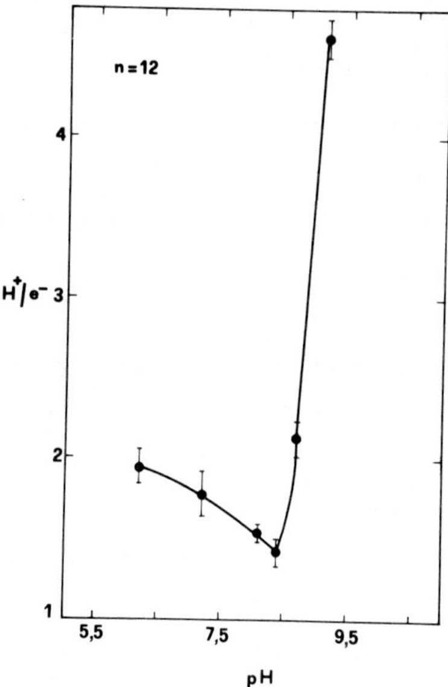

Fig.3. - pH dependence of the H^+/e^- ratio for proton release associated to reduction of exogenous cytochrome c by duroquinol in KCN treated beef-heart mitochondria.

H^+/e^- ratios were obtained at interval of 45 msec and represent the mean of 12 experiments. Vertical bars: S.E.M. Temperature 25°C. For other experimental details see legend to Table I.

The decrease of the H^+/e^- ratio observed going with pH from 6.2 to
8.0, can be explained by inward vectorial hydrogen transfer by b cytochromes. The increase of H^+/e^- ratios to values higher than 2 observed above pH 8.1 is likely to reflect the pH dependence of the proton

conducting component of the pump. In particular it would suggest the existence in complex III of multiple ionizable groups whose pK's are linked to the redox state of a respiratory carrier.

A more precise location of the component(s) involved in the proton pump was provided by the finding that aerobic oxidation of the redox carriers on the oxygen side of the antimycin inhibition site was accompanied by synchronous proton release from mitochondria[24].

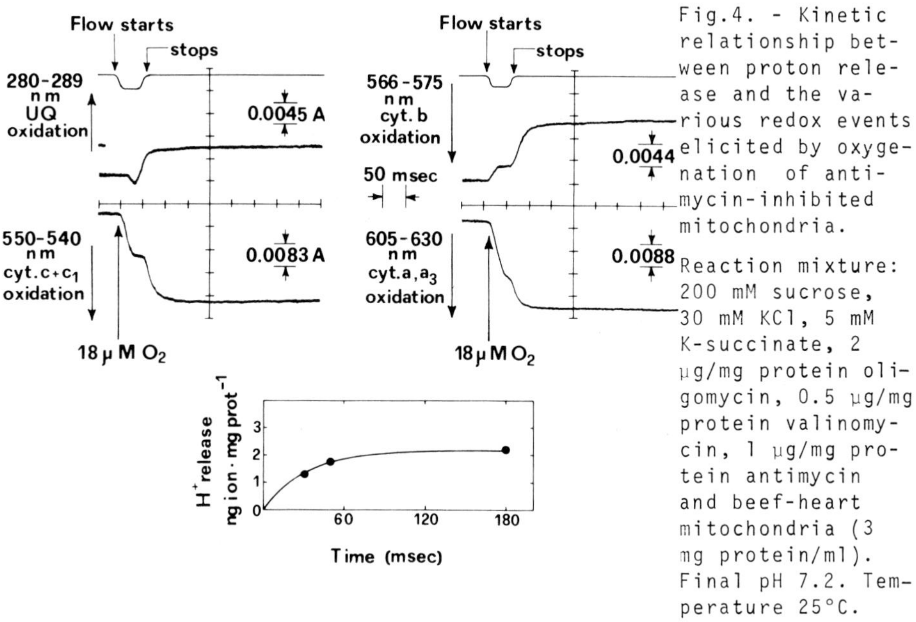

Fig.4. - Kinetic relationship between proton release and the various redox events elicited by oxygenation of antimycin-inhibited mitochondria.

Reaction mixture: 200 mM sucrose, 30 mM KCl, 5 mM K-succinate, 2 μg/mg protein oligomycin, 0.5 μg/mg protein valinomycin, 1 μg/mg protein antimycin and beef-heart mitochondria (3 mg protein/ml). Final pH 7.2. Temperature 25°C.

Fig.4 shows that aerobic oxidation of cytochrome oxidase and cytochrome c was accompanied, in the presence of antimycin, by oxidation of ubiquinone and reduction of b cytochromes. These redox events were accompanied by a fast proton release, synchronous to oxidation of c cytochromes and ubiquinone. The antimycin insensitive oxidation of ubiquinol wasn't the only cause of the proton release. In fact more than 2 gion H^+ were released per mol ubiquinol oxidized. Thus there exists on the oxygen side of the antimycin inhibition site a component of the cytochrome system which is directly involved in the proton pump.

The H^+/e^- stoichiometries measured in the mitochondrial membrane and in complex III and complex IV phospholipid vesicles locate such a component in the b-c_1 complex[1,6,8]. On the other hand Wikstrom has proposed the existence of an analogous component in the cytochrome oxidase complex[10]. This would mediate net transport across the membrane of 1 H^+ per e^- transferred from cytochrome c to oxygen. These possibilities were verified by measuring the H^+/e^- stoichiometry for proton release from mitochondria associated to antimycin insensitive aerobic oxidation of terminal carriers (Table II).

Table II. - STOICHIOMETRY OF H^+ RELEASE AND OXIDATION OF REDOX CAR-RIERS CAUSED BY OXYGEN PULSES OF ANAEROBIC ANTIMYCIN-INHIBITED BEEF-HEART MITOCHONDRIA.

For Experimental details and additions see legend to Fig.4.

		Control	+NEM 30 nmoles mg prot.
		30 msec.	
H^+ release (ngions)		1.22	1.31
Respiratory carrier oxidation (nmoles)	a,a_3	1.07	0.98
	Cu	1.07	0.98
	c,c_1	0.68	0.64
	Fe-S (g=1.90)	0.34	0.32
	$\frac{1}{2}$ QH_2	0.23	0.19
	Total e^- flow	3.39	3.11
H^+/Total e^- flow		0.36	0.42
(Fe-S + $\frac{1}{2}$ QH_2)		0.57	0.51
H^+/(Fe-S + $\frac{1}{2}$ QH_2)		2.14	2.57

If cytochrome oxidase acted as a proton pump the ratio of H^+ release to the total amount of electron transfer to oxygen from all the components oxidized in the presence of antimycin should be equal to 1.

The total amount of electron flow was computed by summing up the moles of cytochrome a, a_3 oxidized and an equivalent number of copper atoms, the moles of c cytochromes, the moles of the Fe-S protein of complex III oxidized, this is taken as half the amount of $c+c_1$, and half the moles of quinol oxidized, which correspond to the moles of b cytochromes reduced. It can be seen (Table II) that the H^+/e^- ratio so measured amounted, both in the absence and presence of NEM,to 0.4. On the other hand the H^+/e^- ratio calculated by dividing the amount of H^+ released by the sum of the Fe-S protein (or c_1) and half the moles of quinol oxidized was around 2.

These observations favour the view that complex III but not complex IV is capable of pumping protons. Furthermore they show that the site of proton pumping is located between the antimycin binding site and cytochrome c.

Further informations were obtained by testing the effect of 2-heptyl-4-hydroxy-quinoline-N-oxide (HQNO) on the antimycin-insensitive redox and proton transfer reactions (Table III).

Table III. - EFFECT OF HQNO ON THE STOICHIOMETRY OF H^+ RELEASE AND OXIDATION OF REDOX CARRIERS CAUSED BY OXYGEN PULSES OF ANAEROBIC ANTIMYCIN-INHIBITED BEEF-HEART MITOCHONDRIA.

For Experimental details and additions see legend to Fig. 4.

		Control	+HQNO 77 nmoles / mg prot.
		30 msec.	
H^+ release (ngions)		0.82	0.21
Respiratory carrier oxidation (nmoles)	a,a_3	0.55	0.58
	Cu	0.55	0.58
	c,c_1	0.33	0.35
	Fe-S (g=1.90)	0.16	0.17
	$\frac{1}{2}\ QH_2$	0.12	0.06
	Total e^- flow	1.71	1.74
H^+/Total e^- flow		0.48	0.12

As expected HQNO had no effect on the oxidation of c cytochromes and cytochrome oxidase, it however depressed oxygen induced cytochrome b reduction and, more markedly, the antimycin insensitive H^+ release. This resulted in a marked decrease of the H^+/e^- ratio, which calculated on the basis of the total electron flow to oxygen amounted only to 0.1.

Thus the proton pumping activity of the cytochrome system is linked to redox events in complex III which are insensitive to antimycin but are inhibited by HQNO.

DISCUSSION.

The fact that two protons are translocated across the mitochondrial membrane per 1 electron flowing from quinols to oxygen has been explained by Mitchell with a protonmotive Q cycle[23]. This cycle can explain the decrease of the H^+/e^- ratio that we have observed going with the pH from 6.2 to 8.1. However it cannot explain the sharp subsequent rise of this ratio to values higher than 2, observed when the pH was increased to more alkaline pH's[19] (see Fig. 3).

Kröger has proposed another scheme which also accounts for the translocation of 2 protons across the membrane per 1 electron transferred through the Q region of the chain[25]. His mechanism proposes that quinone radicals are formed in the reduction of Q ·by the dehydrogenases and oxidation of QH_2 by an electron carrier of the $b-c_1$ complex. The radicals so formed rapidly dismutate. Transmembrane proton translocation would be the result of the fact that at neutral pH's the radical is mostly present as deprotonated anion. Since the percentage of the semiquinone radical existing as anion increases sharply as the pH is raised from 6 to 8, it can be calculated from the ionization curve of the semiquinone (Fig.1) that, according to this mechanism, the H^+/e^- ratio should raise, going from pH 6 to 8, from 1.5 to 2. This however is just the opposite of what is experimentally observed (Fig. 3).

It is therefore apparent that the proton-pumping activity of the $b-c_1$ complex cannot be accounted for by the redox reactions of quinones, neither as they are visualized by the Mitchell Q cycle, nor as proposed by Kröger.

What presented is on the other hand consistent with the vectorial Bohr mechanism that we have proposed for the proton pumping activity

of the b-c$_1$ complex[1,13,19]. According to this mechanism (see Fig.5)
conformational changes, induced in the protein by electron arrival to
and removal from the metal of a redox carrier, result in the opening
of a channel which conducts H$^+$ from the M to the C side of the membrane. It can be conceived that reduction of the electron-accepting
center(s) causes the pK values of the aminoacid residues lined along
the channel to shift to higher values going from the M to the C side
and that oxidation is accompanied by decreases of the pK of the acidic group(s) exposed at the C surface.

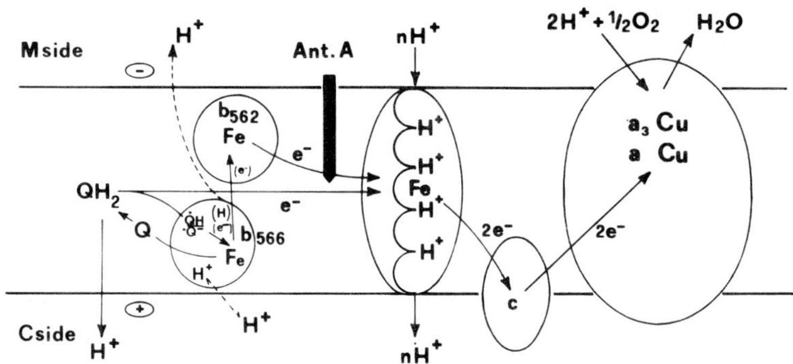

Fig. 5.- Model for the protonmotive function of the cytochrome system.

From a more general point of view it can be noted that this scheme
provides a model for the redox proton pump which considers stereochemical events in the apoproteins of electron carriers as directly involved in the coupling between redox processes and vectorial proton translocation. It is however the combination of these stereochemical events and linkage phenomena with the specific and asymmetric orientation of the redox centers associated to the large lipoproteins polymers of the cytochrome system that appears to endow the redox system
with the capacity of converting redox scalar energy into an osmotic
force.

REFERENCES

1. Papa, S.(1976) Biochim. Biophys. Acta 456, 39-84.

2. Mitchell, P. (1972) in Mitochondria/Biomembranes,(Van Den Bergh,
   S.G., Borst,P., Van Deenen, L.L.M., Reimersma,J.C., Slater,E.C.and
   Tager,J.M.eds.),Vol.28,pp.353-370,North Holland Elsevier,Amsterdam.

3. Mitchell,P.(1966) Chemiosmotic Coupling in Oxidative and Photosyn-

thetic Phosphorylation,Glynn Research Ltd.,Bodmin, Cornwall.

4. Mitchell,P.(1971) in Energy Transduction in Respiration and Photosynthesis (Quagliariello,E., Papa,S. and Rossi, C.S.,eds.),
   pp.123-149, Adriatica Editrice, Bari.

5. Papa, S., Guerrieri, F.and Lorusso, M.(1974) in Dynamics of Energy-Transducing Membranes (Ernster,L., Estabrook, R.W.and Slater,
   E.C.,eds.), pp.417-432,Elsevier Publishing Co., Amsterdam.

6. Papa, S., Lorusso, M. and Guerrieri, F.(1975) Biochim. Biophys.
   Acta 387, 425-440.

7. Lawford, H.G. and Garland, P.B.(1973) Biochem.J. 136, 711-720.

8. Hinkle, P. and Leung, K.H. (1974) in Membrane Proteins in Transport and Phosphorylation (Azzone,G.F., Klingenberg, M.E., Quagliariello, E. and Siliprandi, N.,eds.),North Holland Publishing Co.,
   Amsterdam, pp.73-78.

9. Guerrieri,F. and Nelson, B.D. (1975) FEBS Lett.54, 339-342.

10. Wikstrom, M.K.F. (1977) Nature 266, 271.

11. Rieske, J.S. (1976) Biochim. Biophys. Acta 456, 195-247.

12. Erecinska, M., Wilson, D.F. and Miyata,Y. (1977) Arch. Biochem.
    Biophys. 177, 133-143.

13. Papa, S., Guerrieri, F., Lorusso, M., Simone, S. (1975) Biochimie
    55, 703-716.

14. Straub,J.P. and Colpa Boonstra, J.P. (1962) Biochim. Biophys. Acta
    60, 650-652.

15. Urban, P.F. and Klingenberg, M. (1969) Eur.J.Biochem. 8, 519-525.

16. Prince, R.C. and Dutton, P.L. (1976) FEBS Lett. 65, 117-119.

17. Bridge, N.K. and Porter, G. (1958) Proc.R.Soc. 244, 276.

18. Papa,S., Lorusso,M., Guerrieri,F. and Izzo,G. (1975) in Electron
    Transfer Chains and Oxidative Phosphorylation (Quagliariello, E.,
    Papa,S., Palmieri, F., Slater, E.C., Siliprandi, N.,eds.), North
    Holland Publishing Co., Amsterdam, pp. 317-327.

19. Papa,S., Guerrieri,F., Lorusso,M., Izzo,G., Boffoli,D. and Capua-
    no,F. (1977) in Biochemistry of Membrane Transport (Semenza,G. and
    Carafoli,E.,eds.), Springer Verlag, Berlin,New York, pp. 502-519.

20. Wikstrom, M.K.F. (1973) Biochim. Biophys. Acta 301, 155-153.

21. Rottenberg, H. (1975) Bioenergetics 7, 61-74.

22. Baum, H., Rieske, J.S., Silman, H.I. and Lipton, S.H. (1976) Proc.
    Natl. Acad. Sci. U.S. 57, 798-805.

23. Mitchell, P. (1976) J. Theor. Biol. 62, 327-367.

24. Papa, S., Guerrieri, F. and Lorusso, M. (1974) Biochim. Biophys.
    Acta 357, 181-192.

25. Kröger, A. (1976) FEBS Lett. 65, 278-280.

TRANSMEMBRANE ELECTRON TRANSPORT AND
ENERGY CONSERVATION IN CHLOROPLASTS

Achim Trebst
Dept. of Biology, Ruhr University

Photosynthetic electron flow by chloroplasts is coupled to ATP
formation with a stoichiometry of about 1.3 ATP per 2 electrons
transported from water to $NADP^+$. The characterisation of a second
energy conserving site connected with photosystem II and the water
splitting reactions with P/e_2 ratio of 0.6 - 0.7 made clear that
two (native) energy conserving sites - with broken P/e_2 ratio each-
combine to yield the total P/e_2 ratio of 1 to 1.7 in non cyclic
photophosphorylation. Furthermore a concept of artificial energy
conservation has been developed to explain coupling of certain
sequences of the electron flow system which are not coupled to ATP
formation in the intact system, but are when complemented by artificial donor systems [1,2].

The notion of vectorial electron flow, field formation and proton translocation across the thylakoid membrane (across a functional
barrier not identical with the membrane width) and the built up of
a proton motive force has been able to explain the experimentel results on the coupling in photosynthesis[3]. However, the chemiosmotic

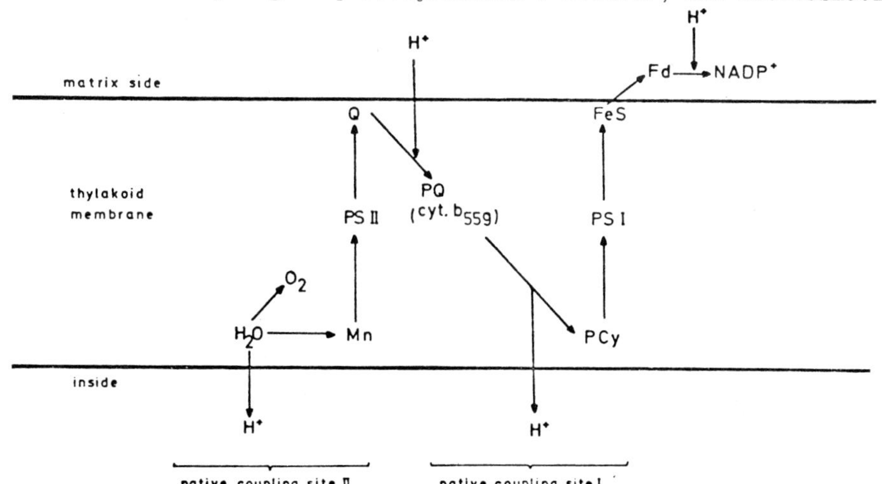

Vectorial electron flow across the thylakoid membrane and
two energy conserving sites connected with either photosystem

coupling mechanism is still not generally accepted. This is partly, because there now seems to be conflicting evidence as to the location in the thylakoid membrane and the accessibility of crucial electron carriers like plastocyanin. Support for vectorial electron flow, however, comes from the properties of artificial electron donor or acceptor systems. This is because the dependence of the coupling of the donor system to ATP formation on lipophilicity and a hydrogen carrying property of the actual mediator does suggest transmembrane electron flow connected on to either photosystem I or photosystem II. The effect of internal pH on donor systems supports an out/in transport of reducing equivalents and protons into a delocalised (inner) space.

The main evidence is that in artificial donor systems for photosystem I DCPIP but not DCPIP-sulfonate or PMS but not PMS-sulfonate is able to donate electrons to photosystem I in intact thylakoid vescicles[4]. Secondly the hydrogen carrying phenylendiamine DAD as an electron donor is coupled to ATP formation, whereas TMPD is not. Similarly the indamines are electron donors and are coupled to ATP formation, except for the fully alkylated derivative which – as TMPD – does not liberate a proton when oxidized[5]. Thirdly the effect of uncouplers on the rate of electron flow of a donor system – stimulation of an indophenol donor system, but not of the phenylendiamine donor system and independent of coupled ATP formation – indicates an internal pH effect on the donor oxidation in the inner thylakoid space[3].

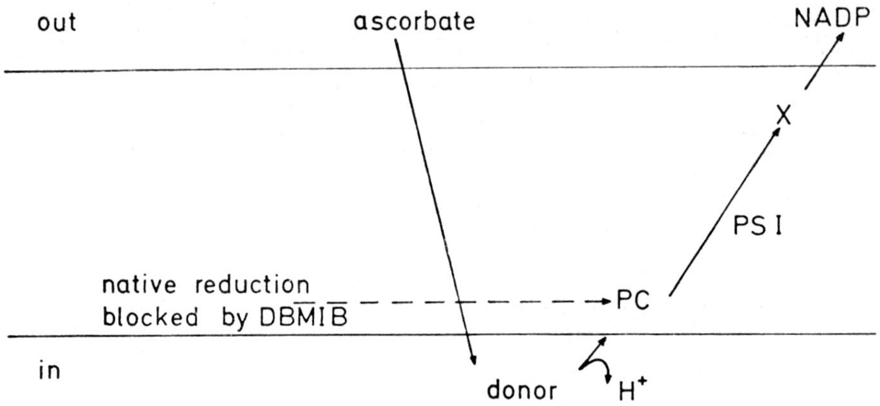

Transmembrane redoxreaction in artificial donor systems for photosystem I, coupled to ATP formation only, when the donor is a hydrogen carrying compound

In artificial donor systems for photosystem II also energy conservation and ATP formation is only observed, if the donor can
carry hydrogen across the membrane i.e. benzidine or catechol, but
are not coupled in the case of tetramethylbenzidine, ferrocyanide
or iodine[6].

In both artificial donor systems for either photosystem I or II
as well as in most cyclic electron flow systems the energy conserving site consists of an electrongenic photosystem and a proton
translocation via the transmembrane shuttle of the lipophilic hydrogen carrying donor[3]. An artificially imposed pH gradient is
built up, delocalised in the inner space of the thylakoid. As is
well known from Jagendorf's experiments, such a pH gradient does
yield ATP. In the donor systems any of the native coupling sites
visualised as proton release in the water splitting reactions or
in the oxidation of plastohydroquinone on the inside of the membrane do not nessessarly participate[1]. Therefore we have termed
this a transmembrane donor system coupled to an artificial energy
conservation[1]. Indeed, this energy conservation might be even more
artificial than we thought. It is conceivable in view of present
reservations as to the chemiosmotic coupling mechanism, that in native energy conservation no delocalised pH gradients are formed nor
transmembrane electron flow occurs, but rather proton pumps into
localised domains. Therefore those doubting should not use electron
donor systems or cyclic photophosphorylation systems with artificial
cofactors to study the mechanism of native coupling sites.

An artificial electron acceptor system is added in photoreductions by photosystem II, when DBMIB or other inhibitors block electron flow to photosystem I. The questions arises whether the acceptor is reduced by plastoquinol at the inside or outside of the membrane i.e. whether a transmembrane flow in → out exsists[7]. The reduction side inside or outside is irrelevant for the coupling ratio
of most mediator is used, in either case the P/e_2 ratio would be
0.6. But in the case of acceptors which do not carry hydrogen across
the membrane, the P/e_2 ratio in the case of inside redcution should
be twice the ratio of the case of outside reduction. The experiment
values unfortunately do not show a significient differenc, when TMPD
vs. phenylenediimine are compared[8]. On the other hand the effect of
internal pH on the rate of electron flow argues for an inside reduction[7]. Just opposite to the observation in donor systems for
photosystem I discussed above quinoid acceptors for photosystem II

are not affected by uncouplers whereas phenylendiimine acceptors systems are strongly inhibited[9], indicating an internal ph effect on the protonisation of the acceptor system[1,3].

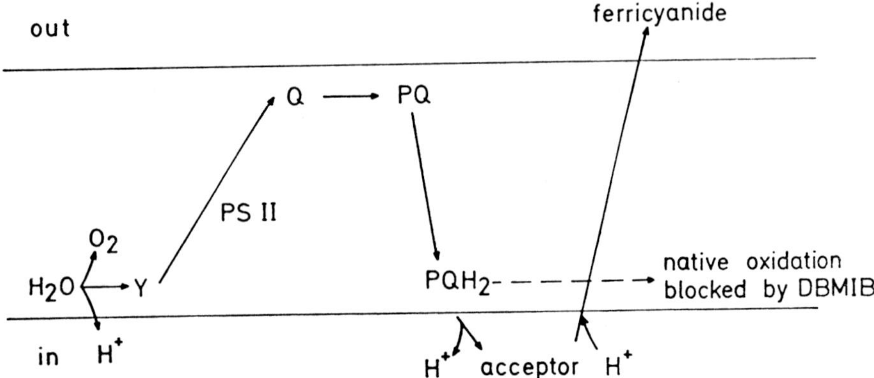

Photoreduction of ferricyanide by photosystem II via an artificial acceptor, which could be reduced via plastoquinol either on the inside (as indicated) or outside of the membrane.

In addition to a transmembrane bypass of an inhibitos site there is evidence for an internal as well as an external bypass[7]. In the absence of substrate amounts of ascorbate or ferricyanide TMPD is able to bypass the inhibition by DBMIB, valinomycin, bathophenanthroline and EDAC of electron flow dependent on both photosystem I and II. By reconnecting plastohydroquinone and plastocyanin oxygen evolution and NADP reduction are restored. This places the inhibition site of these compounds between plastoquinone and plastocyanin. DCMU and $HgCL_2$ inhibition on the other hand is not reversed by TMPD bypass[10].

The P/e_2 ratio of the reconstituted TMPD bypass in a DBMIB inhibition system is approaching one[7,10]. This argues for an internal bypass and an inside oxydation as well as reduction of TMPD because only in this case a P/e_2 ratio of one is expected. In a transmembrane bypass the P/e_2 ratio should be 0.5, because only the photosystem II dependent energy conserving site contributes. The observed ATP values, just below one indicate only some contribution of a transmembrane TMPD bypass. In this TMPD bypass an energy conservation is due to a native coupling site and not to artificially induced ones.

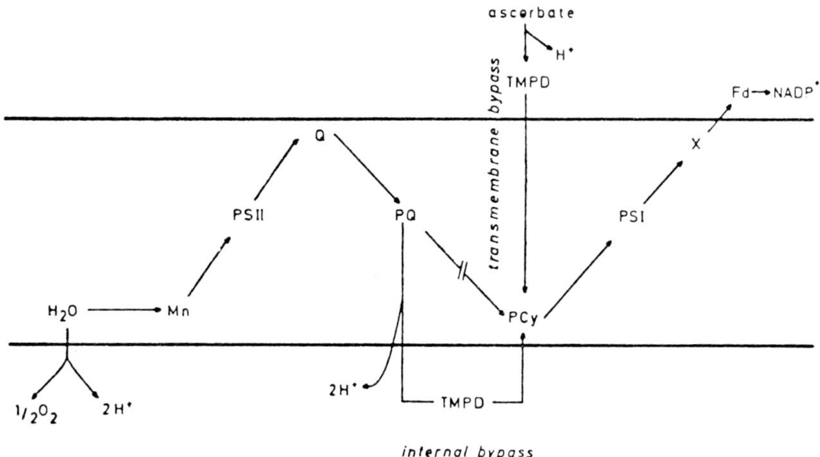

Internal bypass via TMPD of an inhibition
site between plastoquinone and plastocyanin.
An internal bypass is coupled via both energy
conserving sites, whereas a transmembrane TMPD/
ascorbate bypass is not coupled at all

An external bypass occurs, when a hydrophilic cofactor of cyclic
electron flow connects the primary acceptor of photosystem I and
plastoquinone both on the matrix side of the thylakoid. This is the
case in ferredoxin catalyzed photophosphorylation, which is DBMIB
sensitive, suggesting a participation of plastoquinone.

In summary, the existence of transmembrane, internal or external electron transport in chloroplasts supports the notion of vectorial electron flow and of ATP formation via a pH gradient across
the membrane. However, as stated above, these systems may include
artificial energy conserving sites with a transmembrane hydrogen
carrying redox reactions. Such a massive, delocalised proton translocation into the internal space of the chloroplast thylakoid might
not reflect the mechanism in native energy conservation.

This evidence from the work with chloroplasts is of relevance, of
course, for bacterial photosynthesis as well as respiratory electron flow. Also in these systems a transmembrane and an internal
electron flow bypass is well documented. For example: A coupling
of cytochrom oxidase in submitochondrial particles to DAD but not
to TMPD as electron donor is easily explained by the arguments

above and is indication, that there is no coupling site in just the cytochrome oxidase segment [11].

REFERENCES

1 Hauska, G., Reimer, S. and Trebst, A. (1974) Biochim. Biophys. Acta 357, 1.

2 Trebst, A. Reimer, S.,and Hauska, G. (1975) in: Electron Transfer Chains and Oxidative Phosphorylation, ed. E. Quagliariello et al. North Holland Publ. Comp. Amsterdam p. 343.

3 Hauska, G. and Trebst, A., (1977) in: Current Topics in Bioenergetics Vol. 5, ed. D. R. Sanadi, Academic Press N. Y. p. 151.

4 Hauska, G. (1972) FEBS Letters 28, 217.

5 Oettmeier, W., Reimer, S. and Trebst, A. (1974) Plant Science Letters 2, 267.

6 see reviews in Avron, M. (ed) (1975) Proc. 3rd Internat. Congr. on Photosynthesis, Elsevier.

7 Trebst, A. and Reimer, S. (1973) Biochim. Biophys. Acta 325, 546.

8 Selman, B. (1976) J. Bioenerget. and Biomembr. 8, 143.

9 Trebst, A., Reimer, S. and Dallacker, F. (1976) Plant Science Letters 6, 21.

10 Trebst, A. and Reimer, S. (1977) Plant & Cell Physiol. Special Issue p. 201.

11 Hauska, G.,Trebst, A. and Melandri, B. A. (1977) FEBS Letters 73, 257.

© 1977 Elsevier/north-Holland Biomedical Press
Bioenergetics of Membranes. L. Packer et al. ed.

THE DISEQUILIBRIUM BETWEEN STEADY STATE Ca^{++} ACCUMULATION RATIO AND MEMBRANE POTENTIAL IN RAT LIVER MITOCHONDRIA

Giovanni Felice AZZONE, Tullio POZZAN, Marco BRAGADIN
and Stefano MASSARI

C.N.R. Unit for the Study of Physiology of Mitochondria
and Institute of General Pathology, University of Padova, Italy.

It is generally accepted that in steady state mitochondria permeant cations are at electrochemical equilibrium. This assumption leads however to two inconsistencies, one physiological and another experimental:

1) The $\Delta\psi$ calculated on the distribution of K^+ in valinomycin treated mitochondria may reach 200 mV (1-2). This would results in an accumulation ratio for Ca^{++} of about 10^7. Since $\underline{in\ vivo}$ $\left[Ca^{++}\right]_i$ in the matrix is not higher than 10^{-3}M, $\left[Ca^{++}\right]_o$ in the cytosol should be lower than 10^{-10}M. This is incompatible with the maintenance of most Ca^{++} dependent enzyme activities.

2) If all permeant cations are distributed in steady state at electrochemical equilibrium the $\Delta\psi$ calculated on their accumulation ratios should be equal. This is however not the case. The $\Delta\psi$ calculated on the basis of the accumulation ratios for permeant cations varies in a large range between 200 and 90 mV (3). The variation is observed both among the univalent and the divalent cations, whether they are measured independently or simultaneously. Particularly striking is the discrepancy between Mn^{++} and Sr^{++}, which provide values of $\Delta\psi$ of about 90-100 mV, and K^+ which provides values of about 180-200 mV (3).

This discrepancy is open to two interpretations: a) that cation^{++} are transported with a single positive charge on it; b) that cation^{++} are not at electrochemical equilibrium in steady state but rather the accumulation ratio depends on the kinetics of cation^{++} influx and efflux.

Are divalent cations transported with one or two positive charges?

Two lines of evidence oppose the view that divalent cations are transported with a single positive charge: one is based on the stoicheometry and the other on the thermodynamics.

The stoicheometric evidence is the following: a) under anaerobic conditions the uptake of Ca^{++} in exchange with K^+ occurs with a stoicheometry of 2 whether in the presence of an excess of K^+ or of

396

Ca^{++}, whether in the presence or in the absence of weak acids (3);
b) under aerobic conditions the Ca^{++} uptake occurs with a stoicheometry of 4 charges/site similarly to the case of K^{+} (4,5); c) that
Ca^{++} may be transported as $\left[(Ca)_2^{4+} \, HPO_4^{2-} \right]^{2+}$ is in contrast with
the observation that the uptake of Pi accompanying the uptake of
Ca^{++} or Mn^{++} is N-ethylmaleimide sensitive (6); furthermore the
Ca^{++}/Pi ratio is also N-ethylmaleimide sensitive (7).

The thermodynamic evidence is the following: if a cation is
transported together with an anion, the equilibrium condition in
steady state is:

$$\Delta \tilde{\mu}_{C^{2+}} + \Delta \tilde{\mu}_{A^-} = 0 \tag{1}$$

and the divalent cation distribution:

$$\Delta \psi = \frac{RT}{F} \left(\ln \frac{\left[C^{2+} \right]_{in}}{\left[C^{2+} \right]_{out}} + \ln \frac{\left[A^- \right]_{in}}{\left[A^- \right]_{out}} \right) \tag{2}$$

Equation (2) predicts that the steady state distribution of C^{2+} is
affected by the anion distribution. An accumulation ratio of 10^4 or
10^3, with a $\Delta \psi$ of 180 mV, implies that the anion concentration in
the matrix be either 10 times lower or equal to that of the outer
space. However the distributions of OH$^-$, Pi and acetate are opposite. That the anion be Cl$^-$ is in contrast with the lack of effect
of Cl$^-$ on the divalent cation distribution.

<u>The role of cation influx and efflux on the steady state accumula-
tion ratio</u>.

Two lines of evidence support strongly the view that the steady
state accumulation ratio of divalent cations is regulated by the
relative rates of cation influx and efflux.

The first line of evidence originates from the observation
that the accumulation ratio increases or decreases parallel to the
increase or decrease of the rate of cation influx.
a) The steady state accumulation ratio increases in the order Sr^{++} <
Mn^{++} < Ca^{++}; this is also the rate of cation uptake under aerobic
conditions (3).
b) Addition of Ruthenium Red to steady state mitochondria after Ca^{++}
uptake causes Ca^{++} release (8,9,10). The process of Ca^{++} release
corresponds to the attainment of a new steady state cation accumulation ratio, the dimension of which depends on the extent of inhibition of the rate of cation influx (11).

c) Addition of inhibitors of the rate of cation uptake such as Ruthenium Red, La^{3+} or Mg^{++} tend to lower the accumulation ratio (3,8).

Figs. 1 and 2 show the effect of Mg^{++} and Ruthenium Red on the extent of discrepancy between the accumulation ratios (R) of K^+ and of Ca^{++}, as measured simultaneously, in valinomycin treated mitochondria.

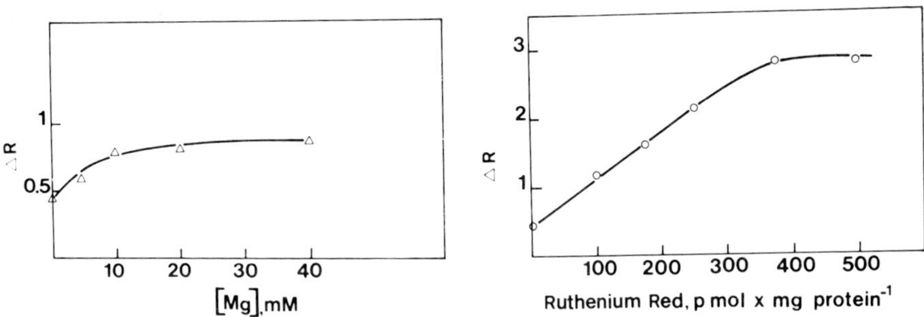

Fig. 1 - Effect of Ruthenium Red on Δ R. Δ R = R (K^+) - R (Ca^{++}), where R K^+ = log K^+ in/$[K^+]$ out and R (Ca^{++}) = 1/2 log $[Ca^{++}]$ in/$[Ca^{++}]$ out.
Medium: 0.2 M Sucrose, 30 mM Tris-Acetate, pH 7.2, 5 mM Succinate-Tris, 2 μM Rotenone, 0.1 γ x mg protein⁻¹ valinomycin, 2 mg/ml mitochondrial protein. Ruthenium Red concentration as indicated.

Fig. 2 - Effect of Mg^{++} on Δ R. Medium as in Fig. 1. $MgCl_2$ as indicated.

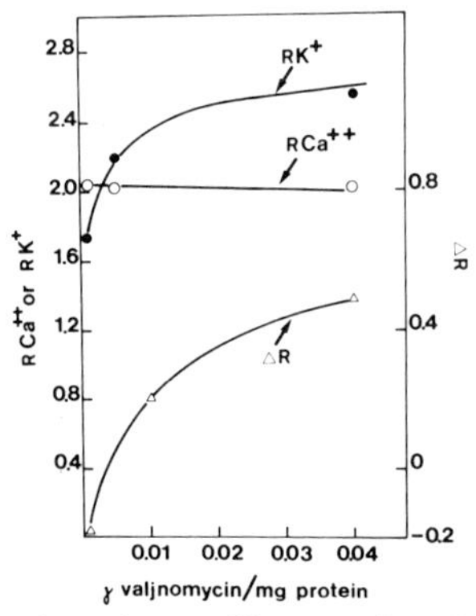

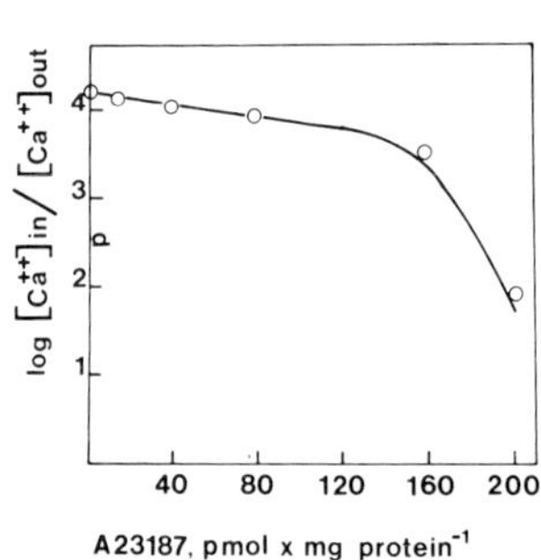

Fig. 3 - Effect of valinomycin on Δ R. Medium as in Fig. 1 except that valinomycin was varied as indicated.

Fig. 4 - Effect of A23187 on Ca^{++} accumulation ratio. Medium: 0.2 M Sucrose, 20 mM Hepes pH 7.2, 10 mM Tris-Acetate, 2.5 mM Succinate-Tris, 2 μM Rotenone, 100 nmol x mg protein⁻¹ $CaCl_2$, 1 mg/ml mitochondrial protein. A23187 as indicated.

The magnitude of the discrepancy ΔR increased with the increase of inhibition of Ca^{++} influx.

Fig. 3 shows an experiment where the rate of K^+ influx was varied by increasing the amount of valinomycin. This resulted in an increase of the K^+ accumulation ratio and no effect on Ca^{++} accumulation ratio: as a consequence ΔR increased at increasing valinomycin concentration.

The other line of evidence originates from the observation that the accumulation ratio decreases or increases parallel to the increase or decrease of the rate of cation efflux.

The steady state Ca^{++} accumulation ratio decreased with the increase of divalent cation ionophore A23187 which catalyzes an electroneutral H^+/cat^{++} exchange, see Fig. 4. The accumulation ratio decreased slightly until the efflux due to A23187 was compensated by a respiratory stimulation and then dropped markedly. Effects similar to those shown in Fig. 4 were observed when increase of the rate of cation efflux was obtained through the addition of nigericin instead of A23187 (on the K^+ distribution).

<u>The dependence of the steady state accumulation ratio on the kinetics of cation transport.</u>

The observations described above are in accord with the view that active transport occurs as two parallel processes of Vinflux and Vefflux.

Then: V net influx = Vinflux − Vefflux   (1)

and in steady state

     V net influx = O when Vinflux = Vefflux   (2)

The steady state distribution of permeant cations does not correspond to $\Delta\psi$ but rather to the parameters determining Vinflux and Vefflux. If Vinflux is an electrical process driven by the proton pump it may be described by:

$$\text{Vinflux} = K_1 \, \Delta\psi \, \frac{\left[\text{cation}^+\right]_i - \left[\text{cation}^+\right]_o \cdot e^{-nF\,\Delta\psi/RT}}{1 - e^{-nF\,\Delta\psi/RT}} \qquad (3)$$

If in equation (3) the force is constant, the rate of influx will depend on K_1, which is a permeability factor reflecting the kinetic constant for the transport of the cations through the membrane. Similarly, Vefflux will depend on K_2 which is the permeability factor for cation efflux; Vnet influx will then be zero depending on the relative values of K_1 and K_2. The steady state distribution will vary according to the values of K_1 and K_2.

Therefore the steady state accumulation ratio (R) is not obtained when R = $\Delta\psi$ but when Vinflux = Vefflux; hence (R) may not reflect the force of the pump. The discrepancy between the various cation accumulation ratios and therefore the disequilibrium between cation accumulation ratios and $\Delta\psi$ depends on the relative rates of cation efflux and influx.

Is the conclusion that the steady state distribution of permeant cations depend on kinetic rather than on thermodynamic parameters in accord with the electrogenic proton pump? Consider an electrogenic H^+ pump operating alone, in the presence of native H^+/cat^+ antiporters or of H^+ leaks, cf Table I, cases 1), 2) and 3). It may be predicted that in case 1) R = $\Delta\psi$ and R be identical for all cations or conditions: in case 2) R < $\Delta\psi$ and R is different for various cations or conditions. In case 3) R = $\Delta\psi$ and R is again equal for all cations or conditions. Furthermore R is dependent on the apparent Km for cation transport in case 2) and not in cases 1) and 3). From the analysis of Table I it appears that the variability of R for the various cations can be explained by the electrogenic H^+ pump through the assumption of native antiporters. The existence of a H^+/K^+ antiporter is indeed one of the basic assumptions of the chemiosmotic hypothesis (12).

<u>Pathway for divalent cation efflux in steady state mitochondria.</u>

Support to the view of a native antiporter catalyzing an electroneutral H^+/cat^{++} exchange is given by the notion that the native divalent cation carrier catalyzes an electrical flux. If under steady state conditions the membrane is under an homogeneous electrical field, negative inside, it is hard to see how the same carrier can catalyze two electrical cation fluxes, one directed inward and another directed outward. The rate of Ca^{++} efflux in steady state mitochondria supplemented with Ruthenium Red can be taken as expression of a Ca^{++} transport independent of the steady state electrical field. Table II analyzes the dependence of the steady state Ca^{++} efflux on ageing, BSA, pH:
a) it is lowest in freshly prepared mitochondria and increases with mitochondrial ageing;

T A B L E 1

R AND $\Delta\psi$ DURING CATION TRANSPORT DRIVEN BY AN ELECTROGENIC PROTON PUMP

Case	Transport processes affecting R		Dependence of R on cat Km	Predicted steady state values for R and $\Delta\psi$	
	Native antiporters	H^+ leak		Relationship between R and $\Delta\psi$	Values of R on various cations
1)	−	−	negligible	$R = \Delta\psi$	identical
2)	+	−	relevant	$R < \Delta\psi$	different
3)	−	+	negligible	$R = \Delta\psi$	identical

T A B L E   II

RATES OF RUTHENIUM RED INDUCED Ca^{++} EFFLUX UNDER VARIOUS CONDITIONS

CONDITIONS	STEADY STATE Ca^{++} EFFLUX nmoles x mg protein^{-1} x min^{-1}
without addition	6
+ BSA	3
pH 6.2	12
pH 7.8	9
ageing (3 hours)	16

Medium: 0.2 M Sucrose, 20 mM Hepes pH 7, 10 mM Tris-Acetate, 2.5 mM Succinate Tris, 5 mM $MgCl_2$, 100 µM Murexide, 100 nmoles x mg prot.$^{-1}$ $CaCl_2$, 2 mg/ml mitochondrial protein. The efflux was induced by adding 400 pmol x mg protein^{-1} Ruthenium Red to steady state mitochondria.

b) it is lowered by the addition of BSA;

c) it is lowest at neutral pH and tends to increase parallel to the shift of pH toward acid or alkaline.

Ageing, BSA, pH shift are generally known as altering the permeability of the mitochondrial membrane to ions rather than as affecting the rate of native antiporters. Fig. 5 shows that the rate of steady state Ca^{++} efflux is correlated with the respiratory control.

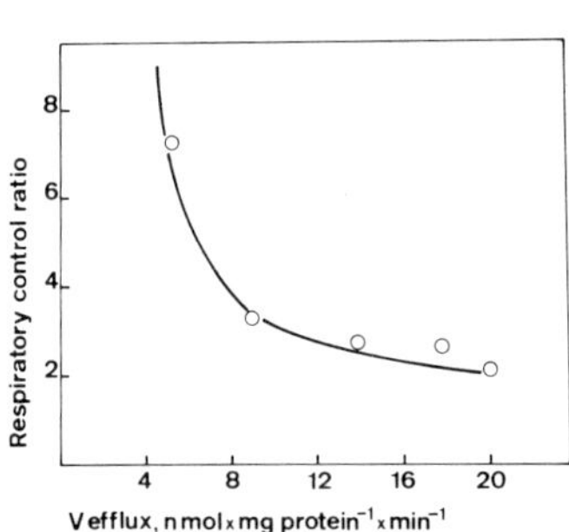

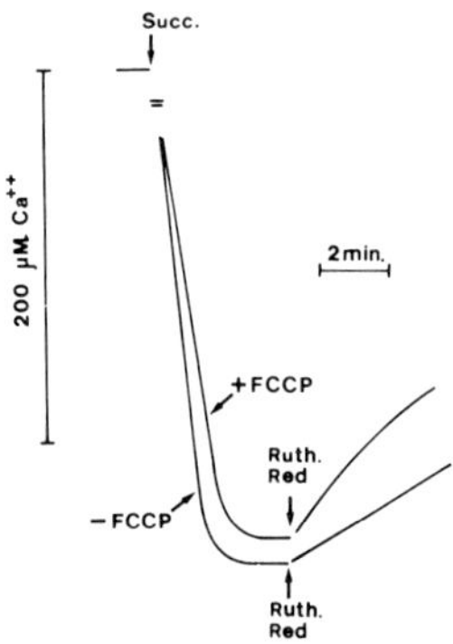

Fig. 5 - Correlation between Ruthenium Red induced Ca^{++} efflux and respiratory control. Medium as in Fig. 4 + 2.5 mM $MgCl_2$ and 100 µM Murexide. Various degrees of respiratory control were obtained by adding BSA, or by varying the time of ageing. The Ca^{++} efflux was induced by adding 400 pmol x mg protein. Ruthenium Red to steady state mitochondria.

Fig. 6 - Enhancement by FCCP of the Ruthenium Red induced Ca^{++} efflux. Medium as in Fig. 4 + 5 mM $MgCl_2$ and 100 µM Murexide. Where indicated: FCCP 22 pmol x mg protein^{-1} and Ruthenium Red 400 pmol x mg protein^{-1}.

The rate of efflux increases hyperbolically with the release of respiratory control. The correlation between cation efflux and the respiratory control suggests a role of an induced H^+ leak.

Indeed Fig. 6 shows that low concentrations of FCCP although had a negligible effect on the extent of Ca^{++} accumulation ratio caused a significant increase of the Ruthenium Red induced Ca^{++} efflux. The capacity of FCCP to enhance the Ruthenium Red induced cation efflux suggests that the process of cation efflux in steady state mitochondria might be due to an electrical H^+ influx coupled to an electrical cation efflux.

Our interpretation is that the apparent electroneutral H^+/cat^{++} exchange is an induced membrane property. The H^+ leak creates local membrane domains with a low electrical field. In these regions the cation concentration gradient drives a $cation^{++}$ efflux through the native electrical carrier. The dimension of the $cation^{++}$ accumulation ratio, which is a function of the relative pump and leak kinetics, will vary considerably among various $cations^{++}$ taken up at different pump rates (11). Furthermore, the accumulation ratio will vary depending on the experimental conditions affecting the pump rate and the conditions of the mitochondrial membrane which determines the leak rate.

Mitochondrial regulation of Ca^{++} gradients in vivo.

As stated in the Introduction, an accumulation ratio of 10^7 in vivo is in contrast with the physiological limits ($10^{-6}M - 10^{-7}M$) of the operation of the sarcotubular system in the muscle cell and for Ca^{++}-dependent enzymatic systems which require a free Ca^{++} concentration between 10^{-5} and 10^{-8} (13). If the accumulation ratio for permeant cations is dependent on the kinetics of influx and efflux the discrepancy between $\Delta\psi$ and divalent cation accumulation ratio in vivo and in vitro is eliminated.

Crompton et al. (14) proposed that this discrepancy is overcome in the heart cell by a Na^+/Ca^{++} antiporter. This fast Na^+/Ca^{++} exchange has been observed only in heart mitochondria and suggested to play a specific role in muscle contraction. However Ca^{++} must be maintained far from equilibrium in all cells. Furthermore, any mechanism of Ca^{++} efflux, without restriction of Ca^{++} influx leads to a marked energy drain. We therefore suggest that the H^+ induced Ca^{++} efflux leads to a reduction of the Ca^{++} accumulation ratio without energy expenditure; this is achieved through a restriction of Ca^{++} influx dependent on the cooperative kinetics of Ca^{++} transport. Since the Ca^{++} efflux is less than 5 nmoles/mg protein^{-1} in tightly coupled mitochondria _in vitro_, it is likely that _in vivo_ the Vinflux is of the same order of magnitude, say less than 2 nmol x mg prot.$^{-1}$ x min^{-1} at cytoplasmic Ca^{++} concentration around 10^{-6}-10^{-7}.

The finding by Vinogradov and Scarpa (15) and by Ackerman et al. (16) that conditions similar to that _in vivo_, i.e. 100 mM KCl and presence of Mg^{++}, render the kinetics of Ca^{++} uptake cooperative with a Km around 50 μM are efficient tools to restrict the rate of Ca^{++} influx to values similar to the rate of efflux at physiological Ca^{++} concentrations. The reduction of the rate of Ca^{++} influx leads even under conditions of a small H^{+} leak, in native mitochondria, to accumulation ratios in the range 10^{3}, i.e. 10^{-3}M in the matrix and 10^{-6} in the cytosol. According to this model the Ca^{++} accumulation ratio under physiological conditions has little relation with the force acting on cation and is rather determined by the kinetic parameters based on the affinity of Ca^{++} with its carrier.

MATERIALS AND METHODS

Rat liver mitochondria were prepared according to standard procedures. Cation$^{+(+)}$ accumulation ratio, were measured isotopically according to Azzone et al. (3). Kinetics of Ca^{++} uptake and release were monitored by following murexide absorbance changes with a double beam. Respiratory control was measured with a Clark electrode.

REFERENCES

1. Rossi, E. and Azzone, G.F. (1969) Europ. J. Biochem. 7, 418-426
2. Mitchell, P. and Moyle, J. (1969) Europ. J. Biochem. 7, 471-484
3. Azzone, G.F., Bragadin, M., Pozzan, T. and Dell'Antone, P. (1977) Biochim. Biophys. Acta 459, 96-109
4. Azzone, G.F. and Massari, S. (1971) Europ. J. Biochem. 19, 97-107
5. Azzone, G.F. and Massari, S. (1973) Biochim. Biophys. Acta 301, 195-226
6. Pozzan, T., Bragadin, M. and Azzone, G.F. (1976) Europ. J. Biochem. 71, 93-99

7. Azzone, G.F., Pozzan, T., Massari, S., Bragadin, M. and Dell'An-
    tone, P. (1977) FEBS Letters, 78, 21-24
8. Puskin, J.G., Gunther, T.E., Gunther, K.K. and Russell, P.R.
    (1976) Biochemistry 15, 3834-3842
9. Sordahl, L.A. (1974) Arch. Biochem. Biophys. 167, 104-115
10. Pozzan, T. and Azzone, G.F. (1976) FEBS Letters 71, 62-66
11. Pozzan, T., Bragadin, M. and Azzone, G.F., submitted for publi-
    cation
12. Mitchell, P. and Moyle, J. (1969) Europ. J. Biochem. 9, 149-155
13. Gompers, B.D. (1976) in Receptors and Recognition (Cuatrecasas,
    P. and Greaves, M.F., eds.) pp. 43-102, Chapman and Hall, London.
14. Crompton, M., Capano, M. and Carafoli, E. (1976) Eur. J. Biochem.
    69, 453-462
15. Vinogradov, A. and Scarpa, A. (1973) J. Biol. Chem. 248, 5527-
    5531
16. Ackerman, K.E.D., Wikstrüm, M.K.F. and Saris, N.E. (1977) 464,
    287-294

Bioenergetics of Membranes. L. Packer et al. ed.

ACID-BASE INDUCED REVERSE ELECTRON FLOW IN CHLOROPLASTS

Yosepha Shahak, Yona Siderer and Mordhay Avron
Biochemistry Department, Weizmann Institute of Science
Rehovot, Israel

INTRODUCTION

The coupled photosynthetic electron transport system has been considered to be
an essentially irreversible process. Recently, a method was developed by which
an energy-dependent reverse electron flow through part of the electron transport
chain was measured in the dark. Rienits et al[1,2] showed that ATPase activity can
drive the flow of electrons uphill towards Q (the primary acceptor of photosystem
11), inducing the reduction of Q and the oxidation of cytochrome f. The reduction of Q was measured by the increase in chlorophyll fluorescence yield under a
very weak measuring light, insufficient to induce the characteristic fluorescence
induction curve[3,4]. According to the chemiosmotic theory[5] transmembrane proton
gradients (Δ pH) serve as intermediate energy pool between electron transport and
ATP formation. By this mechanism the ATP driven reverse electron flow should
occur via the formation of Δ pH by the ATPase. Indeed, an artificially induced
Δ pH (obtained by transition of chloroplasts from acid to base) was also found to
drive reverse electron flow[4,6]. The acid-base driven unlike the ATP driven reverse electron flow is a transient reaction in which the electron donor is the
endogenous reduced plastoquinone pool.

Delayed light emission is commonly considered to originate from the recombination of the oxidized primary donor (Z^+) and the reduced primary acceptor (Q^-) of
photosystem 11[7]. Z^+ and Q^- are usually formed during illumination and are re-
reduced or reoxidized in the dark respectively. Since the dark decay of Q^- is
much faster than that of Z^{+}[8,9], the luminescence observed a few seconds after
illumination is supposedly limited by the amount of Q^-. The reduction of Q, in
this situation, by reverse electron flow may therefore be expected to, and indeed
does induce enhancement of the luminescence. The phenomenon of reverse electron
flow induced luminescence (REFIL) which is driven by acid-base transition of chloroplasts has recently been reported by us[10,11]. This reaction is further discussed in this communication and is compared to reverse electron flow induced Q
reduction (REFIQ).

MATERIALS AND METHODS

Chloroplasts were isolated from lettuce leaves as previously described[11,12].
The procedure for acid-base transition consisted of incubation of chloroplasts in
an acidic medium (acid-stage) for about 2 min at room temperature followed by the
injection of a basic buffer (base stage). The standard reaction mixture used for

the acid stage contained in 2.0 ml: 3 mM succinate, 10 mM $MgCl_2$, 30 mM KCl and chloroplasts containing 40-50 μg chlorophyll; final pH about 5.6. Base stage was achieved by injection of 0.2 ml of either 0.2 M TAPS (tris (hydroxymethyl) methyl amino propane sulfonic acid) or 0.4 M Tris pretitrated to give a final pH of 9.2.

Luminescence and fluorescence were measured in the instruments constructed by Malkin[13,14]. For a more detailed description of the measurements see Shahak et al[11].

RESULTS AND DISCUSSION

The experimental procedure of acid-base induced luminescence and typical results are illustrated in Fig. 1. Following a preillumination period, the chloroplasts were placed in the dark during which their native luminescence decayed essentially completely, the shutter in front of the photomultiplier was then opened to permit observation, the base was injected and immediately a transient of light emission was apparent. DCMU which blocks electron flow between Q and PQ (plastoquinone) inhibited the reaction and so did the absence of preillumination. The electron transport inhibitors NQNO and O-phenanthroline also inhibited the acid base luminescence.

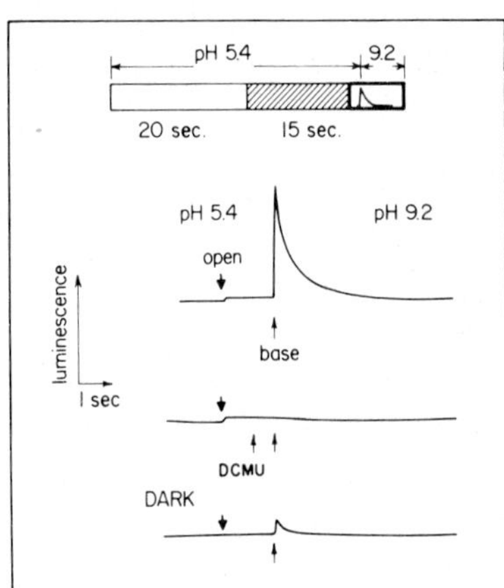

Fig. 1. Proton gradient driven reverse electron flow luminescence. Chloroplasts were incubated in the acidic medium (see Materials and Methods), illuminated for 20 sec, darkened for 15 sec and then the base was injected as described in the scheme. 0.1 ml 0.1 mM DCMU in 10% methanol was injected where indicated.

The triggering of luminescence by acid-base transition is a well known phenomenon[15,16]. Usually the reaction had been executed by transition of chloroplasts from pH 4 to 8.5. The effect of electron transport inhibitors on this triggered luminescence as reported in the literature is contradictory[16-18]. We have found that the inhibition of the acid-base induced luminescence by DCMU, which may represent the involvement of reverse electron flow in the luminescence, depends upon the pH of the acid stage (Figs. 2 and 3 left). When lowering the pH below

about 5.5 the inhibition became less effective, while changing the pH of the base
stage (constant acidic pH-5.6) did not affect the inhibition. It is suggested
that acid-base induced luminescence may occur via three different processes:
(i) a structural change of the membrane which stimulates the recombination reaction[7]; (ii) protonation and deprotonation of Z and QH, the midpoint potentials
of which are pH dependent and which are located at different sides of the membrane

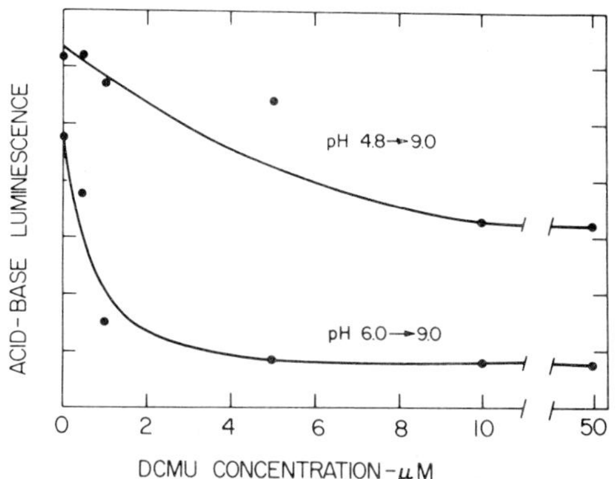

Fig. 2. Inhibition of acid-base induced luminescence by DCMU. Chloroplasts were
illuminated for 15 sec at pH 4.8 or 6.0, as indicated. DCMU and the base were
injected 8.5 and 10 sec after illumination, respectively. The ordinate represents the relative extent of luminescence observed upon injection of the base.

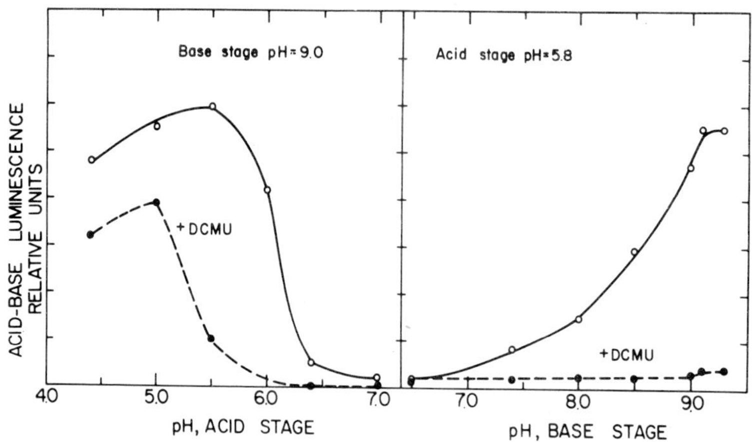

Fig. 3. Dependence of the inhibition by DCMU on the pH of the acid and the base
stages. The reaction mixture was as described in Materials and Methods except for
the indicated different pH values which were adjusted by HCl or KOH. The final
basic pH (9.0) was kept constant when varying the pH of the acid stage (left) or
the acidic pH (5.8) was kept constant when the final basic pH was changed (right).
DCMU (final concentration - 10 μM) and the base were injected 8.5 and 10 sec
after preillumination, respectively.

to give ZH[+] and Q[-] (see ref. 16); (iii) reverse electron flow from plastoquinone.
Only the last process should be inhibited by DCMU. As the acidic pH is lowered
below about pH 5.5 the acid-base luminescence seems to reflect more and more the
first two processes. Thus, further studies on acid-base induced reverse electron
flow were carried out using acidic pH of 5.6-6.0.

Acid-base induced Q reduction has clearly been shown to depend upon the availability of a reduced plastoquinone pool[4,6]. Fig. 4 shows that the same holds for
acid-base induced luminescence. Whereas green preillumination enhanced the reaction (see also Fig. 1), far-red light, which is known to oxidize the electron
carriers located between the two photosystems inhibited the reaction (Fig. 4,
second trace). The addition of DBMIB, an inhibitor of electron transfer between
plastoquinone and cytochrome f[19], did not affect the control reaction (3rd trace)
but did prevent the inhibition by far-red light (lower trace).

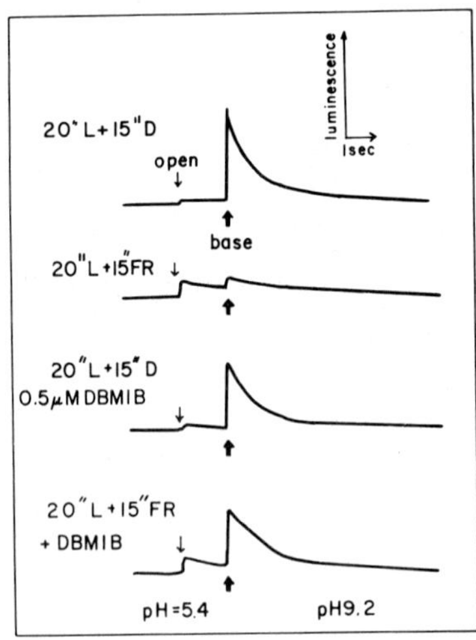

Fig. 4. Effect of far-red light on acid-base induced luminescence. Reaction mixture as described in Materials and Methods except for the addition of 0.25 mM NADP and 93 µg ferredoxin. DBMIB was added, where indicated before preillumination. Preillumination by green light (L) for 15 sec was followed either by 15 sec dark (D) or far red light (FR). The far-red light (10^4 ergxcm^{-2} x sec^{-1}) was filtered through a 730 nm broad-band interference filter. DBMIB was a kind gift from Dr. A. Trebst.

We previously reported[4] that DBMIB inhibited the acid-base induced reduction
of Q and suggested that it oxidizes the plastoquinone. Indeed when prereduced by
ascorbate or dithionite (Fig. 5) DBMIB did not decrease the extent of Q[-] formed
by the acid-base induced reverse electron flow, while the oxidized (untreated)
DBMIB did. In a control experiment both forms of DBMIB inhibited light induced
electron transport from water to NADP to the same extent (data not shown). In
Fig. 4 DBMIB was present during the preillumination which presumably induced its
reduction and thus it did not inhibit the REFIL.

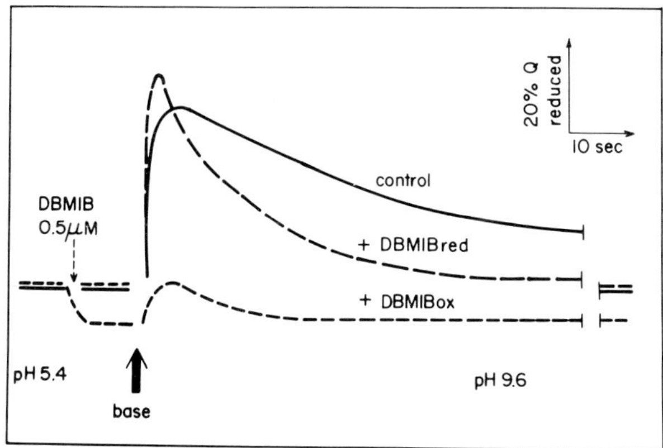

Fig. 5. Effect of DBMIB on acid-base induced Q reduction. Reaction mixture was as described under Materials and Methods except for the bubbling of N_2 to slow down oxidation of the reduced DBMIB (DBMIB red). DBMIB was prereduced by the addition of few crystals of dithionite into a stock solution of 50% ethanol.

The dependence of REFIL on the duration of the dark period which follows the preillumination is given in Fig. 6. The decay of the native luminescence was essentially complete within ten seconds. It may reflect the reoxidation of Q^- which had been formed during preillumination. The decay of the REFIL (Fig. 6 upper curve) was much slower. Since at each point Q^- was formed by reverse electron flow, this decay curve may represent the dark re-reduction of Z^+. In the experiment described in Fig. 7 (dashed line) again the base was injected at different dark intervals after preillumination, but here a single flash was given 3 sec before the base in order to reoxidize the Z. In this experiment the decay was very slow (notice the different time scale). After 10 min dark the extent of REFIL was about half the extent observed 10 sec after preillumination. This decay is suggested to reflect the reoxidation of the plastoquinone pool, since Z^+ is not limiting the luminescence in this case. Indeed the dark decay of REFIQ, which depends only upon the reduction state of the pool, had the same time course (Fig. 7, solid line).

The effect of uncouplers on the two acid-base induced reverse reactions was tested in detail. Uncouplers are known to stimulate the dissipation of the pH gradient across the thylakoid membrane[20]. Some of the uncouplers, like FCCP and S13, are also ADRY agents[21], namely they accelerate the deactivation (reduction) reactions of the oxidized electron donors to photosystem 11. The uncouplers

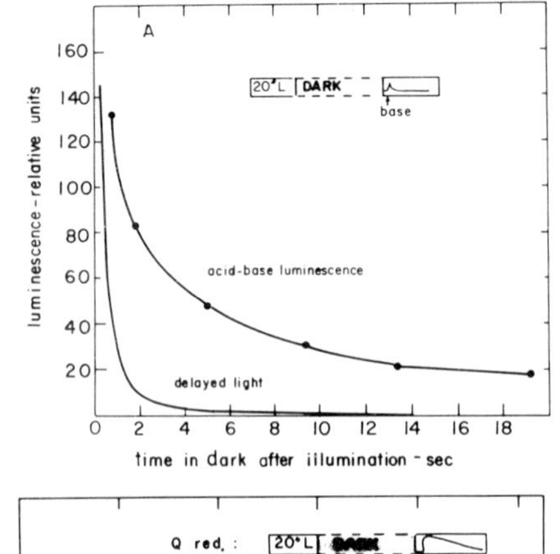

Fig. 6. Decay of acid-base induced luminescence in the dark after pre-illumination. Chloroplasts were incubated at pH 5.7, illuminated for 20 sec and after the indicated dark intervals the base was injected to give a final pH of 9.1.

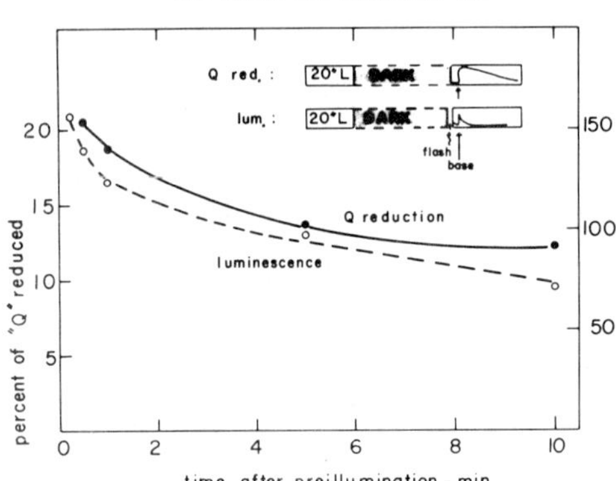

Fig. 7. Effect of the dark period after preillumination on acid-base induced reduction of Q and luminescence. Chloroplasts were illuminated for 20 sec in the acidic buffer (pH 5.8) and then darkened for the indicated period before injection of the base (final pH 9.2). In the luminescence experiment a single flash (half life 100 µsec) was given 3 sec before the base.

tested did not affect the acid-base REFIQ (Fig. 8B) except for the ADRY agents. Under the same reaction conditions all these uncouplers did inhibit the 9-amino-acridine fluorescence changes which follow the induction and dissipation of Δ pH[22] (not shown). The lack of effect of uncouplers on acid-base REFIQ suggests that the maintenance of Δ pH required to drive the burst of electrons backwards is too short for the uncouplers to compete with. The previous finding of lack of requirement for a buffer in the acid stage[6] is compatible with the above suggestion.

It is the pH gradient across the thylakoid membrane that is required for the reverse electron flow and not the change in the pH of the medium. When the transition from the acid to the base stage was performed too slowly (during more than a few seconds) so that Δ pH could not be built-up, reverse electron flow did not occur. Similarly, REFIQ required a period of incubation in the acid. In Fig. 9 chloroplasts were incubated at pH 9.0, then transferred to pH 5.8 and after different intervals in acid were transferred back to based at pH 9.0. When the incubation in acid was too short to permit its equilibration between the inside and outside of the thylakoid, no to very poor reduction of Q was measured upon injection of the base.

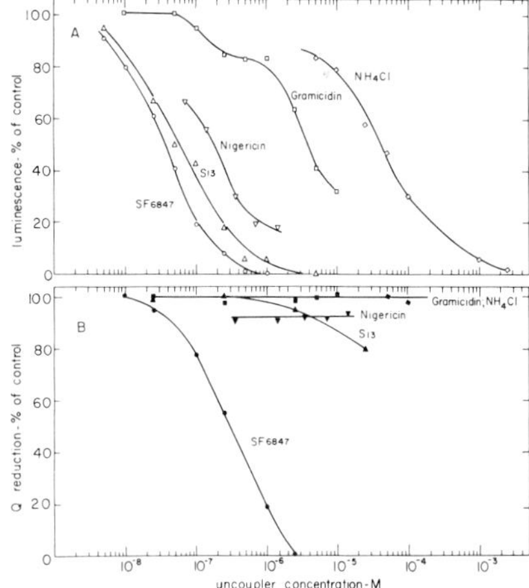

Fig. 8. The effect of uncouplers on acid-base induced luminescence (A) and Q reduction (B). Chloroplasts were illuminated for 20 sec (pH 5.7) and darkened for 60 sec before injection of the base (final pH 9.2). The uncouplers were added before preillumination. In (A) a single flash was given 4.5 sec before injection of the base. In the control of (B) the acid-base transition induced the reduction of 18.3% of Q.

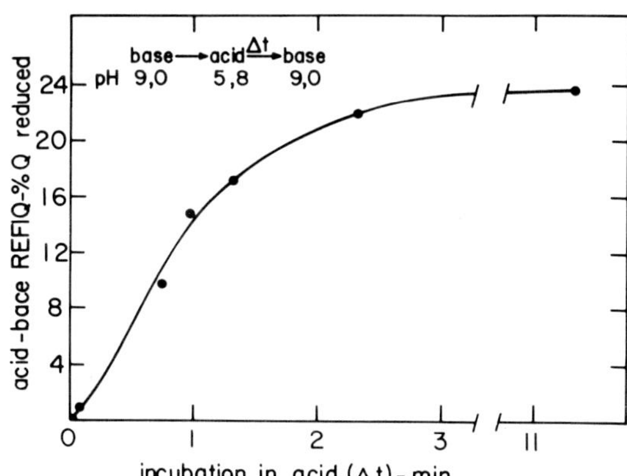

Fig. 9. Dependence of acid-base Q reduction on the duration of incubation in the acid (base-acid-base transition). The initial reaction mixture contained in 2 ml at pH 9.0: 30 mM KCl, 10 mM $MgCl_2$, 5 mM Tris and chloroplasts containing about 40 µg chlorophyll. After 2 min incubation 0.15 ml of 0.2 M succinate were injected to give pH 5.8, and after the indicated period 0.2 ml of 0.4 M Tris were injected to give final pH of 9.0. The 20 sec preillumination ended up 1 min before injection of the base.

The inhibition of the reverse reduction of Q by the ADRY agents FCCP (not shown), SF 6847 and S13 (Fig. 8B) may be due to their ability to catalyze the reduction of the oxidized primary donor of photosystem 11 (here called Z^+) by one of the components of the electron transport chain beyond Q^{21}. Such a cycle will bring about the oxidation of plastoquinone and thus inhibit the reverse electron flow. If this explanation holds then reduction of Z^+ by another external reductant should prevent the ADRY agent from closing the cycle and as a consequence prevent the inhibition of REFIQ. Indeed, the addition of ascorbate and hydroquinone (Fig. 10) as well as ascorbate and p-phenylenediamine (not shown) partially prevented the inhibition of REFIQ by FCCP. The uncoupling of cyclic and non cyclic photophosphorylation by FCCP which was measured under similar conditions was not affected by the addition of ascorbate and hydroquinone (data not shown).

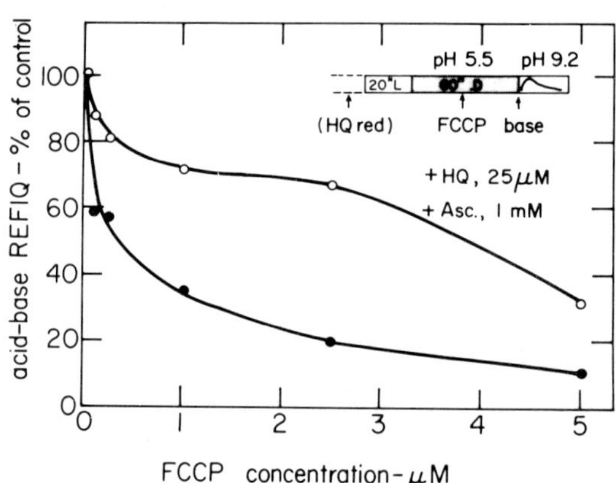

Fig. 10. The effect of ascorbate-hydroquinone on the inhibition of acid-base REFIQ by FCCP. Chloroplasts were illuminated for 20 sec and the base was injected after 1 min dark. FCCP was added about 30 sec before base. Ascorbate and hydroquinone were added (uuper curve) before preillumination.

Acid-base induced luminescence was inhibited by all the uncouplers tested (Fig. 8A). Since REFIL is presumably driven by the reduction of Q and the latter is not sensitive to non-ADRY uncouplers, there might be an additional effect of the acid-base transition in the luminescence mechanism. That additional effect should require a longer maintenance of Δ pH, and thus is sensitive to uncouplers.

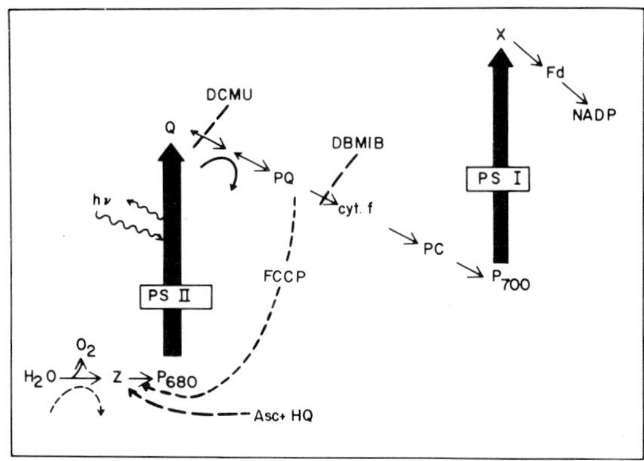

Fig. 11. Scheme of the photosynthetic electron transport system including the suggested coupling site between Q and plastoquinone and the cycle induced by the ADRY agent FCCP.

In Fig. 11 the main conclusions are schematically summarized. We suggest the location of a reversible coupling site between Q and plastoquinone. When light is absorbed by the photosystem 11 reaction centers, electron transfer from Q to PQ is coupled to the translocation of protons into the thylakoid. On the other hand, the induction of Δ pH in the dark can drive reverse electron flow from PQ to Q and on to the reaction centers inducing, under appropriate conditions the emission of light. DCMU inhibits the reverse reaction but DBMIB, which inhibits electron transfer beyond PQ, does not. The ADRY agent FCCP induces the oxidation of plastoquinone while ascorbate and hydroquinone prevent this effect.

REFERENCES
1. Rienits, K. G., Hardt, H., and Avron, M. (1973) FEBS Lett. 33, 28-32.
2. Rienits, K. G., Hardt, H., and Avron, M. (1974) Eur. J. Biochem. 43, 291-298.
3. Malkin, S. and Kok, B. (1966). Biochim. Biophys. Acta 126, 413-432.
4. Shahak, Y., Pick, U., and Avron, M. (1976) in Proc. 10th FEBS Meeting (Densuella, P. and Michelson, A. M., eds.) Vol. 40 p. 305, Elsevier, Amsterdam.
5. Mitchell, P. (1968) Chemiosmotic Coupling and Energy Transduction. Glynn Research, Bodmin.
6. Shahak, Y., Hardt, H., and Avron, M. (1975) FEBS Lett. 54, 151-154.
7. Lavorel, J. (1975) in Bioenergetics of Photosynthesis (Govindjee, ed.) p. 319, Academic Press, New York.
8. Papageorgiou, G. (1975) in Bioenergetics of Photosynthesis (Govindjee, ed.) p. 219, Academic Press, New York.
9. Renger, G. (1973) Biochim. Biophys. Acta 314, 320-402.

414

10. Avron, M., Pick, U., Shahak, Y., and Siderer, Y. (1976) in <u>Structure of Bio-logical Membranes</u> (Abrahamsson, S. and Pascher, I., eds.) p. 25, Plenum, New York.

11. Shahak, Y., Siderer, Y., and Avron, M. (1977) in <u>Photosynthetic Organelles</u>, <u>Structure and Function</u> (S. Miyachi et al., Eds.) p. 115-127, Japanese Society of Plant Physiol., Japan.

12. Avron, M. (1961) Anal. Biochem. 2, 535-543.

13. Malkin, S. and Hardt, H. (1971) in Proc. 2nd Int. Cong. Photosynthesis (Forti, G., Avron, M. and Melandri, A. eds.) p. 253, W. Junk N. V., The Hague.

14. Malkin, S. and Michaeli, G. (1971) in Proc. 2nd Int. Cong. Photosynthesis (Forti, G., Avron, M. and Melandri, A. eds.) p. 149, W. Junk N. V., The Hague.

15. Mayne, B. D. and Clayton, R. K. (1966) Proc. Nat. Acad. Sci. 55, 494-497.

16. Kraan, G. P. B., Amesz, J., Velthuys, B. R., and Steemers, R. G. (1970) 223, 129-145.

17. Miles, C. D. and Jagendorf, A. T. (1969) Arch. Biochem. Biophys. 129, 711-719.

18. Hardt, H. and Malkin, S. (1972) Biochem. Biophys. Res. Comm. 46, 668-676.

19. Trebst, A., Harth, E. and Draber, W. (1970) Z. Naturforsch. 25b., 1157-1159.

20. Jagendorf, A. T. (1975) in Bioenergetics of Photosynthesis (Govindjee, ed.) p. 413, Academic Press, New York.

21. Renger, G. (1973) Biochim. Biophys. Acta 314, 390-402.

22. Schuldiner, S., Rottenberg, H., and Avron, M. (1972) Eur. J. Biochem. 25, 64-70.

415

Bioenergetics of Membranes. L. Packer et al. ed.

HOW DO CYTOPLASMICALLY SYNTHESIZED PROTEINS BECOME
INCORPORATED INTO MITOCHONDRIA? SOME STUDIES OF THE
PROBLEM USING THE ISOZYMES OF ASPARTATE AMINOTRANSFERASE

Ersilia Marra[+], Shawn Doonan[++], Cecilia Saccone[+] and
Ernesto Quagliariello[+]
[+]Istituto di Chimica Biologica, Università di Bari, Italy
[++]Department of Chemistry, University College London, England

INTRODUCTION

Studies of mitochondriogenesis have so far been directed mainly towards
characterization of the intramitochondrial machinery for protein synthesis and
to identification of the translation products of the mitochondrial genome[1,2]. It
is clear, however, that the majority of mitochondrial proteins, and in particular
those of the matrix compartment, are coded by nuclear DNA and synthesised on
cytoribosomes. The suggestion[3] that nuclear mRNA may be imported into
mitochondria and translated on the mitochondrial ribosomes appears to be
untenable[4]. Given that this is so, then a major outstanding problem of
mitochondriogenesis is how cytoplasmically synthesised proteins are directed
specifically towards the mitochondrion and how such proteins are either inserted
into existing membrane structures[5,6] or pass through the mitochondrial membranes
into the intermembranal or matrix spaces.

In the case of yeast, a partial solution to those problems may be afforded by
the identification of cytoplasmic ribosomes associated specifically with the
outer membrane of mitochondria from growing cells[7]. It is suggested that these
ribosomes are involved in synthesis of mitochondrial proteins and that the
product polypeptide chains are transferred directly to the mitochondria. It
remains to be shown, however, that only mitochondrially associated ribosomes are
indeed specifically involved in synthesis of mitochondrial proteins and, if so,
which proteins they produce. Furthermore in contrast with the previously
mentioned hypothesis, in Neurospora Crassa the existence of extramitochondrial
pools of proteins which are transported into the mitochondria has been shown[8].
With higher eukaryotes, specific association between mitochondria and cytoplasmic
ribosomes has not been demonstrated although claims have been made for a special
class of membrane associated cytoplasmic ribosomes involved in the synthesis of
mitochondrial proteins in rat liver[9,10]. Similarly, in an extensive study of the
biosynthesis of glutamate dehydrogenase in rat liver[11], it was shown that the
enzyme was synthesised on cytoplasmic ribosomes, but that these ribosomes
sedimented in the microsomal fraction not as mitochondrially associated particles.

It is clear from the brief survey given above that identification of the
sites of synthesis of cytoplasmically produced mitochondrial proteins, the

factors directing them towards the mitochondrion and mechanisms of insertion into
or passage through mitochondrial membranes are largely unexplored areas of
mitochondriogenesis. We have chosen to study some aspects of this problem with
respect to the enzyme aspartate aminotransferase (E.C. 2.6.1.1). This seemed to
be a good choice for study for several reasons. The enzyme exists in higher
eukaryotes as distinct cytoplasmic and mitochondrial isozymes[12]. Recent work has
shown that, although the isozymes differ markedly in many respects, they are
nonetheless homologous proteins[13]. In the case of rat liver, studies of the
subcellular distribution of the isozymes have shown that the mitochondrial
isozyme accounts for 80% of the total activity; more importantly the localization
of the isozymes was unique in that no cytoplasmic form was found in the
mitochondria or vice versa[14]. It is, however, clear from experiments with somatic
cell hybrids that the mitochondrial isozyme is coded by the nuclear genome[15] and
hence synthesised on cytoplasmic ribosomes. Hence in the case of aspartate
aminotransferase there exist two structurally related isozymes, both synthesised
cytoplasmically, but only one of which is capable of translocation into the
mitochondrion, and more specifically to the matrix compartment[16]. What are
factors responsible for the discrimination between the isozymes and how does the
translocation occur?

In the present paper we show that the mitochondrial but not the cytoplasmic
isozyme has an affinity both for intact mitochondria and for the inside of the
inner membrane; this effect is, however, extremely sensitive to the ionic
strength of the incubation medium and is unlikely to be of importance in vivo.
More interestingly, we have shown using two different analytical methods that
the mitochondrial isozyme can pass from solution into the mitochondrion in vitro
in a process that does not require the presence of other cellular components.
The cytoplasmic isozyme, on the other hand, is completely unable to penetrate
the mitochondrial membrane. It is clear that the use of the cytoplasmic isozyme
provides an in-built control in experiments designed to study the biogenesis and
localisation of the mitochondrial form. This is an obvious advantage of using an
isozyme pair for experiments of this type.

BINDING OF ASPARTATE AMINOTRANSFERASE ISOZYMES TO VARIOUS SUBCELLULAR STRUCTURES

An obvious possible origin of the specificity of localisation of the isozymes
in their respective subcellular compartments is that the mitochondrial but not
the cytoplasmic form has an affinity for mitochondrial membranes; this could be
either for the outer membrane and represent a preliminary stage in transfer of
the enzyme from ribosomes to the matrix or for the inner membrane to retain the
enzyme once translocation had occurred. These possibilities were tested by
measuring binding of both isozymes to intact mitochondria and to sonic
("inside-out") submitochondrial particles. The results are given in Table 1. Also
included are data on the binding to a rat liver microsomal fraction and to red

cell ghosts. It can be seen that the mitochondrial isozyme binds strongly to all
these systems whereas the cytoplasmic isozyme binds to none of them. A similar

TABLE 1

BINDING OF ISOZYMES TO SUBCELLULAR STRUCTURES

Rat liver mitochondria were prepared by standard methods[17]. The microsomal
fraction was obtained from the post-mitochondrial supernatant by centrifugation
at 105.000 xg for 1 h. Submitochondrial particles were prepared by sonication of
a suspension of mitochondria in 250 mM-sucrose, 10 mM-Tris, 1 mM-ATP, 15 mM-$MgCl_2$
at pH 7.4 and were recovered by centrifugation at 105.000 xg for 30 min. Red
cell ghosts were prepared by lysis of human red cells in water for 15 min; the
ghosts were exhaustively washed with water. Pig heart cytoplasmic[18] and
mitochondrial [19] aspartate aminotransferases were purified by previously
published methods. Incubations of mitochondrial and cytoplasmic isozymes (40 µg
and 60 µg respectively) with subcellular fractions were carried out in 250 mM
-sucrose at 37°C for 30 min. Samples were centrifuged and the pellets resuspended.
Enzyme activity was assayed [14] in supernatants and in resuspended pellets.
Control incubations with enzymes alone and with subcellular fractions alone were
carried out where appropriate.

Subcellular fraction	Amount of Protein (mg)	Percentage of Added Enzyme Bound	
		Mitochondrial	Cytoplasmic
Intact Mitochondria	3.1	95	5
Submitochondrial Particles	2.6	86	3
Microsomes	3.1	90	2
Red Cell Ghosts	0.4	85	5

pattern of results was obtained with different concentrations of isozymes and
amounts of subcellular fractions. It should be noted that pig heart isozymes were
used in these experiments, but available evidence (see below) suggests that the
rat liver isozymes behave similarly.

Binding of mitochondrial aspartate aminotransferase to both submitochondrial
particles and to intact mitochondria was measured as a function of ionic
strength of the incubation medium. The results for submitochondrial particles
are given in Table 2. It is clear that the binding is very dependent on ionic
strength and is reduced to approximately one half maximum at a concentration of
KCl of 12 mM. Similar results were obtained with intact mitochondria when
incubated with various concentrations of sucrose and KCl such as to keep the
osmolarity of the incubation mixture constant. Binding was essentially abolished
by 20 mM KCl. Here, however, analysis of the data was complicated by the fact
that the total aspartate aminotransferase activity in the incubation mixtures
appeared to increase with increasing concentrations of KCl. We do not understand
the origin of this phenomenon but it is worth noting that a similar KCl-dependent

TABLE 2

BINDING OF MITOCHONDRIAL ISOZYME TO SUBMITOCHONDRIAL PARTICLES AS A FUNCTION OF
IONIC STRENGTH

The submitochondrial particles were prepared as described in Table 1. Particles
(1.5 mg) were incubated with enzyme (20 µg) for 30 min. at 37°C in 250 mM-sucrose
containing various concentrations of KCl. Enzyme activity was measured in
supernatants and resuspended pellets after centrifugation.

KCl concentration (mM)	0	12	23	46	68
Percentage of Added Enzyme Bound	81	44	9	5	3

increase in activity was observed when the mitochondrial isozyme was incubated
with lecithin liposomes; it may be that lipids plus KCl activate the enzyme.

The results in Table 1 suggest that mitochondrial aspartate aminotransferase
binds indiscriminately to membrane structures. This lack of specificity would not
necessarily rule out binding as a contributary factor in specific localization
of the enzyme since if synthesis occurred on ribosomes associated with
mitochondria then transfer to the outer mitochondrial membrane and thence to the
matrix could occur without the enzyme coming into contact with other structures
to which it might bind. A serious objection to this hypothesis, however, is the
fact that binding is abolished at an ionic strength considerably less than that
known to obtain in intracellular fluids. It seems to us, therefore, that the
binding of the mitochondrial isozyme to membranes at low ionic strength is a
purely electrostatic phenomenon with no physiological relevance. The selectivity
between the isozymes is understandable on this basis since at neutral pH the
mitochondrial isozyme is positively charged whereas the cytoplasmic isozyme is
negative. Hence the origin of the localization of the isozymes must be sought
elsewhere.

The results obtained in the present work are consistent with the repeatedly
reported observation that release of endogenous aspartate aminotransferase from
damaged mitochondria requires solutions of high ionic strength[12,14,20] even
though it is a matrix enzyme [16]. Waksman and Rendon [21,22] have claimed that
mitochondrial aspartate aminotransferase changes its localization in response to
the functional state of the mitochondrion and to the presence of salts of
intermediary metabolites. The results given here suggest strongly that the
phenomenon observed by Waksman and Rendon[21,22] was binding and release of the
enzyme in response to changing ionic strength and that it is very unlikely to
have any physiological importance.

SELECTIVE PERMEABILITY OF MITOCHONDRIA TO ASPARTATE AMINOTRANSFERASE ISOZYMES AS MEASURED BY A FLUORIMETRIC METHOD

It seemed to be of interest to examine the possibility that intact mitochondria may be selectively permeable to mitochondrial aspartate aminotransferase _in vitro._ For this, a method was required for assay of the enzyme levels inside mitochondria. The principles of the method, details of which have already been published[23], are as follows. Mitochondria are treated with rotenone and sodium arsenite to block oxidation of intramitochondrial NAD(P)H by the electron transport system and by the 2-oxoglutarate dehydrogenase complex respectively. Aspartate is added to the suspension and, after allowing time for the intramitochondrial level to reach a maximum, 2-oxoglutarate is added. Transport of the latter substance to the matrix allows the following linked reactions to occur:

$$\text{aspartate} + \text{2-oxoglutarate} \xrightarrow{\;1\;} \text{oxaloacetate} + \text{glutamate}$$

$$2 \left\{ \begin{array}{l} \nearrow \text{NAD(P)H, H}^+ \\ \searrow \text{NAD(P)}^+ \end{array} \right.$$

$$\downarrow$$

$$\text{malate}$$

Reaction 1 is catalysed by internal aspartate aminotransferase and step 2 by endogenous malate dehydrogenase. The rate of oxidation of NAD(P)H, measured fluorimetrically, provides a measure of the intramitochondrial aspartate aminotransferase level provided that neither the rate of entry of 2-oxoglutarate nor the activity of malate dehydrogenase is limiting. Detailed results showing that these processes are not limiting and that the method provides a true measure of aspartate aminotransferase activity have already been presented[23].

Fig. 1 shows the initial rates of decrease of fluorescence for a typical sample of mitochondria compared with the values obtained when the mitochondria were preincubated with either cytoplasmic or mitochondrial rat liver aspartate aminotransferases. It is clear that incubation with the cytoplasmic isozyme did not result in an increase in the level of intramitochondrial enzyme activity. Incubation with the mitochondrial isozyme on the other hand gave rise to a concentration dependent increase in the intramitochondrial level of activity up to a maximum of 60% above that observed in the absence of added enzyme. It seems necessary to conclude that, under the conditions of incubation used, the mitochondrial isozyme can pass freely from solution into its site of intra-mitochondrial action, this is, the matrix space[16].

Failure to detect an increase in rate of oxidation of NAD(P)H after incubation of the mitochondria with cytoplasmic isozyme is important from several points of view. Firstly, it shows that the effect observed with the mitochondrial isozyme was neither an artefact due to enzyme activity remaining in the

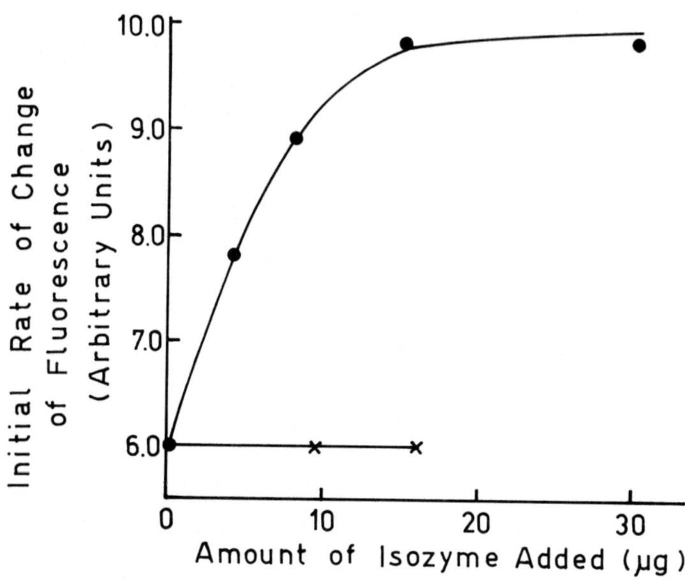

Fig. 1 Levels of intramitochondrial aspartate aminotransferase activity after
incubation of intact mitochondria with aspartate aminotransferase isozymes.
Mitochondria (3.0 mg of protein) in 1 ml 250 mM-sucrose, 20 mM-Tris/HCl, 1 mM
-EGTA, pH 7.4 were preincubated for 3 min at 22°C with 2 µg rotenone and 2 mM
-sodium arsenite. Aspartate was added to 9.6 mM and after a further 1 min,
2-oxoglutarate was added to 3 mM. The initial rate of decrease of fluorescence
of NAD(P)H was measured using an Eppendorf 1101 M photometer. When the effect
of externally added enzyme was to be studied, the enzyme was added to the
suspension one min before the addition of aspartate.
(X) cytoplasmic isozyme; (●) mitochondrial isozyme.

incubation mixture nor a reflection simply of damage to the mitochondria, for

example increased permeability of the membranes due to swelling; Rendon and

Packer[24] have recently shown that mitochondrial swelling under some circumstances

is accompanied by release of malate dehydrogenase from the matrix. Both of the

potential sources of artefacts mentioned would give increased rates with both

isozymes; this provides a clear example of the desirability of using the

cytoplasmic isozyme as a control in experiments with the mitochondrial form.

Secondly, the specificity of the permeation process indicates that the process

is inherent in the structures of the mitochondrial membranes and of the

mitochondrial isozyme. Clearly further study of this phenomenon may allow

elucidation of the origins of the specificity of transport of proteins into

mitochondria, and hence of the specificity of subcellular localization as well

as throwing light on the mechanism of the permeation process per se.

STUDIES WITH RADIOLABELLED ISOZYMES

The fluorimetric method described above is clearly not well suited for examination of the effect of variables such as temperature of incubation, presence of possible inhibitors, or chemical modifications of the protein on the permeability of mitochondria to mitochondrial aspartate aminotransferase. We decided therefore to investigate the use of radiolabelled isozymes for this purpose. The label was introduced by reduction of the enzyme-cofactor linkage with $NaBH_4$. Such reduced isozymes are catalytically inactive but immunochemically indistinguishable from the native forms (Doonan and Porter, unpublished). Hence it seemed likely that the three dimensional structures of the reduced forms may not be significantly different from those of the native enzymes and that the property of selective permeability would be retained.

The technique to be adopted was simply to incubate mitochondria with radiolabelled isozyme and, after removal of the supernatant by centrifugation, to determine the amount of radioactivity remaining in the pellet. For this approach to be successful, however, it was clearly necessary to employ conditions under which radioactivity associated with the mitochondrial pellet arose from enzyme that had entered the mitochondrion rather than from enzyme that had bound to outside of the organelles. This problem did not arise with the fluorimetric assay method since the procedure assays only enzyme present in the mitochondrial matrix.

Early experiments described above had shown that native pig heart mitochondrial enzyme binds to rat liver mitochondria in solutions of low ionic strength but that external binding is abolished at higher ionic strength; the cytoplasmic isozyme did not bind appreciably. Experiments with the homologous system (rat liver mitochondria and native rat liver isozymes) showed that a similar situation pertained, but it was possible to demonstrate that with the mitochondrial isozyme at high ionic strength a small, reproducible amount of the added enzyme was removed from the supernatant but was not bound externally to the mitochondria. By using Triton X-100 unmask the internal enzyme activity it was shown that the enzyme removed from the supernatant at high ionic strength had passed inside the mitochondria. The same pattern of events could be observed using the labelled enzymes. These results will not be discussed in detail here, but suffice it to say that by the appropriate choice of buffer concentration (20 mM-Tris HCl, pH 7.4) it is possible to ensure that radioactivity associated with the mitochondrial pellet after incubation with labelled enzyme is due to internal enzyme and not to material bound externally. This clearly provides a much more flexible assay system than that described previously and one that we intend to use to study further features of the translocation of mitochondrial enzyme to the interior of the mitochondria. The assay is also much more readily quantitated.

Some initial results obtained using the radioactivity assay are as follows. The amount of labelled enzyme entering the mitochondria was found to show a normal hyperbolic dependence on external enzyme concentration. At a saturating concentration of external enzyme, the amount found to enter the mitochondria was approximately 1 µg enzyme/mg mitochondrial protein. The uptake was rapid and appeared to be complete in the time taken to add the enzyme to the mitochondrial suspension and then separate the mitochondria by centrifugation (time of the order of 1 min). Of particular interest was a set of experiments in which the entry of labelled mitochondrial aspartate aminotransferase was measured as a function of temperature. The results are shown in Fig. 2 as a plot of log (radioactivity incorporated) against the reciprocal of the absolute temperature. It can be seen that plot shows a break at $1/T = 3.41 \times 10^{-3}$ K^{-1} i.e. at a temperature of about 20°C. This corresponds well with the temperature at which

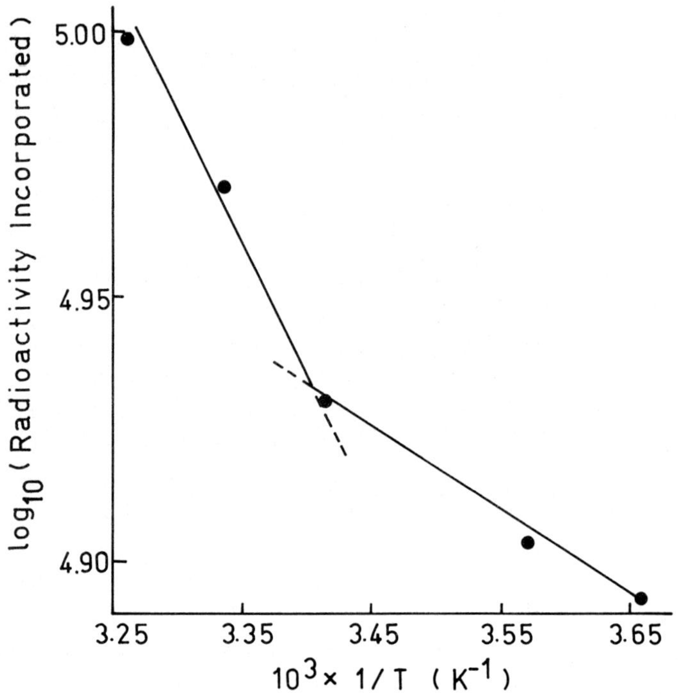

Fig. 2 Dependence of amount of labelled enzyme incorporated into mitochondria on temperature. Mitochondria (2.5 mg of protein) were incubated with mitochondrial aspartate aminotransferase (29 µg radioactivity 556.000 cpm) in 1 ml of 250 mM -sucrose, 20 mM-Tris, 1 mM-EGTA, pH 7.4 at the temperature indicated. Mitochondria were recovered by centrifugation at 4°C and the supernatant removed. The resulting pellet was taken up in 300 µl 5% Triton X-100; insoluble material was removed by centrifugation and an aliquot of the supernatant counted for radioactivity.

breaks are observed in Arrhenius plots for the activities of various enzymes of the inner membrane of rat liver mitochondria and can be correlated with the melting temperature of the lipids in the membrane[25]. Hence in the present case it appears that the amount of enzyme incorporated depends on the state of fluidity of the inner membrane. It would be of considerable interest to repeat this experiment with mitochondria having inner membranes of various lipid compositions.

SUMMARY

Mitochondrial, but not cytoplasmic, aspartate aminotransferase binds to mitochondria at low ionic strength. The effect is abolished at higher ionic strength and is hence unlikely to be of importance in directing newly synthesised enzyme to the mitochondrion. Two different analytical methods have been devised that can be used to measure uptake of mitochondrial aspartate aminotransferase into mitochondria in vitro. The process is selective, in that the cytoplasmic isozyme is completely excluded, and does not appear to require the presence of other subcellular components. The amount of mitochondrial isozyme incorporated seems to depend on the fluidity of the inner membrane. Further studies of the process may throw light on the processes of incorporation of matrix enzymes into mitochondria in vivo.

REFERENCES

1. Schatz, G. and Mason, T.L. (1974) Ann. Rev. Biochem., 43, 51.

2. (1974) The Biogenesis of Mitochondria (Kroon, A.M. and Saccone, C. eds.), Academic Press, New York.

3. Swanson, R.F. (1971) Nature New Biol., 231, 31.

4. Mahler, H.R. and Davidowicz, K. (1973) Proc. Natl. Acad. Sci., U.S., 70, 111.

5. Forde, B.G. et al. (1976) J. Cell. Sci., 21, 329.

6. Ainsworth, P.J. et al. (1974) Bioenergetics, 6, 135.

7. Kellems, R.E. et al. (1974) in "The Biogenesis of Mitochondria" (Kroon, A.M. and Saccone, C. eds.), Academic Press, New York, p. 511.

8. Hallermayer, G. and Neupert, W. (1975) in "Genetics and Biogenesis of Chloroplasts and Mitochondria" (Bucher, TH., Neupert, W., Sebald, W. and Werner S. eds), North-Holland, p. 807.

9. Gaitskhoki, V.S. et al. (1974) FEBS Letters, 43, 151.

10. Shore, G.C. and Tata, J.R. (1977) J. Cell Biol., 72, 726.

11. Godinot, C. and Lardy, H.A. (1973) Biochemistry, 12, 2051.

12. Boyd, J.W. (1961) Biochem. J., 81, 434.

13. Doonan, S. et al. (1974) FEBS Letters, 49, 25.

14. Baumber, M.E. and Doonan, S. (1976) Int. J. Biochem., 7, 119.

15. van Heyningen, V. et al. (1974) in "The Biogenesis of Mitochondria" (Kroon, A.M. and Saccone, C. eds.), Academic Press, New York, p. 231.

16. Matlib, M.A. and O'Brien, P.J. (1975) Arch. Biochem. Biophys., 167, 193.

17. Keingenberg, M. and Slenczka, W. (1959) Biochem. Z., 331, 486.

18. Banks, B.E.C. et al. (1968) Europ. J. Biochem., $\underline{5}$, 528.

19. Barra, D. et al. (1976) Europ. J. Biochem., $\underline{64}$, 519.

20. Boyde, T.R.C. and Hiu, C.F. (1972) Biochem. Biophys. Res. Commun., $\underline{46}$, 231.

21. Waksman, A. and Rendon, A. (1971) Biochem. Biophys. Res. Commun., $\underline{42}$, 745.

22. Waksman, A. and Rendon, A. (1974) Biochimie, $\underline{56}$, 907.

23. Marra, E. et al. (1977) Biochem. J., $\underline{164}$, 685.

24. Rendon, A. and Packer, L. (1976) Mitochondria: Bioenergetics, Biogenesis and Membrane Structure (Packer, L. and Gòmez-Puyou, A. eds.) Academic Press, New York, p. 151.

25. Raison, J.K. (1973) J. Bioenerg., $\underline{4}$, 285.

Bioenergetics of Membranes. L. Packer et al. ed.

A SINGLE CARRIER TO TRANSPORT OXALOACETATE
INTO RAT LIVER MITOCHONDRIA

Salvatore Passarella, Ferdinando Palmieri
and Ernesto Quagliariello
Department of Biochemistry, University
of Bari, Italy

INTRODUCTION

The transport of oxaloacetate mediated by carrier has been recently tested in rat liver mitochondria. Gimpel et al. as result of the research concluded that oxaloacetate is transported only by the dicarboxylate carrier[1]. In agreement with this outcome it was found that oxaloacetate causes Pi efflux via dicarboxylate carrier; but it should be noted that oxaloacetate is transported with much more effectiveness by the oxoglutarate carrier[2].

In this work we have studied the mechanism of the Pi efflux induced by oxaloacetate in order to define the way how the transport of oxaloacetate occurs in rat liver mitochondria.

MATERIALS AND METHODS

(^{32}P)Phosphoric acid and 2-oxo-(5-^{14}C)glutaric acid (sodium salt) have been supplied by the Radiochemical Center (Amersham, England), rotenone by F.P. Penich and Co. (New York), malic enzyme, N-ethylmaleimide, oligomycin, mersalyl and oxaloacetate have been provided by Sigma, NADP^{+} by Boehringer, 2-butylmalonic acid by Aldrich; phthalonic acid was a gift of L. Troisi of the Organic Chemistry Department, Bari.

Loading of mitochondria with metabolites: Rat liver mitochondria were isolated as previously described[3,4], using a medium consisting of 0.25 M sucrose, 1 mM EGTA^{+} and 20 mM Tris-HCl, pH 7.25. The mitochondria (40–50 mg of protein) were incubated at 20°C in 10 ml medium consisting of 100 mM KCl, 1 mM EGTA, 20 mM Tris-HCl, pH 7.0, 1 µg/ml of rotenone and 2 mM Pi plus 5 µg/ml of oligomycin or 2 mM oxoglutarate plus 1 mM arsenite. The mitochondria were left in incubation for 2 min before adding 1.5 mM N-ethylmaleimide then 1 min later, 50 ml of ice cold medium was appended. When oxoglutarate was used only 50 ml of ice cold medium was added after 2 min incubation. The mitochondria were washed in the medium without metabolite and then suspended (40–50 mg of mitochondrial protein). The intramitochondrial anion was labelled by adding to the mitochondrial suspension carrier free ^{32}Pi or (^{14}C)oxoglutarate (in the ratio of 1 µCi/ml each).

$^{+}$Abbreviations used: EGTA, (ethylene-bis(oxyethylenenitrilo))tetraacetic acid; HEPES, 4-(2-hydroxyethyl)-1-piperazineethanesulfonic acid.

Measurement of the metabolite efflux: Pi or oxoglutarate loaded mitochondria (about 2 mg protein) were incubated at 8°C in 1.0 ml of standard medium consisting of 0.2 M sucrose, 10 mM KCl, 20 mM HEPES$^+$-Tris, pH 7.0, 1 mM MgCl$_2$, 1 μg rotenone, 3 μg of oligomycin, and 1 mM N-ethylmaleimide (when Pi loaded mitochondria were used) or 1 mM arsenite (when oxoglutarate loaded mitochondria were used). After 1 min the assay was started by the addition of unlabeled substrate and stopped at the time indicated by rapid addition of 20 mM butylmalonate (as regard Pi) or 20 mM phenylsuccinate (as regard oxoglutarate). After rapidly centrifuging the mitochondria in an Eppendorf microcentrifuge for 2 min, the radioactivity in the pellet and in the supernatants was measured as described previously[2,5].

How to find out malate in the reaction mixture: The malate trapping system consists of 250 μM NADP$^+$ and 0.1 unit of malic enzyme. Before the experiments malic enzyme was dialysed in 100 mM Tris-HCl buffer, pH 7.0. No loss in the enzyme activity was found in this procedure.
Mitochondria (about 2 mg protein) were incubated in 1 ml of   standard medium at 30°C and the changes in fluorescence were measured at the Eppendorf photometer equipped with the fluorescence apparatus. The additions were made as described in the figures at the following concentrations: 0.25 mM oxaloacetate (OAA), 250 μM NADP$^+$, 0.1 unit malic enzyme (M.E.) 45 μM mersalyl (Mers.) 1 mM phthalonate (Phtha.).
Protein were measured with a modified biuret method[6].

RESULTS

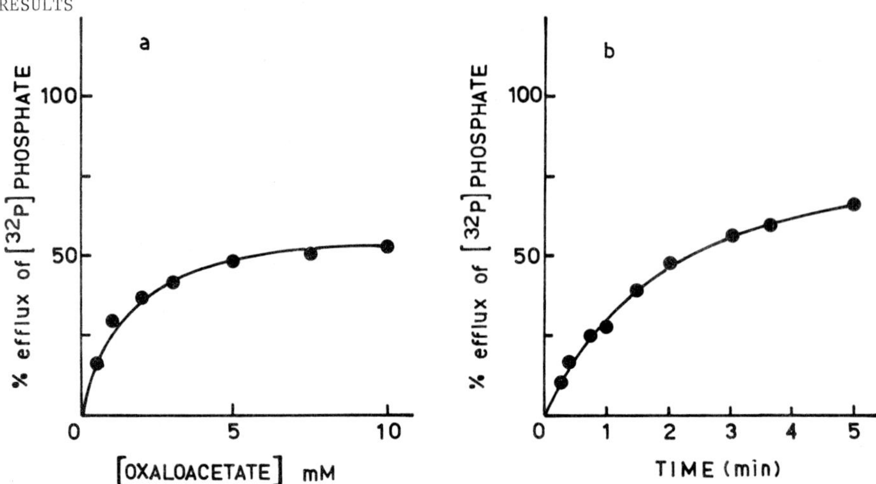

Fig. 1. (a) Efflux of Pi induced by added oxaloacetate. ^{32}Pi loaded mitochondria were incubated in the presence of oxaloacetate for 2 min at 8°C (see Materials and Methods). Mitochondrial protein was 1.6 mg. (b) Time course of the Pi efflux induced by 5 mM oxaloacetate (see Materials and Methods). Mitochondrial protein was 2 mg.

In Fig. 1a. the dependence of the Pi efflux on increasing concentrations of oxaloacetate is reported. The efflux, measured in 2 min, occurs via the dicarboxylate carrier, in fact N-ethylmaleimide, specific inhibitor of Pi own carrier[7,8], is present in the reaction mixture. Half maximum exchange is found at 1 mM oxaloacetate concentration. On the other hand no significant difference is found in the Pi efflux by rising oxaloacetate concentration from 5 to 10 mM. The same results have been attained in several experiments.

In Fig. 1b. the time course of the Pi efflux induced by 5 mM oxaloacetate is shown. The Pi efflux increases with time, without reaching equilibrium after 5 min, when the percentage of efflux is 70%. But surprising enough no evident increase of the Pi efflux has been found by varying oxaloacetate concentration from 5 to 10 mM whereas the Pi efflux induced by 5 mM oxaloacetate increases after 2 min; it should be noted that the Km of the Pi efflux induced by oxaloacetate, ranges between 6.5 and 10 mM[2]. In order to clear up this point, the possible involvement of the oxoglutarate carrier on the Pi efflux has been tested.

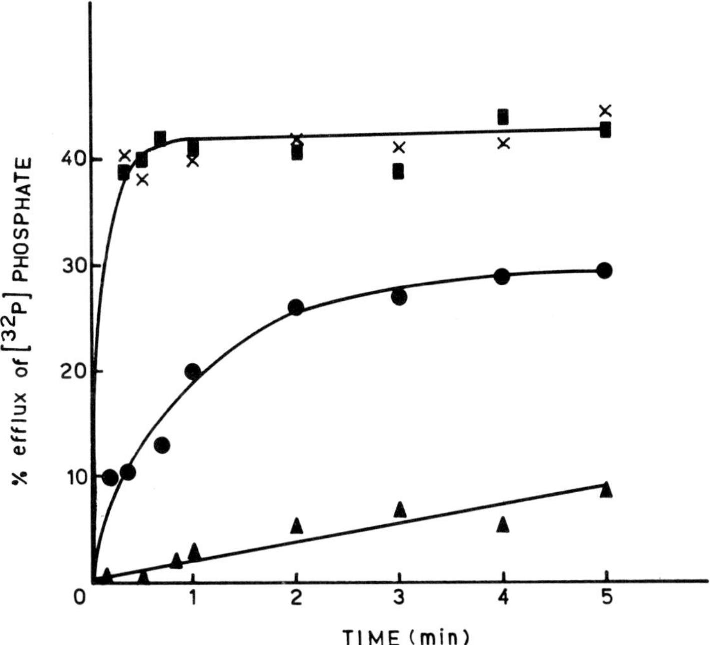

Fig. 2.   Effect of phthalonate on the time course of the Pi efflux induced by Pi and oxaloacetate. [32]Pi loaded mitochondria were incubated in the presence of 1 mM Pi (■,✗) or 1 mM oxaloacetate (●,▲). ✗,△ , 0.5 mM phthalonate was added together with the substrates (see Materials and Methods). Mitochondrial protein was 2 mg.

Phthalonic acid has been recently proposed by Meijer et al.[9] as a rather specific inhibitor of the oxoglutarate carrier in rat liver mitochondria. In the reported

428

experimental conditions this compound competitively inhibits the oxoglutarate/
oxoglutarate exchange with a low Ki (40 µM). Fig. 2 shows the time course of the
Pi efflux induced by 1 mM Pi and 1 mM oxaloacetate in Pi loaded mitochondria, in
the absence or in the presence of 0.5 mM phthalonate. Pi/Pi exchange occurs at
high rate and equilibrium is reached after 1 min. No inhibition by 0.5 mM
phthalonate is found. On the other hand, the Pi efflux induced by oxaloacetate
appears to be slower and equilibrium is reached after 4 min. Phthalonate strongly
inhibits this efflux which suggests that the oxoglutarate carrier activity is
involved in the Pi efflux.

The inhibition caused by phthalonate on the Pi efflux due to oxaloacetate can
be tentatively explained with the hypothesis described in the Scheme I.

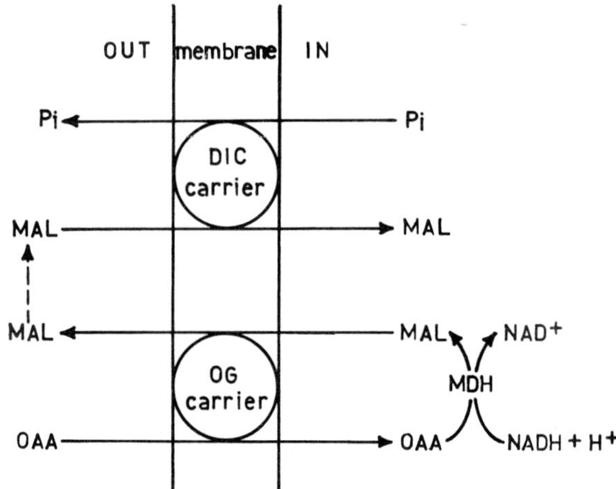

Scheme I: Mechanism of Pi efflux induced by oxaloacetate. Abbreviations used:
OAA, oxaloacetate; MAL., malate; MDH, malic dehydrogenase; Pi, inorganic
phosphate; OG oxoglutarate; DIC., dicarboxylate.

Oxaloacetate enters the mitochondria in exchange with endogenous oxoglutarate or
malate via the oxoglutarate carrier. In the mitochondrial matrix oxaloacetate is
reduced by means of the malic dehydrogenase and of the intramitochondrial NADH.
Malate comes out in exchange with further oxaloacetate and may be taken up via
the dicarboxylate carrier in exchange with Pi.

In order to check this assumption the appearance of malate in the reaction
mixture after the addition of oxaloacetate was tested by measuring the increase of
fluorescence due to the $NADP^+$ reduction caused by malate in the presence of malic
enzyme according to

$$\text{MALATE} + \text{NADP}^+ \xrightarrow{\text{MALIC ENZYME}} \text{PYRUVATE} + CO_2 + \text{NADPH} + H^+$$

A typical experiment is reported in Fig. 3a. Oxaloacetate was added to fresh mitochondria in the presence of rotenone, causing rapid oxidation of the intramitochondrial NADH. The addition of 250 µM $NADP^+$ and 0.1 unit of malic enzyme produces an increase of fluorescence. This datum suggests that malate comes out from the matrix after oxaloacetate permeation; in the experiment it was verified that no change in the oxidation state of $NADP^+$ occurs in the absence of oxaloacetate after the addition of $NADP^+$ and malic enzyme. It is generally accepted that malate is a substrate for the dicarboxylate, tricarboxylate and oxoglutarate carrier[10]. It was reported that no inhibition of the oxaloacetate/ malate exchange is found in the presence of benzene 1,2,3-tricarboxylate, the specific inhibitor of the tricarboxylate carrier[11,12]. This suggests that no oxaloacetate/malate exchange can be mediated by this carrier[2].

In order to check if the malate efflux is mediated by the dicarboxylate carrier the rates of NADH oxidation and $NADP^+$ reduction were tested with respect to their sensitivity to a mersalyl concentration able to inhibit only the dicarboxylate carrier (Fig. 3b, 3c). The initial rate of NADH oxidation is not affected by mersalyl, in agreement with previous results [13], however the -SH reagent inhibits the extent of the oxidation. Besides malic enzyme is irreversibly inhibited by mersalyl. After the incubation of the mitochondrial suspension with a solution of mersalyl, oxaloacetate is added and this -SH reagent is removed by appending 1 mM cysteine in the absence or in the presence of the malic enzyme and of $NADP^+$. Cysteine restores the rate of NADH oxidation; if afterwards $NADP^+$ and malic enzyme are added, a reduction of $NADP^+$ occurs at a rate similar to that obtained without addition of mersalyl; on the contrary, if $NADP^+$ and malic enzyme are added together with cysteine, high inhibition of NADH oxidation is found, likely due to the simultaneous reduction of $NADP^+$ which later on appears evident. This finding suggests that malate was present in the reaction mixture just before $NADP^+$ and malic enzyme addition, which proves that malate is not transported by the dicarboxylate carrier.

The effect of 1 mM phthalonate was also tested in order to check if malate efflux induced by oxaloacetate was due to the oxoglutarate carrier (Fig. 3d, 3e). Phthalonate added together with oxaloacetate strongly inhibits the rate of NADH oxidation. Little $NADP^+$ reduction is found when malic enzyme is added. On the other hand, when phthalonate is added together with malic enzyme and $NADP^+$ the rate of $NADP^+$ reduction does not vary significatively, however the extent of reduction decreases. These results suggest that malate efflux is mediated by the oxoglutarate carrier. It should be noted that the malic enzyme activity is not affected by phthalonate under the same experimental conditions.

$NADP^+$ reduction after the penetration of oxaloacetate was also tested with oxoglutarate loaded mitochondria in the presence of 1 mM arsenite which inhibits the oxoglutarate dehydrogenase. Fig. 3f shows that the initial rate of NADH oxidation is higher than that obtained with fresh mitochondria, however no $NADP^+$

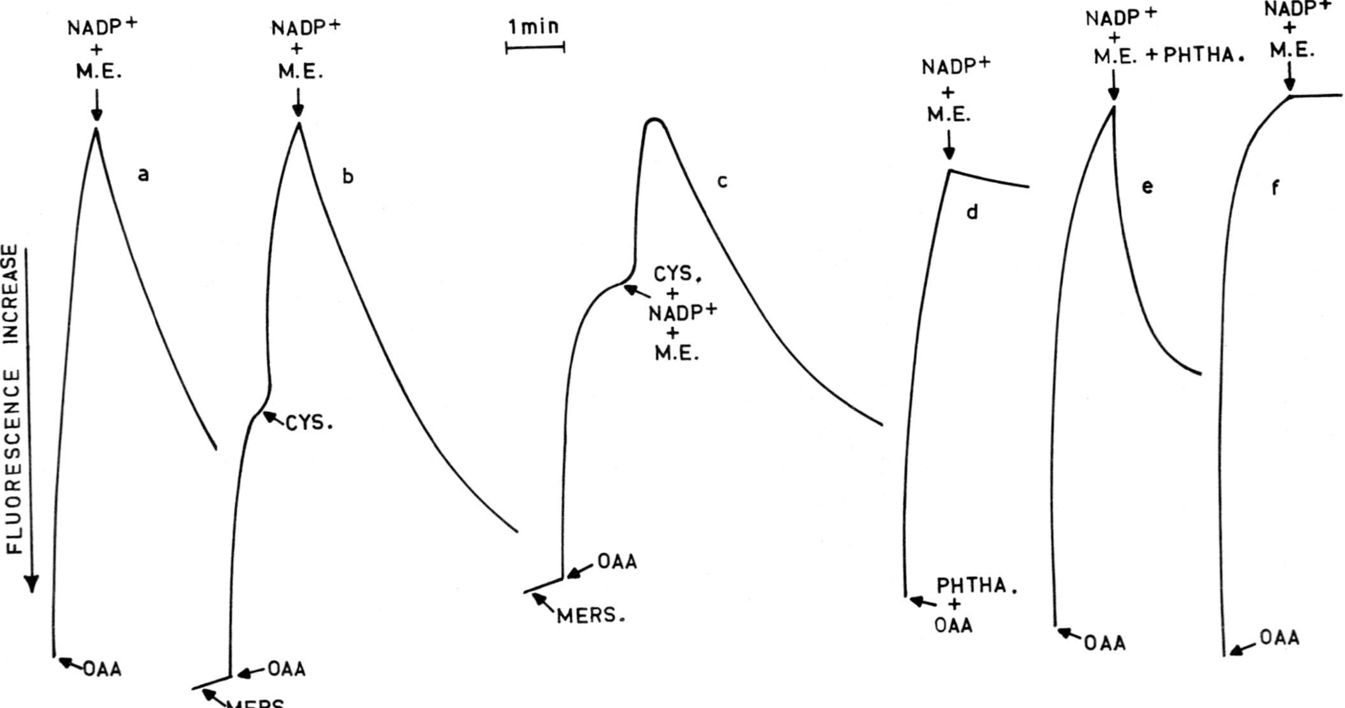

Fig. 3. Pointing out the malate efflux induced by oxaloacetate addition to fresh (a–d) or oxoglutarate loaded (f) rat liver mitochondria and its sensitivity to mersalyl (b,c) and phthalonate (d,e) (see Materials and Methods). Mitochondrial protein was 1.9 mg (a–e) and 1.5 mg (f).

reduction is found when malic enzyme and NADP$^+$ are added. This suggests that no malate is present in the reaction mixture.

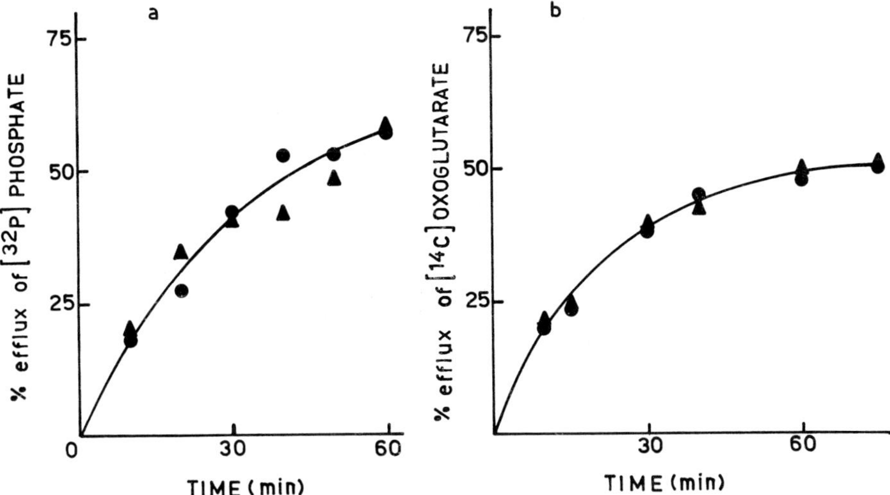

Fig. 4. Effect of NADP$^+$ and malic enzyme on the time course of Pi/Pi (a) or oxaloacetate/oxoglutarate (b) exchanges ^{32}Pi or ^{14}C-oxoglutarate loaded mitochondria were incubated in the presence of 1 mM Pi (a) or 1 mM oxaloacetate (b) in the absence ($\bullet$) or in the presence ($\blacktriangle$) of 250 μM NADP$^+$ and 0.1 unit of malic enzyme (see Materials and Methods). Mitochondrial protein was 1.3 (a) and 1 (b) mg.

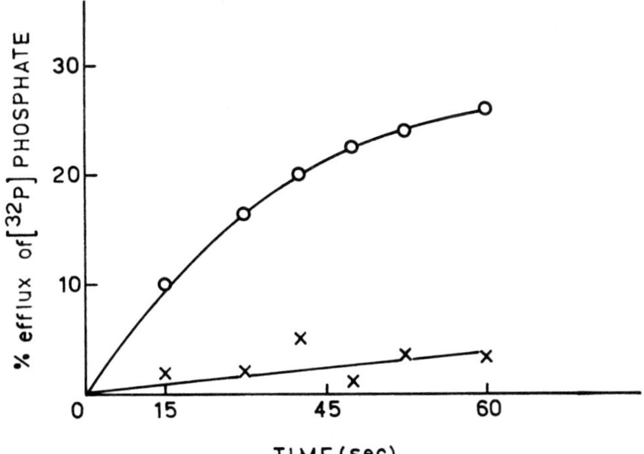

Fig. 5. Effect of NADP$^+$ and malic enzyme on the time course of Pi efflux induced by oxaloacetate. ^{32}Pi loaded mitochondria were incubated in the presence of 1 mM oxaloacetate (O) for the time indicated at 8°C (see Materials and Methods). $\times$, 250 μM NADP$^+$ and 0.1 unit of malic enzyme were added together with oxaloacetate. Mitochondrial protein was 1.95.

In Fig. 4 are reported the time courses of the Pi (a) and oxoglutarate (b) efflux induced by 1 mM Pi and 1 mM oxaloacetate respectively. The effect of 250 µM NADP[+] and 0.1 unit of malic enzyme has been also tested. No inhibition by NADP[+] and malic enzyme was found in both the reactions. This suggests that the presence of the malate trapping system does not affect the activity of the dicarboxylate (4 a) and of the oxoglutarate carrier (4 b) . In Fig. 5 the time course of the Pi efflux induced by 1 mM oxaloacetate is reported in the absence or in the presence of 250 µM NADP[+] and 0.1 unit of malic enzyme. The Pi efflux induced by oxaloacetate is strongly inhibited by NADP[+] and malic enzyme.

DISCUSSION

The experimental data reported in this work strongly suggest that no uptake of oxaloacetate occurs in rat liver mitochondria via the dicarboxylate carrier. In the presence of N-ethylmaleimide, Pi is transported across the mitochondrial membrane bound to the specific site of the dicarboxylate carrier (for ref. see 13). We have recently proved that oxoglutarate, which per se neither has effect on Pi/Pi exchange in Pi loaded mitochondria, nor promotes Pi efflux, completely inhibits Pi efflux induced by oxaloacetate; i.e. when the oxoglutarate carrier does not transport oxaloacetate no Pi efflux occurs. This finding is confirmed by using phthalonate, which acts as an impermeable inhibitor of the oxoglutarate carrier. This oxoacid, in fact, strongly inhibits Pi efflux induced by oxaloacetate. It was reported that phthalonate inhibits also the dicarboxylate carrier although with less affinity than that shown for the oxoglutarate carrier (2 mM v.s. 25-40 µM)[9]. The phthalonate affinity for the dicarboxylate carrier however, does not account for the found inhibition of a possible direct oxaloacetate/Pi exchange. In fact the theoretical percentage of inhibition calculated considering Vmax = 50 µmoles/min x g protein, oxaloacetate Km = 6.5 mM, phthalonate Ki = 2 mM, phthalonate concentration 0.5 mM, is about 15-20%, on the contrary the rate of Pi efflux is inhibited about 90%.

In order to make clear the process of Pi efflux, involving the oxoglutarate carrier, the fluorescence measurements can be usefully discussed. When oxaloacetate is added to fresh mitochondria malate efflux occurs, as proved by NADP[+] reduction; on the other hand, if oxoglutarate loaded mitochondria were used, no reduction resulted; this datum suggests that oxoglutarate is the preferred counteranion. It should be noted that the intramitochondrial oxoglutarate promotes oxaloacetate uptake, i.e. the rate of NAD(P)H oxidation. Malate efflux occurs only via the oxoglutarate carrier, as confirmed by mersalyl insensitivity, and by the inhibition shown by phthalonate added together with oxaloacetate. When no phthalonate is present, malate appears in the reaction mixture and the addition of phthalonate to NADP[+] and malic enzyme, only curbs the extent of NADP[+] reduction. This compound in fact inhibits further malate efflux.

The involvement of the oxoglutarate carrier in the Pi efflux induced by oxaloacetate as well as the malate appearance in the reaction mixture are finally correlated by the evidence that the malate trapping system ($NADP^+$ plus enzyme) strongly inhibits Pi efflux induced by oxaloacetate, without affecting the dicarboxylate and oxoglutarate carrier activity. This finding proves that in rat liver mitochondria malate/Pi exchange occurs when oxaloacetate is added and this exchange is due to the malate efflux as result of the addition of oxaloacetate. Hence direct OAA/Pi exchange is to be excluded and we can state that the oxoglutarate carrier is the single carrier used for the oxaloacetate translocation.

This conclusion may explain the reason why the Pi efflux fails to increase at rising oxaloacetate concentration above 5 mM; this concentration in fact is somewhat saturating for the oxoglutarate carrier[2], so that no significant increase of malate efflux is found at higher concentrations, on the other hand if further time is allowed for the uptake of oxaloacetate the Pi efflux increases above the reported percentage. In not reported experiment the ability of phthalonate in inhibiting the initial rate of malonate/oxaloacetate exchange, in malonate loaded mitochondria, was tested. The oxoacid completely collapses the malonate efflux induced by oxaloacetate, producing little inhibition on malonate/malonate exchange. This suggests that no malonate efflux is caused by oxaloacetate via dicarboxylate carrier. Besides oxaloacetate causes the efflux of sulphate from the mitochondrial matrix definitively sensitive to phthalonate. Sulphate is proved to act as a substrate only for the dicarboxylate carrier[14]. This datum confirms that oxaloacetate is able to cause efflux of substrates of the dicarboxylate carrier in a process in which the oxoglutarate carrier plays an important role as revealed by the sensitivity of sulphate efflux to phthalonate. These achievemnts confirm the existence in rat liver mitochondria of a single carrier, i.e. the oxoglutarate carrier, which catalyzes oxaloacetate/malonate exchange. Thus the experimental data suggest the reliability of the reported scheme, showing that a single carrier i.e. the oxoglutarate carrier transports oxaloacetate in rat liver mitochondria.

REFERENCES

1. Gimpel, G. A., De Haan, E. J., and Tager, J. M. (1973) Biochim. Biophys. Acta 292, 582-591.
2. Passarella, S., Palmieri, F., and Quagliariello, E. (1977) Arch. Biochem. Biophys. 180, 160-168.
3. Klingenberg, M., and Slenczka, W. (1959) Biochem. Z. 331, 486-517.
4. Tyler, D.D., and Gonze, J. (1967) in Methods in Enzymology (Estabrook, R. W., and Pulman, M. E., eds.), Vol. 10, pp. 75-77, Academic Press, New York.
5. Palmieri, F., Quagliariello, E., and Klingenberg, M. (1970) Eur. J. Biochem. 17, 230-238.

6. Kröger, A., and Klingenberg, M. (1966) Biochem. Z. 344, 317-336.

7. Meijer, A. J., Groot, G. S. P., and Tager, J. M. (1970) FEBS Lett. 8, 41-44.

8. Robinson, B. H. (1970) FEBS Lett. 11, 200-204.

9. Meijer, A. J., Van Woerkom, G. M., and Eggelte, T. A. (1976) Biochim. Biophys. Acta 430, 53-61.

10. Klingenberg, M. (1970) in Essays in Biochemistry (Campbell, P. N., and Dickens, F., eds.), Vol. 6, pp. 119-159, Academic Press, London, New York.

11. Robinson, B. H., Williams, G. R., Halperin, M. L., and Leznoff, C. (1970) Eur. J. Biochem. 15, 263-272.

12. Robinson, B. H., Williams, G. R., Halperin, M. L., and Leznoff, C. (1971) Eur. J. Biochem. 20, 65-71.

13. Passarella, S., and Quagliariello, E. (1976) Biochimie 58, 989-1001.

14. Crompton, M., Palmieri, F., Capano, M., and Quagliariello, E. (1974) Biochem. J. 142, 127-137.

EVIDENCE FOR LOCALIZED ENERGY TRANSFER IN RECONSTITUTED SUBMITOCHONDRIAL PARTICLES

Kerstin Nordenbrand, Torill Hundal, Christine Carlsson, Gabriella Sandri[*] and
Lars Ernster
Department of Biochemistry, Arrhenius Laboratory, University of Stockholm,
Stockholm, Sweden

A problem of considerable interest in current studies of membrane bioenergetics
(see ref. 1 for review) is whether different energy-transducing units - electron-
transport catalysts, ATPases, and various ion translocators - interact only by way
of a transmembrane electrochemical gradient, in accordance with the chemiosmotic
hypothesis (2), or whether there are instances of localized interactions between
such units within the membrane. Earlier studies in this laboratory (3-5) have led
to the conclusion that the ATP-driven nicotinamide nucleotide transhydrogenase
reaction catalyzed by submitochondrial particles involves a localized interaction
between ATPase and transhydrogenase in the mitochondrial inner membrane. This
conclusion was primarily based on repeated observations (3-6), using various types
of submitochondrial particles, that the ATPase and ATP-driven transhydrogenase
activities were inhibited with equal efficiencies by the ATPase-inhibitor protein
of Pullman and Monroy (7), under conditions when the ATPase was not rate-limiting
for the ATP-driven transhydrogenase. In interpreting these findings it was assumed that, under the conditions employed, the ATPase inhibitor was not dissociated
from the ATPase - an assumption that was recently verified by Ferguson et al. (8).

More direct evidence for a localized interaction between ATPase and transhydrogenase was obtained in preliminary experiments (4,5) in which beef-heart EDTA particles[**] depleted of ATPase inhibitor and F_1 by treatment with Sephadex and urea
(ESU particles) were reconstituted by adding increasing amounts of purified F_1 and
the restorations of oligomycin-sensitive ATPase and ATP-driven transhydrogenase
were followed. Fig. 1 shows such an experiment. Addition of increasing amounts
of purified beef-heart F_1 to ESU particles resulted in increasing activities of
reconstituted, oligomycin-sensitive ATPase activity until a plateau was reached at

---

[*]Recipient of a travel fellowship from Stiftelsen Blanceflor Boncompagni-Ludovisi,
född Bildt. Permanent address: Istituto di Biochimica, Università di Trieste,
Trieste, Italy.

[**]Abbreviations: EDTA particles, submitochondrial particles prepared in a medium
containing EDTA; ES particles, EDTA particles treated with Sephadex G-50; ESU
particles, ES particles treated with urea; TH, nicotinamide nucleotide transhydrogenase; OS-ATPase, oligomycin-sensitive ATPase; AI, ATPase inhibitor; ANS, 8-ani-
linonaphthalene-1-sulfonate; DCCD, N,N'-dicyclohexyl carbodiimide; FCCP, carbonyl
cyanide trifluoromethoxy phenylhydrazone; Nbf-Cl, 4-chloro-7-nitrobenzofurazan;
PMS, phenazine methosulfate; TMPD, N,N,N',N'-tetramethyl-p-phenylene diamine.

436

about 0.1 mg F_1 per mg ESU particle protein; this value is similar to recent estimates of the F_1 content of native submitochondrial particles (12,12A). It may also be
noted that the maximal oligomycin-sensitive ATPase activity was between 50 and 60
units per mg F_1, which was only slightly below the (oligomycin-insensitive) ATPase
activity of the purified F_1 used in these experiments, appr. 75 units/mg protein.

It may be seen in Fig. 1 that, in accordance with results already reported

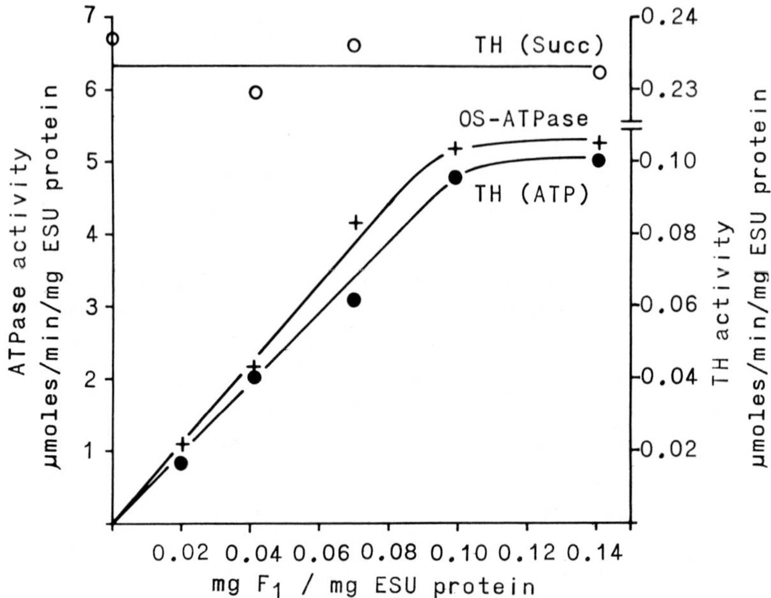

Fig. 1.   Effects of F_1 on the oligomycin-sensitive ATPase (OS-ATPase) and the
ATP- and succinate-driven transhydrogenase (TH) activities of ESU particles.
ESU particles were prepared from EDTA particles (9) by treatment with Sephadex and
urea as described by Racker and Horstman (10).   Beef-heart F_1 was purified
according to Horstman and Racker (11).   ESU particles (10 mg protein/ml) were
preincubated with F_1 in the proportions indicated in a medium consisting of
0.25 M sucrose, 10 mM Tris sulfate, pH 7.5, 2 mM EDTA for 10 min at 25°C.   Aliquots of 1-5 µl and of 50-250 µl, respectively, were used for the assay of ATPase
and TH activities.   ATPase activity was assayed in a medium containing 30 mM
potassium acetate, 25 mM Tris acetate, pH 7.5, 3 mM magnesium acetate, 0.2 mM
NADH, 1.7 µM rotenone, 1 mM phospho(enol)pyruvate, 50 µg lactate dehydrogenase,
50 µg pyruvate kinase, 1 mM FCCP, 3 mM ATP, in a final volume of 3 ml.   The oxidation of NADH was followed spectrophotometrically at 340 nm.   OS-ATPase activity
was determined as the difference in ATPase activity in the absence and presence
of 5 µg oligomycin.   TH activity was assayed in a medium containing 50 mM Tris
acetate, pH 7.5, 170 mM sucrose, 0.1 mM NADH, 0.2 mM NADP$^+$, 150 µg alcohol dehydrogenase, 8 mM hydrazine, 75 mM ethanol, 1.7 µM rotenone, 5 mM MgCl$_2$, and either
3 mM ATP and 0.3 µg oligomycin/mg ESU particle protein or 5 mM succinate and 5
µg rotenone/mg particle protein, in a final volume of 3 ml. The reduction of NADP$^+$
was followed spectrophotometrically at 340 nm. The reaction rates were corrected for
the nonenergy-linked TH activity measured prior to the addition of ATP or succinate.
Temperature, 25°C.

briefly (4,5), the ATP-driven transhydrogenase activity of the ESU particles increased parallel to the oligomycin-sensitive ATPase activity as a function of added F_1, and the two activities reached a plateau at the same amount of added F_1. The ATP-driven transhydrogenase activity (expressed as nanomoles NADH + $NADP^+$ converted into NAD^+ + NADPH per min and mg particle protein) was approximately 50 times lower than the oligomycin-sensitive ATPase activity (expressed as nanomoles ATP hydrolyzed per min and mg particle protein) at all levels of added F_1. Since the conversion of 1 mole of NADH + $NADP^+$ to NAD^+ + NADPH through the ATP-driven transhydrogenase reaction requires the expenditure of 1 mole of ATP (13), it follows that the ATPase activity was in large – appr. 50-fold – excess of the ATP-driven transhydrogenase. Fig. 1 also shows that the energy-linked transhydrogenase activity driven by succinate as the energy source was, as expected, independent of added F_1. This activity was about twice as high as the ATP-driven transhydrogenase. It may thus be concluded that neither the ATPase nor the transhydrogenase was rate-limiting for the ATP-driven transhydrogenase activity. The experiments described in the following have been performed in an attempt to define the rate-limiting step in the ATP-driven transhydrogenase reaction of this reconstituted system.

Ferguson et al. (14) have recently shown that covalently bound, irreversible, ATPase inhibitors such as Nbf-Cl or DCCD diminish the extent of the ATP-induced ANS-fluorescence enhancement of submitochondrial particles. In contrast, reversible inhibitors of ATPase, e.g. replacement of ATP by the less efficient substrate ITP, diminished the initial rate, but not the final extent, of the ATP-induced ANS response. These findings gave strong support to earlier indications (15-21) that the energy-linked ANS response is an expression of a localized event (e.g. a local charge-separation) in the inner mitochondrial membrane, rather than a bulk transmembrane potential (22,23), and indicated that each ATPase molecule may be able to energize primarily only a limited area within the membrane. In the light of these findings it was of interest to compare the ATP-induced ANS response with the ATPase and ATP-driven transhydrogenase activities of the F_1-reconstituted ESU particles. Such a comparison is shown in Fig. 2. Clearly, the extent of the ATP-induced ANS response parallelled the ATPase and ATP-driven transhydrogenase activities as functions of the added amount of F_1 per mg ESU particle protein. These data thus seem to be consistent with the conclusion that what limits the ATP-driven transhydrogenase activity at any given level of F_1 is the relative area of the membrane and thereby the relative number of transhydrogenase molecules that can be energized through the ATPase reaction.

To substantiate this conclusion it was important to eliminate the possibility that the ATP-driven transhydrogenase activity in the reconstituted particles is limited by a general energy leak through the membrane. It is known (13,24-26) that EDTA particles and, consequently, ES and ESU particles derived from them, are

438

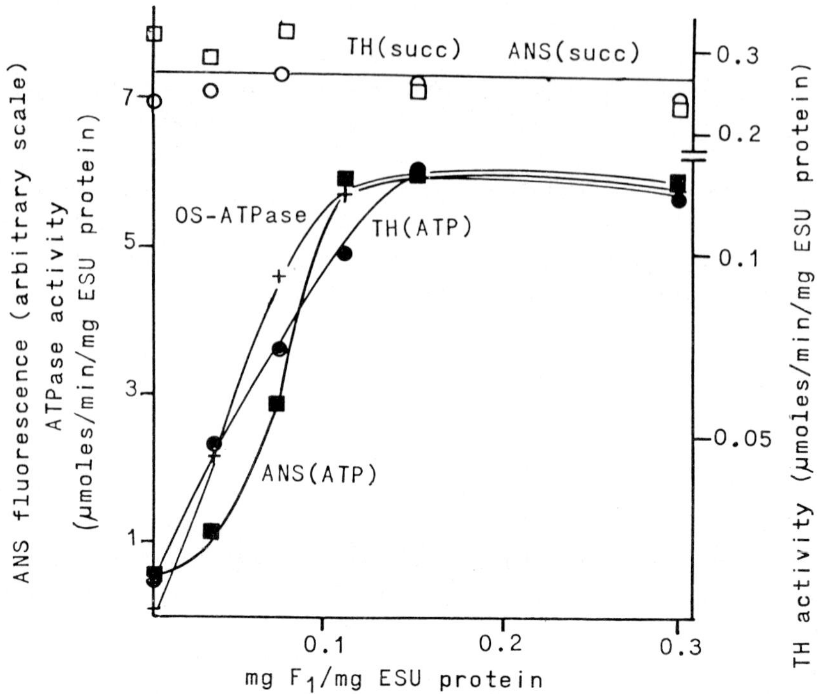

Fig. 2. Effects of F_1 on the oligomycin-sensitive ATPase (OS-ATPase), the ATP-
and succinate-driven transhydrogenase (TH) activities, and the extents of ATP- and
succinate-induced ANS-fluorescence enhancements of ESU particles. Preincubation
of ESU particles with varying amounts of F_1 and assays of ATPase and TH activities
were as in Fig. 1. The ANS-fluorescence enhancement was measured as earlier described (19,20). The assay mixture contained 10 mM Tris-acetate, pH 8.8, 180 mM
sucrose, 20 μM ANS, 0.8 mg ESU particle protein, F_1 as indicated, and either 5 mM
succinate, 1.7 μM rotenone and 5 μg oligomycin/mg particle protein, or 3 mM ATP,
3.3 mM $MgSO_4$ and 0.3 μg oligomycin/mg protein, in a final volume of 3 ml. Fluorescence was measured at 470 nm, with an excitation wavelength of 366 nm. After
reaching steady state following the addition of succinate or ATP, 1 μM FCCP was
added, and the decrease of the extent of fluorescence was taken as a measure of
the succinate - or ATP-induced ANS response. Temperature, 27°C.

virtually uncoupled, indicating that their membranes possess an extensive energy

leak, probably a proton leak. A large portion of this leak seems to proceed spe-

cifically via the "membrane sector" - probably the proton translocator - of the

ATPase system, since oligomycin induces a high degree of respiratory control in

these particles (13,24-26); the amount of oligomycin needed for maximal respira-

tory control is larger in F_1-depleted than in F_1-containing particles (4,27),

suggesting that the removal of F_1 increases the leak. In the case of ATP-driven

reactions, the situation is more complex (13,24,25); low concentrations of oligo-

mycin stimulate these reactions, by diminishing the leak, and high concentrations

inhibit them, by inhibiting the ATPase. This situation prevailed in the systems

here studied, and, in fact, the ATP-driven transhydrogenase activity and the ex-

tent of the ATP-induced ANS response in the experiments shown in Figs. 1 and 2
were measured in the presence of a maximally stimulating concentration of oligomycin. As may be seen in Fig. 3, both parameters were maximal at an oligomycin

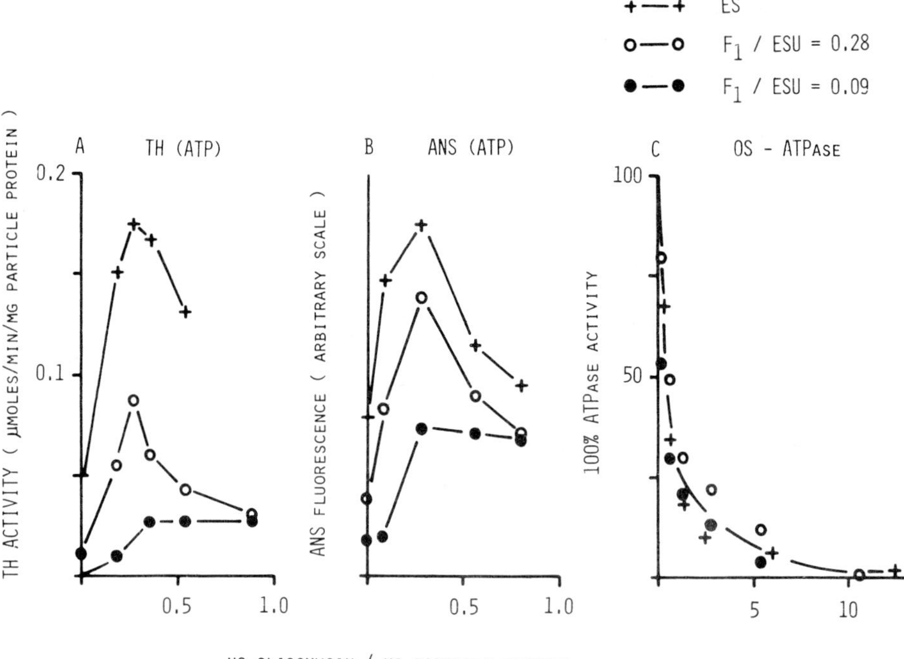

Fig. 3. Oligomycin titrations of the ATP-driven transhydrogenase activity, the
extent of ATP-induced ANS-fluorescence response and oligomycin-sensitive ATPase
(OS-ATPase) activity of ES and reconstituted ESU particles. Assay conditions as
in Figs. 1 and 2.

concentration of 0.3 μg per mg particle protein both in the original ES particles
and in the reconstituted ESU particles, the latter tested at either "saturating"
or "non-saturating" levels of F_1. At higher concentrations of oligomycin, both
parameters decreased, markedly in the ES and the fully F_1-saturated ESU particles,
and less so in the partially F_1-saturated ESU particles. The ATPase activities of
the three types of particles exhibited equal oligomycin titers and were as expected only partially inhibited (by about 50 %) by 0.3 μg oligomycin per mg particle
protein.

Although these results were consistent with the conclusion that the ATP-driven
transhydrogenase activity and the ATP-induced ANS response of the oligomycin-supplemented, F_1-deficient particles were not limited by a general energy leak, they
were clearly insufficient to prove this conclusion because of the dual

effect of oligomycin on these parameters. A more promising approach seemed possible by testing the effect of an uncoupler, FCCP, on the two ATP-dependent energy-linked reactions of the reconstituted particles. It has been shown (20) that
increasing the capacity of the energy-generating system in submitochondrial particles results in a decreased sensitivity of energy-requiring reactions to uncouplers. For example, the FCCP sensitivity of the ascorbate + PMS-supported ANS response of EDTA particles has been found to decrease markedly when the rate of oxygen consumption is enhanced by increasing the concentration of PMS (20). Similarly, as shown in Fig. 4, the succinate-driven transhydrogenase activity of ES particles became increasingly FCCP-sensitive as the rate of succinate oxidation, and

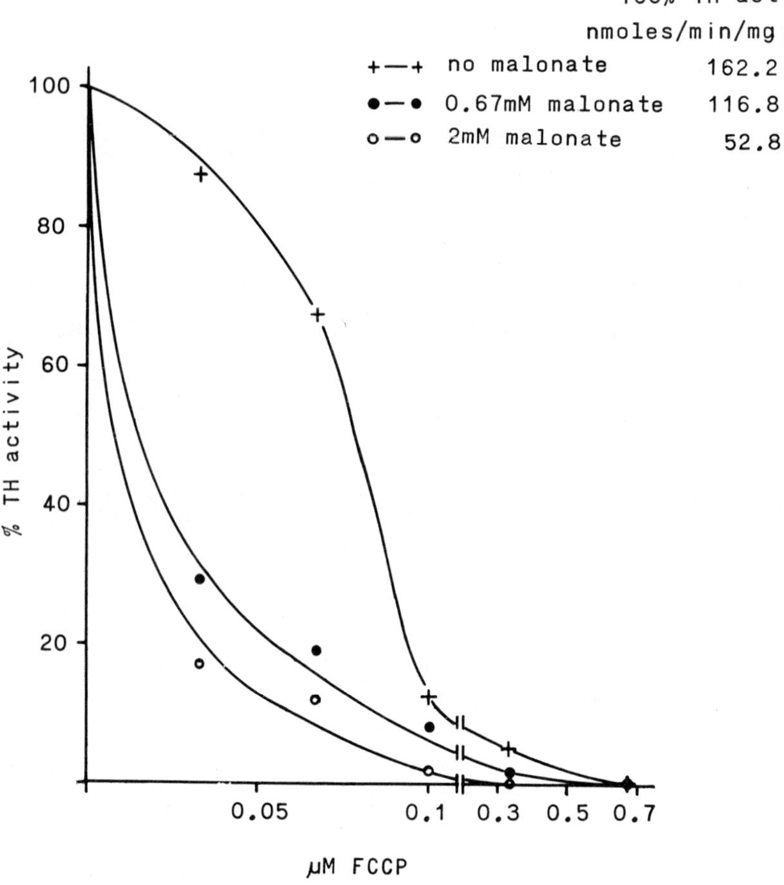

Fig. 4. Effect of malonate on the FCCP sensitivity of succinate-driven transhydrogenase (TH) reaction in ES particles. Assay conditions were as in Fig. 1.

thereby the rate of the succinate-driven transhydrogenase, was diminished by increasing concentrations of malonate. It would appear, thus, that the more limiting the "bulk" energy supply (available, for example, as a transmembrane electrochemical gradient), the more sensitive the system becomes to an energy leak induced by an uncoupler.

In the experiment shown in Fig. 5A, the succinate-driven transhydrogenase activity of ES particles was diminished by the addition of malonate to match the ATP-driven transhydrogenase activity. It may be seen that the succinate-driven reaction was more sensitive to FCCP than the ATP-driven reaction. Similar results were obtained with tightly-coupled Mg^{++}-ATP particles treated according to Van de Stadt et al. (28) to remove ATPase inhibitor, in which the ATP-driven transhydrogenase activity was close to the succinate-driven activity in the absence of

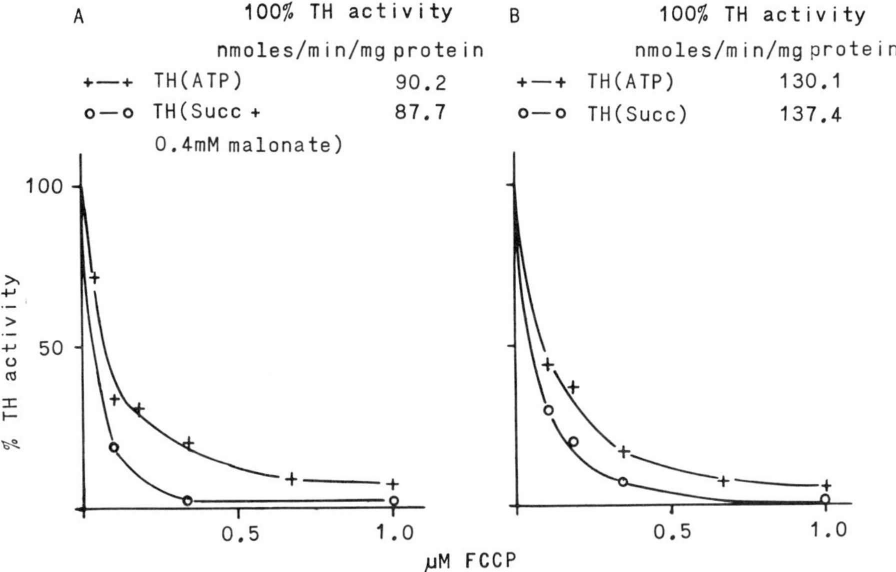

Fig. 5. Comparison of FCCP sensitivities of the ATP- and succinate-driven transhydrogenase (TH) reactions of (A) ES particles and (B) Mg^{++}-ATP particles treated according to Van de Stadt et al. (28) to remove ATPase inhibitor. Assay conditions were as in Fig. 1, except that no oligomycin was added in experiment B.

malonate (Fig. 5B). It appears from these data that in these systems, with their endogenous complements of F_1, the "bulk" energy supply is not rate-limiting for the ATP-driven transhydrogenase reaction.

Fig. 6 compares the FCCP sensitivities of the ATP-driven transhydrogenase activities and of the extents of ATP-induced ANS response of reconstituted ESU

ESU PARTICLES + F_1

□   TH (ATP) 0.32 MG F_1 / MG ESU
■   TH (ATP) 0.04 MG F_1 / MG ESU
○   ANS FLUORESCENCE 0.32 MG F_1 / MG ESU
●   ANS FLUORESCENCE 0.04 MG F_1 / MG ESU

EDTA PARTICLES

▲   TH (ATP)
+   SUCC $\longrightarrow$ NAD$^+$ (ATP)

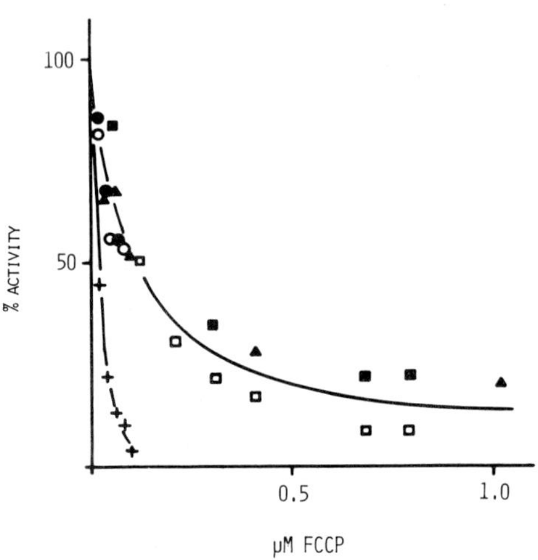

Fig. 6. Comparison of FCCP sensitivities of the ATP-driven transhydrogenase (TH) reaction and the ATP-induced ANS-fluorescence response of reconstituted ESU particles at different levels of saturation with F_1. Assay conditions were as in Figs. 1 and 2. For comparison the FCCP sensitivities of the ATP-driven transhydrogenase (TH) and succinate-linked NAD$^+$ reduction (succ $\rightarrow$ NAD$^+$) of EDTA particles are also indicated; these data are quoted from ref. 3.

particles at saturating and non-saturating levels of F_1. In all cases, the FCCP sensitivities were virtually equal. In addition, they were closely similar to the FCCP sensitivity of ATP-driven transhydrogenase activity of the parent EDTA particles. On the other hand, the FCCP sensitivities of these reactions were markedly lower than that of the ATP-driven succinate-linked NAD$^+$ reduction catalyzed by the same EDTA particles which, as earlier shown (3), is limited by the energy supply. These results strongly indicate that the ATP-driven transhydrogenase activity and the extent of the ATP-induced ANS response of the reconstituted ESU particles at either saturating or non-saturating levels of F_1 are not limited by the actual supply of energy but are most probably limited by the number of F_1 molecules available to locally energize the membrane.

It was of interest to investigate other F_1-dependent reactions of the reconstituted ESU particles. The rates of oxidative phosphorylation (with succinate as substrate) and ATP-driven succinate-linked NAD^+ reduction were too low to allow reliable estimates. Apparently, the particles were too leaky to achieve appreciable rates of these thermodynamically relatively unfavourable reactions (i.e., the forward reaction of coupling site II and the reverse of coupling site I) even in the presence of an optimal concentration of oligomycin. However, the oligomycin-supplemented ESU particles did catalyze a well-measurable, F_1- and ATP-dependent reduction of cytochrome b by ascorbate and TMPD, involving a reversal of coupling site II. The measurement of this reaction is illustrated in Fig. 7, which shows that addition of ascorbate + TMPD to F_1-reconstituted ESU particles in the absence of ATP resulted in an increase in the steady-state level of

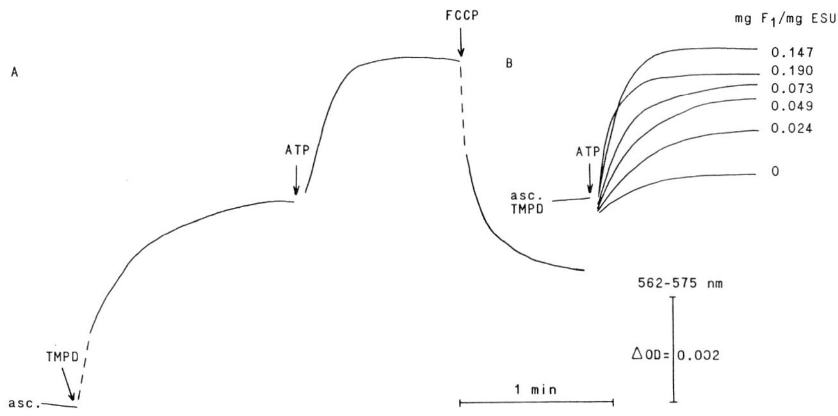

Fig. 7. Measurement of ascorbate + TMPD- and ATP-induced increase in the level of cytochrome b reduction in ESU particles, reconstituted by preincubation with F_1. In (A), F_1 was added at a level of 0.147 mg/mg particle protein; in (B), the level of added F_1 was as indicated. Preincubation was performed as described in Fig. 1. The assay mixture contained 50 mM Tris-acetate, pH 7.5, 170 mM sucrose, 2 mM KCN, 3.3 mM $MgSO_4$, 0.4 µg oligomycin, and 1.11 mg ESU particle protein (+ the indicated amount of F_1) in a final volume of 1 ml. When indicated, 3.3 mM ascorbate, 0.3 mM TMPD, 3 mM ATP and 1 µM FCCP were added. Temperature, 27°C.

reduced cytochrome b (measured at 562-575 nm) and that this level was further increased upon the addition of ATP; FCCP abolished the ATP-induced cytochrome b reduction. As shown in Fig. 8A, the extent of the ATP-induced cytochrome b reduction reached maximum at a level of added F_1 of about 0.1 µg per mg particle protein, i.e., at the same F_1 level as was found saturating for the ATP-driven transhydrogenase activity and the ATP-induced ANS response (cf. Figs. 1 and 2); the ATP-independent reduction of cytochrome b by ascorbate + TMPD was, as expected, also independent of added F_1. The effect of oligomycin on the ATP-induced cytochrome b reduction (Fig. 8B) was likewise similar to those found with the ATP-

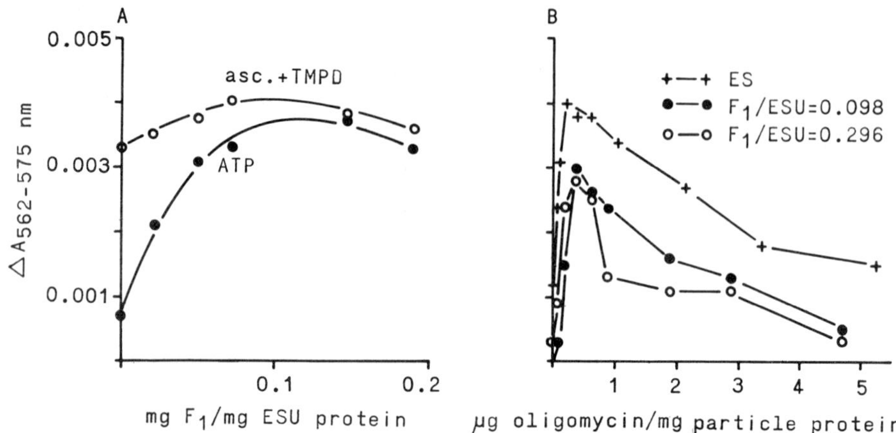

Fig. 8.   (A) Effect of F_1 on the extents of ascorbate + TMPD-induced and ATP-induced reduction of cytochrome $\underline{b}$ in ESU particles.   Measurements were made in the presence of 0.4 µg oligomycin/mg particle protein.
(B) Effect of oligomycin on ATP-induced increase in the extent of cytochrome $\underline{b}$ in ES particles (0.95 mg protein) and in ESU particles (1.08 mg protein) at different levels of F_1.
Other assay conditions were as in Fig. 7.

driven transhydrogenase activity and the ATP-induced ANS response (cf. Fig. 3). These preliminary results are thus consistent with the possibility that also coupling site II, i.e., the cytochrome $\underline{b}$-$\underline{c}_1$ complex, might interact with the ATPase system in a localized fashion.

In conclusion, the present data seem to provide further support for the idea of a localized interaction between the mitochondrial ATPase and nicotinamide nucleotide transhydrogenase, and, perhaps also between other membrane-associated energy-transducing catalysts, including the cytochrome $\underline{b}$-$\underline{c}_1$ complex.   Studies with purified, reconstitutively active preparations of ATPase (29), transhydrogenase (30) and cytochrome $\underline{b}$-$\underline{c}_1$ complex (31), now in progress in our laboratory, may throw further light on the existence and molecular mechanisms of these interactions.

This work has been supported by a grant from the Swedish Natural-Science Research Council.

REFERENCES

1. Boyer, P.D., Chance, B., Ernster, L., Mitchell, P., Racker, E. and Slater, E.C. (1977) Ann. Rev. Biochem. 46, 955-1026.

2. Mitchell P. (1966) Chemiosmotic Coupling in Oxidative and Photosynthetic Phosphorylation, Glynn Research, Bodmin, Cornwall.

3. Ernster, L., Juntti, K. and Asami, K. (1973) J. Bioenergetics 4, 149-159.

4. Ernster, L. (1975) FEBS Symp. 40, 253-276.

5. Ernster, L., Asami, K., Juntti, K., Coleman, J. and Nordenbrand, K. (1976) in Structure of Biological Membranes (S. Abrahamsson and I. Pascher, eds.) Nobel Symp. 34, 135-156, Plenum Press, New York.

6. Asami, K., Juntti, K. and Ernster, L. (1970) Biochim. Biophys. Acta 205, 307-311.

7. Pullman, M.E. and Monroy, G.C. (1963) J. Biol. Chem. 238, 3762-3769.

8. Ferguson, S.J., Harris, D.A. and Radda, G.K. (1977) Biochem. J. 162, 351-357.

9. Lee, C.P. and Ernster, L. (1967) Meth. Enzymol. 10, 543-548.

10. Racker, E. and Horstman, L.L. (1967) J. Biol. Chem. 242, 2547-2551.

11. Horstman, L.L. and Racker, E. (1970) J. Biol. Chem. 245, 1336-1344.

12. Vàdineanu, A., Berden, J.A. and Slater, E.C. (1976) Biochim. Biophys. Acta 449, 468-479.

12A Ferguson, S.J., Lloyd, W.J. and Radda, G.K. (1976) Biochem. J. 159, 347-353.

13. Lee, C.P. and Ernster, L. (1966) B.B.A. Library, vol. 7, 218-236.

14. Ferguson, S.J., Lloyd, W.J. and Radda, G.K. (1976) Biochim. Biophys. Acta 423, 174-188.

15. Azzi, A., Chance, B., Radda, G.K. and Lee, C.P. (1969) Proc. Natl. Acad. Sci. U.S. 62, 612-619.

16. Azzi, A. (1969) Biochem. Biophys. Res. Commun. 37, 254-260.

17. Brocklehurst, J.R., Freedman, R.B., Hancock, D.J. and Radda, G.K. (1970) Biochem. J. 116, 721-731.

18. Datta, A. and Penefsky, H.S. (1970) J. Biol. Chem. 245, 1537-1544.

19. Nordenbrand, K. and Ernster, L. (1971) Eur. J. Biochem. 18, 258-273.

20. Ernster, L., Nordenbrand, K., Lee, C.P., Avi-Dor, Y. and Hundal, T. (1971) in Energy Transduction in Respiration and Photosynthesis (E. Quagliariello, S. Papa, C.S. Rossi, eds.) pp. 57-87, Adriatica Editrice, Bari.

21. Radda, G.K. and Vanderkooi, J. (1972) Biochim. Biophys. Acta 265, 509-549.

22. Jasaitis, A.A., Kuliene, V.V. and Skulachev, V.P. (1971) Biochim. Biophys. Acta 234, 177-181.

23. Bakker, E.P. and Van Dam, K. (1974) Biochim. Biophys. Acta 339, 157-163.

24. Lee, C.P. and Ernster, L. (1965) Biochem. Biophys. Res. Commun. 18, 523-529.

25. Lee, C.P. and Ernster, L. (1968) Eur. J. Biochem. 3, 391-400.

26. Lee, C.P., Ernster, L. and Chance, B. (1969) Eur. J. Biochem. 8, 153-163.

27. Ernster, L., Nordenbrand, K., Chude, O. and Juntti, K. (1974) in Membrane
    Proteins in Transport and Phosphorylation (G.F. Azzone, M.E. Klingenberg,
    E. Quagliariello and N. Siliprandi, eds.) pp. 29-41, North-Holland, Amsterdam.
28. Van de Stadt, R.J., de Boer, B.L. and Van Dam, K. (1973) Biochim. Biophys.
    Acta 292, 338-349.
29. Glaser, E., Norling, B. and Ernster, L., this volume.
30. Rydström, J., Kanner, N. and Racker, E. (1975) Biochem. Biophys. Res. Commun.
    67, 831-839.
31. Gellerfors, P. and Nelson, B.D. (1975) Eur. J. Biochem. 52, 433-443.

Bioenergetics of Membranes. L. Packer et al. ed.

CONFORMATIONAL CHANGE, ATP GENERATION AND TURNOVER RATE OF THE CHLOROPLAST ATPase ANALYZED BY ENERGIZATION WITH AN EXTERNAL ELECTRIC FIELD

H. T. Witt, E. Schlodder and P. Gräber

Max-Volmer-Institut für Physikalische Chemie und Molekularbiologie
Technische Universität Berlin
Straße des 17. Juni 135, 1000 Berlin 12, Germany

This work has two main aspects. On the one hand it is shown that in respect to phosphorylation excitation by an external electric field is equivalent to excitation by light. On the other hand the possibility to exchange the light-induced generator by an external electric generator is used to ask for the relation between (a) the fraction of ATPases which change their conformation, (b) the amount of ATP generated, (c) the rate of turnover and (d) the extent of the electric field strength and transmembrane voltage respectively.

By the light-driven vectorial electron transfer an electric potential difference is generated across the thylakoid membrane. Subsequent protolytic reactions with the separated charges at the outer and inner membrane surface lead to a proton translocation into the inner space of the thylakoid and to the formation of a pH gradient across the membrane. Numerous experiments gave evidence that the efflux of these protons via the ATPase - which is driven by the field and the pH gradient - is coupled with phosphorylation (1,2).

If this mechanism of phosphorylation which was proposed by Mitchell (3) is correct, in principle phosphorylation should be possible with transmembrane electric fields generated by an external energy source. In other words it should be possible to replace the light-induced electric generators - $Chl-a_I$ and $Chl-a_{II}$ - by an external voltage source.

Last year we have demonstrated that in this way phosphorylation is indeed possible (4). It shall be pointed out briefly in which way this method has been realized because the technique is the basis for the following new results.

In Fig.1 the principle of the external electric field method is outlined.

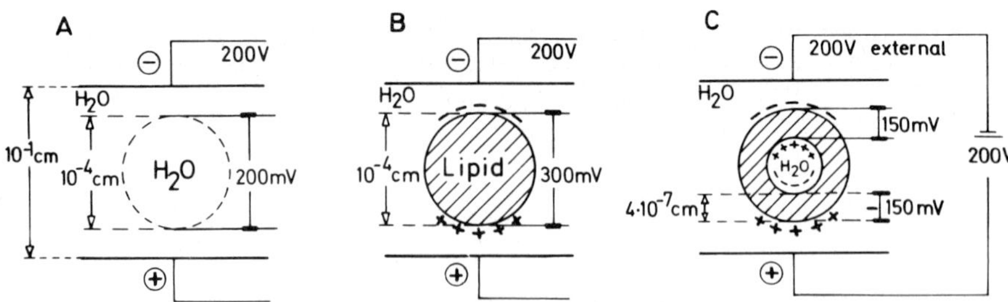

Fig.1. Principle of the external electric field method.

In the left hand scheme A two macroscopic electrodes with a distance of 1 mm are depicted. An aqueous solution is placed between these electrodes and 200 Volts are applied. If we imagine within the solution a sphere of water with a diameter which is thousand times smaller than the distance of the electrodes the voltage across this sphere is of course thousand times smaller than the total voltage, i.e. 200 mV. If we regard a sphere of lipid instead of the sphere of water as shown in the center B of Fig.1, the non-conducting lipid gives rise to a charge accumulation in the water at the outer surface of the sphere. Therefore, the voltage across the sphere is higher and can be calculated to be at the poles 300mV.If we replace the inner space of the sphere of lipid by water, this corresponds to a vesicle with a non-conducting membrane. This is shown in the right hand scheme C. The voltage across the whole sphere is, of course, still 300 mV but the distribution of counter ions inside splits up the voltage into two parts: 150 mV across the membrane at the upper half of the vesicle and 150 mV across the lower half. The generated transmembrane field has at one half of the vesicle the same direction as the light-induced field, i.e. positive inside and negative outside. In the scheme this is the upper half. For this reason the yield of two external voltage pulses has to be compared with one light pulse if the yield is related to the same amount of chloroplast.

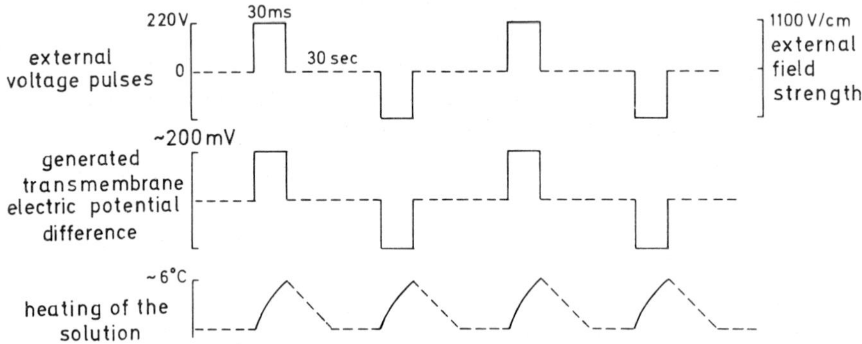

Fig.2. Experimental conditions of the external electric field method.

The experimental conditions were chosen as shown in Fig.2. The electrodes had a distance of 2 mm. We applied pulses of 220 Volts to the chloroplast suspension. The polarity of the voltage pulses was changed after each pulse. The pulse duration was 30 ms. These external pulses generate in the case of suspended thylakoids an electric potential difference of about 200 mV across the membrane (see below). The current induced by the external voltage pulse produces a heating of the solution from 5° to 11^OC. To allow a reverse of the heating of 6^OC the time between the pulses was 30 sec. (A more refined cuvette with electrodes having a distance of 22 mm is described in (4).

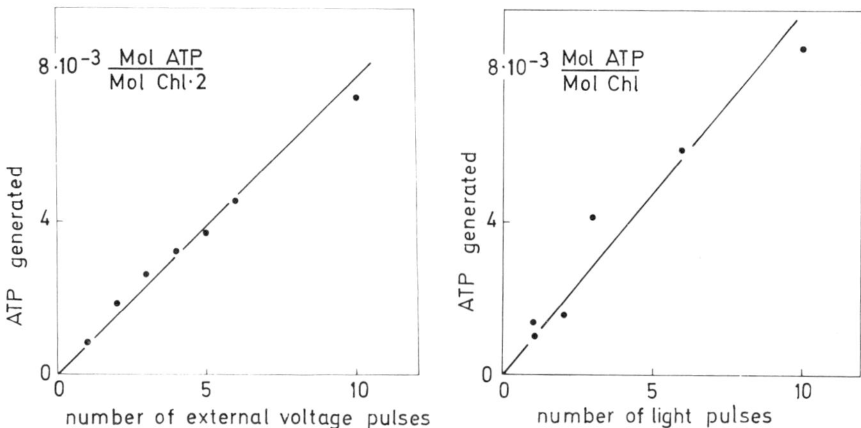

Fig.3. Left: ATP generation as a function of the number of external voltage pulses (duration 30 ms). Right: ATP generation as a function of the number of saturating light pulses (duration 30 ms).

Under such conditions ATP is generated in the dark as shown in Fig.3 on the left hand side. The amount of ATP generated increases linearly with the number of external voltage pulses. On the right hand side in Fig.3 we compared this yield with the amount of ATP generated in saturating light pulses of the same duration. Because the yield in both cases is nearly the same we concluded that phosphorylation generated by light is equivalent to phosphorylation in the dark generated by an external electric field (4).

After this has been clarified and checked by other types of experiments (4) we asked in the following if conformational changes of the ATPase can be observed in the dark when external voltage pulses are used for energization.

For detection of the conformational change of the ATPase we used the energy dependent release of tightly bound adenine nucleotides which were labelled with ^{14}C. This method is well known and was introduced by others (5,6,7). Because in our experiments about one nucleotide is bound per coupling factor, we identify the number of nucleotides released with the number of ATPases which have carried out a conformational change. In the following these ATPases are characterized as "active" ATPases.

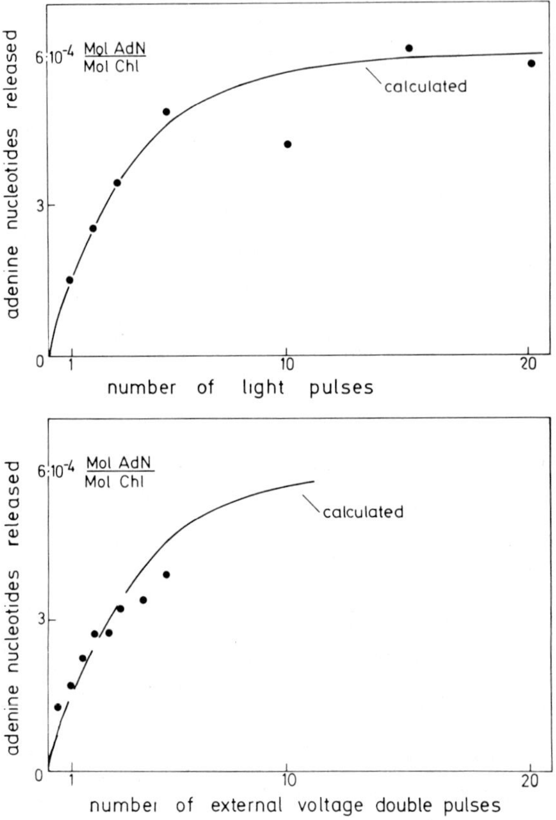

Fig.4. <u>Bottom</u>: Amount of adenine nucleotides released as a function of the <u>number</u> of external voltage pulses. <u>Top</u>: Amount of adenine nucleotides released as a function of the <u>number</u> of saturating light pulses.

In Fig.4 on the bottom the amount of nucleotides released, i.e. the amount of ATPases which have changed their conformation, is depicted as a function of the number of the external voltage double pulses.On the top the same was measured as a function of the number of light pulses. The similarity of both curves in the common range up to 10 pulses again shows that excitation by an external electric field in the dark is equivalent to excitation by light.

This has been proved in a third experiment which is represented in Fig.5. In the experiment shown on the bottom the amount of nucleotides released is measured as a function of the external electric field strength which is proportional to the extent of the electric potential difference induced by the external field across the thylakoid membranes.

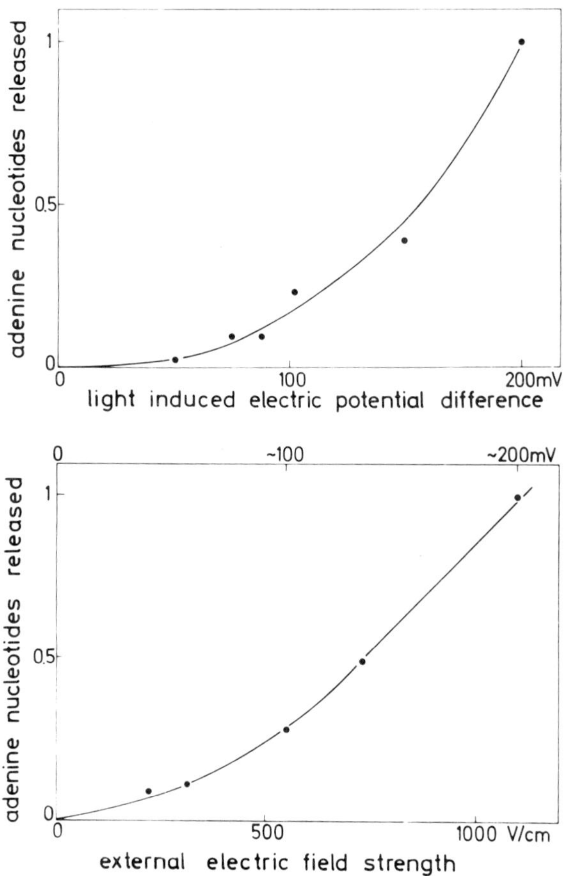

Fig.5. <u>Bottom</u>: Amount of adenine nucleotides released as a function of the external electric field strength and transmembrane electric potential difference induced. <u>Top</u>: Amount of adenine nucleotides released as a function of the transmembrane potential difference induced by light(Relative Units).

In the experiment shown on the top the amount of nucleotides released is measured as a function of the transmembrane electric potential induced by light. (In this case the scale of the electric potential difference on the top of Fig. 5 was obtained from the simultaneously measured electrochromic absorption changes (2)). The functional dependence on the transmembrane electric potential and field strength respectively is similar in both cases.

By the outlined results we have shown in three different ways that in respect to phosphorylation excitation by an external field in the dark is equivalent to excitation by light. We conclude that with both types of excitation the energization of the membrane must be similar. Therefore, the external electric field strength of 1100 V/cm must have induced about the same overall transmembrane voltage as the light pulse, i.e. about 200 mV (s. Fig.5 top and bottom). (This value is in the range with that which can be calculated theoretically from the external field strength and the shape of the thylakoids (4)).

We turn now to the key experiment. We asked for the correlation between the conformational change of the ATPase and phosphorylation, i.e. we compare directly the number of ATPases which change their conformation with the number of ATP molecules generated. In this experiment we prefer excitation by the external electric field instead by light for the following reasons.

In the light always a transmembrane potential plus a pH gradient is produced. Both components depend in a different way on the light intensity and neither the potential nor the pH gradient depends linearly on the light intensity. However, if the transmembrane voltage is induced by an external voltage pulse, the pH gradient is zero. Furthermore, the transmembrane voltage is directly proportional to the external voltage. Lastly the period for which the membrane is energized is identical with the duration of the external pulse (because practically no ion gradients are generated across the membrane). These features of the external electric field method are necessary for the analysis of the results presented in Fig.6.

On the bottom of Fig.6 it is shown what was already demonstrated in Fig.5, the non-linear increase of the amount of nucleotides released as a function of the external electric field strength and the transmembrane electric potential induced. In the experiment shown on the top in Fig.6 we have measured under the same conditions the amount of ATP generated as a function of the external field strength.

The essential point is the comparison between the amount of ATP generated and the amount of nucleotides released. The surprising result is that about six times more ATP is generated than nucleotides released, i.e. the amount of ATP generated is about six times larger than the amount of ATPases which change into the active conformation.

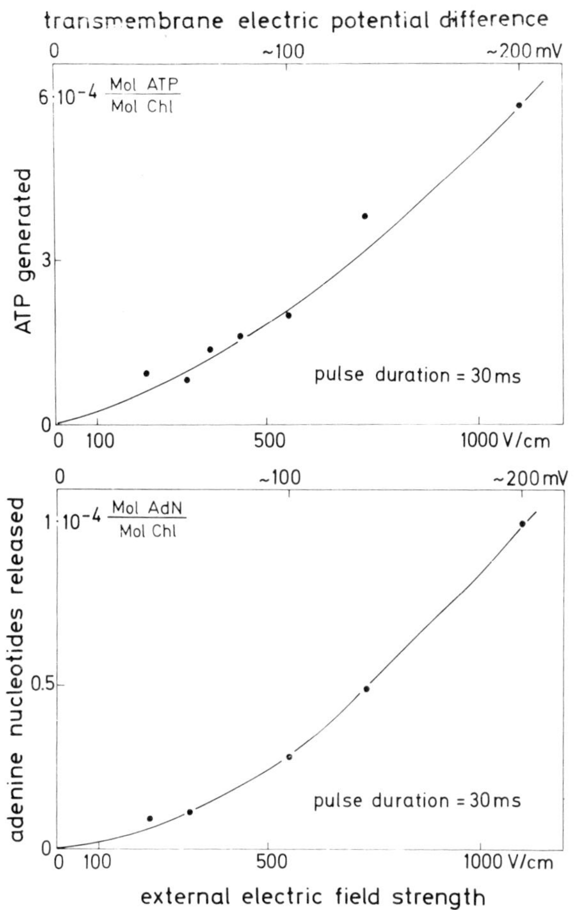

Fig.6. Top: Amount of ATP generated as a function of the external electric field strength and transmembrane electric potential difference induced. Bottom: Amount of adenine nucleotides released as a function of the external electric field strength and transmembrane electric potential difference induced.

This discrepancy can be explained by the assumption that each active ATPase generates about six ATP, i.e. each active ATPase carries out six turnovers during the pulse duration of 30 ms. Six turnovers in 30 ms mean firstly that the turnover time of the ATPase is about 5 ms. Because the ATP/AdN ratio is nearly constant at different electric potentials (s.Fig.6) it can be concluded secondly that the turnover time is independent of the electric potential across the membrane.

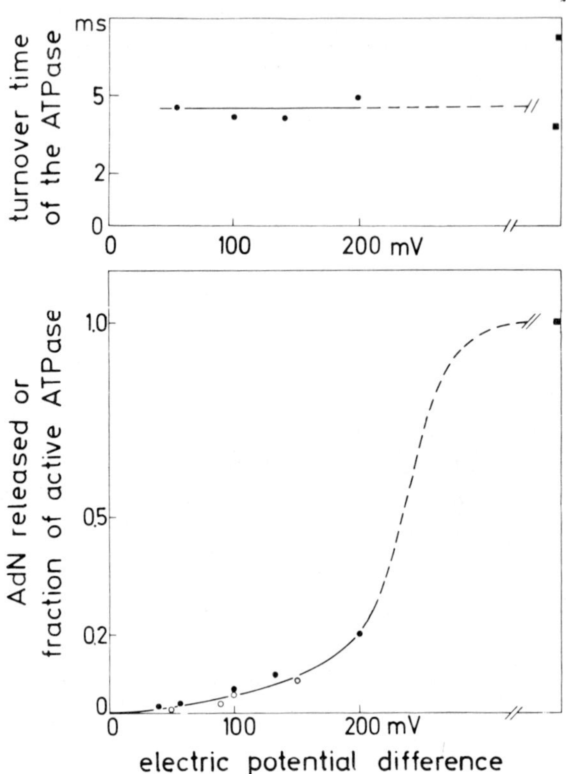

Fig.7. Top: Turnover time of the ATPase as a function of the trans-membrane electric potential difference. Bottom: Fraction of the active ATPase as a function of the transmembrane electric potential difference.

These conclusions are summarized in Fig.7. The solid lines indicate the range of our measurements (up to 200 mV). On the top it is depicted what has been concluded from the measurement in Fig.6, namely that the turnover time of the ATPase is about 5 ms and independent on the transmembrane voltage. On the bottom again the amount of nucleotides released is depicted (s.Fig.5), i.e. the fraction of active ATPases as a function of the transmembrane potential.

What is expected for both events at potentials higher than 200 mV? Because we have observed that the turnover time is constant up to 200 mV we assume that this is valid also at higher potentials (see dotted curve in Fig.7 top). In respect to the fraction of active ATPase the maximal value cannot be higher than one, that means that the extrapolation of the solid curve must end up at one(e.g. in a shape as indicated by the dotted curve in Fig.7 bottom).

The predictions and consequences of the mechanism proposed by these two curves are in accordance with data obtained by independent experiments of other authors:

1. Smith and Boyer (8) have shown that at higher energization (Δ pH = 4.4) the fraction of active ATPases is one. Their result is represented by the quadratic points in Fig.7 bottom.

2. From this work (8) also two turnover times can be read out. These times - represented by the two quadratic points in Fig.7 top - are in fair agreement with our prediction indicated by the dotted curve.

3. Our measurements indicate that at a transmembrane potential of 200 mV only about 10 to 20 % of the ATPases are active (s. Fig.7 bottom). This means that at a higher potential where all ATPases are active the maximal rate of phosphorylation should be 5 to 10 times higher. This is the case. We have measured at 200 mV a rate of 30 mM ATP/Chl·s (this can be calculated from the amount of ATP generated in Fig.1 in a pulse of 30 ms duration). This value is also the usual rate in continuous light. The maximal rate ever obtained, for example in cyclic phosphorylation, is indeed about 300 mM ATP/Chl·s (10) instead of 30 mM ATP/Chl·s.

It has been reported that the release of adenine nucleotides and phosphorylation depends in the same way on different parameters, i.e. on light intensity, on pH_{out}, on various inhibitors (9) and on the electric potential difference (this work). This supports our assumption (s.above) that the conformational change observed here is involved in the mechanism of phosphorylation and is not a side reaction of energization. However, our results do not allow to distinguish between a conformational change as a catalytic step (i.e. without transmitting energy) or as energy transmitting step.

From our results the following can be concluded: (a) Conformational change and activation of the ATPase and phosphorylation are not caused directly by the light-induced electron transport because they can be observed in the dark in the external voltage pulse. (b) The conformational change and activation is not only caused by an increase of H^+ concentration inside the thylakoid because they occur also in the external voltage pulse where no increase of the H^+ concentration in the inner thylakoid space takes place. (c) One possibility for the conformational change and activation of the ATPase by the field may be an indirect action of the electric field by producing a local proton gradient across the ATPase. However, also a direct interaction of the field with electric properties on the ATPase could be discussed.

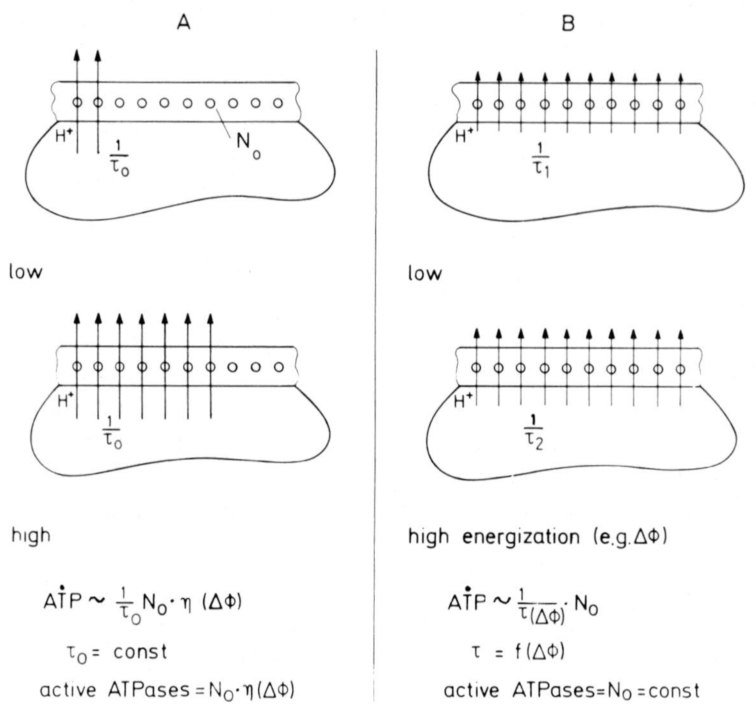

$$\overset{\bullet}{ATP} \sim \frac{1}{\tau_0} N_0 \cdot \eta \, (\Delta\Phi)$$

$$\tau_0 = const$$

$$active\ ATPases = N_0 \cdot \eta(\Delta\Phi)$$

$$\overset{\bullet}{ATP} \sim \frac{1}{\tau_{(\Delta\Phi)}} \cdot N_0$$

$$\tau = f(\Delta\Phi)$$

$$active\ ATPases = N_0 = const$$

Fig. 8. Two types of mechanism – A and B – of the cooperation between the turnover time and activation of the ATPases at low and high energization. Details see text.

The proposed mechanism derived from our results is illustrated in the left hand scheme A of Fig. 8. The cycles represent the ATPases within the membrane. The total number is N_0. The active ones are marked with arrows. The length of the arrows indicate the reciprocal of the turnover time. A membrane at low and high energization, i.e. at low and high electric potential is shown (s. top and bottom in Fig. 8). The fraction of active ATPases increases at higher energization whereas the turnover time is constant and always minimal (indicated by a constant length of the arrows). This scheme corresponds to the equations also indicated on the left hand side. The rate of ATP formation, $\overset{\bullet}{ATP}$, is proportional to the product of the rate constant, that is one over the turnover time, i.e. $1/\tau_0$, times the number of active ATPases which is a function of the potential, $N_0 \cdot \eta \, (\Delta\phi)$. To point out these features we compare the outlined mechanism with the other extreme B shown on the right hand side in Fig. 8. The turnover time decreases at higher energization as indicated by the different lengths of the arrows, $1/\tau$ ($\Delta\emptyset$) but the number of active ATPases is constant, N_o. From our experiments we conclude that the left hand scheme A is realized.

It should be mentioned that we have evidence that the fraction of active ATPases is not fixed to a special ensemble but migrates statistically between all ATPases within about 3 seconds (11).

The mechanism A outlined in Fig.8 left has e.g. the following advantage. The time for gathering n protons in order to synthesize one ATP is minimal and constant and thereby independent on the light intensity. This is not realized by the mechanism B depicted in Fig.8 right and, therefore, gives rise to problems at low intensities.

Our further investigation is focussed to the question in which way the electric field or energization in general activates a certain fraction of ATPases.

This work has been supported by grants of the Deutsche Forschungsgemeinschaft and by the Commission of the European Communities.

References

1. Jagendorf, A.T. (1975) in Bioenergetics of Photosynthesis (Govindjee, ed.) pp. 413-492, Academic Press, New York.

2. Witt, H.T. (1975) in Bioenergetics of Photosynthesis (Govindjee, ed.) pp. 493-554, Academic Press, New York.

3. Mitchell, P. (1966) Biol. Rev. 41, 445-502.

4. Witt, H.T., Schlodder, E. and Gräber, P. (1976) FEBS Lett. 59, 272-276.

5. Harris, D.A. and Slater, E.C. (1975) Biochim. Biophys. Acta 387, 335-348.

6. Boyer, P.D., Smith, D.J., Rosing, J. and Kayalar, C. (1975) in Electron Transfer Chains and Oxidative Phosphorylation (Quagliariello, E. et al., eds.) pp. 361-372, North-Holland Publ. Co., Amsterdam.

7. Strotmann, H., Bickel, S. and Huchzermeyer, B. (1976) FEBS Lett.61, 194-198.

8. Smith, D.J. and Boyer, P.D. (1976) Proc. Natl. Acad. Sci. USA 73, 4314-4318.

9. Bickel-Sandkötter, S. and Strotmann, H. (1976) FEBS Lett. 65, 102-106.

10. Avron, M. (1960) Biochim. Biophys. Acta 40, 257-272.

11. Gräber, P., Schlodder, E. and Witt, H.T. (1977) Biochim. Biophys. Acta, in press.

Bioenergetics of Membranes. L. Packer et al. ed.

THYLAKOID MEMBRANE SURFACE CHARGES IN RELATION

TO PROMPT AND DELAYED CHLOROPHYLL FLUORESCENCE

J. Barber
Department of Botany, Imperial College,
London S.W.7. U.K.

INTRODUCTION

Perhaps the most striking feature of the photosynthetic apparatus of oxygen
evolving organisms is the extent of the thylakoid membrane surface area in
relation to the volume of the intrathylakoid and stromal compartments. In the
case of higher plants the amount of thylakoid membrane in the chloroplast can vary
considerably and no precise general value of the surface area/volume ratios can be
given. However it is possible to give a rough estimate of these values using relevant published figures. Heldt (1) has estimated the intrathylakoid and stromal
volumes of spinach chloroplasts to be 3.3 and 23µl/mg chl. respectively while I
have used a thylakoid surface area value of 16.7 cm^2/µg chl. (ref.2). These figures give surface area to volume ratios of 5×10^6 and 7×10^5 for the intrathylakoid and stromal compartments respectively. In this paper I want to discuss the
important implications which arise from these very large surface area/volume
ratios with regard to ionic levels at the chloroplast membrane surfaces and their
relation to the control of prompt and delayed chlorophyll fluorescence yields.

SURFACE CHARGES AND DOUBLE LAYER CONCEPTS

When isolated thylakoid membranes are placed in a conducting solution with a
pH above 4.3 and subjected to an electric field they move towards the positive
electrode indicating that they carry an excess of negative charges on their outer
surface. As Fig.1 shows, when the pH is dropped below 4.3 the chloroplasts change
their direction of electrophoretic flow indicating that under these acid conditions
the negative charges are neutralised and that the membranes now carry an excess
positive charge. Thus this shows, in agreement with other work (3,4,5) that at
physiological pH the thylakoid membrane surface is negatively charged. The negative charges are most probably due to carboxyl groups of the membrane proteins
(4,5).

The existence of a large negatively charged membrane surface area relative to
the intrathylakoid and stromal volumes will greatly influence the distribution of
ions within the chloroplast and particularly at the thylakoid surface. In fact
the negatively charged surface will attract cations so as to maintain electroneu-
trality but because of the chemical potential gradient created the cations will
remain in a diffuse layer adjacent to the membrane surface (6,7). The cation
composition of this diffuse layer will depend on the electrolyte composition of
the medium in which the membrane is suspended and may be quite different to the

460

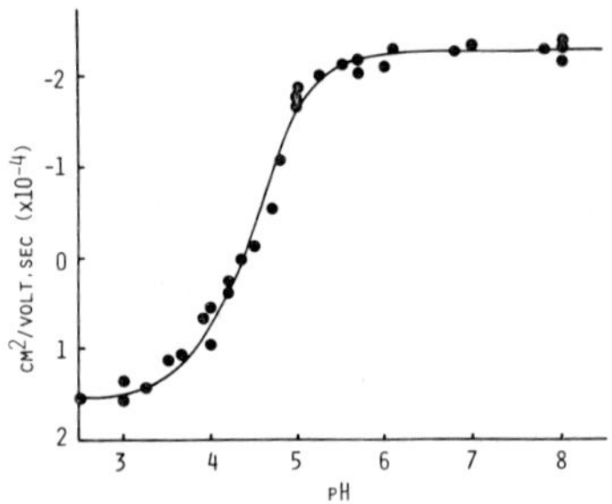

Fig.1. Electrophoretic mobility of
isolated thylakoid membranes as a function
of pH with ionic strength of 0.02 using
KCl in 6% sorbitol. Measurements made on
a Zeiss Cytopherometer Ph40.
(Nakatani and Barber, unpublished results)

composition of the bulk solution. The theory of electrical double layers can be
complex but the classical approach of Gouy (8) and Chapman (9) with its various
assumptions (see ref.7) seems adequate for application to biological membrane
surfaces where the surface charge densities are reasonably low (10). The Gouy
Chapman theory links the Boltzmann and Poisson equations and leads to an expression which relates the electric field strength ($d\psi/dx$) with any point x from the
membrane surface having a potential (ψ) relative to the bulk solution.

$$\frac{d\psi}{dx} = \pm\, 2\left(\frac{2\pi RT}{\varepsilon}\right)^{\frac{1}{2}} \left| \sum_i C_{i\alpha} \left(\exp\left(\frac{-Z_iF\psi}{RT}\right) -1\right)\right|^{\frac{1}{2}} \qquad \ldots.1$$

where F is the Faraday, R is the Gas Constant, ε is the permittivity of water,
T is the absolute temperature and C_α is the bulk concentration of an ion having
a charge Z.

Since the Gauss equation relates the electric field to the surface charge
density on the membrane (q): i.e. $d\psi/dx = 4\pi q/\varepsilon$, equation 1 can be rewritten:

$$q = \pm\, \left| \frac{RT\varepsilon}{2\pi} \sum_i C_{i\alpha} \left(\exp\left(\frac{-Z_iF\psi_0}{RT}\right) -1\right)\right|^{\frac{1}{2}} \qquad \ldots.2$$

where ψ_0 is the potential when $x = 0$ (i.e. surface potential) relative to the bulk
solution ($x = \alpha$).

In principle knowing the ionic composition of the bulk solution and the value
of q, it is possible to calculate the potential (ψ), the electric field ($d\psi/dx$)
and the space charge density ($d^2\psi/dx^2$) as a function of distance from the membrane
surface. Moreover by using the Boltzmann equation:

$$C_{ix} = C_{i\alpha}\, \exp\left(\frac{-Z_iF\psi}{RT}\right) \qquad \ldots.3$$

it is possible to calculate the ionic composition at any point x in the diffuse layer.

CATION INDUCED CHANGES IN THE YIELD OF CHLOROPHYLL AND 9-AMINOACRIDINE FLUORESCENCE

Some years ago it was shown by Homann (11) and Murata (12,13) that the addition of metal cations to isolated thylakoid membranes increases the steady-state chlorophyll fluorescence yield. In these experiments the chloroplasts had been pretreated with 3-(3,4-dichlorophenyl)-1,1-dimethyl urea (DCMU) and the yield changes were attributed to alterations in the membrane conformation rather than to changes in the redox state of the photosystem two traps. More recently Gross and Hess (14) found that cation induced fluorescence changes are more complex showing antagonistic effects between low concentrations of mono- and divalent cations. The essential features of the cation induced chlorophyll fluorescence changes are shown in Fig.2.

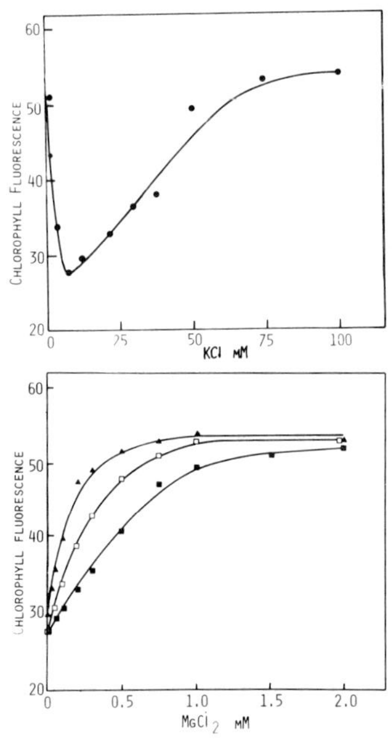

Fig.2. Effect of cations on chlorophyll fluorescence from isolated spinach thylakoid membranes treated with 10^{-5}M DCMU. The suspending medium contained 0.1M sorbitol brought to pH 7.0 with Tris base.

a) upper curve, concentration curve for monovalent cations (K^+).

b) lower curve, competitive effect of K^+ on the Mg^{2+} stimulated increase in fluorescence yield using membranes which had been washed with 0.5mM Na-EDTA prior to experimenting. The suspension contained the following KCl concentrations: closed triangles 1mM, open squares 3mM and closed squares 10mM. Data from ref.6.

In Fig.2(a) the thylakoids have been suspended in a cation free medium and the fluorescence level is high. However the introduction of low levels of K^+ brings the fluorescence to a level which corresponds to the starting level of the experiments reported by Homann (11) and Murata (12,13). As shown, further addition of K^+ under these conditions increases the yield back to the higher level. A similar increase in fluorescence is also seen on adding divalent cations but the concentration required is much lower (see Fig.2(b)). A large range of monovalent cations cause essentially the same result as Fig.2(a) including all the alkali metal cations and organic cations like choline and lysine. Also, for the divalent cation effect there is little or no specificity between the alkaline earth cations and organic cations like lysyl-lysine which acts equally as well as Mg^{2+}. The initial starting level for the type of experiment shown in Fig.2(a) depends on whether the isolated thylakoids have been exposed to monovalent cations. For instance, washing the preparation with 0.5mM Na-EDTA greatly reduces the initial yield of fluorescence and under these conditions very low levels of divalent cations are needed to restore the fluorescence to the high yield. Actually the concentration of divalent cations required to induce maximum fluorescence is determined by the level of monovalent cations in the suspending medium as shown in Fig.3(b) (also see ref.15).

To understand the mechanism controlling these cation induced fluorescence changes it is important to note the main features:

   i) Not related to changes in ionic or in osmotic strength.

  ii) There is a significant difference between monovalent and divalent cations but little dependence on the nature of cations within these two charged groups.

 iii) Antagonism between low levels of monovalent and divalent cations.

  iv) Antagonism between low and high levels of monovalent cations.

   v) Independence of the nature of the anion.

 .vi) No obvious relationship with the structure, lipid solubility and coordination number of the cations.

All these observations point to an electrical rather than a chemical effect and can be explained in terms of the diffuse double layer concepts presented above. Chlorophyll fluorescence studies and other observations (16-19) suggest that in the intact chloroplast Mg^{2+} acts as the main cation in the double layer. Thus when the thylakoid membranes are carefully isolated into a cation free medium we can assume that there is a significant amount of Mg^{2+} at the surface and the bulk concentration of the divalent cation is very low, say 10^{-6}M. Under these conditions the chlorophyll fluorescence yield is high. Introduction of monovalent cations (e.g. K^+) lowers the yield and at the same time will reduce ψ_o and bring about K^+/Mg^{2+} exchange at the surface. Further addition of K^+ decreases ψ_o even more and will continue to displace Mg^{2+} from the diffuse layer. However under these conditions the fluorescence rises. This decrease in ψ_o and the monovalent/ divalent exchange at the membrane surface can be calculated using modified forms of equations (2) and (3), (see legend of Fig.3). Such calculations are shown in

Fig.3(a) and 3(b) but they do not explain the "dip" effect seen with fluorescence. To account for this phenomenon it is necessary to plot the total diffusible positive charge (space charge density) at the membrane surface as a function of the bulk monovalent cation level. In Fig.3 the surface charge density q has been taken as 2.5μ coulombs cm^{-2} (1 electronic charge per 640Å^2), since this gives a dip in the calculated curve corresponding to the concentration of monovalent cations which give the minimum fluorescence yield (5-10mM).

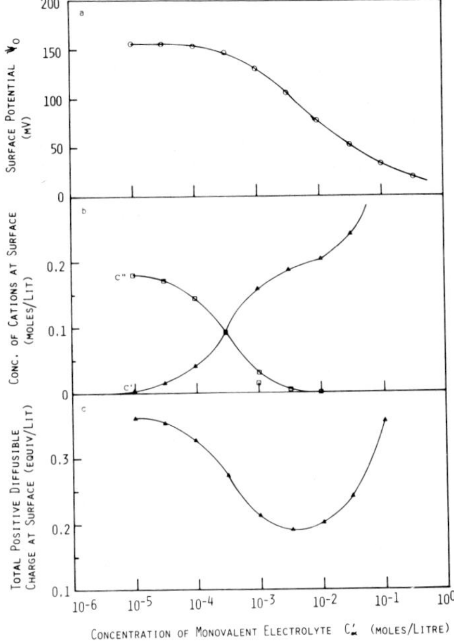

Fig.3

a) Surface potential ψ_0 calculated from the following expression derived from Eq.2 for a membrane having a surface charge density (q) of 2.5μ coulombs cm^{-2} bathed in a solution containing Z-Z type monovalent and divalent salts at bulk concentrations of C'_α and C''_α (see ref.7).

$$2C''_\alpha \cosh^2\left(\frac{F\psi_0}{RT}\right) + C'_\alpha \cosh\left(\frac{F\psi_0}{RT}\right)$$
$$- \left(2C''_\alpha + C'_\alpha + \frac{q}{2A}\right) = 0 \qquad ....4$$

where $A = \left(\frac{RT\epsilon}{2\pi}\right)^{1/2}$, $c''_\alpha = 10^{-6}M$ and C'_α is varied.

b) Concentrations of monovalent C'_0 and divalent C''_0 at the membrane surface calculated using Eq.3.

c) Positive space charge density at membrane surface.

As discussed in a previous paper (ref.7) this approach can explain the various antagonistic and competitive effects seen between monovalent and divalent cations on fluorescence.

This concept can be taken one step further by deriving the following expression for mixed electrolytes from equation 1.

$$\frac{d\psi}{dx} = \pm \left(\frac{8\pi RT}{\epsilon}\right)^{1/2} \left| 4C' \sinh^2\left(\frac{F\psi}{2RT}\right) + 4C''_\alpha \sinh^2\left(\frac{F\psi}{RT}\right) \right|^{1/2} \qquad ....5$$

Taking $C'' = 10^{-6}M$ as before, it is possible to generate the curves shown in Fig.4 by numerical methods using a computer and again taking q = 2.5μ coulombs/cm^2. As shown, the electric field ($d\psi/dx$) shows a maximum as C' is increased and the effect extends out into the diffuse layer for some distance. In contrast the minimum seen in the positive space charge density curves only occurs within a few

Ångstroms of the membrane surface. At this stage it is not clear whether the electric field or the space charge density changes bring about the cation induced fluorescence effects although it is not difficult to visualize that electrical effects of this type near the membrane surface can bring about changes in the interactions between the fixed negative charge of the membrane proteins and thus induce conformation and the associated chlorophyll fluorescence changes.

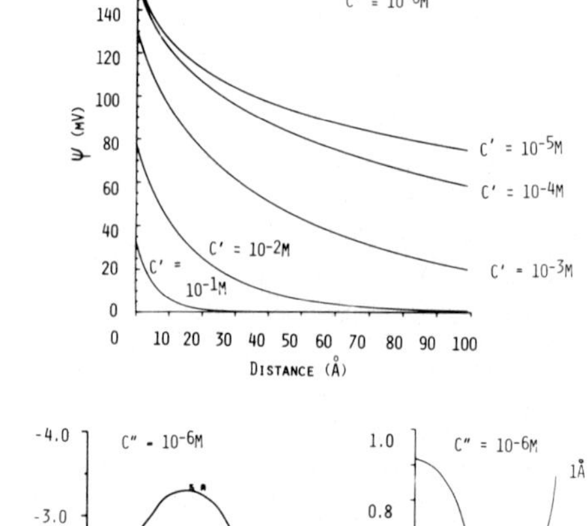
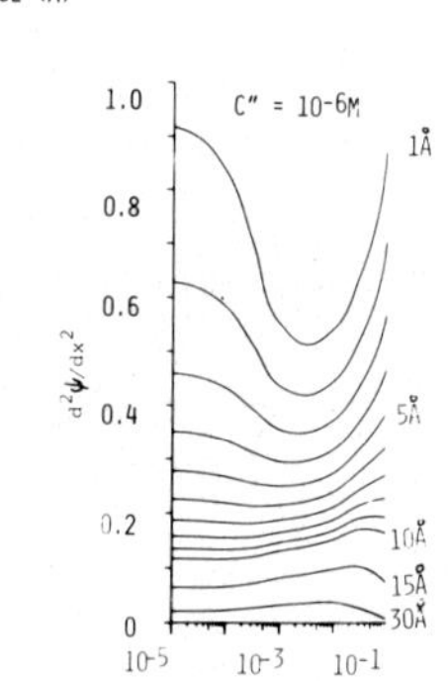

Fig.4 Computer generated curves obtained by carrying out numerical integration of Eq.5 showing how

a) the potential ψ,

b) the electric field $(\frac{d\psi}{dx})$ and

c) the space charge density $d^2\psi/dx^2$

varies as a function of distance from the membrane surface for different monovalent salt levels and with $C''_\alpha = 10^{-6}$M (from ref.7).

It is possible to probe the double layer of the thylakoid membranes using 9-aminoacridine (9-AA) as a fluorescent monovalent cation. When thylakoid membranes are added to a cation free medium containing 9-AA at 25µM there is a quenching of the fluorescence which does not involve any shifts in its absorption or emission spectra and occurs in the absence of a high energy state (20). The degree of 9-AA fluorescence lowering is a function of the amount of membrane added and probably reflects concentration quenching due to the attraction of the positively charged dye into the diffuse layer. Addition of cations to this mixture induces an increase of 9-AA fluorescence which, like the chlorophyll fluorescence changes, are very dependent on the charge carried by the cation and essentially independent of the associated anion. As Fig.5 shows the order of effectiveness is polyvalent>divalent>monovalent and virtually no variation was found within the

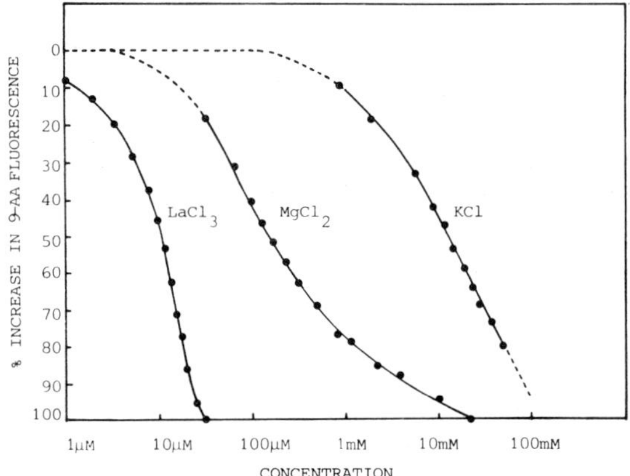

Fig.5. The effectiveness of mono-, di- and tri-valent cations in releasing the quenching of 9-AA fluorescence by chloroplasts (from ref.20).

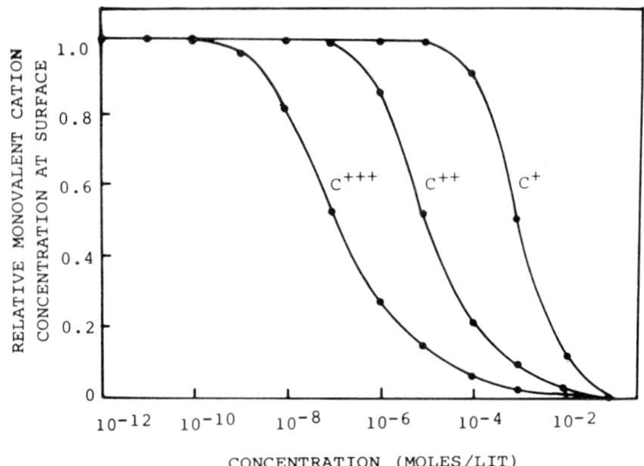

Fig.6. Theoretical curves showing the displacement of a monovalent cation (C'_o) from the membrane surface by additions of monovalent, divalent, and trivalent cations. The membranes are assumed to carry a net negative charge of 2.5μ coulombs cm^{-2} and to be suspended in a medium containing an initial bulk concentration of monovalent cations at 1mM. The values of C'_o have been obtained using Eq.3 and the ψ_o values calculated from the appropriate expressions for mixed electrolytes.

i) mixtures of monovalent salts, $C'_{1\alpha}/C'_{2\alpha}$, $4(C'_{1\alpha} + C'_{2\alpha})\,\sinh^2\dfrac{F\psi_o}{2RT} - \dfrac{q^2}{A^2} = 0$ ...6

ii) monovalent/divalent mixture C'_α/C''_α , Eq.4 in Fig.3 legend.

iii) monovalent/trivalent mixture C'_α/C'''_α

$$4C'''_\alpha\,\cosh^3\left(\frac{F\psi_o}{RT}\right) + (C'_\alpha - 3C'''_\alpha)\,\cosh\left(\frac{F\psi_o}{RT}\right) - \left(C'_\alpha + C'''_\alpha + \frac{q}{2A^2}\right) = 0 \qquad ...7$$

groups of monovalent and divalent cations tested. Furthermore, like the cation induced chlorophyll fluorescence changes, choline and lysine were as effective as alkali metal cations, and lysyl-lysine was almost as effective as alkaline earth metal cations.

It appears that 9-AA can act as a diffusible monovalent cation which increases its fluorescence when displaced from the diffuse electrical layer adjacent to the thylakoid membrane. The differential removal of a monovalent cation from the surface by cations carrying different charges can be predicted using expressions for mixed electrolytes derived from Eq.2 (see Fig.6).

The above calculations have used a value of $q = 2.5\mu$ coulombs/cm^2 or 1 electron charge per $(25\text{Å})^2$. The electrophoretic mobility data shown in Fig.1 can be used to directly estimate the value of q. For the ionic strength employed the following equation can be used to calculate the zeta potential (ζ) which is the potential difference between the surface of shear and the bulk of the liquid

$$\zeta = \frac{4\pi\Omega\mu}{\epsilon} \qquad \qquad \dots\dots 8$$

where Ω is the viscosity of the suspending medium and μ is the electrophoretic mobility. Using a value μ determined at pH 7.0 of 2.3×10^{-4} cm^2/volt.sec. gives a calculated zeta potential of about -38mV. Making the crude assumption that $\zeta = \psi_0$ and bearing in mind the membranes were suspended in 20mM KCl then using Eq.2 it can be calculated that $q = 1.33\mu$ coulombs/cm^2 or 1 electronic charge per $(35\text{Å})^2$. However, as pointed out by Haydon (10) the various assumptions made in this type of calculation for charged membrane systems inevitably result in an under-estimation of the surface charge and the assumed value of 2.5μ coulombs/cm^2 is therefore not unreasonable. Moreover it should also be pointed out that changes in ionic conditions of the suspending medium will affect the number of hydrogen ions in the diffuse layer and this, together with changes in the amounts of adsorbed bound cations (in the Helmholtz layer) will mean that q is variable and dependent on the cationic constituency of medium even at neutral pH. No allowance for changes in q have been made in the calculations presented in Figs.3 and 6.

THE RELATIONSHIP BETWEEN THE YIELD FACTORS FOR PROMPT AND DELAYED FLUORESCENCE

Lavorel first pointed out the possible importance of the prompt chlorophyll yield on the intensity of delayed fluorescence emission from photosynthetic systems (21). By analogy with the relation between incident light intensity I and prompt chlorophyll fluorescence intensity F; $F = \phi_F I$, where ϕ_F is the prompt fluorescence yield, Lavorel suggested that delayed fluorescence intensity L should be described by a similar relationship:

$$L = \phi_{DF} J \qquad \qquad \dots\dots 9$$

where J is the rate of chlorophyll singlet formation and ϕ_{DF} is the fluorescence yield of the chlorophylls through which the delayed fluorescence exciton migrates. To understand the relationship between ϕ_{DF} and ϕ_F has important implications since it would reveal information about the extent of exciton migration away from the trap from which it originated (22,23). Unfortunately earlier studies designed to investigate the relationship between ϕ_{DF} and ϕ_F have not been entirely satisfactory since there are a number of factors which can effect J and which must be controlled (see 22,23). A useful expression to describe J was formulated by Crofts et al(24) based on the results of several workers and can be written:

$$J = (Z^+ChlQ^-)\ k'\ \nu\ exp\ \left| -(E_{ac} - \Delta p)/kt \right| \qquad \ldots\ldots 10$$

where (Z^+ChlQ^-) is the concentration of the charge transfer complex generated in the reaction centre of photosystem two (PS2), thought to act as the precursor for delayed fluorescence, k' is a constant containing entropy terms, ν is a frequency factor, E_{ac} is the activation energy for the back reaction and Δp is the high energy state expressed as Mitchell's proton motive force given by

$$\Delta p = \Delta \psi + 2.303\ \frac{RT}{F}\ \Delta pH \qquad \ldots\ldots 11$$

where $\Delta \psi$ is the electrical gradient and ΔpH is the pH gradient across the membrane.

To study the relationship between ϕ_F and ϕ_{DF} we have used the cation induced changes in ϕ_F described in the above sections. The intensity of prompt and 1msec delayed fluorescence has been monitored using a phosphoroscope previously described (2). The isolated chloroplasts were treated with 5.5 x 10^{-5}M DCMU to keep Q fully reduced and also with 10^{-7}M gramicidin to reduce Δp to zero. Under these conditions the intensity of the delayed emission is reduced by more than 50 times. When the fluorescence yield is increased by the addition of cations there is a concomitant increase in the 1msec steady state delayed fluorescence yield. When the changes in the total prompt fluorescence yield ϕ_F (including both the variable ϕ_V and the background fluorescence corresponding to dark adapted fully open traps ϕ_O) is used then there is a close relationship between the change in the two yield factors. This relationship is shown in Fig.7 for the Mg^{2+} induced changes.

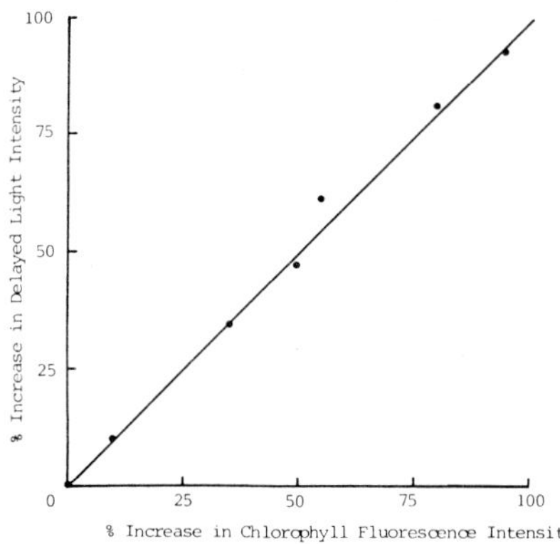

Fig.7. Comparison of the Mg^{2+} induced change in the yield of prompt and 1msec delayed fluorescence from pea chloroplasts showing that $\phi_{DF} = \phi_F$ where $\phi_F = \phi_V + \phi_O$ (see text).
The chloroplasts were suspended in 0.33M sorbitol and 0.01M HEPES brought to pH7.6 with Tris base and treated with 5.5x10^{-5}M DCMU and 10^{-7}M gramicidin.

(Data by Barber, Mauro and Lannoye, in press)

CONCLUSION

In this paper I have emphasised the importance of the diffuse electrical layer adjacent to the negatively charged thylakoid membrane in controlling the yield of prompt fluorescence. Within the functioning chloroplast, proton pumping and secondary ion movements almost certainly affect the surface charge densities on both sides of the thylakoid membrane and consequently alter the nature of the associated diffuse electrical layers. These physiologically induced changes will bring about alterations in membrane conformation which can be seen as changes in chlorophyll fluorescence reflecting changes in spillover of energy between the two photosystems (16,25). I have also presented evidence, based on the cation induced effects, that an exciton created by a back reaction in a PS2 trap is able to rapidly migrate into the light harvesting chlorophyll antenna system and has the same properties as an exciton generated by direct light absorption. Such a conclusion suggests that the open reaction centre generated by the back reaction is either not a perfect trap or that there is sufficient lag before it is able to trap again.

ACKNOWLEDGEMENTS

Support for the work has come from the Science Research Council and the EEC Solar Energy Research and Development Programme. The work has been carried out in collaboration with various colleagues, J. Mills, H. Nakatani, G.F.W. Searle, A. Telfer, A. Love, R. Lannoye and S. Mauro.

REFERENCES

1. Heldt, H.W., Werdan,K., Milovancev, M. and Geller, G. (1973) Biochim.Biophys. Acta 314, 224-241.

2. Barber, J. (1972) Biochim. Biophys. Acta 275, 105-116.

3. Nobel, P.S. and Mel, H.C. (1966) Arch. Biochem. Biophys. 113, 695-702.

4. Berg, S., Dodge, A., Krogmann, D.W. and Dilley, R.A. (1974) Plant Physiol. 53, 619-627.

5. Davis, D.J. and Gross, E.L. (1975) Biochim. Biophys. Acta 387, 557-567.

6. Barber, J. and Mills, J. (1976) FEBS Lett. 68, 288-292.

7. Barber, J., Mills, J. and Love, A. (1977) FEBS Lett. 74, 174-181.

8. Gouy, G. (1910) Ann. Phys. (Paris) Serie (4) 9, 457-468.

9. Chapman, D.L. (1913) Phil. Mag. 25, 475-481.

10. Haydon, D.A. (1961) Biochim. Biophys. Acta 50, 450-457.

11. Homann, P.H. (1969) Plant Physiol. 44, 932-936.

12. Murata, N. (1969) Biochim. Biophys. Acta 189, 171-181.

13. Murata, N., Tashiro, H. and Takamiya, A. (1970) Biochim. Biophys. Acta 197, 250-256.

14. Gross, E.L. and Hess, S. (1973) Arch. Biochem. Biophys. 159, 832-836.

15. Vandermeulen, D.L. and Govindjee (1974) Biochim. Biophys. Acta 368, 61-70.

16. Barber, J. (1976) In "The Intact Chloroplast", Vol.1 of Topics in Photosynthesis. Ed. J. Barber, pp89-134. Pub. Elsevier, Amsterdam.

17. Telfer, A., Barber, J. and Nicolson, J. (1975) Biochim. Biophys. Acta 396,
301-309.

18. Portis, A.R. and Heldt, H.W. (1976) Biochim. Biophys. Acta 449, 434-444.

19. Krause, G.H. (1977) Bicohim. Biophys. Acta 460, 500-510.

20. Searle, G.F.W., Barber, J. and Mills, J.D. (1977) Biochim. Biophys. Acta
in press.

21. Lavorel, J. (1968) Biochim. Biophys. Acta 153, 727-730.

22. Lavorel, J. (1975) In "Bioenergetics of Photosynthesis", Ed. Govindjee,
pp.223-317. Pub. Academic Press.

23. Malkin, S. (1977) In "Primary Photosynthetic Processes", Vol.2 of Topics in
Photosynthesis, Ed. J. Barber, Pub. Elsevier, Amsterdam.

24. Crofts, A.R., Wraight, C.A. and Fleischman, D.E. (1971) FEBS Lett. 15, 89-100.

25. Butler, W.L. and Kitajima, M. (1975) Biochim. Biophys. Acta 396, 72-85.

Bioenergetics of Membranes. L. Packer et al. ed.

^{31}P HIGH RESOLUTION NMR STUDIES OF BIOENERGETICS IN E. COLI

R. G. Shulman, G. Navon[*], S. Ogawa, T. Yamane
T. R. Brown, K. Ugurbil, P. Glynn and H. Rottenberg
Bell Laboratories
Murray Hill, New Jersey 07974

The improved sensitivity and resolution of modern Nuclear
Magnetic Resonance (NMR) spectrometers have recently enabled
meaningful measurements to be made on living tissue and cells. In
this brief article we show how NMR measurements of the ^{31}P
metabolites can follow _in vivo_ cellular processes which are
fundamental to the flow and storage of information. For coherence
we have chosen examples from our recent studies of E. coli for
which detailed descriptions are being published elsewhere.[1-3] All
spectra reported here have been obtained with a Bruker HX-360 NMR
spectrometer operating at 145.7 MHz for the ^{31}P absorption, in the
Fourier Transform mode. The NMR spectra were accumulated from free
induction decays of the ^{31}P nuclei with data acquisition times of
0.6 seconds and the spectra were obtained by summing at least 50
spectra.

Figure 1 shows the NMR spectra in the vicinity of the P_i peaks
in suspension of E. coli cells.[1] The position of the P_i peak
changes with pH so that the position is used as a measure of the
pH. Because the sample has been prepared in the cold the cells are
not perfectly run down at the beginning of the experiment so that
the internal pH_{in} is 6.73 while the external pH_{out} was 6.13.
Spectra were accumulated for 3 minute periods, the top spectrum
was taken between 7 and 4 minutes before oxygen bubbling started.
Note that in the next spectrum both internal and external pH's have
become more acidic as a result of glycolysis. After the onset of
oxygen bubbling pH_{in} rises within a few minutes to ∿7.55 where it
remains for about one hour depending upon conditions. Separate
measurements of the rate of respiration indicate that in these dense
suspensions (∿50 mg of protein/ml or ∿6x10^{11} cell/ml) the rate of
oxygen consumption per cell is 6 to 10 times slower than it is with

---

[*]Present Address: Department of Chemistry, Tel-Aviv University
Ramat-Aviv, Tel-Aviv, Israel

an unlimited oxygen supply, but that increasing the oxygen supply
several fold does not change the measured pH values.[2] In
accompanying experiments it was shown that pH_{in} of energized E. coli
reached ~7.5 for pH_{out} values between 6.0 and 8.0, in accord with
previous pH measurements using the distribution of the weak acid
DMO across the plasma membrane.[4]

One particularly valuable aspect of measuring the pH by the NMR
technique is that the intrinsic time resolution is the time it takes
to accumulate one free induction decay or ~1 second. This time

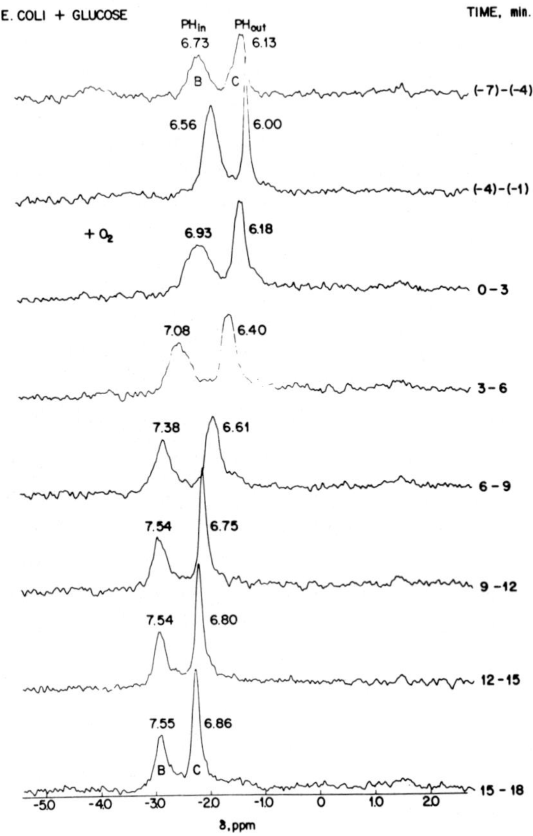

Fig. 1. 145.7 MHz ^{31}P NMR spectra of a suspension
of E. coli cells grown in M-9 medium with glucose
as the carbon source. After centrifugation
the pellet (~1 ml) was suspended in 1 ml of M-9
medium with added 3mM glucose, 50mM MES, 50mM
HEPES and 20 mM phosphate at pH 6.0 and 20°C.

resolution is realized in the spectra shown in Figure 2 where the
free induction decays are synchronized with a three second long
pulse of bubbling oxygen.[2] The bubbling was intentionally vigorous
so as to disturb the magnetic field homogeneity, and it can be
seen that the inorganic phosphate NMR peaks are very broad during
and immediately after the bubbling, but become sharp a few seconds
after the bubbling stops. Note that with the time sequence employed
of 3 seconds bubbling followed by 15 seconds off the pH remains
constant. This shows that the response time of the pH_{in} of E. coli
to loss of oxygen is slower than 15 seconds and in fact separate
experiments show that it is of the order of a few minutes.[2]

In addition to monitoring the pH, the ^{31}P NMR can also be used
to follow the concentrations of metabolites and the effects of drugs.

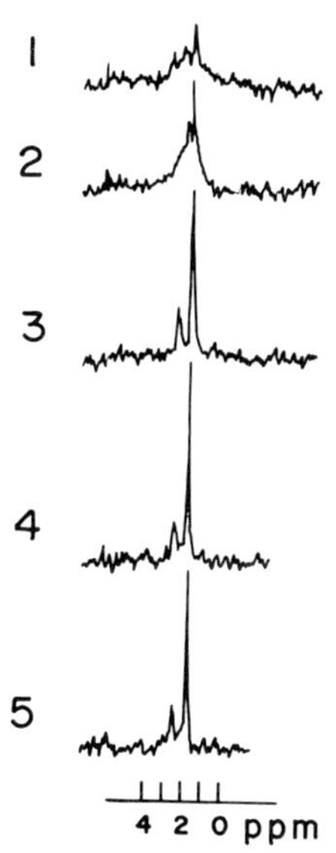

Fig. 2. ^{31}P NMR spectra of
orthosphosphate peaks in a
suspension containing 6×10^{11} E.
coli cells/cc. Pure oxygen was
bubbled for 3 seconds and turned
off for 15 seconds. The five
spectra were obtained from 90°
radio frequency pulse synchronized
with the bubbling as shown. Each
spectrum is a sum of 50 individual
free induction decays.

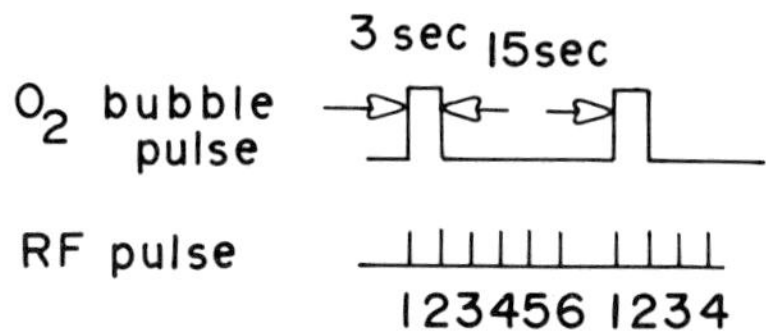

Figure 3 shows the full NMR spectra in the phosphate region of a suspension of anaerobic E. coli. Each spectrum is a two minute accumulation centered at the time shown. Six minutes before the introduction of glucose the top spectrum shows that the cells are run down, $\Delta pH = pH_{in} - pH_{out} = 0$. A weak phosphomonoester peak is observed to the left of the intense P_i peak. To the right (higher fields) near +10.5 ppm is the NAD^+ peaks while just slightly higher at ∿12 ppm is the weaker UDPG peak. Two minutes after addition of glucose the sugar phosphates in the phosphomonoester region and the terminal phosphate of nucleotide diphosphates (NDP) at +5 ppm increase in intensity. At 4 minutes both NTP and NDP signals can be seen near 5 ppm (the NTP is at a slightly lower field) and the β

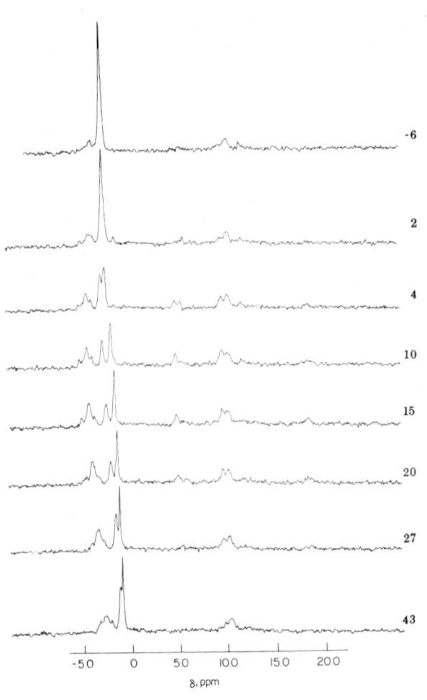

Fig. 3. E. coli, MRE 600 were grown in M9 medium with 10mg/l succinate as the carbon source. Cells were harvested at mid-log phase, washed twice and resuspended in cold medium (10 mM Na_2HPO_4, 10 mM KH_2PO_4, 100 mM PIPES, 50 mM MES, 85 mM NaCl, pH 7.4). At time zero, glucose was added to the NMR samples up to a concentration of 50 mM. All measurements were made at 20°C.

phosphate peak of NTP at +18 ppm is seen. The sugar phosphate region is beginning to show structure, and the main peak is identified as fructose-1,6-diphosphate (FDP). The P_i peak is beginning to split showing a non vanishing value of ΔpH. As the NTP signal increases and subsequently decreases it is accompanied

by a rise and fall in ΔpH. These results, and similar results which
we have published,[1] seemed in agreement with one of Mitchell's
hypothesis i.e., the ΔpH was created by the ATP being hydrolyzed
by an ATPase, with subsequent proton translocation. If this were
true then an ATPase inhibitor should prevent the ATP hydrolysis and
not allow the ATP created during glycolysis to create a ΔpH. These
results are in fact obtained in Figure 4 where the only difference
from Figure 3 is that the E. coli were incubated in 2 mM DCCD
before the NMR experiments started. The two differences noted are
first the NTP builds up faster and lasts longer in the presence of
DCCD and second the ΔpH created is almost zero (note the shoulder at
16 minutes in Figure 4 compared to the large splittings between
pH_{in} and pH_{out} observed in Figure 3.) In addition to being consistent with this proposed pathway for ATP hydrolysis it is important to
note that ATP lasts longer and builds up faster in the DCCD treated
cells. This suggests that the ATP hydrolysis is an appreciable
shunt for the use of ATP.

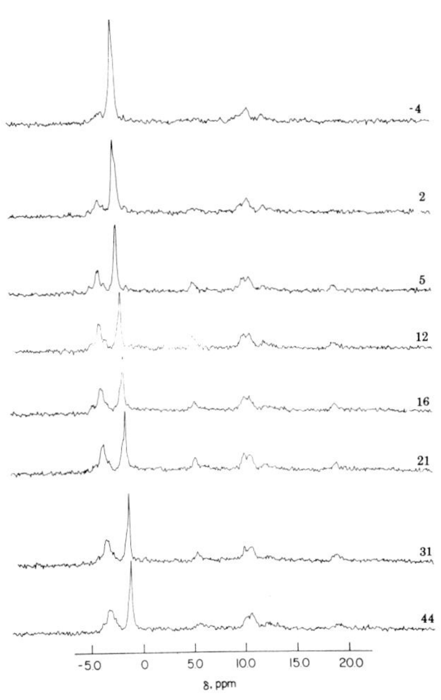

Fig. 4. Same conditions as
Figure 3 except that prior to
the experiments shown the
cell suspension was incubated
for 15 minutes at 20°C in the
presence of 2 mM DCCD.

These experiments have been selected to show some of the potential for ^{31}P NMR measurements of bioenergetic paths in intact cells. More complete description of these results as well as similar experiments in rat liver mitochondria are being presented elsewhere.

REFERENCES

1. Navon, G., Ogawa, S., Shulman, R. G. and Yamane, T. Proc. Natl. Acad. Sci. USA 74, 888 (1977).
2. Ogawa, S., Shulman, R. G., Glynn, P., Yamane, T. and Navon, G. Biochem. Biophys. Act. (to be published).
3. Ugurbil, K., Shulman, R. G. and Glynn, P. (to be published)
4. Padan, E., Zilberstein, D. and Rottenberg, H. Eur. J. Biochem. 63, 533 (1976).
5. For a review see Harold, F. M. Current Topics Bioenergetics 6, 89 (1977).

                                        477
Bioenergetics of Membranes. L. Packer et al. ed.

AN INVESTIGATION INTO THE DIFFERENT POPULATIONS OF CAROTENOIDS IN CHROMATO-
PHORES FROM RHODOPSEUDOMONAS SPHAEROIDES AND CAPSULATA.

Marc Symons, Christine Swysen and Christiaan Sybesma
Biophysical Laboratory, Vrije Universiteit Brussel,
Pleinlaan 2, 1050 Brussels
Belgium.

INTRODUCTION

Light-induced absorption changes in the 400-500 nm spectral region
in cells and all cell-free preparations attributed to carotenoid bandshifts, have been studied extensively (1-12). There is a general agreement that these absorption changes reflect the generation of an electrical potential across the
thylakoid or chromatophore membrane (4-8). However, there is still no unambiguous experimental evidence to support a specific electrochromic mechanism that
quantitatively would explain the phenomena. The original interpretation that a-
bout 10% of the absorption bands of Rhodopseudomonas sphaeroides shifts by about
10 nm to the red upon illumination could not be reconciled with any theory of
electrochromism.

Moreover, no appreciable shift in the isosbestic point has been observed. To explain this, Amesz et al (9) assumed that an increase in light intensity results in an increase in the number of carotenoid molecules exhibiting
a band shift which is independent of the light intensity. Another theory (11)
holds that the larger part of the absorbance changes originates from an electrochromic change in the extinction coefficient.

Recent experimental evidence (20,21) has shown, however, that there
are two pools of carotenoids. One of these pools (comprising 20 to 40% of the
carotenoid content) shifts over about 1.7 nm upon illumination; the other pool
containing the rest of the carotenoids does not show any change in absorption
upon illumination.

RESULTS

Rhodopseudomonas sphaeroides : Part of the evidence was obtained by computer a-
nalysis of the light-induced absorption difference spectra.
Fig. 1 shows a carotenoid absorption difference spectrum obtained from point by
point measurements of the light-induced absorption changes of a chromatophore preparation of the G1C mutant of Rhodopseudomonas sphaeroides
(this mutant contains neurosporene as the predominant carotenoid). Details on
the preparation and experimental equipment can be found in (21). The spectrum
was corrected for non-carotenoid absorbance changes by substracting the residual

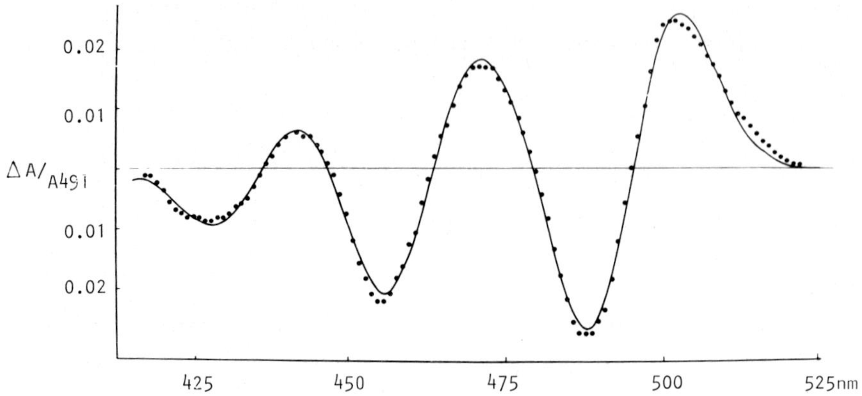

Fig. 1 Light-induced and computer fitted carotenoid absorption difference spectra of chromatophores from the GlC mutant of _Rps. sphaeroides_. Chromatophores at a concentration of 21 μM (determined by using the extinction coefficient of BChl _in vivo_ given by Clayton (31)) were suspended in 10 mM glycyl glycine and 10 mM choline chloride, pH 7.8. Open circles : experimental absorption difference spectrum obtained by point by point measurements and corrected for non-carotenoid absorbance changes (see text). Solid line : fit by the DIFSPC computer program, giving the parameters compiled in table 1.

spectrum after addition of 50 μm gramicidin. Gramicidin at this concentration is known to abolish electrochromic band shifts by neutralising the membrane potential (9.20).

We have tried to fit this spectrum with a theoretical difference spectrum obtained by subtraction of an convoluted spectrum of three Gaussian bands from a similar spectrum in which the bands are shifted over a (variable) distance $\Delta\lambda_i$.
We also allowed for a change in oscillator strength of the band $\Delta\varepsilon_i$. Each band is characterized by the fraction of total absorption, α_i, which is involved in the shift, its center wavelength, λ_i, and the e^{-1}-half bandwidth ω_i.
For a detailed discussion of the program, see (21).

In order to find an unambiguous fit for the difference spectrum, it was necessary to experimentally determine the magnitude of the light-induced band shifts. If one assumes that the form of the spectrum responsible for the light-induced absorption changes is identical to that of the bulk carotenoids,

the shift of only one band needs to be measured. This assumption seems to be a
reasonable one, since all the bands of the total spectrum originate from one molecule. According to Labhart (13) the bands can be assumed to result from a vibrational splitting of one electronic transition.

The electrochromic shift of the red-most band appeared to be most
ammenable to measurement. The shift was determined from plots of the wavelength
of the isosbestic point versus the maximum absorbance change (cf. Fig. 2).

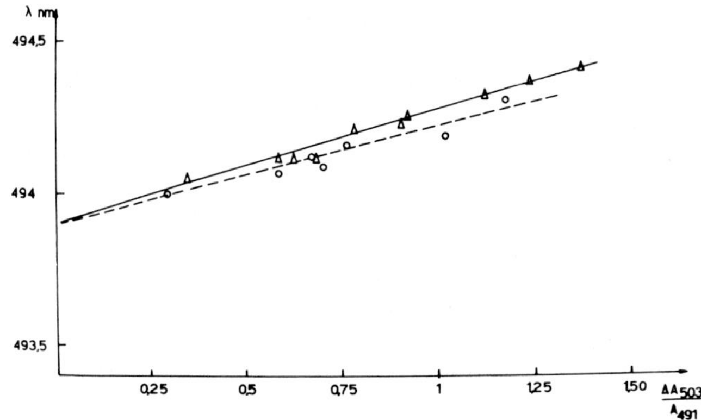

Fig. 2 Plots of the localization of the isosbestic point vs the maximum light-
induced change in absorption at different intensities of actinic light
for two preparations of Rps. sphaeroides G1C chromatophores. Conditions
as given in Fig. 1. Triangles : chromatophore preparation after 2x1
minute sonication; circles : chromatophores after 2x0.5 minutes sonication.

It can be shown that for shifts not exceeding 5 nm this relation should be linear
The wavelengths of the isosbestic point were then extrapolated to zero absorbance change. Typical values of the shift obtained this way were $\Delta\lambda_i$ = 1.5 $\pm$ 0.3nm
(for the long wavelength band). Values for the shift of the middle band are
some 30% lower than those for the red-most one. This indicates once more that
the spectrum which is responsible for the light-induced changes has the same
shape as the bulk carotenoid spectrum.

Taking $\Delta\lambda_1$ = 1.75 nm the band parameters for the population which
shifts in the light are given in table 1. Compared with the center wavelengths
of the bulk carotenoid peaks (see table 2) the results clearly indicate that the
carotenoid molecules responsible for the light-induced absorption changes form a
pool which is different from the major part of the carotenoid content. Thus we
can conclude that there are at least two populations of the same carotenoid,
which in the G1C-mutant is predominantly neurosporene (96%), see (10).

TABLE 1

Parameters of the population undergoing an electrochromic shift.

Band	Centre	e^{-1}-width	Fraction (% of tot. abs.)	Electrochromic extinction change	Shift
1	494.5 nm	10.5 nm	19 %	− 0.6 %	1.75 nm
2	463 nm	10.3 nm	19 %	− 1.3 %	1.4 nm
3	436.5 nm	8.9 nm	19 %	− 0.8 %	1.0 nm

TABLE 2

Parameters of the population remaining unshifted.

Band	Centre	e^{-1}-width	Fraction (% of tot. abs.)	Centra of total abs. spectrum corrected	non-corrected
1	489.5 nm	10.9 nm	84 %	490.5 nm	490.5 nm
2	459.5 nm	12.0 nm	85 %	459 nm	459.5 nm
3	431.5 nm	9.6 nm	89 %	431.5 nm	433.5 nm

The absorption spectrum of these two moietes were computed by another
program. This was done by subtracting the sum of the three Gaussian curves
coming from the first analysis from the absorption spectrum of the chromatophores suspension. The latter was corrected for non-carotenoid absorption by using
a normalized absorption spectrum of the carotenoidless mutant R-26 of Rps.
sphaeroides. The remainder was analyzed again in the three Gaussian curves.
The result of this analysis is given in Fig. 3.
The parameter values of the remaining three bands are given in table 2. The
table also contains the wavelength of maximum absorption of the chromatophores,
corrected for non-carotenoid absorption together with the non-corrected values.
Obviously, there is discrepancy between the two values : the correction appears
to be necessary in order to calculate good estimations for the distance between
the two pools.

A calibration of the light-induced bandshifts in terms of membrane
potential was carried out by the method indicated by Jackson and Crofts (9) :
KCl pulses were administered in the presence of 1µM valinomycin. An average
calibration value of 140 mV per nm shift was determined, so that the shift men-
tionned in table 1 correspond to membrane potentials of about 240 mV.

Rhodopseudomonas capsulata : Experiments carried out with Rps. capsulata wild
type chromatophores lead qualitatively to the same conclusions that there are
two spectrally different pools of carotenoids. The field sensitive pool contains
about 25% of the total carotenoids, the distance between the two pools however
appears to be about 10 nm; this is twice as large compared with Rps. sphaeroides
GIC chromatophores. The light-induced band shifts are about the same as for

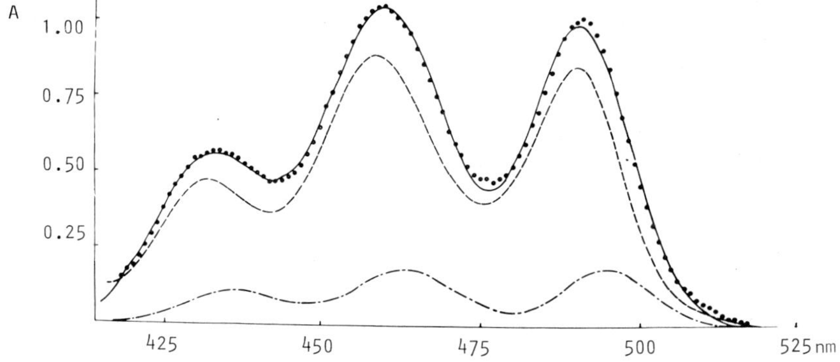

Fig. 3 Analysis of the <u>in vivo</u> carotenoid absorption spectrum. Dashed-dotted
line : spectrum of the "shifting" carotenoid pool (data from table 1);
dashed line : spectrum of the "remaining" carotenoid pool; dots : measured <u>in vivo</u> absorption spectrum corrected for non-carotenoid absorption
(see text); solid line : FITCAR computer fit of the sum of the "shifting"
spectrum and the "remaining" spectrum, giving the parameters compiled in
table 2.

sphaeroides, but the peak values of the potentials measured were substantially
larger (up to 400 mV).

A feature commonly observed in Rps. sphaeroides and capsulata is that
the shape and location of the total carotenoid spectrum change upon partly deterioration of the membrane. The bands become broader and show a blue shift of up
to 2 nm, depending on the degree of deterioration (Symons et al, unpublished results). The origin of these changes remains obscure until now.

In Rps. capsulata a very rapid change in the spectrum of the chromatophores was observed : it appeared to be that the difference spectrum, after a
few minutes actinic illumination, was shifted by about 0.5 nm to the blue.
However, the total absorption spectrum showed only a shift of maximum 0.2 nm, to
this kind of deterioration : we can conclude that these changes in the absorption
spectrum are predominantly caused by a shift of the pool which is light sensitive
These results substantiate the fact that there are at least two spectrally different populations of carotenoids, as shown by the computer analysis of the <u>Rps.</u>
<u>sphaeroides</u> spectra.

<u>DISCUSSION</u>

In order to explain the linear dependence of the bandshifts on the
field strength (contrary to what would be expected for a non-polar molecule) one
has to assume, as pointed out by Schmidt et al (15), that the carotenoids, at
least those which exhibit a light-induced shift, are subjected to a permanent
potential (cf. 22).

482

If the thickness of the membrane, the polarizability difference $\alpha_g - \alpha_e$ between the ground and excited state and the angle (θ) between the molecular axis and both the permanent electrical field (F_p) and the light-induced field (F_a) are kwown, the magnitude of the permanent field as well as the shift produced by the permanent field can be calculated from a relation given by Schmidt (15).

$$\Delta\nu = \frac{1}{2hc} (\alpha_g - \alpha_e) (F_p \cos \theta_p + F_a \cos \theta_a)^2$$

Taking 76 Å for the thickness of the membrane (16), 780 $Å^3$ for the polarizability of neurosporene (19), and 45° for the orientation of the permanent field as well as the carotenoid molecules in respect to the membrane surface (30) the permanent field has a strength of 7×10^6 V/cm. This would cause a displacement of the carotenoid absorption bands of 27 nm to the red, which is about the shift observed when carotenoids are extracted from the Rps. sphaeroides G1C chromatophores (22).

Considering the differences concerning the calibration value and distance between the two populations between Rps. sphaeroides and capsulata, it seems difficult to us to explain the wavelength shift between the two pools only by the strength of the permanent field. So we think that both populations are subjected to a local permanent field, both, however, in a different way, perhaps as a result of different orientation and/or localization.

REFERENCES

1. Smith, L. and Ramirez, J. (1960) J. Biol. Chem. 235, 218-225.
2. Clayton, R.K. (1963) Proc. Natl. Acad. Sci. 50, 583-587.
3. Vredenberg, W.J. and Amesz, J. (1966) Biochim. Biophys. Acta, 126, 244-253.
4. Emrich, H. B., Junge, J. and Witt, H. T. (1969), Z. Naturf. 24b, 1144-1146.
5. Reich, R., Scheerer, R., Sewe, K.-U. and Witt, H. T. (1976), Biochim. Biophys Acta 449, 285-294.
6. Borisevich, G. P., Kononenko, A. A., Venediktov, P. S., Verkhoturov, V. N. and Rubin, A. B. (1975), Biofizika, 20, 254-258.
7. Jackson, J. B. and Crofts, A. R. (1969) FEBS letters 4, 185-189.
8. Jackson, J. B. and Crofts, A. R. (1971) Eur. J. Biochem. 18, 120-130.
9. Amesz, J., 't Mannetje, A. H. and de Grooth, B. G. (1973) Abstr. Symp. Prokaryotic Photosynth. Org. Freiburg, pp. 34-35.
10. Holmes, N. G. and Crofts, A. R. (1977) Biochim. Biophys. Acta 459, 492-483.
11. Conjeaud, H. and Michel-Villaz, M. (1976) J. Theor. Biol. 62, 1-16.
12. De Grooth, B. G. and Amesz, J. in Abstracts of "International Conference on the primary electron transport and energy transduction in photosynthetic bacteria" Brussels, September 1976.

13. Labhart, H. (1957) J. Chem. Physics 27, 957-965.

14. Matsuura, K. and Nishimura, M. (1977) Biochim. Biophys. Acta 459, 483-491.

15. Schmidt, S. Thesis Technical University, (1973) Berlin.

16. Ueki, T., Kataoka, M. and Mitsui, T. (1976) Nature 262, 809-810.

17. Breton, J. (1974) Biochem. Biophys. Res. Comm. 59, 1011-1017.

18. Clayton, R.K. (1963) in Bacterial Photosynthesis (Gest, H., San Pietro, A. and Vernon, L. P., eds.), p. 498, Antioch Press, Yellow Springs, Ohio.

19. Labhart, H. (1961) Helv. Chim. Acta 44, 457-460.

20. De Grooth, B. G. and Amesz, J. (1977), Biochim. Biophys. Acta, in press

21. Symons, M., Swysen, C. and Sybesma C., (1977) Biochim. Biophys. Acta, in press.

22. Holmes, N. G., Thesis University of Bristol, (1975).

CHANGE OF LECITHIN AGGREGATION DUE TO VALINOMYCIN-LIPID INTERACTION, AND ITS RELEVANCE TO ENERGY CONVERSION

Dieter Walz
Biozentrum, University of Basel, Klingelbergstrasse 70
CH-4056 Basel, Switzerland

INTRODUCTION

Valinomycin's ability to act as a carrier for certain alkali ions in biological and artificial membranes is well-known and its ionophoric properties have been extensively investigated. Obviously the antibiotic has to penetrate into the membrane in order to transport ions across the hydrophobic barrier[1] which implies an interaction of valinomycin with the constituents of the membrane. Less attention has been paid to this aspect although nuclear magnetic resonance studies of artificial lipid bilayers have indicated effects on the lipids molecules due to the interaction with valinomycin[2,3].

Lecithin vesicles whose membranes contain chlorophyll a or b turned out to be a system which can provide additional information about this phenomenon[4,5]. Since the parameters which determine the light absorption process in chlorophyll are sensitive to influences from the surrounding medium (solvatochromism) this pigment is able to reflect certain alterations in its vicinity by a change of the absorption spectrum. It was thus found that valinomycin dissolved in a membrane affects the lipid matrix. This phenomenon, which is not related to the antibiotic's ionophoric properties and hence is not dependent on the presence of potassium ions, offers a possible explanation for several not yet understood effects of valinomycin in biological membranes (e.g. thylakoid membranes).

MATERIALS AND METHODS

Chlorophyll a or b containing lecithin vesicles were prepared as described previously[6]. They have diameters around 200 Å and are bounded by a single bilayer membrane accomodating between 2 and 18 chlorophyll molecules per 1000 lecithins[6]. The absorption spectrum of chlorophyll a and b (Figs. 1 and 3) is typical for the pigments in the monomeric form[7,8].

The difference spectra were measured with an Aminco DW-2 spectrophotometer in split beam mode; sample and reference cuvette contained vesicle suspensions of identical composition except for valinomycin which was present in the sample only (for details see ref. 4). The vesicles were prepared and stored in a solution containing 0.01 M MOPS*, pH 7.2, and LiCl or NaCl with varying amounts of KCl to a total concentration of 0.2 M salt. Different ratios of K^+ concentration inside to outside of the vesicles could then be established by diluting the vesicle stock solution with isotonic buffer solutions having the appropriate KCl content besides LiCl or NaCl.

RESULTS

Experiments with equal K^+ concentrations inside and outside the vesicles: Under this condition, no diffusion potential across the vesicle membranes arises when valinomycin is added to the suspension. Nevertheless, clear-cut difference spectra were observed for both chlorophyll a (Fig. 2) and b (Fig. 4) containing vesicles. These absorbance differences were found to be independent of the type of alkali ions present in the aqueous phase and, therefore, are not related to the alkali ion binding or ionophoric properties of valinomycin. Difference spectra obtained with varying molar ratios of valinomycin to lecithin and using vesicles with different chlorophyll contents showed that the wavelength dependence of the absorbance changes is independent of the sample composition. Hence a quantity $Q(\lambda_o)$ was introduced which represents the intensity of the absorbance changes averaged over the whole wavelength range and corrected for the varying amounts of chlorophyll present in the samples (for details see ref. 4). $Q(\lambda_o)$ could be correlated to the molar ratio of valinomycin to lecithin which indicates a valinomycin-lecithin interaction. A direct interaction of valinomycin with chlorophyll could be excluded since the experimental data did not satisfy the pertinent relations derived for such an interaction[4].

A further analysis of the experimental data yielded the distribution coefficient of valinomycin for the vesicle membrane/water two-phase system ($\sim$20,000) and, with this quantity, the molar ratio of valinomycin dissolved in the membrane to lecithin could be calculated for each sample[4]. $Q(\lambda_o)$ was found to be proportional to this ratio up to a value of 0.0286 and constant above this limit, which thus represents the saturation point for dissolving valinomycin in vesicle membranes (Fig. 5). In a valinomycin-saturated membrane, we then find an average of 35 lecithin molecules per valinomycin which corresponds to two circular layers (see Fig. 5 inset).

Experiments with different K^+ concentrations inside and outside the vesicle: Valinomycin added to a vesicle suspension under this condition has a twofold effect, it elicidates the spectral changes of chlorophyll as described above and induces a diffusion potential across the vesicle membrane which again alters the spectrum of

---

* 1-Morpholinopropane sulfonic acid, pH adjusted with LiOH or NaOH

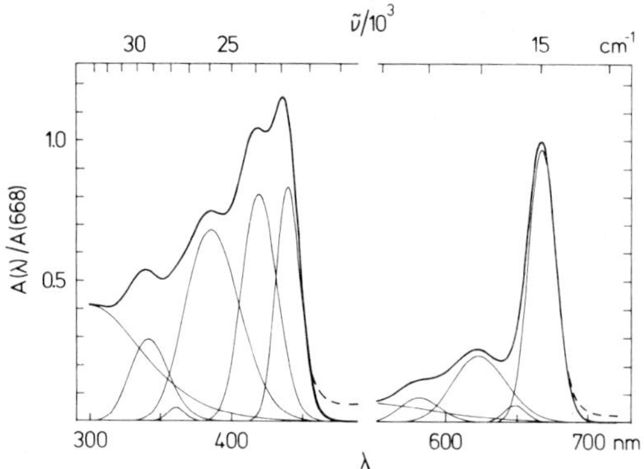

Fig. 1. Absorption spectrum of chlorophyll a incorporated into membranes of lecithin vesicles, A(λ), normalized by the absorbance at the main red peak A(668). The spectrum was deconvoluted into Gaussian components (thin lines) which are substitutes for the vibrational structure of the absorption bands associated with different electronic transitions (see refs. 5,7). The heavy line indicates the spectrum calculated as the sum of the components; it coincides with the measured spectrum (broken line) except for those wavelength ranges where components were neglected. The spectrum in the omitted range (cf. Fig. 2) runs almost parallel to the abscissa as a slightly wavy line. The upper scale indicates the wavenumber, $\tilde{\nu}$.

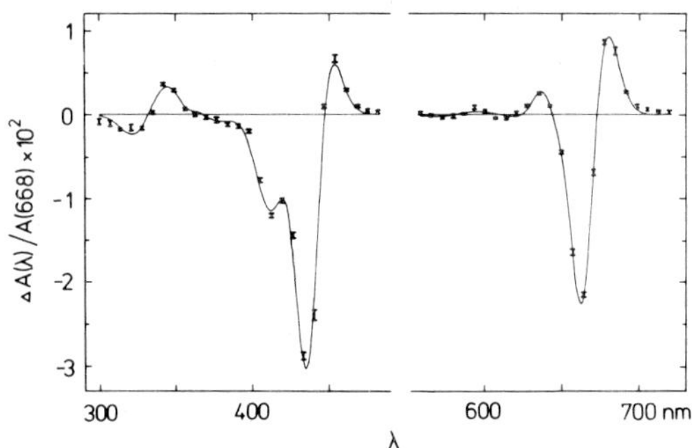

Fig. 2. Absorbance differences, ΔA(λ), induced by valinomycin in the spectrum of chlorophyll a in lecithin vesicles, and normalized by the absorbance of the vesicle suspension at 668 nm, A(668). The experimental data, which were obtained with vesicles having equal K$^+$ concentration inside and outside, are marked by bars representing $\pm$ standard deviation. The curve was calculated using the coefficients resulting from the analysis of the difference spectra in terms of a solvatochromism (ref. 5) and with the Gaussian components shown in Fig. 1. No absorbance changes were observed in the omitted wavelength range.

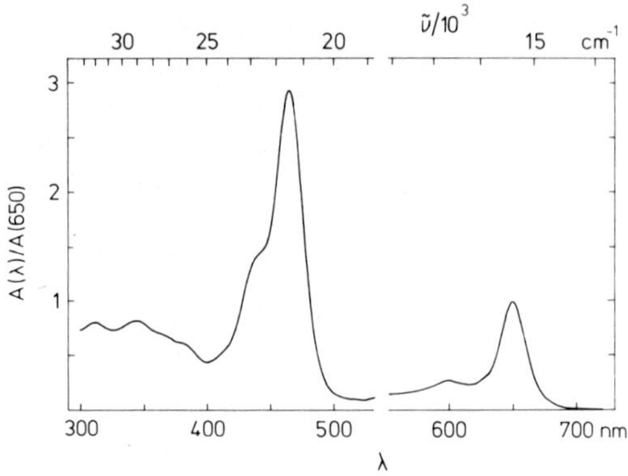

Fig. 3. Absorption spectrum of chlorophyll b in lecithin vesicles, A(λ), normalized by the absorbance at the main red peak, A(650). The upper scale indicates the
wavenumber, $\tilde{\nu}$.

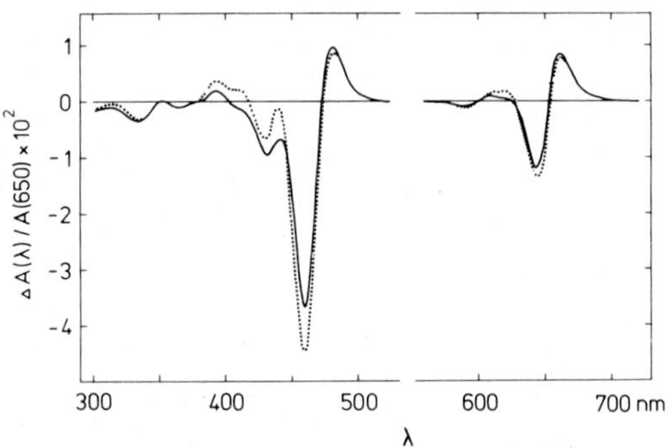

Fig. 4. Valinomycin-induced absorbance differences, ΔA(λ), of chlorophyll b in
lecithin vesicles, normalized by the absorbance of the vesicle suspension at 650 nm,
A(650). The two spectra shown were obtained with samples of identical composition
except for the K^+ concentration which was either the same inside and outside the
vesicles (heavy line) or ten times larger in the suspending medium than inside the
vesicles (dotted line).

chlorophyll due to an electrochromic effect[9]. The wavelength dependence of the
absorbance changes in the presence of an externally applied electric field closely
follows the difference spectrum without a field but the intensities are slightly
different (Fig. 4). Experiments where the polarity of the membrane potential was
opposite to that illustrated in Fig. 4 yielded essentially the same difference
spectra. Similar results were also obtained with chlorophyll a containing vesicles.

Due to the fairly small differences between the valinomycin-induced difference
spectra in the presence and absence of a membrane potential, the electrochromic
part of the absorbance changes can be quantified with low accuracy only. It
appears, however, that it is mainly dependent on the square of the electric field

strength, which was to be expected in view of the mirror symmetrical arrangement of the two monolayers in the membrane. On the other hand, increasing molar ratios of valinomycin (dissolved in the membrane) to lecithin up to half saturation (cf. Fig. 5) at a constant membrane potential yielded increasing electrochromic absorbance changes so that the relative contributions to the difference spectra, i.e. those due to valinomycin and to an electrochromism, respectively, were constant [about 9:1 in terms of $Q(\lambda_o)$].

Role of chlorophyll: A direct interaction of valinomycin with chlorophyll could be excluded, and the measured absorbance changes of chlorophyll upon incorporation of valinomycin into the vesicle membranes can, therefore, only arise from a change of the pigment's environment, viz. the lipid matrix. Hence the state of aggregation of lecithins surrounding a valinomycin molecule should be different from that of lipids in a domain without valinomycin in the sense that the two states represent two slightly different solvents for chlorophyll. The spectral changes can then be understood in terms of a solvatochromism[10], i.e. the influence of the solvent on the absorption spectrum of the solute, a well-known effect for chlorophyll[11]. The analysis of the difference spectra in this respect (cf. Fig. 2, for details see ref. 5) yielded a set of solvatochromic coefficients pertinent to the electronic

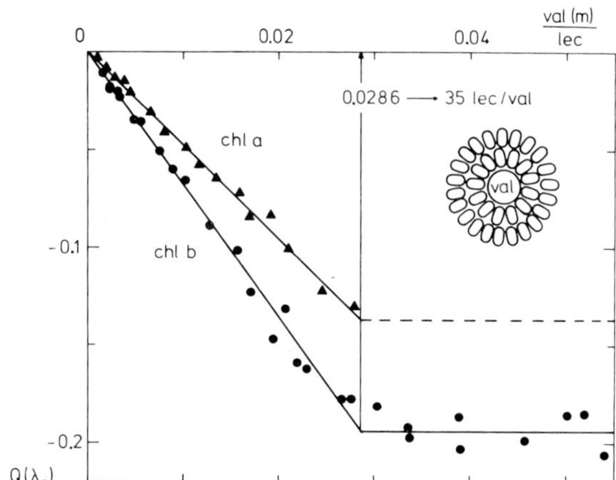

Fig. 5. Dependence of $Q(\lambda_o)$ on the molar ratio of valinomycin dissolved in the membrane to lecithin, val(m)/lec; for the definition of $Q(\lambda_o)$ see text and ref. 4, λ_o = 435nm for chlorophyll a and λ_o = 461nm for chlorophyll b. Experimental conditions limit the range of val(m)/lec-values for chlorophyll a containing vesicles at 0.028 (see ref. 4). However, this limit is higher for chlorophyll b containing vesicles since the extinction coefficient for the Soret band of chlorophyll b is about three times higher than that of chlorophyll a. The inset shows that 35 lecithin molecules (surface area in the membrane ~70 $Å^2$) can be exactly arranged into two circles around valinomycin.

490

transitions which occur in chlorophyll on excitation with light and which are represented by the Gaussian components[7] shown in Fig. 1 for chlorophyll _a_.

An interpretation of the solvatochromic coefficients revealed that a change in electrostatic interaction is the main reason for the alteration of the spectral parameters of chlorophyll[5] which implies a location of the chromophore in the domain of the lipid's polar head groups, in agreement with the findings of other investigators (see e.g. ref. 12). The difference in effective electric field strength was estimated[5] and found to be of the order of 10^7 V/cm which explains the small contributions of membrane potentials corresponding to field strength of at most 10^5 V/cm. The difference in electrostatic interaction for chlorophyll in different lipid aggregations most probably arises from a reorientation of the anisotropic chromophore (i.e. the porphyrin ring) and less from an actual difference of the effective electric field strength. The dependence of the electrochromic absorbance changes due to a constant membrane potential on the amount of valinomycin present in the membrane corroborates the postulated reorientation of the chromophore.

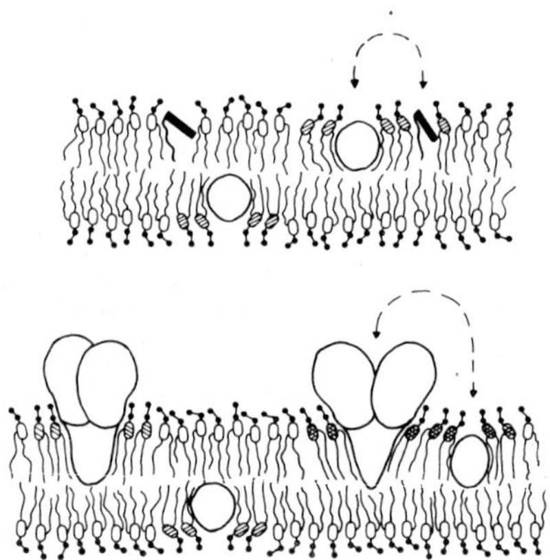

Fig. 6. Schematic representation of a membrane containing chlorophyll (symbolized by the heavy bar with "tail", top) or a protein with boundary lipids (hatched, bottom). Valinomycin dissolved in the membrane (circles) changes the aggregational state of the two layers of lipids surrounding it (as indicated by the hatching different from that of the protein's boundary lipids). Chlorophyll present within these "boundary lipids" of valinomycin adapts to the altered lipid environment by a reorientation of the porphyrin ring (top right). When valinomycin approaches the protein, the two boundary lipid layers interfere and a state of lipid aggregation for the boundary lipids common to both components results (indicated by cross hatching) which differs from that for the isolated components (bottom right). As a consequence, a change in conformation of the protein (and probably also of valinomycin, see ref. 1) occurs thus influencing the protein's function.

DISCUSSION

The present experiments have shown that valinomycin dissolved in a lecithin membrane interacts with the surrounding lipids and alters their state of aggregation (Fig. 6). A saturation for dissolving valinomycin in the membrane is obviously reached when the aggregational state of all the lipids has been changed, which means that the interaction extends over two circular layers of lipids around each valinomycin molecule (cf. Fig. 5). A similar phenomenon was already described for membrane bound proteins and called boundary lipid formation[13].

Chlorophyll incorporated into the membrane acts as a sensor for the aggregational state of the lipids and, therefore, demonstrates that a change in lipid aggregation has also an influence on non-lipid species embedded in the membrane. At present, no evidence seems to rule out the possibility that an altered lipid matrix has an effect on membrane bound proteins as well thus inducing structural and functional changes of these proteins (Fig. 6). Valinomycin has been extensively used as a tool to increase the permeability to K^+ in many biological membranes. Changes in lipid aggregation due to the valinomycin-lipid interaction are liable to occur in every case; the extent of these changes, however, may greatly vary with type and composition of the lipids. Moreover, the diverse proteins found in different membranes are certainly not equally susceptible to an altered lipid phase, and phenomena resulting from affected proteins can be masked by the system's impairment due to the ionophoric action of the antibiotic.

Valinomycin added to suspensions of energy converting organelles induces K^+ dependent effects on ion transport[14-17]. In mitochondria this leads to uncoupling of phosphorylation in a way entirely explainable by the enhanced K^+ permeability[14,18]. In chromatophores, however, valinomycin inhibits electron transport[19,20], and an inhibitory effect on energy transfer was earlier reported for chloroplasts[16,19]. More recent investigations have shown that valinomycin partially inhibits electron flow which, as a consequence, abolishes phosphorylation[17,21,22]; yet a direct influence of valinomycin on the phosphorylating system seems to exist[21]. All these effects are not related to the ionophoric properties of valinomycin, since they occur even in the absence of K^+ ions, but most probably arise from the antibiotic's influence on the lipid matrix. It is conceivable that the altered lipid aggregation hampers the conformational change of CF_1 associated with the activation of the coupling factor in illuminated chloroplasts. Valinomycin's inhibition of the electron transport can be released by an internal TMPD bypass; this locates its point of inhibition between plastoquinone and the acceptor site of photosystem I[21], i.e. between a lipid-like component and a membrane bound protein complex. Again, the altered lipid matrix could induce another conformation of the photosystem I protein complex which is unable to accept electrons from reduced plastoquinone but still able to accept them from reduced TMPD or other artificial electron donor systems to photosystem I[17,21]. A direct effect on the lipid-like component, however, is more plausible; the valinomycin-induced lipid aggregation probably interferes with the mobility of plastoquinone and thus slows down the quinone shuttle accross the thylakoid membrane. This would also explain the slight inhibition of the electron flow from water to artificial electron acceptor systems of photosystem II[21].

Obviously, the above mentioned effects are not detectable when valinomycin is used to establish a K^+-driven diffusion potential across the membranes of not activated chromatophores[23] or to accelerate the decay of the electric field set up by the primary photoreactions in the membranes of chromatophores[23] or chloro-plasts[24]. On the other hand, valinomycin added to purple membrane fragments suspended in water cannot display any ionophoric properties, but the observed quenching of a transient in the photochemical processes of bacteriorhodopsin[25] may be in line with its direct influence on membrane components.

Effects of valinomycin which cannot be rationalized on the basis of the iono-phoric properties alone have also been reported for other biological systems. The most striking example is the influence of this antibiotic on neuroblastoma cells which, in stationary phase of growth, display partially developed electrical pro-perties[25]. The phenomena observed on addition of valinomycin to these cells, viz. a twofold increase in resting potential accompanied by a similar increment in input membrane resistance and a dramatic enhancement of electrical excitability, could not be explained by the increased K^+ conductance alone but required the assumption of an even larger decrease in Na^+ or Cl^- conductance and a change in the rate at which the membrane permeability to Na^+ increases during spike activity. This was taken to indicate a structural rearrangement of the membrane induced by valino-mycin[25] which is conceivable on account of the antibiotic's interaction with lipids. The dual activity of valinomycin should be kept in mind when using it to affect such complex processes as circadian rhythms in plants[27] or algae[28]. Although a certain dependence of these rhythms on the K^+ concentration in the cytoplasm seems to exist, the phase shifts elicitated by valinomycin most probably reflect the antibiotic's interference with the state of the membranes. This is corroborated by the fact that valinomycin reverses the phase shifts induced by small amounts of ethanol[28]. Preliminary experiments with chlorophyll containing vesicles revealed that ethanol induces similar effects on the lipid aggregation as does valinomycin. However, the activities of the two agents are not simply additive but display a complex interrelation which deserves further investigation.

The schematic representation in Fig. 6, which illustrates how the altered lipid aggregation around valinomycin interferes with the boundary lipid layer of a protein, can be generalized. Boundary lipids of different proteins could similarly interfere with each other thus creating an indirect protein-protein interaction mediated and possibly modulated by the lipid phase. A change in conformation of a protein (e.g. due to the binding of an agent) is likely to be followed by a change in aggregation of the boundary lipids which in turn is transferred to the boundary lipids of other proteins in the membrane and thus may affect the conformation and function of these proteins. Owing to such events, a signal received locally by a specialized protein could spread over the whole membrane and elicitate the appropriate responses. An illustrative example for this mechanism is presented by the membranes of neuroblastoma cells (see above); the incorporation of valinomycin into the membranes, i.e. changing the lipid matrix locally, modulates the function of several proteins.

Finally the effects induced by valinomycin in chlorophyll containing vesicles could help for a further elucidation of the mechanisms by which certain probes detect membrane properties. These probes are added to the suspending medium and, like valinomycin, are partially dissolved in the membranes. Their sensor ability arises, like that of chlorophyll, from the susceptability of their absorption or fluorescence spectrum to influences from the surroundings. When such a probe shows a change in fluorescence or absorbance, this could indeed indicate a change in hydrophobicity of that domain (e.g., of a protein) where the probe is presumably located. However, it could be a change in lipid aggregation (e.g. due to a change in protein conformation) to which the probe then responds in a similar way as the chlorophylls do. In addition, the dependence of the solubility of a species in the membrane on the state of lipid aggregation, as demonstrated by valinomycin, could lead to a repartition of the probe not only between the aqueous and the membraneous phase but between different regions within the membrane.

REFERENCES

1. Grell, E., Funck, Th. and Eggers, F. (1975) in Membranes - A Series of Advances (Eisenman, G., ed.) Vol. 3, pp. 1-126, Marcel Dekker, New York and Basel

2. Finer,E.G., Hauser, H. and Chapman, D. (1969) Chem. Phys. Lipids 3, 386-392

3. Hsu, M. and Chan, S. I. (1973) Biochemistry 12, 3872-3876

4. Walz, D. (1976) J. Membrane Biol. 27, 55-81

5. Walz, D. (1977) J. Membrane Biol. 31, 31-64

6. Ritt, E. and Walz, D. (1976) J. Membrane Biol. 27, 41-54

7. Shipman, L. L., Cotton, M. Th., Norris, J. R. and Katz, J. J. (1976)
   J. Amer. Chem. Soc. 98, 8222-8230

8. Lehoczki, E. (1975) Biochim. Biophys. Acta 408, 223-227

9. Kleuser, D. and Bücher, H. (1969) Z. Naturforsch. 24b, 1371-1374

10. Liptay, W. (1969) Angew. Chem. 81, 195-206

11. Seely, G. R. and Jensen, R. G. (1965) Spectrochim. Acta 21, 1835-1845

12. Oettmeier, W., Norris, J. R. and Katz, J. J. (1976) Biochem. Biophys.
    Res. Commun. 71, 445-451

13. Vanderkooi, G. (1974) Biochim. Biophys. Acta 344, 307-344

14. Höfer, M. and Pressman, B. C. (1966) Biochemistry 5, 3919-3925

15. Jackson, J. B., Crofts, A. R. and von Stedingk, L.-V. (1968) Eur. J.
    Biochem. 6, 41-45

16. Karlish, S. J. D. and Avron, M. (1971) Eur. J. Biochem. 20, 51-57

17. Telfer, A. and Barber, J. (1974) Biochim. Biophys. Acta 333, 343-352

18. Mitchell, P. (1968) Chemiosmotic coupling and Energy Transduction,
    Glynn Research Lab., Bodmin, Cornwall

19. Keister, D. L. (1970) in Electron Transport and Energy Conservation (Tager,
    T. M., Papa, S., Quagliariello, E. and Slater, E. C., eds.) pp. 469-472,
    Adriatica Editrice, Bari

20. Gromet-Elhanan, Z. (1970) Biochim. Biophys. Acta 223, 174-182

21. Trebst, A. and Reimer, S. (1977) Plant Cell Physiol., in the press

22. Keister, D. L. and Minton, N. J. (1970) Bioenergetics 1, 367-377 .

23. Jackson, J. B. and Crofts, A. R. (1971) Eur. J. Biochem. 18, 120-130

24. Junge, W. and Schmid, R. (1971) J. Membrane Biol. 4, 179-192

25. Slifkin, M. A. and Caplan, S. R. (1975) Nature 253, 56-58

26. Spector, I., Palfrey, C. and Littauer, U. (1975) Nature 254, 121-124

27. Bünning, E. and Moser, I. (1972) Proc. Natl. Acad. Sci. U. S. 69, 2732-2733

28. Sweeny, B. M. (1974) Plant Physiol. 53, 337-342

Bioenergetics of Membranes. L. Packer et al. ed.

COUPLING FACTOR ATPase COMPLEX OF RHODOSPIRILLUM RUBRUM CHROMATOPHORES: ISOLATION AND PURIFICATION OF AN OLIGOMYCIN AND DCCD SENSITIVE ATPase

Zippora Gromet-Elhanan and Rachel Oren
Biochemistry Department, Weizmann Institute of Science, Rehovot,
Israel

SUMMARY

Extraction of Rhodospirillum rubrum chromatophores with Triton X-100 results in solubilization of the whole coupling factor ATPase complex. This complex has been purified by glycerol gradient centrifugation and found to contain about eight different polypeptide subunits. The Triton solubilized as well as the membrane bound complex exhibit both Mg^{2+} and Ca^{2+} dependent ATPase activities that are sensitive to oligomycin, DCCD and venturicidin.

Triton treatment of LiCl depleted chormatophores, which have previously been shown to lack only the β-subunit of the coupling factor complex, results in extraction of an inactive ATPase complex. It is, therefore, concluded that the β-subunit is indispensable for activation of the Triton-solubilized ATPase.

INTRODUCTION

Detailed understanding of the mechanism of electron transport coupled ATP synthesis requires the investigation of the catalytic component, the coupling factors,of the system . The best studied coupling factor ATPase complex is that of mitochondria, where the whole complex has been isolated by detergent[1] and found to be very similar to the membrane bound ATPase in its cation specificity and sensitivity to inhibitors[2]. It has recently been extensively purified[3] and shown to contain about ten different polypeptide subunits.Such a complete detergent soluble complex has not been as yet isolated from photosynthetic systems.

A water soluble ATPase, which contains only five out of the ten subunits of the whole complex has been isolated from mitochondria[2] as well as from chloroplasts[4] and chromatophores[5]. However, this ATPase, unlike the detergent soluble one, has been shown to be different from the membrane bound ATPase. In R. rubrum chromatophores for instance the membrane bound enzyme can be induced by both Ca^{2+} and Mg^{2+} and is sensitive to oligomycin[6],whereas the water soluble ATPase can be induced by Ca^{2+} but is inhibited by Mg^{2+} and is resistant to oligomycin[7]. In the present communication the purification and properties of a Triton X-100 solubilized ATPase from R. rubrum chromatophores are described.

METHODS

The growth of R. rubrum cells and the isolation and storage of chromatophores were as previously described[8-10]. Bacteriochlorophyll was determined in vivo using the absorbance

496

coefficient given by Clayton[11]. Chromatophores were extracted by Triton X-100 as outlined by
Oren and Gromet-Elhanan[12] and by LiCl as described by Gromet-Elhanan[13]. Protein was
determined according to Lowry et al.[14].

ATPase activity was assayed by following the hydrolysis of $(\gamma-{}^{32}P)$ATP which was prepared
according to the method of Avron[15]. The reaction mixture contained in a final volum of 1 ml:
10 mM HEPES-NaOH, pH 8.0, 4 mM ATP containing 2×10^5 cpm $(\gamma-{}^{32}P)$ATP and either 2 mM
$CaCl_2$ or 4 mM $MgCl_2$. The reaction was started by addition of either chromatophores (10 μgr
bacteriochlorophyll) or Triton solubilized enzyme (20 μg protein) and stopped by the addition
of cold TCA to a final concentration of 5%. The released $({}^{32}P)$ Na_2HPO_4 was separated by the
isobutanol-benzene procedure[16].

RESULTS

We have recently reported that Triton extraction of R. rubrum chromatophores resulted in
solubilization of an oligomycin senstive ATPase[17]. The enzyme was concentrated by ultrafiltra-
tion through Diaflo XM-300 and further purified on a glycerol gradient. As illustrated in Fig. 1,
ATPase activity could be induced by either Ca^{2+} or Mg^{2+} and both activities appeared in the
same protein fractions, indicating that they belong to the same enzyme complex. The effect
of increasing cation concentrations on the ATPase activity of the membrane bound and Triton
soluble ATPases are compared in Fig. 2. Although the Triton ATPase, like the membrane bound
one, could be induced by either Ca^{2+} or Mg^{2+} it was found to be much more sensitive to an
excess of free Mg^{2+}. At equimolar concentrations of Mg^{2+} and ATP the membrane-bound
ATPase retained about 80% of its activity whereas the Triton soluble one was inhibited over 90%.
On the other hand, excess of free Ca^{2+} had a very little inhibitory effect in both types of
ATPase. A fixed concentration of 2 mM Mg^{2+}, 4 mM Ca^{2+} and 4 mM ATP was therefore used
in all further investigations.

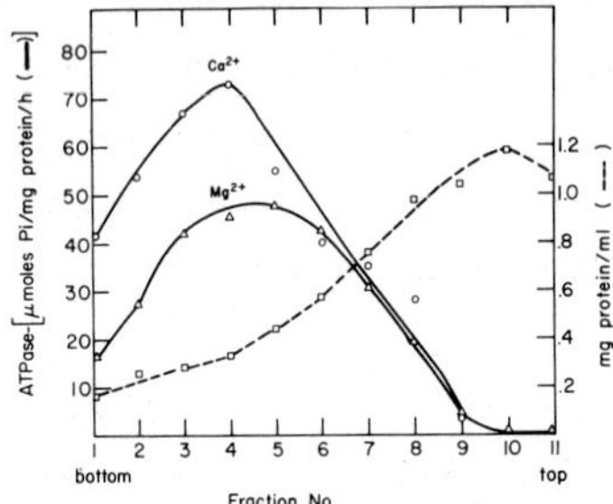

Fig. 1. Distribution profile of Triton X-100 solubilized R. rubrum ATPase on a glycerol gradient.
The Triton extract (4 ml) was applied to 32 ml of 5-15% glycerol gradient containing 5 mM
HEPES-NaOH,pH 8.0, and 0.1% Triton X-100. The gradients were centrifuged 15 hours at
26,000 rpm in the SW 27 Beckman rotor. 0—0, Ca^{2+} dependent ATPase activity; △—△ Mg^{2+}
dependent ATPase activity; □----□, protein concentration.

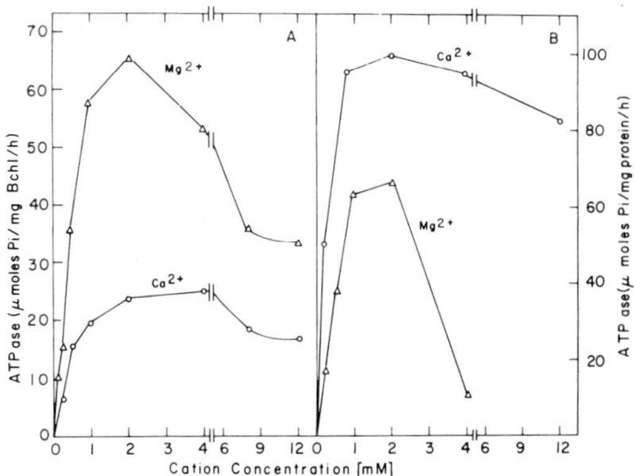

Fig. 2. The dependence of membrane-bound and Triton soluble ATPase activity on the concentration of divalent cation. The reaction mixture was as described under Methods, except that the cation concentration was varied as indicated.
A. ATPase activity of the membrane bound enzyme. B. Activity of the soluble enzyme.

The sensitivity of ATPase to oligomycin was similar when tested in the membrane bound and in the Triton soluble state (Fig. 3). Both Mg^{2+} dependent ATPase activities were completely inhibited by 20 μM oligomycin. This is in contrast with the water soluble ATPase which was shown to be resistant to oligomycin[7]. DCCD which, like oligomycin, was found to inhibit both the membrane-bound and detergent solubilized ATPase in mitochondria[18], was an effective inhibitor of both types of ATPase also in R. rubrum (Fig. 4). Here the Triton soluble ATPase was somewhat more sensitive than the membrane bound one with 50% inhibition being obtained with 3 and 7 μM DCCD respectively. Another antibiotic that was found to inhibit both types of

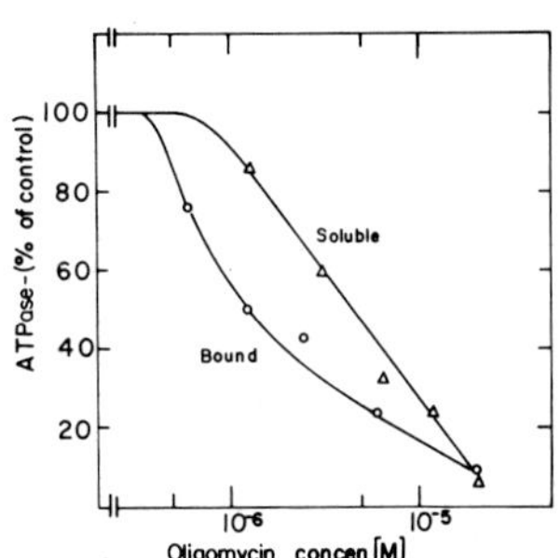

Fig. 3. The effect of oligomycin on the membrane bound and Triton soluble Mg^{2+}-dependent ATPase. 0—0 , membrane bound ATPase (control activity 180 μmoles ATP hydrolyzed/mg bacteriochlorophyll/h). △——△, soluble ATPase (control activity 38 μmoles ATP hydrolyzed/ mg protein/h).

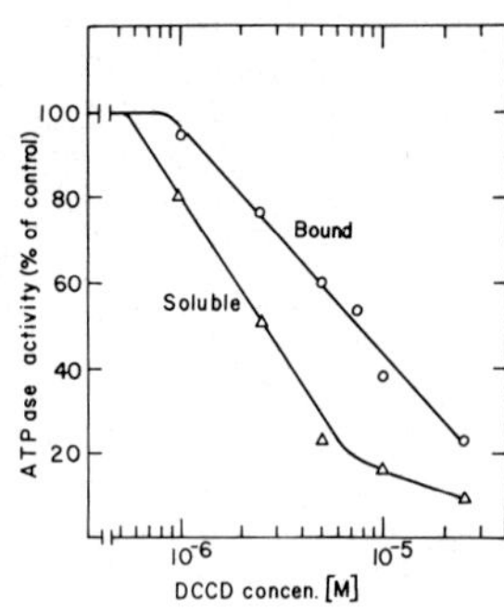

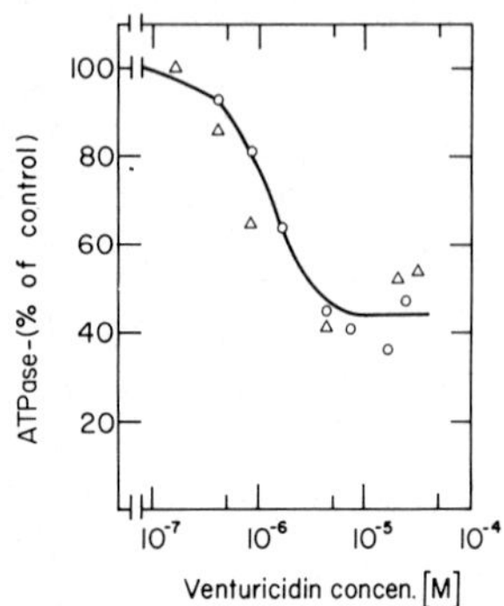

Fig. 4. The effect of DCCD on the membrane bound and on Triton soluble Mg^{2+}-dependent ATPase. The control activities were the same as in Fig. 3.

Fig. 5. The effect of venturicidin on the membrane bound and Triton soluble Mg^{2+}-dependent ATPase. The control activities were the same as in Fig. 3.

ATPase to a similar extent is venturicidin (Fig. 5) but its inhibition levelled off at around 50%.

It has previously been reported[19] that extraction of R. rubrum chromatophores by LiCl released only the β-subunit of the coupling factor. Although the removed β-subunit had by itself no ATPase activity, its removal caused complete loss of both Mg^{2+} and Ca^{2+}-dependent ATPase activity of the membrane. This subunit seems therefore to be essential for membrane-bound ATPase activity. In order to check its function in the Triton soluble ATPase, LiCl-depleted R. rubrum chromatophores were extracted by Triton X-100 and the ATPase activity of this extract was compared with that of a Triton extract of coupled chromatophores. As illustrated

Table 1: ATPase activity of coupled and LiCl-depleted chromatophores and of their Triton X-100 extracts

System assayed	ATP hydrolysis dependent on	
	Mg^{2+}	Ca^{2+}
	μmoles Pi released/mg protein/h	
Coupled chromatophores	3.1	1.2
LiCl-depleted chromatophores	0.2	0
Triton extract of coupled chromatophores	12.4	6.0
Triton extract of LiCl-depleted chromatophores	1.0	0

The Triton extraction of LiCl depleted membranes was performed by the same procedure used for coupled membrane extraction[12].

in Table 1, neither Mg^{2+} nor Ca^{2+} induced ATPase activity was present in the Triton extract of LiCl-depleted chromatophores.

DISCUSSION

The ATPase activity of R. rubrum photosynthetic membranes has been shown to be dependent on divalent cations, with both Ca^{2+} and Mg^{2+} being active, and both activities being inhibited by oligomycin[6]. The data summarized above demonstrate that Triton X-100 extraction releases all these activities from R. rubrum chromatophores (Figs. 1 and 3). Moreover the membrane bound and Triton solubilized Mg^{2+} ATPase exhibit a similar sensitivity not only to oligomycin but also to DCCD (Fig. 4) and venturicidin (Fig. 5) which have not been previously tested in chromatophores.

The water soluble ATPase, which has been previously isolated from R. rubrum chromatophores is specific for Ca^{2+} and inhibited by Mg^{2+} [7]. The Triton solubilized ATPase is activated by both Mg^{2+} and Ca^{2+}, but an excess of free Mg^{2+} causes a stronger inhibitory effect on it than on the membrane bound ATPase (Fig. 2). These results suggest that the membrane and to some degree its solubilized components protect the enzyme from the inhibitory effect of an excess of free Mg^{2+}.

The finding that Triton extraction of LiCl depleted R. rubrum chromatophores does not release an active ATPase (Table 1), indicates that the β-subunit released by LiCl is indispensable not only for the membrane bound but also for Triton X-100 solubilized ATPase activity.

The above results demonstrate that from chromatophores, as well as from mitochondria, detergents can extract the whole coupling factor ATPase complex which is very similar to the membrane bound one in its cation specificity and sensitivity to inhibitors.

REFERENCES

1. Kagawa, Y. and Racker, E. (1966) J. Biol. Chem. 241, 2467-2474
2. Pederson, P.L. (1975) J. Bioenerg. 6, 243-275
3. Serrano, R., Kanner, B.I. and Racker, E. (1976) J. Biol. Chem. 251, 2453-2461
4. Nelson, N. (1976) Biochim. Biophys. Acta 456, 314-338
5. Melandri, B.A. and Baccarini-Melandri, A. (1976) J. Bioenerg. 8, 109-119
6. Johansson, B.C., Baltscheffsky, M. and Baltscheffsky, H. (1971) in: Proc. 2nd Int. Cong. Photosynthesis (Forti, G., Avron, M. and Melandri, B.A., eds.) Vol. 2, pp 1203-1209 Dr. W. Junk N.V. Publishers, The Hague.
7. Johansson, B.C., Baltscheffsky, M., Baltscheffsky, H., Baccarini-Melandri, A. and Melandri, B.A. (1973) Eur. J. Biochem. 10, 109-117
8. Briller, S. and Gromet-Elhanan, Z. (1970) Biochim. Biophys. Acta 205, 263-272
9. Gromet-Elhanan, Z. (1970) Biochim. Biophys. Acta 223, 174-182.

500

10. Gromet-Elhanan, Z. (1972) Eur. J. Biochem. 25, 84-88

11. Clayton, R.K. (1963) In: Bacterial Photosynthesis, eds. Gest, H., San Pietro, A. and
    Vernon, L.P., pp 495-500 Antioch Press, Yellow Springs, Ohio.

12. Oren, R. and Gromet-Elhanan , Z. in preparation

13. Gromet-Elhanan, Z. (1974) J. Biol. Chem. 249, 2522-2527

14. Lowry, O.H., Rosenbrough, N.J., Farr, A.L. and Randal, R.J. (1951) J. Biol. Chem.,
    193, 265-275

15. Avron, M. (1961) Anal. Biochem. 2, 535-543

16. Avron, M. (1960) Biochim. Biophys. Acta 40, 257-271

17. Oren, R. and Gromet-Elhanan, Z. (1977) FEBS Lett. 79, 147-150

18. Bulos, B. and Racker, E. (1968) J. Biol. Chem. 243, 3891-3900

19. Philosoph, S., Binder, A. and Gromet-Elhanan, Z. (1977) J. Biol. Chem. in press

STRUCTURAL ORGANIZATION OF OLIGOMYCIN-SENSITIVE ATPase-ATPsynthase
IN PIG HEART MITOCHONDRIAL INNER MEMBRANE

D.C. Gautheron, C. Godinot, H. Maïrouch, B. Blanchy, F. Penin & Z. Wojtkowiak
Laboratoire de Biologie et Technologie des Membranes du CNRS
Université Claude Bernard de Lyon – 69621 Villeurbanne
France

INTRODUCTION

In respiring cells ATP is synthesized through oxidative phosphorylations by an ATPase-ATPsynthase complex coupled to the respiratory chain. This complex in mitochondria is a constitutive part of the inner membrane structure ; Extensive work has been done to define its components and has been recently reviewed[1-5]. One of the main characteristics is that ATP synthesis or hydrolysis is inhibited by oligomycin[6] either in intact inner membrane or in the isolated complete ATPase complex which was thus called oligomycin-sensitive ATPase (OS-ATPase). A soluble F_1-ATPase can be separated from the whole complex, which is no longer sensitive to oligomycin[7]. The oligomycin target is thought to be located in an hydrophobic part of the complex called the membrane factor. A correct association of F_1 to the membrane factor together with one or two other peptides, OSCP[8,9] and F_6[10,11] is necessary to restore the oligomycin inhibition.

Pullman et al[12] demonstrated that *in situ* F_1 is oriented towards the matrix face of the inner membrane ; besides F_1 is associated to a small peptide natural inhibitor[13] susceptible to trypsin attack[14] and only accessible from the matrix face.

The understanding of the mechanism of oxidative phosphorylation requires the knowledge of the organization of the OS-ATPase complex inside the inner membrane.

In the present work, the use of trypsin on pig heart mitoplasts (inner membrane + matrix, devoided of outer membrane) or on sonicated inverted inner membrane vesicles, proves that a peptide fraction susceptible to trypsin attack, and involved in oligomycin sensitivity, is located near the outer surface of the inner membrane.

On the other hand, it is known that reagents of thiols inhibit ATP synthesis and energy recovery while they stimulate ATPase activity in mitochondria (cf review[15]) ; the implication of very reactive thiols in oxidative phosphorylation has been proved in pig heart mitochondria[16]. Here, the effects of thiol reagents (exhibiting differential permeability) on the oligomycin-sensitive ^{32}Pi-ATP exchange of either mitoplasts or inverted vesicles, prove that two different populations of thiols are involved in ATP synthesis, one being accessible from the outer face of inner membrane and the other more easily from the inner face.

The polarity of the membrane fractions is assessed by various techniques.

MATERIALS AND METHODS

Previously described procedures have been used to prepare and test pig heart mitochondria[17], mitoplasts[18], ETP (inverted electron-transfer particles) and soluble F_1-ATPase[19] and to estimate inner membrane marker enzymes[18,19] : cytochrome oxidase, succinate-cytochrome c reductase, NADH oxidase, ATPase, malate dehydrogenase.

To obtain sonicated vesicles, mitoplasts homogenized at 0° in 0.25 M sucrose, 10 mM Tris-HCl, pH 7.4 at 20 mg protein/ml were sonicated 3×10 sec in 5 ml aliquots with a Branson Sonifier B12. Care was taken to prevent temperature from raising above 8° during sonication. Unbroken mitoplasts were spun down at 25,000 g for 10 min and sonicated vesicles were collected after centrifugation at 78,000 g for 90 min and the pellets suspended in 0.25 M sucrose, 10 mM Tris-HCl, pH 7.4, 0°.

Since the heating step often used in the preparation of F_1-ATPase produces a conformational change that masks the cooperativity between the adenine nucleotide sites of the enzyme[20], the accessibility of the thiol groups to thiol reagents[20] and the retention of F_1 by a Sepharose-DTNB column[21], the heating step was always omitted to obtain F_1 as *native* as possible. Thiols reagents were used as previously : DTNB |5,5'-dithio-*bis*-(2-nitrobenzoic) acid|[16] and CPDS |carboxypyridine disulfide = 6,6'-dithiodinicotinic acid|[22].

The Sepharose-cytochrome c columns were prepared by coupling cytochrome c directly to CNBr-activated Sepharose 6 MB according to Cuatrecasas[23].

^{32}Pi-ATP exchanges were estimated according to Conover et al[9].

Trypsin activity was measured using $N\alpha$-p-toluene sulfonyl-L-arginine methyl ester as substrate according to Rick[24] ; the specific activity was 0.25 mmole substrate/min/mg protein at 30°.

Nucleotides and most enzymes were obtained from Boehringer ; trypsin (Type III or XI), oligomycin, soybean trypsin inhibitor and $N\alpha$-p-toluenesulfonyl-L-arginine methyl ester were from Sigma Chemical Co.

RESULTS AND DISCUSSION

<u>Polarity of particles</u>. It was assessed by three different methods.

Cytochrome c stimulation of marker enzymes : cytochrome c has been shown to be located on the outer surface of the inner mitochondrial membrane[26,27]. It can be easily removed by hypotonic or salt treatment[25]. During our preparation of mitoplasts, the swelling step in hypotonic phosphate buffer removes most of the cytochrome c from its sites ; thus, if the membrane remains normally oriented, the addition of external cytochrome c should stimulate the activity of cytochrome c-dependent marker enzymes. In contrast if the vesicles are inverted, no stimulation should be observed upon addition of cytochrome c if it cannot reach its binding sites. Table 1 shows that NADH oxidase is stimulated by cytochrome c externally added to mitoplasts, while in sonicated inverted particles where NADH is freely accessible, cytochrome c does not increase the rate of oxidation. The activities of succinate-cytochrome c reductase and cytochrome oxidase are low in inverted vesicles since added cytochrome is poorly accessible, while they are high in mitoplasts ; only the addition of Lubrol WX to inverted vesicles significantly increased the activity, by allowing cytochrome c to reach its sites. These results agree with those of Huang et al[28] on beef heart mitochondrial particles and indicate that mitoplasts have kept their original polarity while sonicated vesicles are mostly inverted.

TABLE 1

CYTOCHROME C – DEPENDENT ACTIVITIES OF MARKER ENZYMES

IN MITOPLASTS AND INVERTED VESICLES

Purified mitoplasts, normally oriented, were obtained by differential centrifugation after swelling of mitochondria in 10 mM phosphate K buffer, pH 7.4 according to Maîsterrena et al[18]. Inverted vesicles were prepared by sonication of mitoplasts as described in Materials and Methods.

NADH oxidation was measured by oxypolarography ± 0.1 mM cytochrome c

Cytochrome c reduction was estimated spectrophotometrically at 550 nm

Cytochrome oxidase activity was measured polarographically ± Lubrol WX 1 mg/ mg protein.

	NADH oxidase cytochrome c		Succinate-cytochrome c reductase	Cytochrome oxidase Lubrol	
	0	+		0	+
	μmoles substrate transformed/min/mg protein				
Mitoplasts	0.20	0.47	0.12	1.36	1.53
Inverted vesicles	0.86	0.88	0.05	0.65	1.51

Retention of mitoplasts on Sepharose-cytochrome c column : The latter conclusion is further demonstrated by affinity chromatography of right-side out particles. When mitoplasts in suspension are applied to a Sepharose-cytochrome c column, most of them are retained as shown in Fig. 1-A ; only a small fraction absorbing at 280 nm is found in the exclusion volume ; high salt concentration must be applied to elute active mitoplasts. This can be explained either if some mitoplasts are partly damaged and inverted, or if some cytochrome c still remains on the surface of mitoplasts. The latter interpretation is the right one since after extensive removal of cytochrome c by treatment of mitoplasts

504

with 0.15 M KCl according to Jacobs and Sanadi[25], the Sepharose-cytochrome c
column retains all the mitoplasts : as shown in Fig. 1-B, the small excluded peak
no longer exists and only the application of high KCl concentration elutes the
mitoplasts. This proves that all the mitoplasts have kept their normal orientation.

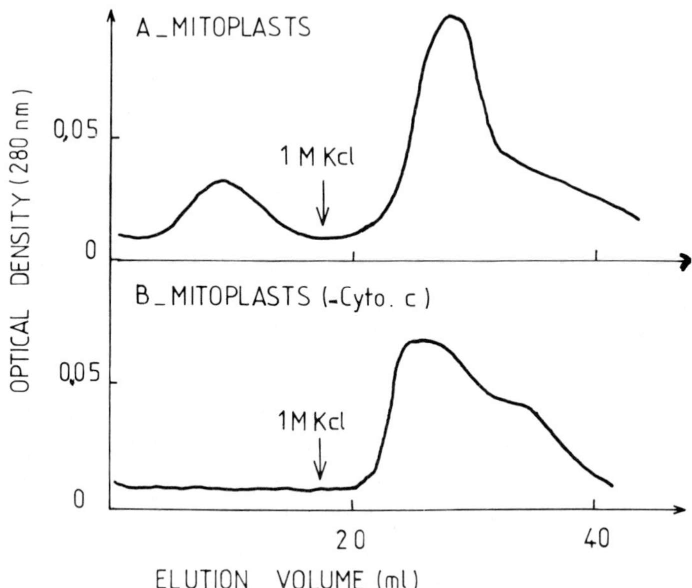

Fig. 1. Retention of mitoplasts on a Sepharose-cytochrome c column. A – MP =
normal mitoplasts. B – MP(-cyto. c) mitoplasts after removal of cytochrome c by
washing the mitoplasts with 0.15 M KCl according to Jacobs and Sanadi[25]. MP
1.9 mg protein and MP(-cyto c) 1.5 mg protein, suspended in 0.25 M sucrose, 10 mM
Tris-HCl, 1 mM EDTA, pH 7.4, were applied on top of Sepharose-cytochrome c
columns (10 × 1 cm) equilibrated with the suspension medium. After washing with
the same medium until O.D. returned to base line, 1 M KCl in the medium was
applied to the column to elute particles retained by cytochrome c.

Action of trypsin on the ATPase activity of mitoplasts and inverted particles.
We see in Fig. 2 that mild trypsin treatments strongly increase the ATPase activity of inverted vesicles while they barely affect the activity of mitoplasts ;
if the membrane preparations are incubated with both trypsin and trypsin inhibitor, no change of activity was observed with either mitoplasts or inverted
vesicles. According to Racker the natural protein inhibitor of F_1-ATPase is

oriented towards the matrix side of inner membrane and very sensitive to trypsin attack[14]. Therefore our results prove that our mitoplasts are right-side out and that, in contrast our sonicated vesicles are mostly inside-out since trypsin can attack the F_1-protein inhibitor and thus increase ATPase activity, only with sonicated vesicles.

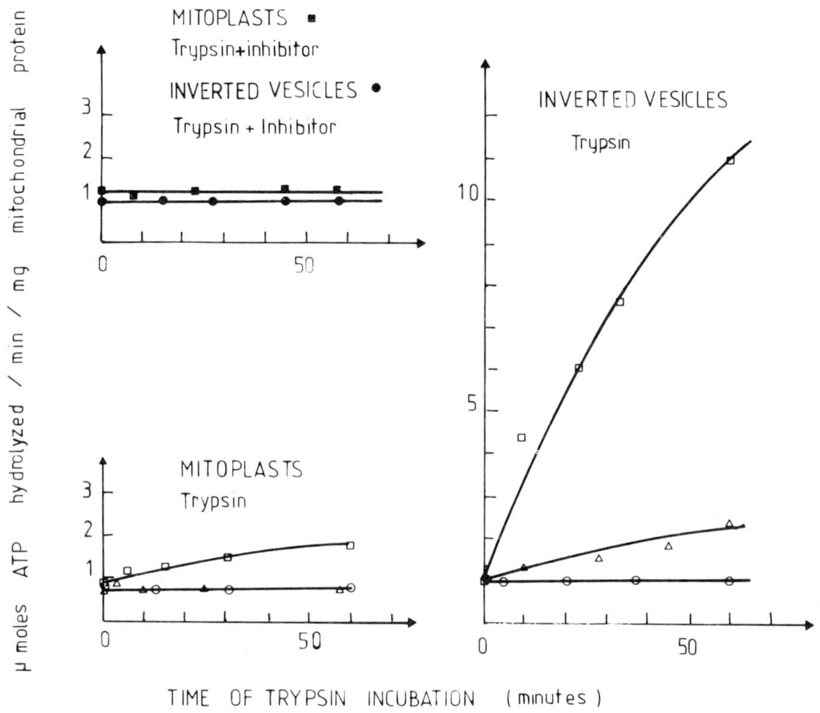

Fig. 2. Effects of mild treatments of trypsin on the ATPase activity of mitoplasts and inverted vesicles. Mitoplasts and inverted vesicles (1 mg protein), prepared as in Table 1, were incubated in 1 ml 0.25 M sucrose, 10 mM Tris-HCl, pH 7.4, at 30° C in the presence of various concentrations (O 2 µg, Δ 20 µg, □ 200 µg) of trypsin, or in the presence of 200 µg trypsin + 1 mg trypsin inhibitor (mitoplasts ■ , inverted vesicles ●). At the indicated times, 10 µl aliquots were removed and tested for ATPase activity spectrophotometrically with pyruvate kinase and lactate dehydrogenase as auxiliary enzymes according to[19].

<u>Effects of trypsin on oligomycin sensitivity of ATPase</u> : We see in Fig. 3, that trypsin decreases the inhibition of ATPase activity by oligomycin both in normal mitoplasts and inverted vesicles and that the presence of the trypsin inhibitor completely prevents this effect. This should not be interpreted as the

result of trypsin digestion, on both outer and inner faces of inner membrane, of a peptide that would be the target site of oligomycin. This interpretation can be retained only in the case of mitoplasts, which have kept their original polarity. In the case of inverted vesicles the loss of oligomycin sensitivity must be due to the solubilization of F_1-ATPase from the membrane. Indeed we see in Fig. 4 that upon trypsin treatment of inverted vesicles, the ATPase activity is progressively solubilized and passes into the supernatant fraction obtained after rapid

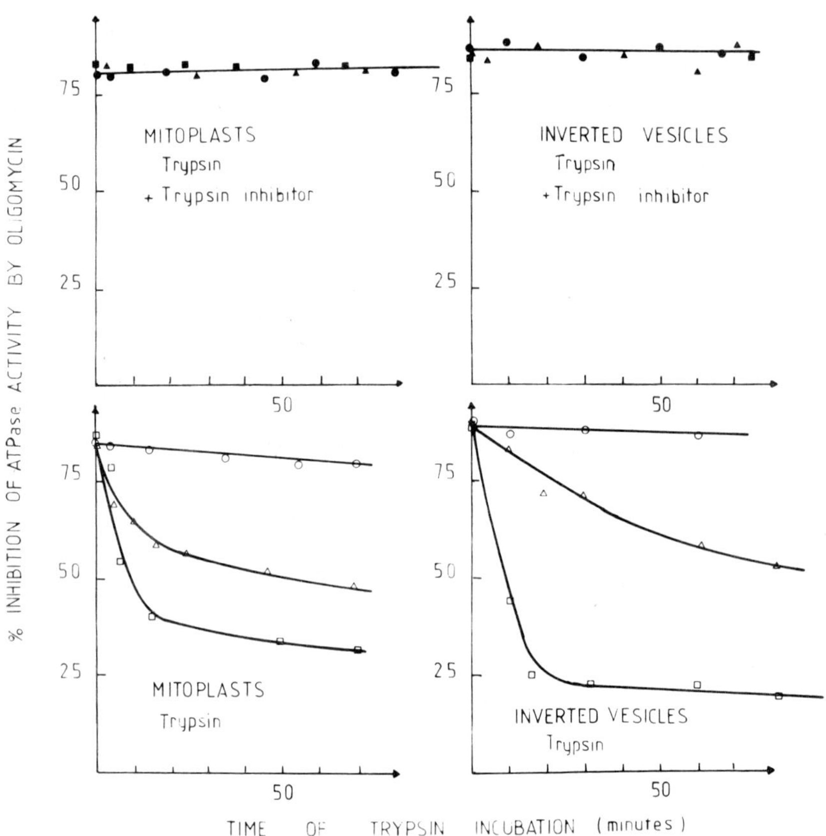

Fig. 3. Effects of trypsin on oligomycin sensitivity of ATPase in mitoplasts and inverted vesicles. Conditions as in Fig. 2. Trypsin concentrations : circles 2 µg, triangles 20 µg, squares 200 µg/ml. Oligomycin 5 µg/ml.

centrifugation of the trypsin-treated inverted vesicles. Hence, the ATPase is no longer sensitive to oligomycin because it is no more associated with the other membrane components of the OS-ATPase complex. Similar results obtained with ETP[29] were explained by the great susceptibility to trypsin of OSCP[8] and F_6[10,11], since these peptides are believed to link F1 to the membrane factor. In contrast, with mitoplasts, no ATPase activity can be detected in the supernatant fraction after

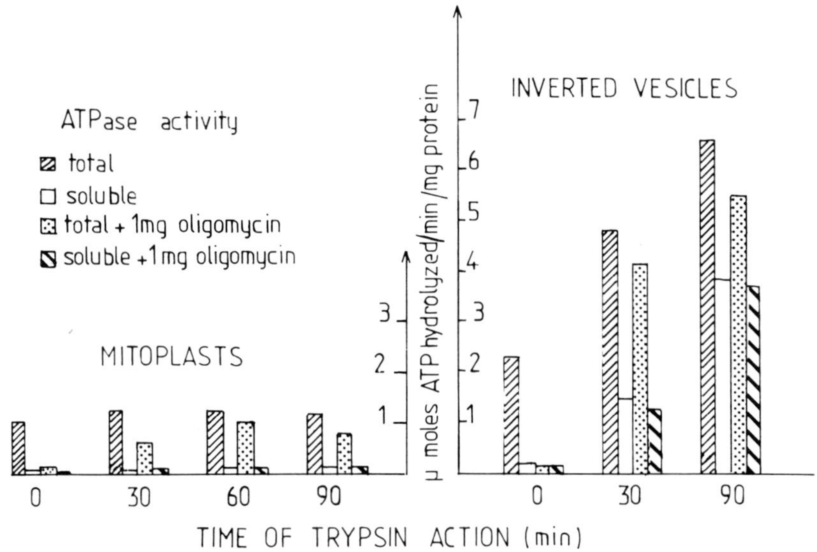

Fig. 4. Comparative solubilization of ATPase activity and loss of oligomycin sensitivity under the influence of trypsin on mitoplasts and inverted vesicles. Mitoplasts and inverted vesicles (1 mg protein) were incubated at 30° in 1 ml 0.25 M sucrose, 10 mM Tris HCl, pH 7.4 in the presence of 200 µg trypsin as Fig. 2. At the indicated times, aliquots were removed, trypsin inhibitor was added ; ATPase activity in the presence or absence of oligomycin was measured either directly on these aliquots or in the supernatant fractions obtained after centrifugation in an Airfuge Beckman for 10 min at 100,000 g. Similar results were obtained with two other experiments.

trypsin treatment and rapid centrifugation. One could argue that trypsin might solubilize the F_1-ATPase within the matrix, hence making it insensitive to oligomycin ; however the fact that trypsin does not greatly increase ATPase activity in mitoplasts, indicates that trypsin does not reach the other face of the membrane (since otherwise it would attack the F_1-natural inhibitor), and is against this argument. Another type of experiment also completely rules out the previous interpretation. The concentration of Lubrol WX able to solubilize the maximal activity of malate dehydrogenase, enzyme reputed to be located on the mitochondria matrix side, was determined ; at this concentration (0.3 % Lubrol WX) the ATPase activity could not be solubilized from the mitoplasts treated with trypsin as shown in Fig. 5. Therefore F_1-ATPase is not solubilized in the matrix and we can conclude that the loss of oligomycin sensitivity is due to the trypsin digestion of peptide(s) located near the outer surface of inner mitochondrial membrane.

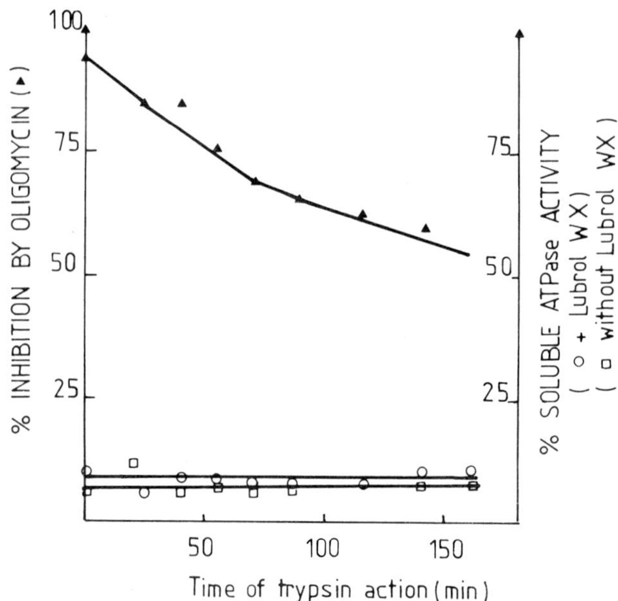

Fig. 5. Lack of solubilization of F_1-ATPase into the matrix during the trypsin attack of mitoplasts which produces a loss of oligomycin sensitivity. Conditions of trypsin attack of mitoplasts as in Fig. 4. At the indicated times two aliquots are removed ; to one of them, 0.3 % Lubrol WX is added to induce the leakage of matrix enzymes as tested on malate dehydrogenase ; after 10 min standing the total ATPase activity is measured and both aliquots of trypsin treated mitoplasts (± Lubrol) are quickly centrifuged in an Airfuge Beckman and the ATPase activity is measured in the supernatant fraction.

It is not possible as yet, to tell if the peptide part damaged by trypsin contains the binding site of oligomycin or if it is only structurally related to oligomycin sensitivity.

The following experiments show that although trypsin has a limited effect on the membrane peptides and permeability, it has some chaotropic effects on the structural organization of the ATPase in the membrane. It is known that the ATPase integrated in the whole membrane complex and the OS-ATPase is not cold-sensitive while after solubilization from the membrane, the F_1-ATPase becomes cold-labile and dissociates into its components[12]. We see in Fig. 6 that after 2 hours treatment of mitoplasts by trypsin, the ATPase activity, which is not released from the membrane, however becomes partly cold-sensitive ; under the trypsin attack, the ATPase activity increases slightly, possibly under a structural uncoupling effect ; the addition of trypsin inhibitor completely stops this increase, but if the mixture is put at 0°, the ATPase activity (tested at 30°) decreases progressively to finally reach its initial level. If F_1 had been solubilized and then denatured by cold-treatment, it would have lost all activity.

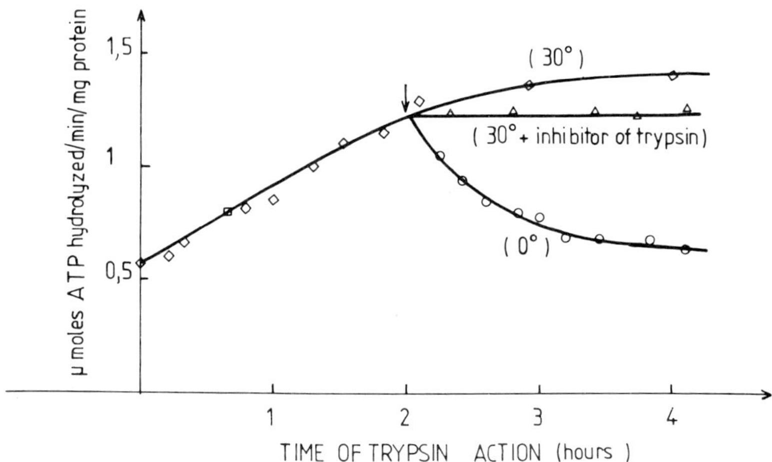

Fig. 6. Effect of cold-treatment on the ATPase activity of mitoplasts after trypsin action. Mitoplasts (1 mg protein) were incubated at 30° in 1 ml 0.25 M sucrose, 10 mM Tris-HCl, pH 7.4, in the presence of 200 μg trypsin. At the indicated times 10 μl aliquots were removed and their ATPase activity measured spectrophotometrically at 30°. After 2 hours trypsin treatment, the mixture was divided into three parts : one was kept at 30°, another was dipped into an ice bucket and trypsin inhibitor was added to the third one which was kept at 30°.

<u>Orientation of thiols involved in oligomycin sensitivity.</u>

Previously the use of DTNB proved the implication of very reactive thiols in energy coupling processes of pig heart mitochondria[16] ; this agent appeared quite permeant since it reacted with many mitochondrial thiols. In contrast, CPDS, a highly charged molecule and thus, a non penetrating thiol agent reacted relatively slowly with pig heart mitochondrial thiols : 3 min $\frac{1}{2}$ reaction and a ratio of 500-600 nmoles CPDS/mg protein, were necessary to completely inhibit the transition state 4/state 3, the ATP synthesis[22] and only the rapid phosphate influx into the ATP-synthase system, via the Pi → OH$^-$ exchange, which appears the only important phosphate transport for ATP synthesis ; besides, CPDS did not inhibit Pi efflux[31]. The combined use of isotopically labelled CPDS and spectrophotometric studies allowed to detect vicinal dithiols involved in the coupling mechanism[32]; the relationship between these strategic vicinal dithiols and the ATP-dependent ones detected by the same method in F_1-ATPase isolated from the same mitochondria (1 vicinal dithiol/mole of F_1), remains to be proved.

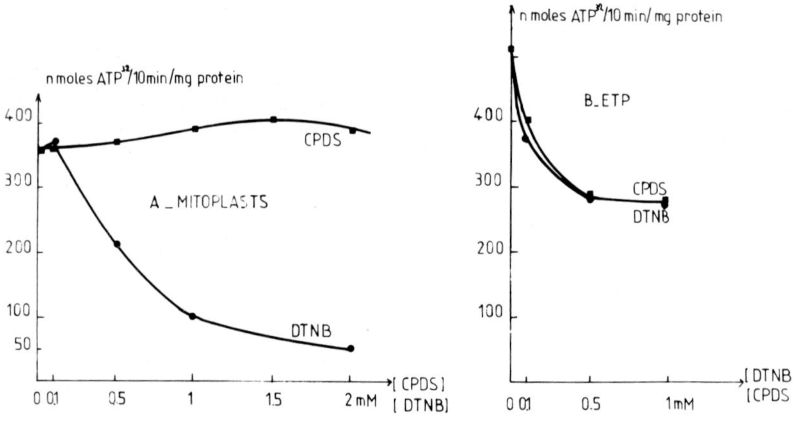

Fig. 7. Comparative effects of CPDS and DTNB on the oligomycin sensitive ^{32}Pi-ATP exchange of mitoplasts and inverted vesicles. Mitoplasts (2.7 mg protein) and ET^D (inverted vesicles, 3.4 mg protein) are incubated in 0.2 ml of 0.25 M sucrose, 10 mM Tris-HCl, pH 7.5 in the presence of various concentrations of CPDS or DTNB, for 30-40 min at 0°. Aliquots of 100 µl are used to estimate ^{32}Pi-ATP exchange : final volume 1 ml, 20 mM phosphate K, 10 mM ATP, 10 mM Mg SO$_4$, 30°, pH 7.4, ^{32}Pi (from the C.E.A. Saclay, France) 1.2×10^6 cpm/assay. The exchange proceeds for 10 min then 100 µl 35 % HClO$_4$ are added to stop the reaction ; ^{32}Pi incorporated into ATP is estimated according to Pullman[30]. The exchange is completely inhibited by oligomycin (1 µg/mg protein).

In Fig. 7 we see that the effects of CPDS and DTNB on the oligomycin-sensitive ^{32}Pi-ATP exchange are different on mitoplasts and inverted vesicles, when the amount of thiol agent is lower than 150 nmoles/mg protein. With mitoplasts, CPDS in such low amounts, does not affect the ^{32}Pi-ATP exchange, while DTNB completely inhibits this exchange. In contrast, CPDS and DTNB, at the same concentration, inhibit with a similar pattern the [32]Pi-ATP exchange of inverted vesicles (ETP), but in a limited way since a plateau of about 50 % inhibition was reached between 30-60 nmoles CPDS or DTNB/mg protein.

These results prove that thiol groups involved in the oligomycin-sensitive [32]Pi-ATP exchange have a dual location on both side of inner membrane. With right-side out particles (mitoplasts) CPDS at such low levels, cannot react with the thiols involved in the phosphate transport to the ATP synthase system[31] while DTNB appears permeant or reactive enough to inhibit phosphate transport[33] and thus the [32]Pi-ATP exchange. In contrast the experiments with inverted particles suggest the participation in the [32]Pi-ATP exchange of another type of thiols easily accessible from the inner face of inner membrane, and reactive to low concentrations of inhibitors ; the partial inhibition (about 50 %) observed, indicates that these thiols would be related to the mechanism of [32]Pi-ATP exchange in the ATP-synthase and not to the transport of phosphate itself. This conclusion agrees with the fact that CPDS did not affect the efflux of phosphate from Pig heart mitochondria[32]. Besides it would be tempting to correlate these thiols accessible from the inner face of inner membrane with the ATP-dependent vicinal thiols accessible in F_1-ATPase, only when this enzyme is in the conformation exhibiting cooperativity between nucleotide sites[20,21].

REFERENCES

1. Senior, A.E. (1973) Biochim. Biophys. Acta 801, 249-277.

2. Penefsky, H. (1974) The Enzymes X, 375-394.

3. Kagawa, Y. (1974) Methods in Membrane Biology, Vol. 1 p. 201-269, ed. by Korn, E.D., Plenum Press.

4. Pedersen, P.L. (1975) Bioenergetics, 6, 243-275.

5. Gautheron, D.C. and Godinot, C. (1977) in "Living systems as energy convertors" R. Buvet and J.P. Massue, eds. North-Holland publishing company, pp. 89-102.

6. Lardy, H.A., Johnson, D. and Mc Murray, W.C. (1958) Arch. Biochem. Biophys. 78, 587-595.

7. Kagawa, Y. and Racker, E. (1966) J. Biol. Chem. 241, 2461-2467.

8. Mac Lennan, D.H. and Tzagoloff, A. (1968) Biochemistry 7, 1603-1610.

9. Conover, T.E., Prairie, R.L. and Racker, E. (1963) J. Biol. Chem. 238, 2831-2837.

10. Russel, L.K., Kinkley, S.A., Kleyman, T.R. and Chan, Sh. P. (1976) Biochem. Biophys. Res. Commun. 73, 434-443.

11. Kanner, B.I., Serrano, R., Kandrach, A. and Racker, E. (1976) Biochem. Biophys. Res. Commun. 69, 1050-1056.

12. Pullman, M.E., Penefsky, H.S., Datta, A. and Racker, E. (1960) J. Biol. Chem. 235, 3322-3328.

512

13. Pullman, M.E. and Monroy, G.C. (1963) 238, 3762-3769.

14. Racker, E. (1963) Biochem. Biophys. Res. Commun. 10, 435-439.

15. Gautheron, D.C. (1973) Biochimie 55, 727-745.

16. Sabadie-Pialoux, N. and Gautheron, D.C. (1971) Biochim. Biophys. Acta 234, 9-15.

17. Godinot, C., Vial, C., Font, B. and Gautheron, D. (1969) Eur. J. Biochem. 8, 385-394.

18. Maïsterrena, B., Comte, J. and Gautheron, D. (1974) Biochim. Biophys. Acta 367, 115-126.

19. Di Pietro, A., Godinot, C., Bouillant, M.L. and Gautheron, D.C. (1975) Biochimie 57, 959-967.

20. Godinot, C., Di Pietro, A. and Gautheron, D.C. (1975) FEBS Letters 60, 250-255.

21. Godinot, C., Di Pietro, A., Blanchy, B., Penin, F. and Gautheron, D.C. (1977) J. Bioenergetics Biomembranes, in the press.

22. Abou-Khalil, S., Sabadie-Pialoux, N. and Gautheron, D.C. (1975) Biochem. Pharmac. 24, 49-56.

23. Cuatrecasas, P. (1970) J. Biol. Chem. 245, 3059-3065.

24. Rick, W. (1974) Methods of Enzymatic Analysis 2nd edition (H.U. Bergmeyer ed.) Vol. 2 pp. 1021-1024, Academic Press, New-York.

25. Jacobs, E.E. and Sanadi, D.R. (1960) J. Biol. Chem. 235, 531-534.

26. Harmon, H.J., Hall, J.D. and Crane, F.L. (1974) Biochim. Biophys. Acta 344, 119-155.

27. Racker, E., Burstein, C., Loyter, A. and Christiansen, R.O. (1970) in "Electron Transport and Energy Conservation" Adriatica Editrice pp. 235-241.

28. Huang, C.H., Keyhani, E. and Lee, C.P. (1973) Biochim. Biophys. Acta 305, 455-473.

29. Hortsman, L.L. and Racker, E. (1970) J. Biol. Chem. 245, 1336-1344.

30. Pullman, M.E. (1967) Methods in Enzymology, Vol. X, pp. 57-60 Acad. Press, New-York.

31. Abou-Khalil, S., Sabadie-Pialoux, N. and Gautheron, D.C. (1975) Biochimie 57, 1087-1094.

32. Sabadie-Pialoux, N., Abou-Khalil, S. and Gautheron, D.C. (1976) J. Microscopie Biol. Cell. 26, 19-24.

33. Tyler, D.D. (1969) Biochem. J. 111, 665-678.

Bioenergetics of Membranes. L. Packer et al. ed.

A QUANTITATIVE STUDY OF THE RECONSTITUTION OF MITOCHONDRIAL OLIGOMYCIN-SENSITIVE ATPase[*]

Elzbieta Glaser, Birgitta Norling and Lars Ernster
Department of Biochemistry, Arrhenius Laboratory,
University of Stockholm, Stockholm, Sweden

INTRODUCTION

Mitochondrial ATPase in native membrane-bound form is sensitive to energy-transfer inhibitors like oligomycin and DCCD and is cold-stable, whereas solubilized mitochondrial ATPase (F_1) is insensitive to these inhibitors and is cold-labile. Several preparations of membrane-bound ATPase (often called oligomycin-sensitive ATPase [OS-ATPase]) from beef heart mitochondria have been described (1-6). They all consist of the soluble F_1 and a membrane portion containing several protein components, some of which are very hydrophobic. One or several of the membrane components apparently confer oligomycin sensitivity and cold stability on the system.

The composition of the membrane part of OS-ATPase is not well characterized for the beef heart system. OS-ATPase from yeast contains four membrane proteins (7) and the preparation from the thermophilic bacterium PS3 has three components (8).

In the present investigation we have studied the composition of beef heart OS-ATPase ("Complex V" of Hatefi et al. [5]) using an SDS-polyacrylamide gel electrophoresis method that gives high resolution of low molecular weight components (9). NaBr-treatment (10,11) of Complex V results in solubilization of F_1 and several other components of the system. The residual portion, called F_0, consists of five major polypeptides which can be used together with isolated F_1 to reconstitute oligomycin-sensitive and cold-stable ATPase activity.

The data reported here concern primarily various quantitative aspects of the reconstituted system, including the binding of F_1 to F_0 and its influence on the oligomycin and DCCD sensitivity, cold-stability and catalytic activity of the bound ATPase. Some data on the proton-translocating ability of F_0 are also presented.

Parts of these results have been reported in a preliminary form (12).

MATERIALS AND METHODS

Complex V was prepared according to Hatefi et al. (5) except for one modification. As starting material submitochondrial particles were used instead of mitochondria. The particles were derived from total beef heart mitochondria by

---

[*]Abbreviations: DCCD, N,N'-dicyclohexyl carbodiimide; OSCP, oligomycin-sensitivity conferring protein; OS-ATPase, oligomycin-sensitive ATPase; SDS, sodium dodecyl sulfate.

514

sonication three times for one minute in a medium consisting of 0.25 M sucrose,
10 mM Tris-acetate, pH 7.5, and with a protein concentration of about 25 mg/ml.
Complex V contained 0.085 mg phospholipid/mg protein. The ATPase activity of
Complex V varied between 3 and 8 μmoles/min/mg protein at 30°C.

Complex V was treated with 3.5 M NaBr according to Tzagoloff et al. (10).
After centrifugation a floating pellet was obtained which was suspended in 0.66 M
sucrose, 50 mM Tris-Cl, pH 7.5, and 1 mM histidine, to a protein concentration of
about 10 mg/ml. It contained 0.22 mg phospholipid/mg protein. This preparation
will be referred to as F_0.

F_1 was prepared by the method of Horstmann and Racker (13). OSCP (14), F_6 (15),
and DCCD-binding protein (16) were prepared as described in the literature.

Reconstitution of OS-ATPase was performed in the following manner: an aliquot
of F_0, suspended in 0.33 M sucrose, 25 mM Tris-Cl, pH 7.5, and 0.5 mM histidine
at a protein concentration of about 5 mg/ml, was incubated with an aliquot of F_1
dissolved in 0.25 M sucrose, 10 mM $Tris-SO_4$, 0.25 mM EDTA, pH 8.0, at a protein
concentration of 8 mg/ml at 25° or 0°C. Incubation time and protein ratios between F_0 and F_1 were varied as described in connection with the individual experiments.

To estimate the binding of F_1 to F_0, samples were centrifuged at 11,400 x g for
10 min at 15°C. The pellet was suspended in 0.25 M sucrose, 10 mM $Tris-SO_4$, and
0.25 mM EDTA, pH 8.0. After recentrifugation, the two supernatants were combined.
The final pellet was suspended in the same medium as above. Concentration of free
F_1 in the supernatant was determined by the method of Lowry et al. (17) and the
amount of F_1 bound in the pellet was calculated as the difference between added
and free F_1. Cold-inactivation of the reconstituted system was measured after
storage overnight at 0°C. Controls were stored at 25°C. Trypsin treatment of
Complex V was performed in a reaction mixture containing 2 mg/ml Complex V protein,
0.25 M sucrose, 10 mM Tris-acetate, pH 7.5, and 30 μg trypsin/mg Complex V at 30°C
for 3 hrs. Trypsin inhibitor was then added in a 4-fold excess to the amount of
trypsin on the protein basis. F_0 from trypsin-treated Complex V was prepared in
the same manner as F_0 from normal Complex V and will be referred to as try-F_0.

ATPase activity was determined spectrophotometrically by coupling the reaction
to the pyruvate kinase and lactate dehydrogenase reactions and following the oxidation of NADH at 340 nm (18). The reaction mixture contained 30 mM potassium acetate, 25 mM Tris acetate, pH 7.5, 3 mM magnesium acetate, 1 mM phospho(enol)pyru-
vate, 50 μg lactate dehydrogenase, 50 μg pyruvate kinase, 0.2 mM NADH and 3 mM ATP,
in a final volume of 3 ml. The temperature was 30°C.

SDS-Tris-glycine polyacrylamide gel electrophoresis (12.5 % acrylamide containing 3 % methylene-bis-acrylamide) according to Laemmli (9) was used routinely.
SDS-phosphate polyacrylamide gel electrophoresis (10 % acrylamide containing 3 %

methylene-b̲i̲s̲-acrylamide) according to Weber and Osborn (19) was performed occasionally for comparison. The gels were stained with Coomassie blue and scanned at
540 nm with a linear scanning attachment connected to a Pye-Unicam spectrophotometer. Molecular weight determinations were made using the following standards:
serum albumin (68,000), pyruvate kinase (57,000), lactate dehydrogenase (36,000),
carbonic anhydrase (29,000), cytochrome c̲ (11,700).

Artificial phospholipid vesicles were prepared by the cholate dialysis method
according to Hinkle and Leung (20) except that 0.1 M Tris-Cl, pH 7.5, was used
instead of phosphate buffer and the medium contained 100 mM or 150 mM KCl. F_0
(0.2 mg F_0/3 mg phospholipids) was added to the liposomes before dialysis.

Protein was determined by the biuret (21) or Lowry e̲t̲ a̲l̲. (17) method.

RESULTS AND DISCUSSION

Composition of Complex V. Fig. 1A shows the polypeptide composition of Complex
V determined by SDS-Tris-glycine polyacrylamide gel electrophoresis according to
Laemmli (9). This gel electrophoresis system visualizes 17 bands as compared to
8-10 bands visible by the SDS-phosphate polyacrylamide method according to Weber
and Osborn (19) (not shown). Other preparations of OS-ATPase have been reported
to contain 8-12 bands (5,6,11), as determined by the method of Weber and Osborn
(19) which apparently gives a lower resolution, especially in the low molecular
weight region.

In Fig. 1A, bands 4, 5, 9, 13 and 17 are F_1 subunits, band 12 is OSCP, and band
15 is probably F_6. These conclusions were reached by isolating F_1, OSCP and F_6 by
reported methods (13-15) and coelectrophoresing them with Complex V (not shown).
Bands 6 and 7 may be core proteins of Complex III (22) which have similar molecular weights. One of the three bands with molecular weights higher than the α subunit of F_1 probably belongs to NADH dehydrogenase, since the preparation shows a
low but significant NADH-ferricyanide reductase activity. The molecular weight of
band 10 is estimated at 29,000, which corresponds to the molecular weight of the
"hydrophobic protein" of Capaldi (23) and seems to be the adenine nucleotide translocator (6). Complex V contains a so-called "uncoupler-binding protein" with a
molecular weight of 20,000-30,000 according to Hatefi e̲t̲ a̲l̲. (5,24); this may
correspond to band 11 or part of band 10.

When Complex V was treated with an amount of C^{14}-DCCD required for complete inhibition (2 nmoles/mg protein) a distinct band of radioactivity appeared on the
gel with a peak corresponding to a molecular weight of about 14,000. This value
is in agreement with the molecular weight of the DCCD-binding protein according to
Stekhoven e̲t̲ a̲l̲. (25). However, in the present case no protein peak was visible
in this range. The amount of DCCD bound, estimated from the area of the radioactive peak, was 0.2-0.5 nmoles/mg Complex V.

Composition of F_0. NaBr treatment of OS-ATPase (10,11) has been reported to
solubilize F_1. When Complex V was treated with 3.5 M NaBr using the same method,

Band no.	Complex V	F_0 derived from Complex V	Possible identity of bands
1	68,000		
2	62,000		
3	58,000		NADH dehydrogenase
4	52,000		F_1 subunit
5	50,000		F_1 subunit
6	47,000		Core protein of Complex III
7	45,000		Core protein of Complex III
8	34,000		
9	32,000		F_1 subunit
10	29,000		ATP-ADP translocator
11	23,000	23,000	
12	21,000	21,000	OSCP
13	11,700		F_1 subunit
14	10,500	10,500	
15	9,400	9,400	F_6
16	8,600	8,600	
17	7,200		F_1 subunit

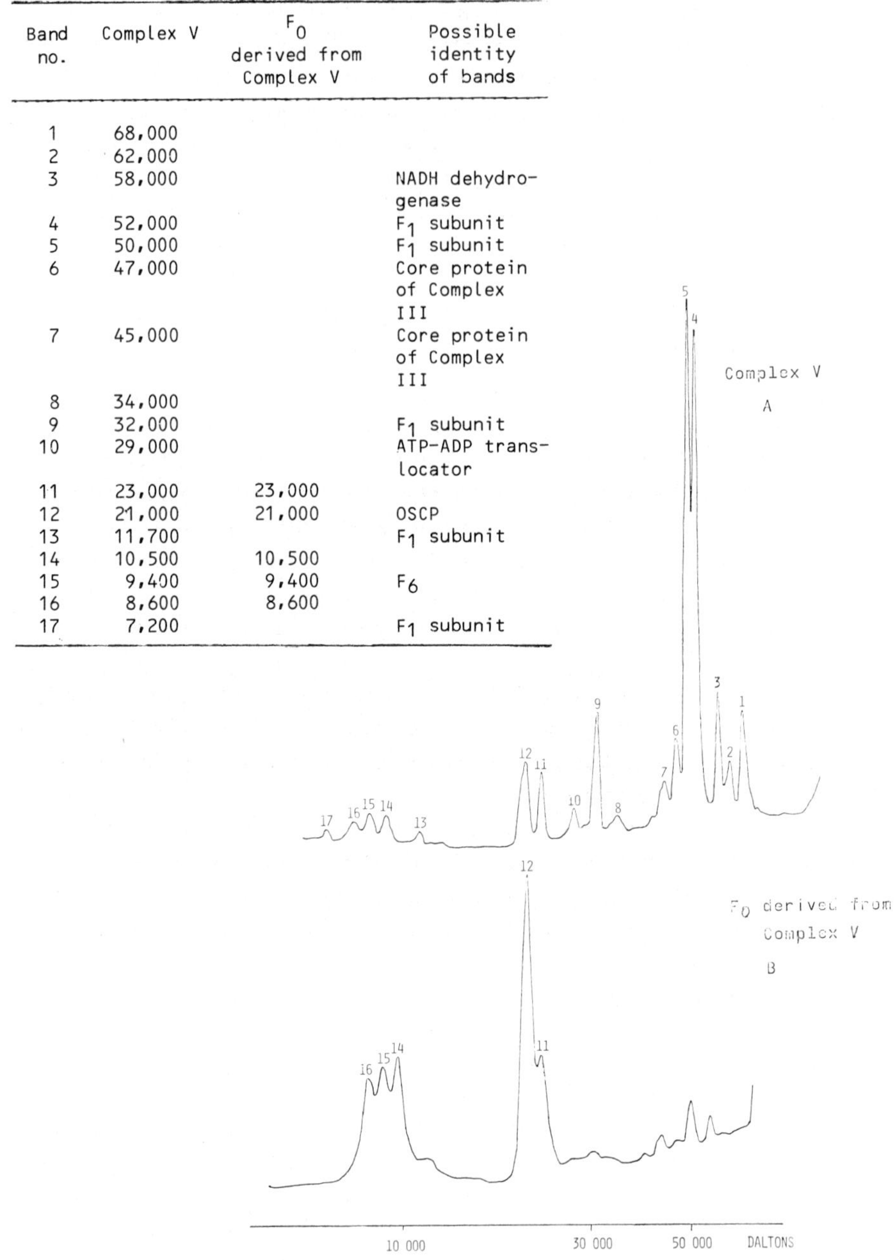

Fig. 1. Polypeptide compositions of Complex V and of F_0 derived from Complex V. SDS polyacrylamide gel electrophoresis was performed according to Laemmli (9).

it was found that not only F_1 but also several other components were solubilized. The NaBr-insoluble pellet is referred to in the following as F_0. The degree of solubilization of various components of Complex V differed from one preparation to another but five major polypeptides were always found in F_0. These are shown in Fig. 1B. The bands which correspond to proteins with molecular weights of 21,000 and 9,500 were identified as described above with OSCP and F_6, respectively. The bands with molecular weights of 23,000, 10,500 and 8,500 are of unknown identity.

Earlier investigations have indicated the requirement for a component, with a molecular weight of 29,000 in the reconstitution of OS-ATPase activity (6,11). This component was only occasionally present in our F_0 preparations. The fate of the DCCD-binding protein upon NaBr treatment of Complex V could not yet be assessed. However, as will be shown below, DCCD did inhibit the reconstituted OS-ATPase and the F_0-induced enhancement of H^+ translocation in reconstituted liposomes.

<u>Reconstitution of OS-ATPase.</u> In the following the conditions for reconstitution of OS-ATPase activity were studied. For this purpose F_0 was incubated with soluble F_1 at 0° and 25^oC. As shown in Fig. 2, soluble F_1 became oligomycin-sensitive to a significant extent already during the first minutes at both temperatures, although the reconstitution was somewhat faster at the higher temperature.

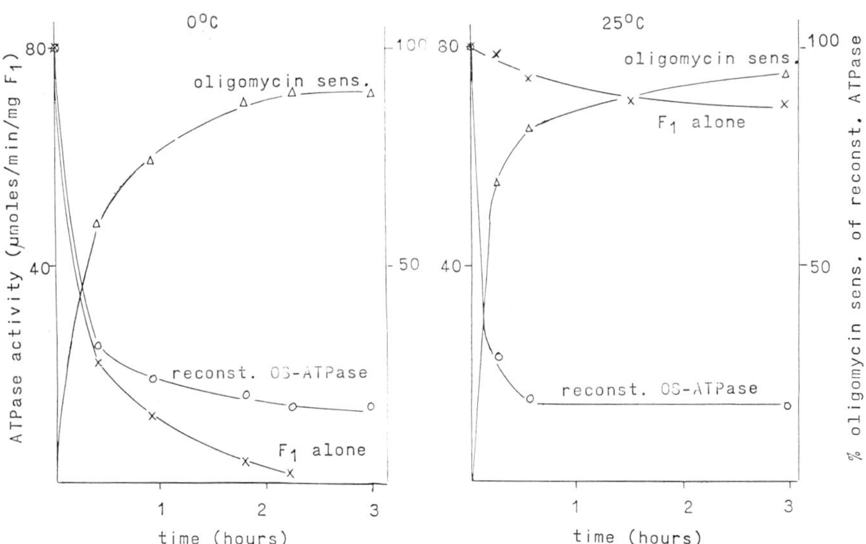

Fig. 2. Reconstitution of OS-ATPase as a function of time and temperature. F_0/F_1 ratio (mg protein/mg protein) was 10.

518

At 0°C, F_1 alone is cold-labile, whereas the reconstituted OS-ATPase is cold-stable since the same activity is obtained for the reconstituted system at 0° and 25°C. This activity was lower (15 µmoles/min/mg F_1) than that of soluble F_1 (80 µmoles/min/mg F_1), a phenomenon that has been reported earlier for membrane bound F_1 (6,26). Addition of micellar soybean phospholipids stimulated the reconstituted ATPase only slightly ($\leqslant$ 30 %)

The final extent of reconstitution of OS-ATPase activity was dependent on the F_0/F_1 ratio (Fig. 3). Maximal extent of oligomycin sensitivity ($\sim$ 85 %) required an F_0/F_1 ratio (mg protein/mg protein) of at least 4; at F_0/F_1 ratios of 2 and 1, the final extent of oligomycin sensitivity remained at about 30 and 15 %, respectively, even on prolonged incubation (Fig. 3A). On the other hand, the decrease

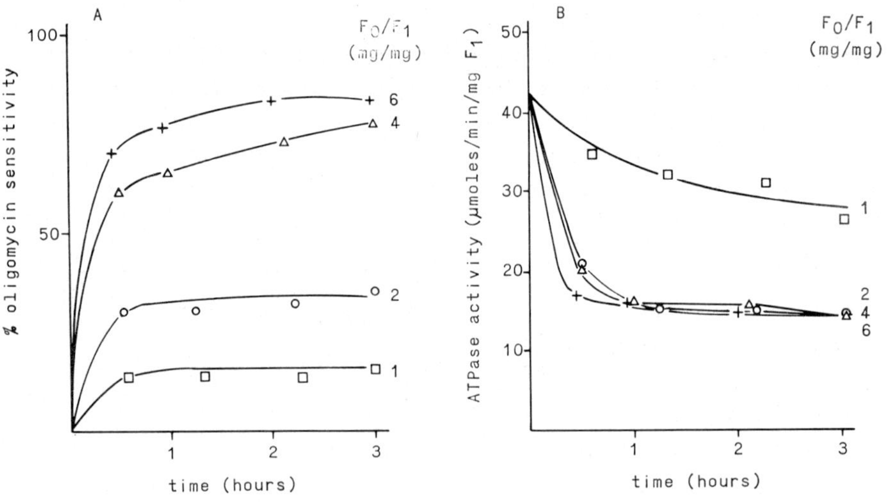

Fig. 3. Reconstitution of OS-ATPase as a function of F_0/F_1 ratio.

in ATPase activity was maximal at an F_0/F_1 ratio of 2, and equal at F_0/F_1 ratios of 2, 4 and 6 (Fig. 3B). This indicates that the decrease in activity is not related to the conferral of oligomycin sensitivity.

Fig. 4 compares the gel-electrophoretic pattern of Complex V with those of the reconstituted systems at F_0/F_1 ratios of 2, 5.5 and 11.5. It may be seen that the pattern obtained at an F_0/F_1 ratio of 5.5 comes closest to that found with Complex V. This F_0/F_1 ratio is in accordance with the 85 % oligomycin sensitivity found with Complex V.

As shown in Fig. 5, the maximally reconstituted OS-ATPase had oligomycin and

DCCD titer similar to those of Complex V. It may be noted that in both systems,
the molar amount of oligomycin needed for maximal inhibition was 2-3 times larger
than that of DCCD.

Differentiation between binding of F_1 and conferral of oligomycin sensitivity.
In the experiment shown in Figs. 6-8, a constant amount of F_0 was incubated with

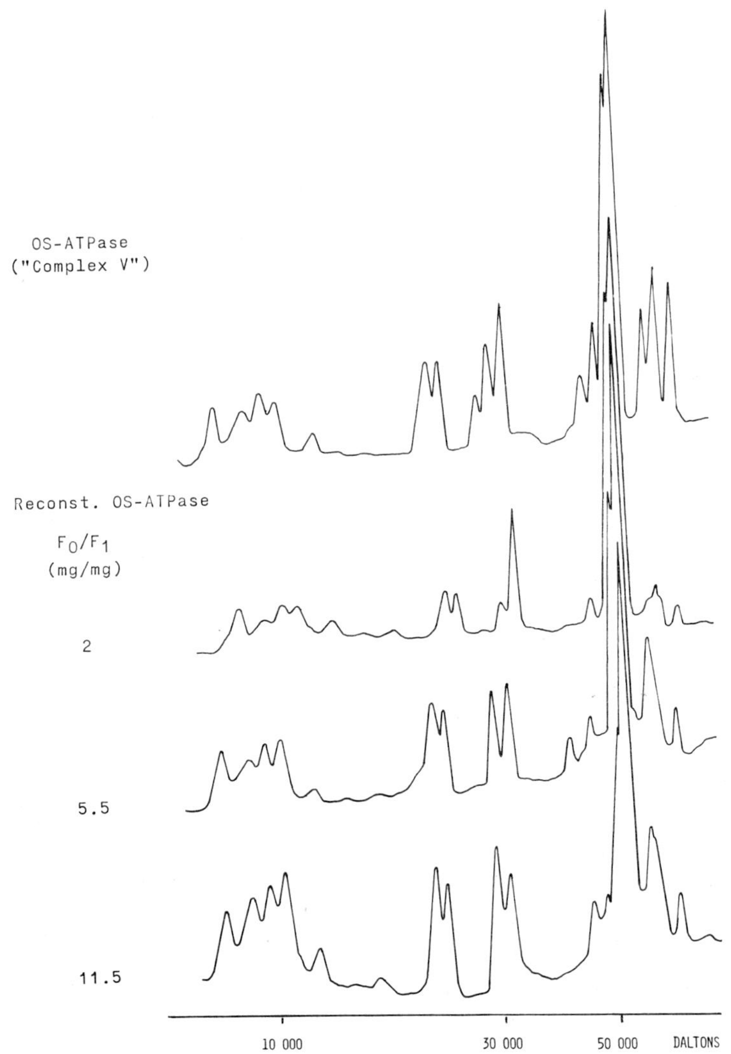

Fig. 4. Polypeptide compositions of Complex V and reconstituted OS-ATPase at
different F_0/F_1 ratios. SDS polyacrylamide gel electrophoresis according to
Laemmli (9).

520

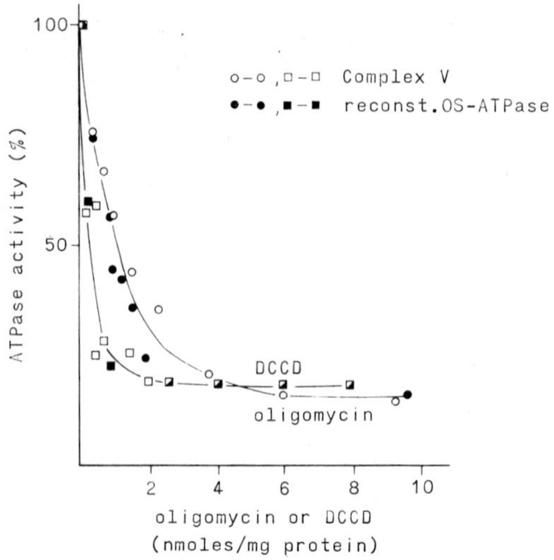

Fig. 5. Oligomycin and DCCD titrations of the ATPase activities of Complex V and reconstituted OS-ATPase. Reconstitution was performed at an F_0/F_1 ratio of 10. Preincubation with DCCD was done with both systems for 4 hrs before measuring ATPase activity.

varying amounts of soluble beef-heart F_1 at $25^{\circ}C$ for 2 hrs. After centrifugation and washing of the pellets, ATPase activity was determined in both the pellets and the supernatants. In addition, the amount of protein in the supernatants was determined as a measure of free F_1. Two incubations were performed, one with normal F_0, and one with F_0 derived from Complex V treated with trypsin (try-F_0). The experimental details are described in Materials and Methods.

As shown in Fig. 6A, all F_1 became bound at or above an F_0/F_1 ratio (mg protein/mg protein) of 2, which was indicated by the absence of protein and of ATPase activity in the supernatants. Below an F_0/F_1 ratio of 2, the supernatants contained F_1 with a specific activity close to that of the added F_1 ($\sim$ 85 µmoles/min/ mg F_1). The oligomycin sensitivity of the bound F_1 increased with increasing F_0/F_1 ratio even beyond the point where all F_1 was bound. This indicates that not all bound F_1 is oligomycin-sensitive and that F_1 binds preferentially to those sites conferring oligomycin sensitivity. The ATPase activity of the bound F_1 was low ($\sim$ 15 µmoles/min/mg F_1) and virtually independent of the degree of oligomycin-sensitivity, suggesting again (cf. Fig. 3B) that the partial inactivation of F_1 by its binding to F_0 is not related to the conferral of oligomycin sensitivity.

The ability of try-F_0 to bind F_1 was similar to that of normal F_0, complete binding again accurring at or above an F_0/F_1 ratio of 2 (Fig. 6B). However, the degree of oligomycin sensitivity of the F_1 bound to try-F_0 was greatly decreased

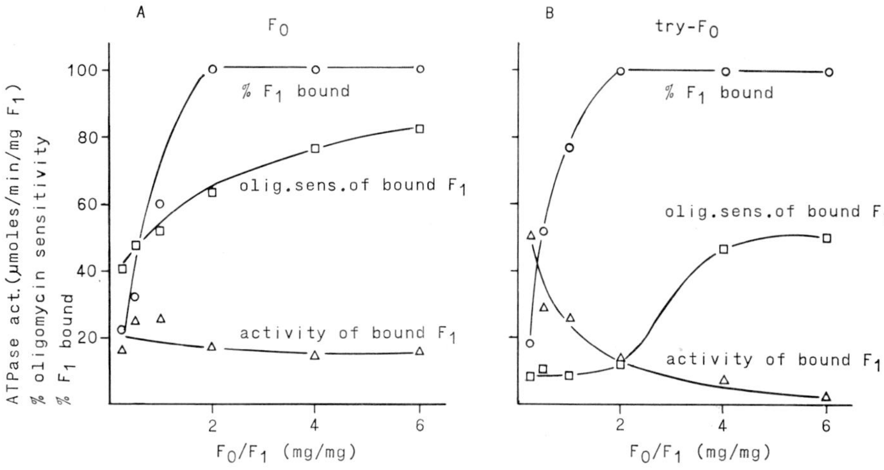

Fig. 6. Binding and oligomycin-sensitivity of reconstituted ATPase as a function of F_0/F_1 ratio. "try-F_0" refers to F_0 derived from trypsin-treated Complex V.

as compared to that of the normal F_0F_1 complex. This shows that trypsin treatment of Complex V modifies F_0 in such a way that the number of oligomycin-sensitivity conferring binding-sites for F_1 decreases without a decrease in the total number of F_1-binding sites. Just as in the case of normal F_0, the degree of oligomycin sensitivity of bound F_1 increased with increasing try-F_0/F_1 ratio, indicating that also in this case F_1 binds preferentially to oligomycin-sensitivity conferring sites. The ATPase activity of the bound F_1 was increasing with decreasing oligomycin sensitivity, suggesting that, in contrast to normal F_0, try-F_0 does not inhibit F_1 when binding it in an oligomycin-insensitive manner.

From the data in Fig. 6 one can calculate the amount of F_1 bound to F_0 at different F_0/F_1 ratios. It may be seen in Fig. 7 that at low F_0/F_1 ratios, this value approaches 0.8 mg F_1/mg F_0 for both normal F_0 and try-F_0. If one assumes a molecular weight of 360,000 for F_1 and 90,000 for F_0 (cf. ref. 27) this means that $360,000/(0.8 \times 90,000) = 5$ moles of F_0 (or try-F_0) are needed to bind 1 mole of F_1; or, in other words, that every fifth F_0 (or try-F_0) is capable of binding one F_1. Since, in the case of normal F_0, the activity of bound F_1 is constant regardless of the degree of oligomycin sensitivity (cf. Fig. 6A), the latter is a direct measure of the amount of F_1 bound in an oligomycin-sensitive manner at any given ratio of F_0/F_1. As shown in Fig. 7, at low F_0/F_1 ratios this value approached 0.3 mg F_1/mg F_0. This would indicate that $0.3/0.8 \cong 40$ % of the total F_1-binding sites on F_0 confer oligomycin sensitivity; or, in other words, that every twelfth or thirteenth F_0 binds F_1 in an oligomycin-sensitive manner. In the case of

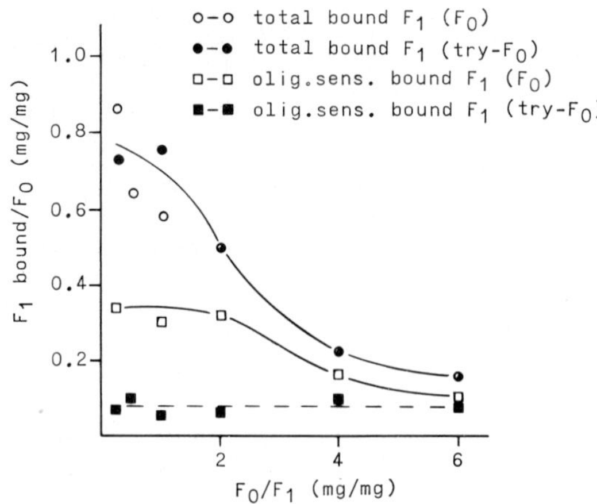

Fig. 7. Extents of binding of F_1 to F_0 as function of F_0/F_1 ratio. "Total" and "olig. sens." refer, respectively, to the total amount of F_1 bound and to the amount bound yielding oligomycin-sensitive ATPase. "try-F_0" refers to F_0 derived from trypsin-treated Complex V. Data derived from Fig. 6.

try-F_0 the amount of F_1 bound in an oligomycin-sensitive manner at low F_0/F_1 ratios approaches the value of 0.06 (Fig. 7), suggesting that $0.06/0.8 \cong 8$ % of the total F_1-binding sites on try-F_0 confer oligomycin sensitivity on F_1, corresponding to one oligomycin-sensitivity conferring binding-site in every 70th try-F_0. Even if this value may be an underestimate (because of the lower ATPase activity of the oligomycin-sensitive try-F_0F_1 complex as compared to the oligomycin-insensitive one [cf. Fig. 6B]), it is evident that try-F_0 contains considerably less oligomycin-sensitivity conferring binding sites as compared to normal F_0.

When the reconstituted pellets were stored overnight at 0° or 25^oC, the ATPase activities of bound F_1 decreased at both temperatures (Fig. 8) but more in the cold than at room temperature. Soluble F_1 was completely cold-inactivated under these conditions. In the case of F_1 bound to normal F_0, the extent of cold-inactivation was constant at all F_0/F_1 ratios, and the degree of oligomycin sensitivity was not affected. In the case of F_1 bound to try-F_0, both the extent of cold-inactivation and the degree of oligomycin sensitivity increased at low F_0/F_1 ratios. These findings indicate that, in the case of normal F_0, the cold-stability of bound F_1 is equal regardless of the degree of oligomycin sensitivity, whereas in the case of try-F_0, F_1 bound in an oligomycin-insensitive manner, is less cold-stable than F_1 bound in an oligomycin-sensitive manner.

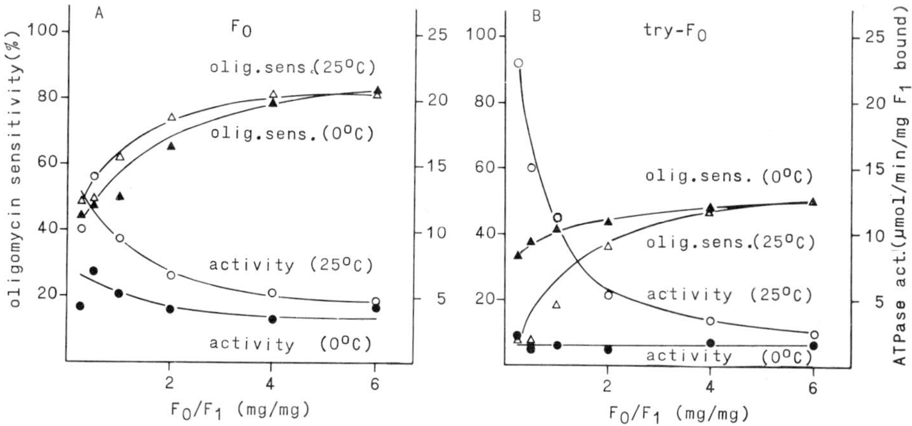

Fig. 8. Effect of storage overnight at 0° and 25°C on the activity and oligomycin sensitivity of reconstituted F_0F_1 complex at different F_0/F_1 ratios. "try-F_0" refers to F_0 derived from trypsin-treated Complex V.

It thus appears that although F_0 and try-F_0 can bind equal amounts of F_1, they differ in their effects on the ATPase activity, oligomycin sensitivity and cold-stability of the bound F_1.

Attempts were made to identify the protein component(s) of F_0 affected by trypsin treatment of Complex V. However, so far no component(s) responsible for the effects here described could be identified. When isolated F_0 was treated with trypsin, the 10,500 molecular weight component disappeared as revealed by gel electrophoresis. This preparation was unable to confer either cold-stability or oligomycin sensitivity on F_1. The ability to bind F_1 could not be assessed because of difficulties in sedimenting the trypsin-treated F_0 by centrifugation.

It is evident from the experiments described above that F_0 derived from Complex V - and probably from other preparations of mitochondrial OS-ATPase as well - has a relatively limited ability to bind F_1 as indicated by the high molar ratios F_0/F_1 needed for reconstitution of oligomycin sensitivity. Furthermore, as revealed by the present data, F_0 can bind F_1 in both an oligomycin-sensitive and an oligomycin-insensitive manner, suggesting that different components of F_0 are involved in the binding of F_1 and the conferral of oligomycin sensitivity. From recent reports by Vàdineanu et al. (28) and Russell et al. (29) it appears that F_6 and OSCP may fullfil these two functions, F_6 being responsible for the binding and OSCP for the conferral of oligomycin sensitivity. On the basis of these reports our results may be rationalized by assuming that only a relatively small proportion of the isolated F_0 (and also of the parent Complex V) contains both F_6 and OSCP, another small proportion contains F_6 but not OSCP, whereas the bulk lacks either F_6 or both components. Treatment of Complex V with trypsin seems to further diminish the number of complete sets of F_0, primarily by decreasing the amount of functional OSCP.

$\underline{F_0\text{-Induced, oligomycin- and DCCD-sensitive proton translocation in liposomes.}}$ Reconstituted OS-ATPase has been shown by Racker and associates (6,30) to catalyze an ATP-dependent electrogenic translocation of protons across artificial membranes. Schipakin $\underline{et}$ $\underline{al}$. (31) have reported that a NaCl-treated OS-ATPase from beef heart mitochondria promoted the K^+ + valinomycin-mediated translocation of protons in liposomes. The proton translocation was slower when the NaCl-treated OS-ATPase was pretreated with DCCD before incorporation into liposomes. Similar experiments with NaBr-treated Complex V (F_0) are shown in Table 1. It may be seen that F_0 did enhance the initial rate of K^+ + valinomycin-induced proton translocation. This effect of F_0 was partially inhibited to varying extents by oligomycin and DCCD. These results are consistent with the conclusion that mitochondrial F_0 contains an oligomycin- and DCCD-sensitive proton translocator. Strong evidence for a similar function of F_0 from the thermophilic bacterium PS3 has been reported by Kagawa and associates (8,32).

TABLE 1

PROTON TRANSLOCATION IN LIPOSOMES CONTAINING F_0 DERIVED FROM COMPLEX V

For preparation of liposomes and incorporation of F_0, see Materials and Methods. The liposomes contained 100 mM KCl in Exp. 1 and 150 mM KCl in Exp. 2. The reaction mixture consisted of 5 mM Tris-Cl, pH 7.5, and 0.2 ml liposomal suspension containing 0.2 mg F_0 protein, in a final volume of 3 ml. The reaction was started by the addition of 0.5 µg valinomycin and the change in pH was followed with a glass electrode. When indicated, 6 nmoles oligomycin or 1 nmole DCCD was present in the reaction mixture. In the case of DCCD this was added to F_0 prior to incorporation into liposomes.

System	Initial velocity of H^+ translocation after addition of valinomycin (ng ions H^+/min)	
	Exp. 1	Exp. 2
Liposomes alone	40	74
Liposomes + F_0	265	550
Liposomes + F_0 + oligomycin	162	320
Liposomes + F_0 + DCCD	75	310

SUMMARY

1. Oligomycin-sensitive ATPase was prepared from beef-heart submitochondrial particles according to Hatefi et al. (5) ["Complex V"]. Treatment of this preparation with 3.5 M NaBr removed F_1 and several other polypeptides. The NaBr-insoluble residue, called F_0, contained 5 major components as revealed by SDS polyacrylamide gel electrophoresis according to Laemmli (9). Two of these were identified as F_6 and OSCP. F_0 also contained the DCCD-binding protein although this was not visible on the gels.

2. The interaction of F_0 with purified, soluble beef-heart F_1 was investigated. F_0 was capable of binding F_1 and conferring oligomycin-sensitivity and cold-stability on its ATPase activity. Furthermore F_0 was found to diminish the specific activity of F_1 ATPase. A comparison of these effects at varying F_0/F_1 ratios shows that F_0 binds F_1 in both an oligomycin-sensitive and an oligomycin-insensitive manner, and that both types of binding involve a conferral of cold-stability and a decrease in specific activity. High F_0/F_1 ratios favoured the oligomycin-insensitive type of binding, indicating that F_1 binds preferentially to oligomycin-sensitivity conferring sites. Treatment of OS-ATPase with trypsin resulted in an F_0 with a decreased proportion of oligomycin-sensitivity conferring binding-sites and a diminished ability to lower the specific activity and cold-lability of F_1.

3. F_0 induced an oligomycin- and DCCD-sensitive enhancement of K^+ + valinomycin-driven proton translocation across the membrane of artificial phospholipid vesicles.

This work has been supported by a grant from the Swedish Natural-Science Research Council. We thank Ilona Belló-Janczak for excellent technical assistance.

REFERENCES

1. Kagawa, Y. and Racker, E. (1966) J. Biol. Chem. 241, 2467-2474.
2. Tzagoloff, A., Byington, K.H. and Mac Lennan, D.H. (1968) J. Biol. Chem. 243, 2405-2412.
3. Kopaczyk, K., Asai, J., Allmann, D.W., Oda, T., and Green, D.E. (1968) Arch. Biochem. Biophys. 123, 602-621.
4. Sadler, M.H., Hunter, D.R. and Haworth, R.A. (1974) Biochem. Biophys. Res. Commun. 59, 804-812.
5. Hatefi, Y., Stiggall, D.L., Galante, Y., and Hanstein, W.G. (1974) Biochem. Biophys. Res. Commun. 61, 313-321.
6. Serrano, R., Kanner, B.J., and Racker, E. (1976) J. Biol. Chem. 251, 2453-2461.
7. Tzagoloff, A., and Meagher, P. (1971) J. Biol. Chem. 246, 7328-7336.
8. Sone, N., Yoshida, M., Hirata, H., and Kagawa, Y. (1975) J. Biol. Chem. 250, 7917-7923.
9. Laemmli, U.K. (1970) Nature (London) 227, 680.

526

10. Tzagoloff, A., Mac Lennan, D.H., and Byington, K.H. (1968) Biochemistry $\underline{7}$, 1596-1602.

11. Capaldi, R.A. (1973) Biochem. Biophys. Res. Commun. $\underline{53}$, 1331-1337.

12. Glazek, E., Norling, B., and Ernster, L. (1976) Abstr. 10th Int. Congress of Biochem., Hamburg, 1976.

13. Horstmann, L.L., and Racker, E. (1970) J. Biol. Chem. $\underline{245}$, 1336-1344.

14. Senior, A.E. (1971) J. Bioenerg. $\underline{2}$, 141-150.

15. Kanner, B.L., Serrano, R., Kandrach, M.A., and Racker, E. (1976) Biochem. Biophys. Res. Commun. $\underline{69}$, 1050-1056.

16. Cattell, K.J., Lindop, C.R., Knight, I.G., and Beechey, R.B. (1971) Biochem. J. $\underline{125}$, 169-177.

17. Lowry, O.H., Rosebrough, N.I., Farr, A.L., and Randall, R.J. (1951) J. Biol. Chem. $\underline{193}$, 265-275.

18. Pullman, M.E., Penefsky, H.S., Datta, A., and Racker, E. (1960) J. Biol. Chem. $\underline{235}$, 3322-3329.

19. Weber, K. and Osborn, M. (1969) J. Biol. Chem. $\underline{244}$, 4406-4412.

20. Hinkle, P. and Leung, K.H. (1974) in Membrane Proteins in Transport and Phosphorylation (Azzone, G.F., Klingenberg, M.E., Quagliariello, E., and Siliprandi, N., eds.), pp. 73-78, North-Holland Publishing Co., Amsterdam.

21. Gornall, A.G., Bardwill, C.I., and David, N.M. (1949) J. Biol. Chem. $\underline{177}$, 751-766.

22. Gellerfors, P., Lundén, M., and Nelson, B.D. (1976) Eur. J. Biochem. $\underline{67}$, 463-468.

23. Capaldi, R.A., Komai, H., and Hunter, D.R. (1973) Biochem. Biophys. Res. Commun. $\underline{55}$, 655-659.

24. Hanstein, W.G. and Hatefi, Y. (1974) J. Biol. Chem. $\underline{249}$, 1356-1362.

25. Stekhoven, F.S., Waitkus, R.F., and van Moerkerk, H.Th.B. (1972) Biochemistry $\underline{11}$, 1144-1150.

26. Hammes, G.G. and Hilborn, D.A. (1971) Biochim. Biophys. Acta $\underline{233}$, 580-590.

27. Senior, A. (1973) Biochim. Biophys. Acta $\underline{301}$, 249-277.

28. Vàdineanu, A., Berden, J.A., and Slater, E.C. (1976) Biochim. Biophys. Acta $\underline{449}$, 463-479.

29. Russell, L.K., Kirkley, S.A., Kleyman, T.R., and Chan, S.H.P. (1976) Biochem. Biophys. Res. Commun. $\underline{73}$, 434-443.

30. Kagawa, Y., Kandrach, A., and Racker, E. (1973) J. Biol. Chem. $\underline{248}$, 676-684.

31. Schipakin, V., Chuchlova, E., and Evtodienko, Y. (1976) Biochem. Biophys. Res. Commun. $\underline{69}$, 123-127.

32. Yoshida, M., Okamoto, H., Sone, N., Hirata, H., and Kagawa, Y. (1977) Proc. Natl. Acad. Sci. USA $\underline{74}$, 936-940.

GLYCOPROTEIN NATURE
OF ENERGY TRANSDUCING ATPASES

J. Manuel Andreu, Vicente Larraga and Emilio Muñoz
Sección de Bioquímica de Membranas,
Instituto de Inmunología y Biología Microbiana,
Velázquez 144, Madrid-6
Spain

SUMMARY

We have determined the sugar composition of highly purified energy – transducing
ATPases (or coupling factors) from two strains of Micrococcus lysodeikticus and spinach
chloroplasts. Form A* of M. lysodeikticus ATPase contains per 340,000 g glycoprotein:
32 mol rhamnose, 39 mol arabinose, 59 mol mannose, 6 mol galactose, 83 mol glucose
and 6 mol glucosamine together with minor amounts of ribose (0–1 mol). Form B of M.
lysodeikticus BF_1 has a less complex carbohydrate composition containing per 370,000 g:
mannose and glucose (56 mol of each monosaccharide), 5–8 mol glucosamine and 5 mol
ribose. On the other hand, chloroplast ATPase is comparatively less enriched in sugars
but reveals the most complex composition. It contains per mol of glycoprotein (325,000 g):
rhamnose (4 mol), fucose (5 mol), ribose (4 mol), arabinose (20 mol), glucose (16 mol),
glucosamine (3 mol) and galactosamine (9 mol). No sialic acid was found to be part of
these proteins. We were unable to detect significant β-elimination of serine residues by
alkaline sodium borohydride treatment of whole ATPases. Preliminary evidence based on
conventional colorimetric procedures suggests that mitochondrial ATPase may also be a
glycoprotein. The possible significance of the glycosilation of these ATPases in their role
as phosphorylating coupling factors and extrinsic membrane proteins is discussed.

INTRODUCTION

Energy-transducing ATPases are complex membrane proteins possessing related overall
functions. Nevertheless, clear differences exist between them with regard to their sub-
structure and topological localization. The Na^+, K^+ -ATPase from plasma membranes
belongs to the category of integral proteins[1] and has been identified as a glycoprotein[2].

---

*Abbreviations: F_1, mitochondrial ATPase or coupling factor 1; CF_1, chloroplast ATPase;
BF_1, bacterial membrane ATPase, in this case referred specifically to that of Micrococcus
lysodeikticus; forms A and B of M. lysodeikticus ATPase, two enzyme forms isolated from
two different strains of the microorganism.

528

The Ca^{2+}-ATPase from sarcoplasmic reticulum may also be categorized as integral or intrinsic membrane protein on the basis of its solubility properties and it appears to have an amphipatic structure[3], but there is no direct demonstration that the polypeptide chain spans the bilayer like the Na^{+} - K^{+} -ATPase does[4]. Interestingly, there is preliminary evidence suggesting that this Ca^{2+} transport protein is a glycoprotein[5]. On the other hand, the proton-translocating ATPases (F_1 coupling factors) from mitochondria, chloroplasts and bacteria were classified as peripheral or extrinsic membrane proteins on the basis of the operational criteria as defined by Singer[6] and their ultrastructural appearance in negative staining[7]. There was also earlier evidence suggesting that F_1 and CF_1 were not likely glycosilated[8].

It has been recently suggested by Guidotti[5] that all the transport processes carried out across eukaryotic membranes should be catalyzed by oligomeric glycoproteins which span the membrane[5]. The little information still available (see for a review ref. 9) on the structure, role and location of glycoproteins isolated from the inner mitochondrial membrane and plasma membranes of bacterial cells raised some doubts about the possible generalization of that postulate.

In the course of our studies on the purified coupling factor from Micrococcus lysodeikticus membranes we observed a series of characteristics like microheterogeneity[10,11] and possible degradation of some of its subunits[10,12] that could not be only explained by accidental exogenous proteolysis[10,11]. We have recently presented evidence to support the notion that this BF_1 factor was a glycoprotein[10,12]. We have extended this study since and in the present report we describe the sugar composition of the ATPases (coupling factors) from two strains of Micrococcus lysodeikticus and chloroplasts as well as evidence suggesting that beef heart mitochondrial ATPase is likely glycosilated. The possible glycoprotein nature of several coupling factors suggests that this is a general property of the proteins involved in phosphorylation. The finding is interesting in connection with the above mentioned proposal by Guidotti[5] but poses some questions concerning the topology and function of glycoproteins in membranes in general and energy-transducing membranes in particular.

MATERIALS AND METHODS

ATPase preparations. The ATPase (forms A[10] and B_A[11]) from Micrococcus lysodeikticus was isolated and purified by polyacrylamide gel electrophoresis as described[10,11]. Crude coupling factor from spinach chloroplasts was prepared according to Lien and Racker[13] and also purified by preparative gel electrophoresis. The CF_1 obtained was more than 98% homogenous as judged by analytical gel electrophoresis, had no detectable tryptophan fluorescence and ATPase activity of 35 μmol substrate transformed $\cdot$ min^{-1} $\cdot$ mg protein^{-1} assayed as described by Lien and Racker[13]. Beef heart mitochondrial F_1 was a generous gift of Drs. A. Gómez-Puyou and Marieta Tuena de Gómez-Puyou. In our laboratory, this F_1 preparation showed to be about 99% homogenous by analytical gel electrophoresis at different pH values and polyacrylamide concentration and had a specific activity of 75 μmol $\cdot$ min^{-1} $\cdot$ mg protein^{-1} assayed as described by Senior and Brooks[14].

Protein was estimated by the method of Lowry et al[15] or by ultraviolet absorption from the molar extinction coefficients reported before[10, 16]. Analytical polyacrylamide gel electrophoresis and gel staining for protein and carbohydrate were carried out with the appropriate protein and glycoprotein controls as described[12]. Total content of neutral sugars was estimated by a micro-scaled orcinol reaction[12, 17].

Amino acid analysis. The analysis was carried out by ion exchange chromatography in a Durrum D 500 autoanalyzer following hydrolysis of approximately 10μg protein samples added of n-leucine as internal standard in 100 μl 6 N HCl for 22h at 110°C.

Alkaline sodium borohydride treatment. In an attempt to identify the eventual products of β-elimination, purified whole BF$_1$ and CF$_1$ (about 200 μg protein) were treated with 30 μl 0.15 N NaOH containing 0.30 M (2.5 mCi) NaB$[^3H]_4$ (Radiochemical Centre, Amersham) for 30 min at 37°C. Mixtures were acidified with acetic acid and taken to dryness with a rotary evaporator. The residues were dissolved in methanol and evaporated six times, dissolved in water and lyophilized. Protein was hydrolyzed as above. The $[^3H]$--labelled samples (100 μg protein) were chromatographed in a Jeol JLC-5 AH amino acid autoanalyzer connected and synchronized to a fraction collector. Aliquots (0.25 ml) of the eluate were counted in Bray's mixture[18] with an Intertechnique SL 32 spectrometer.

Sugar analysis. To analyze the sugar composition of the ATPases hydrolyses were carried out in 1 N H$_2$SO$_4$ (aprox. 2 μg sugar/20 μl) for 4h at 100°C. These conditions appeared to be the optimal ones for the release of sugars in most of the cases analyzed. Hydrolysates were neutralized with BaCO$_3$ and reduction and acetylation performed essentially as described[19] keeping the volumes as small as possible. In some instances, samples were chromatographed twice through Dowex 50 x 8 (H$^+$) to eliminate the residual Tris that could interfere in the chromatograms. Inositol, xylose and galactose were used as internal standards for all sugars, pentoses and hexoses respectively. The alditol acetates were analyzed by gas liquid chromatography on ECNSS-M (Applied Science Laboratories) at 180 or 200°C in a Varian 1440/10 gas chromatograph with a Hewlett Packard 3380 A integrator. Peak assignements were made on the basis of sample coincidence with the relative retention times ($\pm$ 0.01) of standards at the two column temperatures. Sialic acid was assayed with the thiobarbituric acid method of Warren[20].

The small amounts of hexosamines present were estimated from their well separated peaks in the amino acid chromatograms with appropriate corrections for loss by hydrolysis[21].

RESULTS

<u>Chemical composition of two forms of M. lysodeikticus BF$_1$</u>. As estimated by the orcinol reaction, the neutral sugar content of form B of <u>M. lysodeikticus</u> ATPase amounted to 0.09 g sugar/g peptide, a value lower than the 0.12 g sugar/g peptide reported before for form A of this BF$_1$[12]. Similar lower values for form B were obtained with other colorimetric procedures, e.g. 0.11 g sugar/g peptide (form B) with phenol sulfuric acid instead of a 0.19 value for form A[12]. Table I compares the sugar molar composition of these two forms and shows important qualitative as well as quantitative differences. Alditols of each one of the sugars listed were not detected when the analysis was carried out omitting the reduction step with borohydride. In spite of the differences in sugar composition found between the two forms of <u>M. lysodeikticus</u> ATPase, they had a similar amino acid composition, except for minor variations with regard to the glycine and serine content. These differences may account for the presence or absence of some minor subunits[10, 11] rather than reflect a change in the amino acid sequence.

The possibility of an <u>O</u>-glycosidic bond of the sugar moiety to serine residues was explored by attempting the detection of [^{3}H] alanine after alkaline-boro[^{3}H] hydride treatment of the whole protein. Only a small amount of tritium-labelled alanine, <u>i.e.</u> 0.16 residues/mol protein could be estimated. Nevertheless, it is worth noting that the particular conditions of this experiment (see MATERIALS AND METHODS) may not give a good quantitative yield of transformation of serine into [^{3}H] alanine[22].

<u>Sugar composition of CF$_1$</u>. As estimated by the orcinol reaction, the total neutral sugar content of CF$_1$ was 0.06 g sugar/g peptide. The sugar composition of CF$_1$ recorded in Table 1 showed certain similarities with that of form B of <u>M. lysodeikticus</u> BF$_1$ although differences were also evident. The main differences stem from the presence in CF$_1$ of arabinose, 6-deoxyhexoses, galactosamine besides glucosamine and glucose instead of mannose.

It is worth noting that sialic acid did not appear to be a component of BF$_1$ or CF$_1$.

TABLE I

SUGAR COMPOSITION OF THE ATPASES FROM TWO STRAINS OF Micrococcus lysodeikticus AND SPINACH CHLOROPLASTS

Monosaccharides	ATPases or F_1 factors		
	BF_1 form A	BF_1 form B (mol/mol protein[*])	CF_1
Rhamnose	32		4
Fucose			5
Ribose	0-1	5	4
Arabinose	39		20
Xylose			
Mannose	59	56	
Galactose	6		25
Glucose	83	56	16
Glucosamine	6	5-8	3
Galactosamine			9
Sialic acid			
Total	226	124	86

[*]Calculated on the basis of the reported molecular weight values: 340,000, BF_1 form A[10], 370,000, BF_1 form B[11] and 325,000, CF_1[16] taken as the weights of the whole glycoproteins.

Preliminary evidence on the glycoprotein nature of mitochondrial ATPase. Although a detailed analysis of the sugar composition of F_1 is not yet available, we wish to report the preliminary evidence which supports the notion that this coupling factor may also be a glycoprotein. Purified F_1 was freed for contaminating residual saccharose from the isolation procedure by repeated gel filtration on Sepharose 6B and Sephadex G25. The total content of neutral sugars was then estimated by the orcinol reaction[17]. It amounted to 0.03 g sugar/g peptide. Moreover, the protein showed a positive periodic acid-Schiff reaction after polyacrylamide electrophoresis in both normal and sodium dodecyl sulfate gels (Fig. 1).

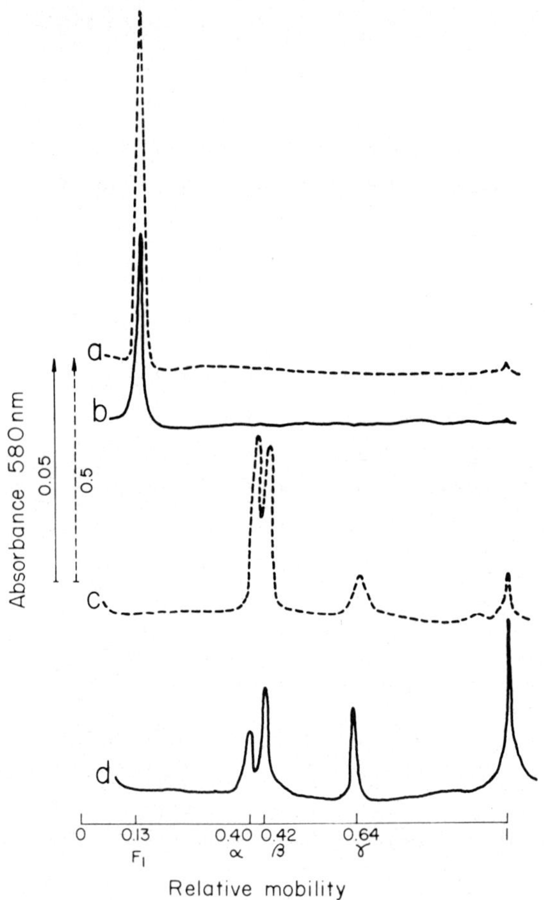

Fig. 1. Polyacrylamide gel electrophoresis and carbohydrate stain of mitochondrial F_1. F_1 was electrophoresed under non-dissociating conditions in 7% acrylamide gel and a) stained for protein (10 µg) and b) stained for carbohydrate (40 µg). F_1 was electrophoresed under dodecylsulfate denaturing conditions with 10% acrylamide gel and c) stained for protein (20 µg) and d) stained for carbohydrate (60 µg).

DISCUSSION

From this work, it seems reasonable to conclude that the glycoprotein nature of the ATPases/coupling factors is a general property rather than a peculiar characteristic of a bacterial ATPase as it might be perhaps envisaged from our previous work[12]. It is worth noting that the present results have confirmed the presence of hexosamines in the three coupling factors (forms A and B of <u>M. lysodeikticus</u> ATPase and CF_1). This apparent contradictory result with previous reports on the lack of hexosamines in F_1 and CF_1[8] may be explained because the relative low content of amino sugars (<u>i</u>. <u>e</u>., less than 0.01 g sugar/g peptide) and the lower sensitivity of the methods employed before. We are aware that the quantitative sugar determination of mitochondrial F_1 can be questioned because it could be attributed to residual bound saccharose. However, the positive periodic acid–Schiff reaction after dodecylsulfate gel electrophoresis suggests a possible covalent link of sugar to the polypeptide chains in F_1.

The presence of rhamnose, ribose and arabinose does not seem to be a common feature in many glycoproteins[23]. Concerning the small amounts of ribose present it could be accounted for by up to 1 – 6 mol of adenine nucleotides bound to the enzyme. Although the present work has confirmed the glycoprotein nature of <u>M. lysodeikticus</u> ATPase and extended this character to other F_1 factors, it has not provided a clear molecular basis to understand their alkali lability and microheterogeneity[10,12,24]. From the results reported here the presence of an <u>O</u>-glycosidic bond to serine appears unlikely. The possibility of a low efficiency in the detection of <u>O</u>-glycosidic linkages in whole BF_1 has been considered (see RESULTS). Work is now in progress to isolate glycopeptides and to attempt β-elimination with these smaller fragments.

In spite of their common glycoprotein character, it is interesting that differences in the sugar moiety of the coupling factors seem to exist both at the quantitative and qualitative level. The differences are particularly striking in the case of ATPases isolated from two closely related strains of the same microorganism. These differences then point to a high degree of specificity for the sugar portion of these energy–transducing proteins. However, this is a subject open to speculation because the exact role of the sugar portion in the coupling factors is largely unknown. It may be postulated that the sugars, like in many other glycoproteins, are located at the exterior of the polypeptide moiety and play a recognition role. This role may mediate the interactions of the coupling factors with their enviroment. However, this possibility would imply a topological orientation of the sugar portion towards the interior of the bacterial cytoplasm or the mitochondrial matrix since it is generally believed that F_1 and BF_1 factors are extrinsic or peripheral membrane proteins facing the internal part of the bilayer[25]. This glycoprotein orientation would be in controversy with that proposed for most of the glycoproteins in biological membranes[26]. As another possibility, the sugars might mediate the interaction of the F_1 factors with the integral components, <u>i</u>. <u>e</u>. the Fo factor, of the proton translocating ATPases. This possibility of orientation would be somehow controversial with the current views and principles of membrane structure because it would propose an interaction between extrinsic and intrinsic membrane proteins through carbohydrates. Although less likely, the possibility also exists that a portion of the F_1 factors spans the bilayer. In this case, the glycoproteins might be topologically orientated towards the exterior mediating the communication between the two faces of the membrane.

As a second alternative, it may be postulated that the sugars play an essential role in the function of the ATPases, i. e. ion binding and/or ion movement. To account for this functional role, the sugars may form or be part of a hydrophilic pore which would in turn regulate the ion movements. This possibility would imply a somewhat unusual location of the sugar moiety buried within the core portion of the ATPase protein part. At this stage, it appears evident that a definitive answer to these questions must await further experimentation which should be specifically oriented to define the funtion and topology of the sugar portion in F_1 energy-transducing proteins.

ACKNOWLEDGMENTS

Part of this work was carried out during a stay of J. M. A. at the Max-Planck-Institut für Immunbiologie (Freiburg, GFR). We are indebted to Drs. B. Kichhöfen and O. Lüderitz from this Institut for laboratory facilities, advice and encouragement. We are particularly indebted to Mr. R. Warth for the amino acid analyses. We thank Prof. F. García-Olmedo for the facilities given for spinach culture and Dr. Alberto Marquet for his help in the preparation of CF_1. We also wish to thank Drs. A. Gómez-Puyou and M. Tuena de Gómez-Puyou for their generous gift of purified mitochondrial ATPase. The work was supported by a short-term EMBO fellowship to J. M. A. and by a grant of the Fondo Nacional para el Desarrollo de Investigación Científica to E. M.

REFERENCES

1. Singer, S. J. (1974) Ann. Rev. Biochem. 43, 805-833.
2. Jorgensen, P. L. (1974) Biochim. Biophys. Acta, 356, 53-67.
3. MacLennan, D. H. (1975) Can. J. Biochem. 53, 251-261.
4. Kyte, J. (1975) J. Biol. Chem., 250, 7443-7449.
5. Guidotti, G. (1976) Trends in Biochem. Sci., 1, 11-13.

6. Singer, S. J. (1976) in Surface Membrane Receptors, Bradshaw, R. A., Frazier, W. A., Merrel, R. C., Gottlieb, D. I. and Hogue-Angeletti, R. A. eds., pp 1-24, Plenum Press, New York and London.

7. Kagawa, Y. and Racker, E. (1966) J. Biol. Chem., 241, 2475-2482.

8. Penefsky, H. S. (1974) in The Enzymes, Boyer, P. D. ed., vol X, pp. 375--394, Academic Press, New York and London.

9. Hughes, R. C. (1976) Membrane Glycoproteins, Butterworths, London, Boston.

10. Nieto, M., Muñoz, E., Carreira, J. and Andreu, J. M. (1975) Biochim. Biophys. Acta, 413, 394-414.

11. Carreira, J., Andreu, J. M., Nieto, M. and Muñoz, E. (1976) Molec. Cell. Biochem. 10, 67-76.

12. Andreu, J. M., Carreira, J. and Muñoz, E. (1976) Febs lett. 65, 198-203.

13. Lien, S. and Racker, E. (1971) in Methods in Enzymology, San Pietro, A. ed., vol XXIII part A, pp. 547-555, Academic Press, New York and London.

14. Senior, A. E. and Brooks, J. C. (1970) Arch. Biochem. Biophys. 140, 257--266.

15. Lowry, O. H., Rosebrough, N. J., Farr, A. L. and Randall, R. J. (1951) J. Biol. Chem., 193, 265-275.

16. Farron, F. (1970) Biochemistry 9, 3823-3828.

17. Winzler, R. J. (1955) in Methods of Biochemical Analysis, Glick, D. ed., vol 2, pp. 279-311, Interscience Pub. Inc., New York.

18. Bray, G. A. (1960) Anal. Biochem. 1, 279-285.

19. Laine, R. A., Esselman, W. J. and Sweeley, C. C. (1972) in Methods in Enzymology, Ginsburg, V. ed., vol XXVIII, part B, pp. 159-167, Academic Press, New York and London.

20. Warren, L. (1959) J. Biol. Chem., 234, 1971-1975.

21. Allen, A. K. and Neuberger, A. (1975) Febs lett. 60, 76-80.

22. Downs, F. and Pigman, W. (1976) in Methods in Carbohydrate Chemistry, Whistler, R. L. and BeMiller, J. N. eds., vol VII, pp. 200-204, Academic Press, New York, San Francisco, London.

23. Glick, M. C. (1974) in Methods in Membrane Biology, Korn, E. D. ed., vol 2, pp. 157-204, Plenum Press, New York and London.

24. Adolfsen, R., McClung, J. A. and Moudrianakis, E. A. (1975) Biochemistry 14, 1727-1735.

25. Oppenheim, J. and Salton, M. R. J. (1973) Biochim. Biophys. Acta 298, 297-322.

26. Bretscher, M. S. and Raff, M. C. (1975) Nature 258, 43-49.

AUTHOR INDEX